IFIP Advances in Information and Communication Technology

718

IFIP Advances in Information and Communication Technology

The IFIP AICT series publishes state-of-the-art results in the sciences and technologies of information and communication. The scope of the series includes: foundations of computer science; software theory and practice; education; computer applications in technology; communication systems; systems modeling and optimization; information systems; ICT and society; computer systems technology; security and protection in information processing systems; artificial intelligence; and human-computer interaction.

Edited volumes and proceedings of refereed international conferences in computer science and interdisciplinary fields are featured. These results often precede journal publication and represent the most current research.

The principal aim of the IFIP AICT series is to encourage education and the dissemination and exchange of information about all aspects of computing.

More information about this series at https://link.springer.com/bookseries/6102

Mieczyslaw Lech Owoc ·
Felix Enigo Varghese Sicily ·
Kanchana Rajaram · Prabavathy Balasundaram
Editors

Computational Intelligence in Data Science

7th IFIP TC 12 International Conference, ICCIDS 2024
Chennai, India, February 21–23, 2024
Revised Selected Papers, Part II

Springer

Editors
Mieczyslaw Lech Owoc
Wroclaw University of Economics and Business
Wroclaw, Poland

Kanchana Rajaram
Sri Sivasubramaniya Nadar College of Engineering
Chennai, Tamil Nadu, India

Felix Enigo Varghese Sicily
Sri Sivasubramaniya Nadar College of Engineering
Chennai, Tamil Nadu, India

Prabavathy Balasundaram
Sri Sivasubramaniya Nadar College of Engineering
Chennai, Tamil Nadu, India

ISSN 1868-4238 ISSN 1868-422X (electronic)
IFIP Advances in Information and Communication Technology
ISBN 978-3-031-69988-7 ISBN 978-3-031-69986-3 (eBook)
https://doi.org/10.1007/978-3-031-69986-3

This Springer imprint is published by the registered company Springer Nature Switzerland AG
The registered company address is: Gewerbestrasse 11, 6330 Cham, Switzerland

Preface

The International Conference on Computational Intelligence in Data Science (ICCIDS) is an annual conference organized by the Department of Computer Science and Engineering, Sri Sivasubramaniya Nadar College of Engineering. The seventh edition of the conference, ICCIDS 2024, took place during February 21–23, 2024. The International Federation for Information Processing (IFIP) has been the organizing body and technical co-sponsor for ICCIDS for the past four years.

ICCIDS 2024 offered a platform for researchers, scientists, engineers, and academics from across the globe to network and exchange their most recent, unique research experiences along with the latest advancements and trends in Artificial Intelligence (AI). The conference theme included major areas of AI application, such as computer vision, natural language processing, speech processing, and integration with IoT and cyber security. The event drew both foreign and national authors who showcased their technical skills through written contributions and presentations.

The conference received 341 submitted papers, of which 21% were accepted for presentation after a double-blind review process in which each submission received 2 or 3 reviews.

The first day of the conference commenced with an inauguration ceremony with the release of the souvenir of ICCIDS 2024. It was followed by two keynote talks. The first talk was on *Localization and Clustering in Wireless Sensor Networks*, by *Saroja P. Kanchi, Kettering University, USA*. The second talk was delivered on *Impact of Fakes on Sustainable Development Actions - Trusted Knowledge as Counteraction* by *Mieczysław Lech Owoc, Wrocław University of Economics & Business, Poland*. Presentations of the accepted papers were organized as five tracks as follows:

1. Applications of AI/ML in NLP
2. Applications of AI/ML in Image Processing
3. Application of AI/ML in KDM, Cloud Computing and Security
4. Data Analytics
5. Applications of ML

There were four parallel tracks, namely 1, 2, 3, & 4, with 6 paper presentations respectively coordinated by the session coordinators with evaluation by the session chairs.

On the second day of the conference, there were two keynote talks, one hands-on session followed by paper presentations. The first keynote talk was on *Self-Supervised Learning: An Efficient Technique to Process Unlabeled Image Data* by *Mayuri Mehta* from *Sarvajanik College of Engineering & Technology, India*. The second talk was on *Gen-AI and the Next Wave of Democratization of IT* by *Kausikram Krishnasayee, Kissflow, India*. There was a Hands-On Session on *AI and XAI-based Use Cases in Healthcare using Python* by *Mayuri Mehta* from *Sarvajanik College of Engineering & Technology, India*. There were two sessions for parallel tracks 2 and 3 with 12 paper presentations each coordinated by the session coordinators with evaluation by the session chairs.

On the third day of the conference, there were two keynote talks followed by paper presentations. The first keynote talk was delivered by *Arun Raj Kumar P.*of *NIT Calicut* on *Securing Tomorrow: Unveiling the Synergy of Computational Intelligence and Data Science in Network Security*. The second talk was on *Cloud-Edge Coopetition: A Win-Win Partnership* by *Zakaria Maamar, University of Doha for Science and Technology, Qatar*. There were four tracks, 1, 2, 4, and 5, with 8, 8, 7, and 8 paper presentations respectively coordinated by the session coordinators with evaluation by the session chairs.

We acknowledge the invaluable assistance of the program and technical committee members. A complete list of track chairs and track PC members can be found in the next pages. Most reviewers provided detailed and constructive comments which were valuable for the authors to continue improving their papers, even if the submissions were not selected for the conference.

Given the high-quality work done by authors, reviewers, and track chairs, we are confident that the ICCIDS 2024 proceedings capture the current state of the art of research in the field of AI and knowledge management and will have significant impact on the research community in the longer term.

February 2024

Mieczyslaw Lech Owoc
Kanchana Rajaram
Felix Enigo Varghese Sicily
Prabavathy Balasundaram

Organization

Chief Patron

Shiv Nadar	SSN Institutions, India

Patron

Kala Vijayakumar	SSN Institutions, India

General Chairs

Annamalai Veiravan	Sri Sivasubramaniya Nadar College of Engineering, India
Radha Sankararajan	Sri Sivasubramaniya Nadar College of Engineering, India
Ramasamy Perumalsamy	Sri Sivasubramaniya Nadar College of Engineering, India

Conference Chairs

Kanchana Rajaram	Sri Sivasubramaniya Nadar College of Engineering, India
Felix Enigo Varghese Sicily	Sri Sivasubramaniya Nadar College of Engineering, India
Prabavathy Balasundarm	Sri Sivasubramaniya Nadar College of Engineering, India

Program Committee

Eunika Mercier-Laurent	University of Reims Champagne-Ardenne, France
Mieczysław L. Owoc	Wrocław University of Economics and Business, Poland
Dominique Verdejo	IFIP Chair, WG 12.13 - AI for Global Security, France
Zahia Guessoum	Université de Reims Champagne-Ardenne, France
Nada Matta	Université de technologie de Troyes, France

Technical Program Committee

Chutisant Kerdvibulvech	NIDA, Thailand
Yanhui Guo	University of Illinois Springfield, USA
Leung, Yiu Wing	Hong Kong Baptist University, China
Kaliappan Ravindran	City College of New York, USA

Mishra Priyanka	VIT Chennai, India
Kanakaraj	Mepco Schlenk Engineering College, India
Palani Shanmugam	Rajalaksmi Engineering College, India
B. Kamala	Sri Sairam Engineering College, India
N. Ramya	Sri Sairam Engineering College, India
A. R. Vasudevan	NIT Calicut, India
Surendiran	NIT Puducherry, India
J. M. Nandhini	Anna University, India
N. Prakash	B.S. Abdur Rahman Crescent Institute of Science and Technology, India
Parthasarathy S.	Thiagarajar College of Engineering, India
S. T. Padmapriya	Thiagarajar College of Engineering, India
N. Gopinath	Sri Sairam Engineering College, India
Sahayam Subin	SNU Chennai, India
Dhalayakumar	SNU Chennai, India

Session Coordinators

S. Angel Deborah	Sri Sivasubramaniya Nadar College of Engineering, India
Y. V. Lokeswari	Sri Sivasubramaniya Nadar College of Engineering, India
V. Balasubramanian	Sri Sivasubramaniya Nadar College of Engineering, India
S. Raghvendra Kumar Sakali	Sri Sivasubramaniya Nadar College of Engineering, India
S. Lakshmi Priya	Sri Sivasubramaniya Nadar College of Engineering, India
M. Saritha	Sri Sivasubramaniya Nadar College of Engineering, India
S. Rajalakshmi	Sri Sivasubramaniya Nadar College of Engineering, India
Nilu R. Salim	Sri Sivasubramaniya Nadar College of Engineering, India
S. Saraswathi	Sri Sivasubramaniya Nadar College of Engineering, India
A. Beulah	Sri Sivasubramaniya Nadar College of Engineering, India
S. M. Dhannya	Sri Sivasubramaniya Nadar College of Engineering, India
K. R. Sarathchandran	Sri Sivasubramaniya Nadar College of Engineering, India

Session Chairs

Saroja P. Kanchi	Kettering University, USA

Kishore B.	Mahalingam College of Engineering & Technology, India
Mieczysław Lech Owoc	Wrocław University of Economics and Business, Poland
Mayuri Mehta	Sarvajanik College of Engineering & Technology, India
B. Latha	Sri Sairam Engineering College, India
Ayesha	VIT University, India.
Noor Mahammad S. K.	IIITDM Kancheepuram, India
Kaja Mohideen	B.S. Abdur Rahman Crescent Institute of Science and Technology, India
Rahul Raman	IIITDM Kancheepuram, India
Arun Raj Kumar P.	NIT Calicut, India
Sriram M.	SDE1, Samsung Ecommerce, India

PC Chairs

Mieczysław Lech Owoc	Wrocław University of Economics and Business, Poland
Dominique Verdejo	IFIP Chair, WG 12.13 - AI for Global Security, France
Zahia Guessoum	University of Reims Champagne-Ardenne, France
Nada Matta	University of Technology of Troyes, France
Eunika Mercier	University of Reims Champagne-Ardenne, France
Mirnalinee Thanganadar Thangathai	Sri Sivasubramaniya Nadar College of Engineering, India
Kanchana Rajaram	Sri Sivasubramaniya Nadar College of Engineering, India
Felix Enigo Varghese Sicily	Sri Sivasubramaniya Nadar College of Engineering, India
Prabavathy Balasundarm	Sri Sivasubramaniya Nadar College of Engineering, India

Sponsors

EVOKE MEDIA

JB SOFT SYSTEM
Where Expertise meets Excellence

ssn

Contents – Part II

Applications of AI/ML in KDM, Cloud Computing & Security

Data Analytics

Applications of ML

Contents – Part I

Applications of AI/ML in KDM, Cloud Computing & Security

Healthify App Using Blockchain with Cloud

S. Ponmani, M. Sushil Mane(✉), and A. Vinita Piola

Department of Computer Science, Rajalakshmi Engineering College, Thandalam, Chennai, Tamil Nadu, India

{ponmani.s,200701268,200701290}@rajalakshmi.edu.in

Abstract. In the modern healthcare industry, hospitals still use outdated data management systems for patient data, and the way patient health records are kept and safeguarded does not reflect our technological advancements in this field. Blockchain technology can completely transform the healthcare sector by enhancing the security and storage of medical data and records. This project uses cloud computing to implement blockchain technology in the health sector. If we have an internet connection, cloud storage enables data access on any device, at any time, and from any location. In the healthcare industry, blockchain offers advantages like accurate patient care, and real-time data access. This project proposes hosting a blockchain code on a cloud network to develop a prototype app for the healthcare sector. By utilizing cloud infrastructure, the solution can adapt to the evolving needs of the healthcare industry while meeting regulatory requirements.

Keywords: Cloud · Blockchain · Consortium Blockchain · Healthcare · Ethereum Blockchain · Gossip Protocol · Permissioned · Permissionless Blockchain

1 Introduction

Hospitals still rely on outdated systems to manage data and the current methods of storing and protecting health records in the healthcare industry don't reflect our advancements in this field. Blockchain technology can advance healthcare by improving the security and storage of data and records. This project aims to integrate technology into the healthcare industry using cloud computing. With an internet connection cloud storage allows for data access, on any device at any time and from anywhere. In healthcare blockchain offers benefits such as patient care and real time access to data among others. To develop a prototype application for the healthcare industry this project proposes storing code on a cloud network. By leveraging cloud infrastructure, the system can meet standards and adapt to the evolving needs of the healthcare business.

2 Related Works

The study provides a methodology that aids in the identification, classification, and mapping of pertinent publications in the field. Database searches are conducted via the search protocol. It provides an understanding of the research by categorizing and mapping the

Published by Springer Nature Switzerland AG 2024
M. L. Owoc et al. (Eds.): ICCIDS 2024, IFIP AICT 718, pp. 3–16, 2024.
https://doi.org/10.1007/978-3-031-69986-3_1

chosen publications. It has limited prototype evaluation and depth of analysis is very low [6]. The process discusses different consensus algorithms. These procedures are employed to approve and append transactions to the blockchain. It guarantees the data which is stored on the blockchain is secure and intact. It focuses less on putting the suggested techniques into practice [8].

The main infrastructure for managing EHR (electronic health record) data in the process is blockchain technology, namely Hyperledger Fabric. It provides functions like privacy protection, identity management, and a modular design appropriate for use in medical settings. Because it makes use of a permissioned blockchain network, users are recognized and dependable entities. It has limited prototype evaluation and less focus on research progress [1]. The concept of OCBS (Off-Chain Blockchain Systems) and its potential use in healthcare data management are examined in this study. With on-chain and off-chain system architecture, it also offers a blockchain framework that protects patient privacy for medical data. It avoids getting into the technical details of a specific mechanism or algorithm and instead concentrates on the conceptual framework and design considerations for integrating blockchain technology into health sector data management. It investigates potential uses of OCBS for healthcare data management to get around the limitations and challenges of traditional blockchain techniques. It lacks comprehensive technical specifications and implementation examples for the proposed OCBS framework [11].

The procedure goes through difficulties and fixes for utilizing blockchain technology on IoT. This process can be used as an initial point for additional research and as a reference in future theses and academic papers on related topics. It provides less information and uses fewer methodologies [15]. The study talks about blockchain technology, which makes use of searchable cryptography and attribute-based access control (ABAC). This scheme incorporates searchable encryption techniques to enable safe keyword searching over encrypted data. Consequently, authorized users can search for keywords without revealing their actual values. ABAC can be used to define rules for access control according to environment attributes, permissions, subject (user), and object (resource). The drawback is that implementing the recommended strategies would be too difficult and impractical [19].

The main technology employed in this study is blockchain. In particular, the authors use a permissioned Blockchain framework designed for business applications called Hyperledger Fabric 2.0. Smart contracts are used by Hyperledger Fabric to speed up transactions and communications within the healthcare system. It describes how the gossip protocol allows nodes inside the blockchain network to exchange information. The absence of information about the security and privacy precautions included in the model is a major limitation of this study [16]. The process discusses how the healthcare industry is becoming more and more in need of blockchain and IoT technologies. IoT networks are critical for facilitating communication between patients and healthcare professionals by providing services like data analysis and remote health monitoring. Processing different types of data gathered from IoT devices and sensors, tracking patient records in real-time, and managing massive amounts of data are among the challenges [9].

The research study talks about secure verification mechanisms (SBVM) on cloud authentication servers. This server maintains the security of the cloud environment and oversees verifying the identity of each cloud user. A cryptographic technique is employed to safeguard data in cloud environments. This paper explains how to compute hash values for evidence using the hashing algorithm. This algorithm generates keys using the search optimization algorithm. It is employed to increase the efficiency of the main generation system through optimization. A thorough comparison with other digital techniques is not provided by the research [20]. The study uses the edge and fog computing concepts to process and analyze data collected from Internet of Things sensors. Edge computing concentrates on locating and storage resources close to the places where data is generated, whereas fog computing extends this concept to a distributed infrastructure between Edge devices and the cloud. Reduced bandwidth usage and real-time data processing are made possible by this. It lacks originality and has methodological limitations [3].

This procedure provides an answer for anonymous patient data in digital medical records. The method suggested makes use of an open-architecture, open-source cloud storage service for file sharing, simplifying data access. NoSQL databases make it feasible to retrieve data efficiently and quickly. It is the Data Anonymization Module (DAM) job to make patient data anonymous. However, there are no details about the implementation of the recommended solution [2]. The study mentions the safe transfer of medical data through attribute-based encryption. Access control can be based on the values of data attributes and user attributes thanks to a cryptographic technique called ABE (Attribute-based Encryption). It uses cryptography techniques to stop unauthorized access to patient health records maintained on an external cloud. It does not go into specific cryptographic algorithms, but it discusses the use of cryptography for security. Steganography is the process of concealing data within other non-secret data, such as data concealed within images or other files. While it is mentioned to benchmark against other systems, a detailed comparative analysis is not given. Readers may find more detailed comparisons with existing solutions useful [12].

This process analyses blockchain, a decentralized, immutable ledger for data and transaction recording. Common mechanisms include blockchain network development and consensus algorithms. A distributed system of nodes maintains and verifies the blockchain. Decentralization enhances trust and security. Blockchain-based access control systems ensure that authorized users can write, alter, and read data which is on the blockchain. This is necessary to protect data integrity and privacy. Cryptographic techniques are widely used to protect data in the blockchain, both during transmission and storage. The disadvantage of this strategy is that non-experts might find it difficult to fully understand the blockchain concept due to the complex algorithms involved [17]. To manage electronic medical records and other healthcare data, the article presents a conceptual architecture for a medical blockchain platform. Data lakes, blockchain, user groups, and smart contracts are all included in the architecture. Its technical details and outcomes are restricted [13].

The Hyperledger Fabric blockchain is used by the authors to construct their framework for consent management. The process of granting and cancelling access permissions is automated by the authors using smart contracts. The technical difficulties encountered during the implementation phase are acknowledged in this study [5]. This study uses blockchain mainly for security, smart contracts, and decentralization. It enables decentralized and IoT networks by removing the need for central authorities. This study fails to introduce new ideas, methods, or findings and simply uses the existing knowledge [7].

To maintain data consistency across the blockchain network, the process makes use of raft algorithm as a consensus mechanism for consortium blockchain. Fault tolerance is improved by this mechanism, which stops system crashes caused by broken node servers or issues with network communication. The implementation of the proposed mechanism may be challenging and require a high degree of technical expertise. Healthcare institutions may face challenges in implementing and integrating this technology [10].

Blockchains that are private and permitted can be used to limit access to only those who are authorized, protecting user privacy, and adhering to laws such as HIPAA. Blockchain uses cryptographic hashing to guarantee data immutability and integrity. Every data block has a distinct hash of the one before it, forming a chain of blocks that makes data changes in one block necessitate changes in all other blocks, hence making the chain extremely secure. It doesn't add anything noteworthy to the body of knowledge already in existence. Depending on its focus, the study might not address every facet of blockchain technology, omitting crucial factors [14].

The study suggests utilizing blockchain technology as an essential part of architecture. Blockchain is used in connected homes to store data that is shared by IoT devices. It ensures the integrity and privacy of the data gathered from wearables by offering a safe and unchangeable method of data storage. It makes mention of using a private blockchain, which limits access to the blockchain network to known users and is permission-based. It also talks about fog computing, which is done with k3's Kubernetes distribution and Raspberry Pi single-board computers. In a smart home setting, fog computing is used for data processing and aggregation. There are no novel or creative concepts presented in this study [18]. The study discusses distributed ledgers, cryptographic techniques, and consensus algorithms. Contracts that have their terms directly encoded into code are known as smart contracts. This study has limited scope and inadequate research design [4].

S. No.	Author	Year	Inference	Limitations
1	Alevtina Dubovitskaya [1]	2020	Hyperledger Fabric, Permissioned Blockchain This choice enhances control, privacy, and security compared to permissionless blockchains like Bitcoin or Ethereum	• Complexity • Scalability • Regulatory Compliance

(*continued*)

(*continued*)

S. No.	Author	Year	Inference	Limitations
2	Antonio Celesti [2]	2020	Cloud Storage, NoSQL Database, Data Anonymization Module (DAM), Web Server Gateway Interface (WSGI)	• Limited Implementation Details • Limited Discussion of Privacy and Security
3	Bouthaina Dammak [3]	2022	Edge computing, fog computing, LoRaWAN	• Lack of Performance Metrics • Limited Scalability Discussion
4	CH. V. N. U Bharathi Murthu [4]	2020	Blockchain technology, Smart Contracts, Privacy Protection, Performance analysis	• Complexity • Rapid Technological Changes
5	Cornelius C. Agbo [5]	2020	Hyperledger Fabric, Consent Transactions, Smart Contracts, Permissioned Blockchain	• Technical challenges • Lack expertise in blockchain technology
6	Cornelius C. Agbo [6]	2019	Systematic Mapping Study Process, Search Protocol, Data Extraction and Analysis, Classification and Mapping, Identification of Research Gaps	• Limited Prototype Evaluation • Depth of Analysis • Focus on Research Progress
7	Dinh C. Nguyen [7]	2020	Smart Contracts, Decentralization, Deep Reinforcement Learning, Internet of Things (IoT)	• Lack of Originality • Inadequate Research Design • Limited scope • Complexity and Accessibility
8	Emeka Chukwu [8]	2020	Consensus algorithms are methods that are employed in the blockchain to verify and insert transactions	• Limited Focus • Potential Bias

(*continued*)

(*continued*)

S. No.	Author	Year	Inference	Limitations
9	Jameel Almalki [9]	2022	The paper discusses the growing demand for IoT and blockchain technologies across various sectors, emphasizing their importance in healthcare. IoT networks are vital for connecting healthcare providers and patients, offering services like remote health monitoring and data analysis	• Limited Experimental Data • Scalability Challenges
10	Jing Zhang [10]	2020	Consensus Mechanism (Raft Algorithm), Privacy-Preserving Data Sharing, Asymmetric Encryption	• Complexity • Scalability issues
11	Ken Miyachi [11]	2021	Off-Chain Blockchain Systems (OCBS)	• Lack of Technical Details
12	Kirtirajsinh Zala [12]	2022	Attribute-based Encryption (ABE), Cryptography, Steganography	• Lack of Specific Details • Limited Comparative Analysis
13	Lei Hang [13]	2019	MediBloc, MedRec, MediLedger, HealthCoin, Connecting Care	• Limited Technical Details • Absence of Results
14	Lei Hang [14]	2022	Data Integrity and Immutability, Consensus Mechanisms, Smart Contracts, Privacy and Permissioning	• Lack of Novelty • Limited Scope
15	Md Ashraf Uddi [15]	2021	It focuses on talking about the difficulties and solutions associated with implementing blockchain technology within IoT ecosystem. Depending on which blockchain platform or network is being discussed, different consensus methods can be employed	• Lack of Specifics • Limited Information

(*continued*)

(*continued*)

S. No.	Author	Year	Inference	Limitations
16	Nihar Ranjan Pradhan [16]	2022	Hyperledger Caliper, Gossip Protocol, Transaction Ordering Schemes, Multi-Hosted Testbed	• Limited Discussion on Privacy • Limited Comparison • Real-World Implementation Challenges
17	Sanjeev Kumar Dwivedi [17]	2021	Blockchain technology, Decentralization, Smart Contracts, access control and authentication, data encryption	• Resource Intensive • Complex
18	Solomon Hopewell Kembo [18]	2020	Private Blockchain, Hyperledger Fabric, IoT Subsystem, Fog Computing, Data Privacy	• Lack of Originality • Methodologic al Limitations • Limited Generalizability
19	Yingwen Chen [19]	2021	Consortium Blockchain, Searchable Encryption, Attribute-Based Access Control (ABAC), Trapdoor Generation, Privacy Preservation	• Complexity • Performance Overhead
20	Yunus Khan [20]	2021	Cloud authentication server with secure verification mechanism, CB-EL GAMAL algorithm, SHA-3 algorithm	• Complexity • Lack of Comparative Analysis • Resource intensive

3 Proposed System

The proposed solution by the Heathify app combines blockchain and cloud technologies to build a robust healthcare platform. Encrypted data is stored on a blockchain with decentralized identity management and secure registration, authentication, and data entry by users. First, the Healthify app, a user-friendly mobile application, is created. The app will have two logins: a patient login and a doctor login. Patients provide their medical information when they log in, and that data is then encrypted using the AES algorithm. Blocks will be created from this encrypted data using blockchain technology. These blocks are protected by cloud environments provided by InterPlanetary File System (IPFS), guaranteeing their safe storage. IPFS (InterPlanetary File System) is a decentralized peer-to-peer storage system, whereas AWS is a centralized cloud storage system. As a result, IPFS is being used. Using their login details, the doctor makes a request using the Healthify app. First, the patient receives the request of the doctor. The blocks in the cloud are decrypted and sent to the app when the patient gives permission, allowing the doctor to view the patient's medical history. In an emergency, the doctor can submit an emergency request to obtain the patient's information directly if the patient is unable to give permission to the request (Fig. 1).

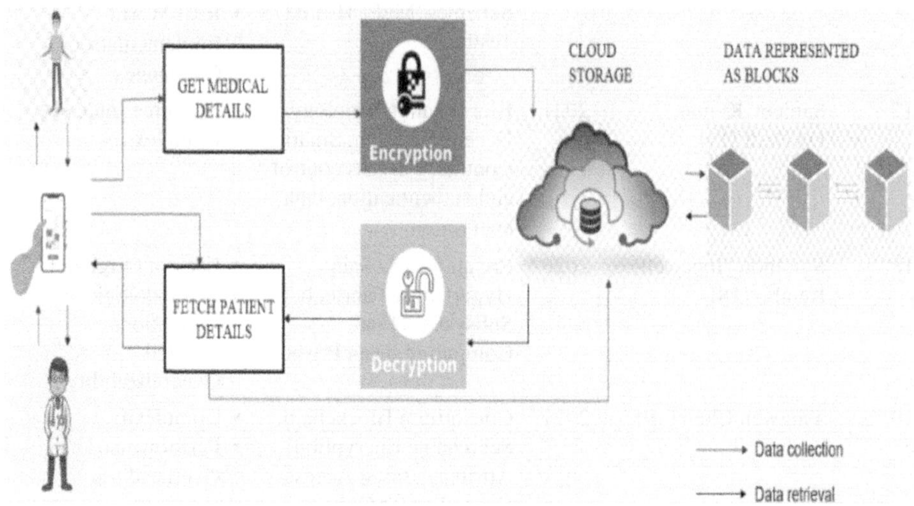

Fig. 1. Architecture Diagram of Proposed System

4 Methodologies Used

Healthify Application Design

The Healthify app is designed using React. The application consists of two logins: one for patient and the other for the doctor. In the patient login, the patient can login using their email id and password and upload their medical details. In the doctor's login, the doctor can enter their login details and view the patient's medical details.

Data Encryption

The core of the system is encryption because the data needs to be stored securely. This module aims to keep the data secure by encrypting the medical details of the patient. Encryption is done using AES algorithm.

Cloud Infrastructure

The patient details are securely uploaded to the IPFS cloud, ensuring confidentiality using AES encryption. Files present in IPFS cloud are distributed across multiple nodes, reducing reliance on centralized servers, and making the network more resilient to censorship and outages.

Blockchain Integration
Within the IPFS system, the file will be hashed, generating a distinct hash key. The blockchain will thereafter store this hash key.

Data Access
When a doctor requires access to the patient's medical history for examination purposes, they will initiate a request to the patient. The patient may decide to approve and decline the request. Upon acceptance, the doctor will retrieve the hash key associated with the patient's file from the blockchain.

5 Results

In the patient login, the patient will first login using their username and password. After successful login, the patient can upload their medical details, which will be encrypted and stored in IPFS which further generates a hash key value. This hash key is stored in the blockchain. In the doctor login, the doctor will ask the patient for permission to access their medical details when necessary for an examination. The request is up to the patient to accept or refuse. When accepted, the doctor will obtain from the blockchain the hash key linked to the patient's record. Using this hash key, the doctor will access

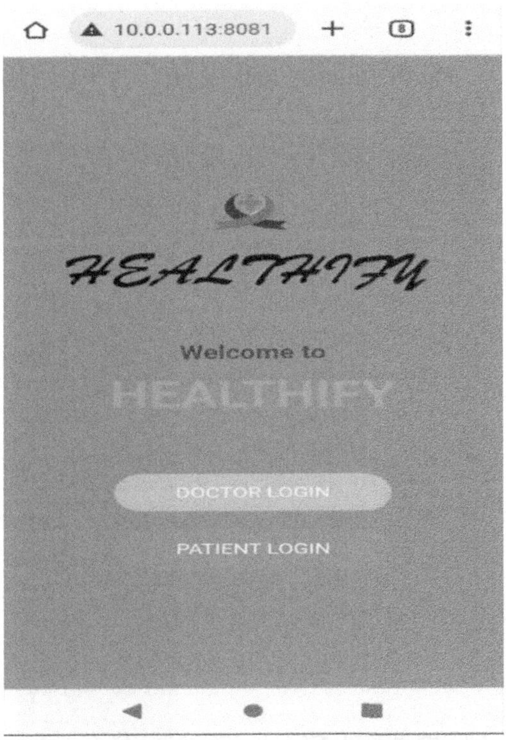

Fig. 2. Healthify App Login Page

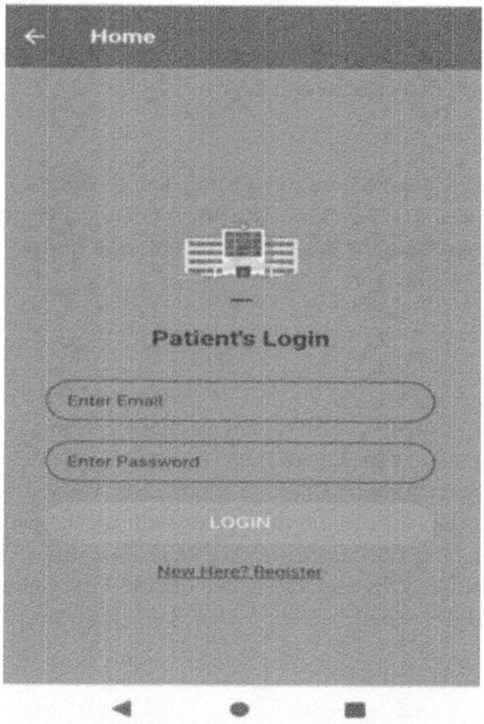

Fig. 3. Healthify App Patient Login Page

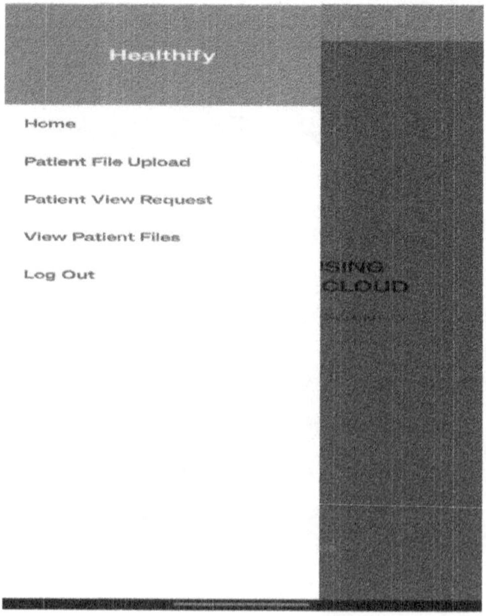

Fig. 4. Patient Login Dashboard

Fig. 5. Patient Medical Details Upload Page

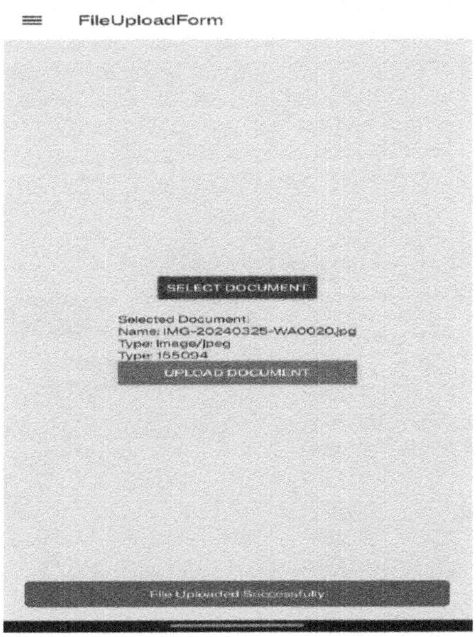

Fig. 6. Patient File Upload

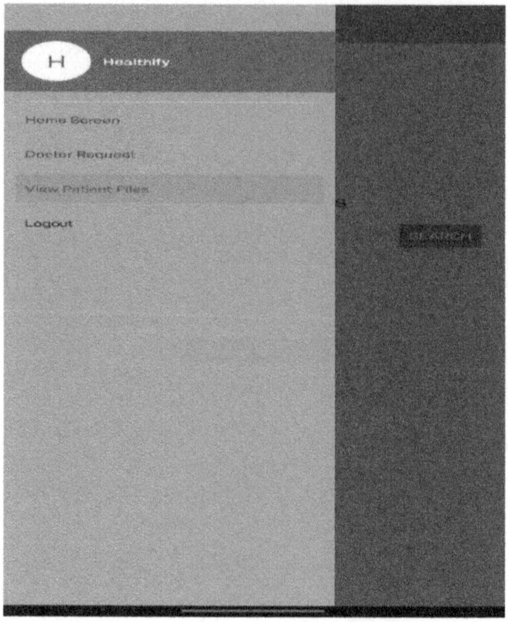

Fig. 7. Doctor Login Dashboard

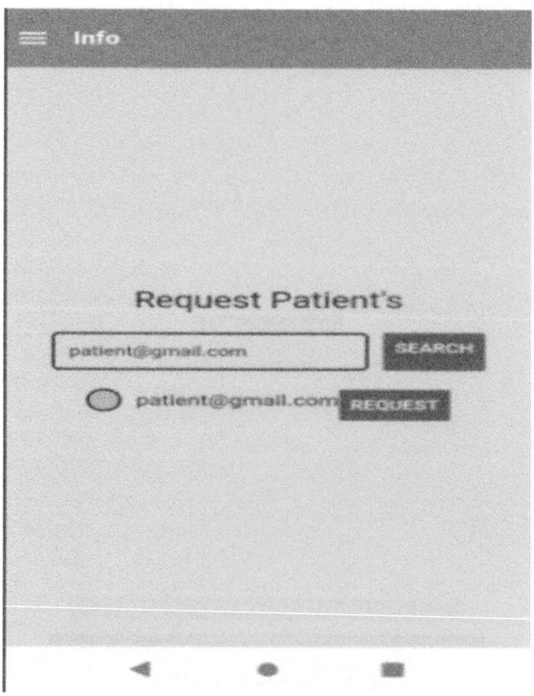

Fig. 8. Doctor requesting patient details

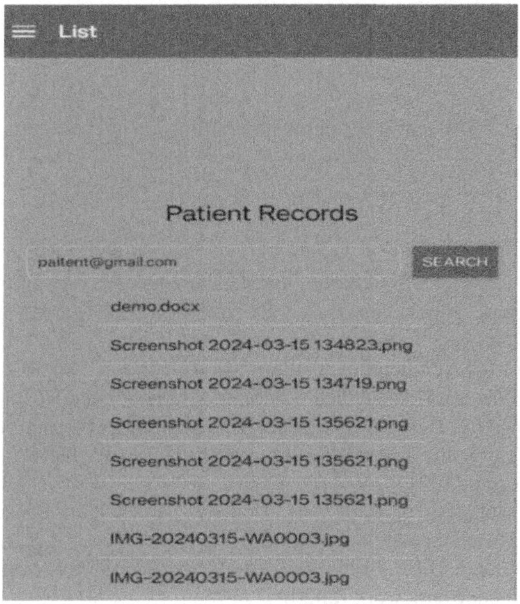

Fig. 9. Doctor viewing patient records

the encrypted file from the IPFS cloud. The file will then be decrypted using the secret key provided by the patient. With the decrypted file in hand, the doctor can review the patient's medical history, aiding in their examination and treatment decisions (Figs. 2, 3, 4, 5, 6, 7, 8 and 9).

6 Conclusion

The concept of using a cloud network to host a blockchain code to develop a prototype application has the power to completely transform the healthcare sector. While hosting a blockchain-based application in the healthcare sector has numerous advantages, it's crucial to consider the disadvantages as well. These include the requirement for widespread industry acceptance, privacy difficulties, scalability challenges, and legal compliance. Developing an application for the healthcare sector with blockchain and cloud computing might improve patient care, data security, and operational efficiency. The gossip protocol allows clients and servers to communicate with each other peer-to-peer. It is used to establish far faster connections to blockchain technology. On a permissioned blockchain, data may only be accessed and retrieved by authorized users. The blocks that are produced will be safely kept in an IPFS-style cloud computing architecture. This ensures that data may be accessed from anywhere at any time. As the healthcare sector grows, this tactic may completely transform how data is managed and shared, eventually benefiting patients as well as healthcare providers.

Acknowledgement. The author wishes to express sincere gratitude to our mentors, professors, and peers for their unwavering support, insightful feedback, and encouragement throughout the duration of this research project. Their guidance and mentorship have been invaluable in shaping the direction and outcomes of this study.

Disclosure of Interest. The author declares no conflicts of interest pertaining to this research endeavor.

References

1. Dubovitskaya, A., Baig, F., Shukla, R., Xu, Z., Zambani, P.S., Swaminathan, A.: ACTION-EHR: Patient-Centric Blockchain-Based Electronic Health Record Data Management for Cancer Care (2020)
2. Celesti, A., Ruggeri, A., Galletta, A., Villari, M., Romano, A., Fazio, M.: Blockchain-Based Healthcare Workflow for Tele-Medical Laboratory in Federated Hospital IoT Clouds (2020)
3. Dammak, B., Turki, M., Baklouti, M., Mars, R., Dhahbi, A., Cheikhrouhou, S.: LoRaChain-Care: An IoT Architecture Integrating Blockchain and LoRa Network for Personal Health Care Data Monitoring (2022)
4. Murthy, C.V.B., Shri, M.L., Kadry, S., Lim, S.: Blockchain Based Cloud Computing: Architecture and Research Challenges (2020)
5. Agbo, C.C., Mahmoud, Q.H.: Design, and Implementation of a Blockchain-Based E-Health Consent Management Framework (2020)
6. Agbo, C.C., Mikael Eklund, J., Mahmoud, Q.H.: Blockchain Technology in Healthcare: A Systematic Review (2019)
7. Nguyen, D.C., Seneviratne, A., Pathirana, P.N., Ding, M.: Integration of Blockchain and Cloud of Things: Architecture, Applications and Challenges (2020)
8. Chukwu, E., Garg, L.: A Systematic Review of Blockchain in Healthcare: Frameworks, Prototypes, and Implementations (2020)
9. Almalki, J., et al.: Enabling Blockchain with IoMT Devices for Healthcare (2022)
10. Zhang, J., Li, Z., Liu, C., Tan, R.: Design and Application of Electronic Rehabilitation Medical Record (ERMR) Sharing Scheme Based on Blockchain Technology (2020)
11. Miyachi, K., Mackey, T.K.: hOCBS: a privacy- preserving blockchain framework for healthcare data leveraging an on-chain and off-chain system design (2021)
12. Zala, K., Thakkar, H.K., Jadeja, R., Singh, P., Kotecha, K., Shukla, M.: PRMS: Design and Development of Patients' E-Healthcare Records Management System for Privacy Preservation in Third Party Cloud Platforms (2022)
13. Hang, L., Choi, E., Kim, D.-H.: A Novel EMR Integrity Management Based on a Medical Blockchain Platform in Hospital (2019)
14. Hang, L., Zhang, L., Chen, C., Yang, J.: Blockchain for applications of clinical trials: taxonomy, challenges, and future directions (2022)
15. Uddin, Md.A., Stranieri, A., Gondal, I., Balasubramanian, V.: A survey on the adoption of blockchain in IoT: challenges and solutions (2021)
16. Pradhan, N.R., et al.: A Novel Blockchain-Based Healthcare System Design and Performance Benchmarking on a Multi-Hosted Testbed (2022)
17. Dwivedi, S.K., Roy, P., Karda, C., Agrawal, S., Amin, R.: Blockchain-Based Internet of Things and Industrial IoT: A Comprehensive Survey (2021)
18. Kembo, S.H., Mukorera, B.T., Mpofu, P., Jacques, S.: Blockchain-Secured Health Wearables in Smart Homes Utilising Raspberry Pi Web of Things Gateways (2020)
19. Chen, Y., Xue, G., Zhou, H., Meng, L.: A Blockchain-Based Medical Data Sharing Mechanism with Attribute-Based Access Control and Privacy Protection (2021)
20. Khan, Y., Verma, S.: An Intelligent Blockchain and Software-Defined Networking-Based Evidence Collection Architecture for Cloud Environment (2021)

A Systematic Review of Various Deep Learning Techniques for Network Intrusion Detection System

A. N. Sasikumar(✉) and Sheeba S. Lilly

Department of Computer Science and Engineering, SRM Institute of Science and Technology, Ramapuram Campus, Chennai 600 089, Tamil Nadu, India
{sa8082,lillyshs}@srmist.edu.in

Abstract. The collective frequency and intricacy of attacks on computer networks remain a threat to information security within computer schemes. In order to solve this, researchers exploited network intrusion detection systems (NIDSs) to safe guard networks and information. Because of its lively nature of malware and continuous change in attacks, these systems normally recognize intrusions by the examination of network traffic. This survey demonstrates several Deep Learning (DL) approaches for the identification and categorization of unexpected attacks. The constant variations in the activities of the network make it essential to examine several datasets by dynamic and static techniques. Here 25 research papers are analyzed and surveyed. A complete assessment of many DL classifiers was exposed on various benchmark malware datasets. First, it offers a classification of computer intrusions with an explanation of categorized methods. Next, a general outline of the NIDS is described with its basic features. Then, the classification of NIDS based on their categorized methods is explained. After that, the challenges faced by the categorized methods are deliberated in the section of research gaps and issues. Lastly, the analysis in the survey is performed in terms of publication year, category analysis, tools used and performance metrics.

Keywords: Intrusion Detection System · Host-based IDS · Signature-based IDS · Anomaly detection-based IDS · Deep Learning

1 Introduction

An intrusion detection system (IDS) is a platform of software or hardware which keeps tracking on the activity of system or network and identifies the signs of malicious activity. At present, almost every computer system in today's use function in an environment of a network which always leaves them open to a wide range of destructive actions. IDS have already developed as one of the significant elements in computer security as the number of intrusion incidences is rising steadily over a period of time. The necessity of Machine Learning (ML) is to efficiently differentiate between malicious and benign system processes which have never been greater due to the fast change of logging data

Published by Springer Nature Switzerland AG 2024
M. L. Owoc et al. (Eds.): ICCIDS 2024, IFIP AICT 718, pp. 17–35, 2024.
https://doi.org/10.1007/978-3-031-69986-3_2

[1]. The purpose of IDSs is to avoid intrusions while safeguarding the data, unauthorized access and programs of computer systems. The IDSs are able to categorize both extrinsic and intrinsic intrusions into a computer network's organization and it activates an alarm if a network compromises the security intimidates [27]. The main goal of IDS is to identify a wide range of intrusions with unknown and known attacks. It is also utilized to discover and adjust to new attacks and to recognize intrusions in a speedy pattern [5]. The objective of network intrusion detection is to recognize doubtful offensives and take suitable action to protect the network from destruction by minimizing financial costs [10].

DL has been broadly examined and has substantial success in the fields such as image identification, natural language processing and forecasting of weather [28]. A DL method displays the capabilities of outstanding learning for managing difficult data processing due to its non-linear nature. In recent years, the fast improvement of parallel computing has delivered a foundation of hardware for DL approaches [11]. The attackers are increased by the advancement of Information Technology. Furthermore, these attacks have intensely extended and can cause considerable loss in economy to the administrations grounded on the computer systems. Moreover, there are various methods such as IDSs, signature-based systems and anomaly-based systems which are offered to neutralize these intrusions [8]. IDSs have been recently seen as a substantial growth in the utilization of DL approaches like Convolutional Neural Networks (CNN) [29], Deep Convolutional Generative Adversarial Network (DCGAN), and deep long-short-term memory recurrent neural network [30]. Better classification performance is made probable by the ability of DL's which is utilized to extract a many illustrative feature subset over several levels [9]. DL methods have also been employed in NIDS because it was found that they could efficiently address the issues in the fields of image classification, object recognition and text classification. The widely held mechanism using DL approaches concentrates on the areas of anomaly-based intrusion detection and dimensionality reduction [19].

The intention of this survey is to assess many DL methods exploited for NIDS. The classification of other techniques is Host-based methods, signature-based methods, network-based methods and anomaly detection-based methods. The estimation is completed by recognizing several tools utilized performance measures, datasets, categories and year of publication. The disadvantages that are attained in the paper are also described.

The survey is well-organized in the existing way and is revealed as follows: the categorization of several methods of NIDS is explained in Sect. 2. The general outline of NIDS is demonstrated in Sect. 3. The description and literature survey of NIDS is illuminated in Sect. 4.The research gaps and issues for noticing the difficulties are displayed in Sect. 5. The analysis of datasets, evaluation metrics, tools and publication year are mentioned in Sect. 6. Lastly, Sect. 9 determines the conclusion of survey.

2 Categoriation of Network Intrusion Detection System

This section demonstrates the approaches of NIDS which are characterized as Host-based methods, network-based methods, signature-based methods, anomaly detection-based methods, and other methods. Figure 1 establishes the categorization of the Network intrusion detection system.

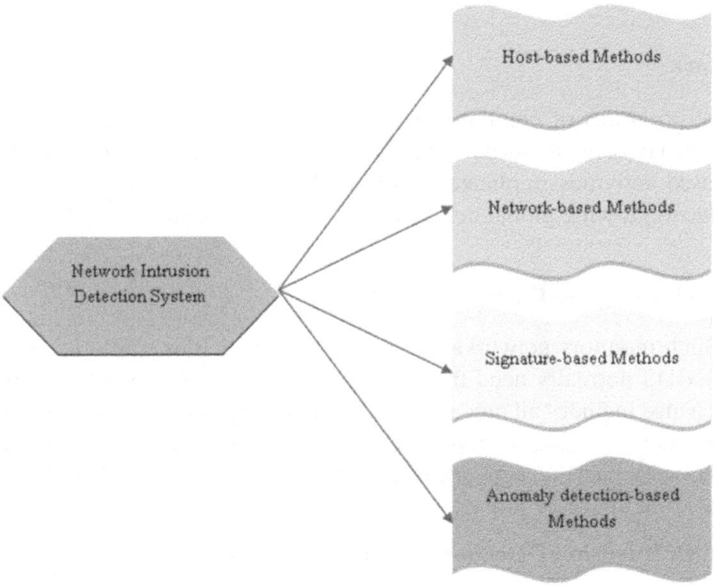

Fig. 1. Categorization of Network Intrusion Detection System

3 General Outline with Steps

In this section, the general outline of NIDS is deliberated in detail and its structural representation of NIDS was explained as follows:

3.1 Internet

The Internet and computer systems need prevention to safeguard against virus attacks and intrusions as more and more sensitive data is being stored and made available via online networks. Computer systems are protected from attackers employing security tools which include IDS, anti-virus software, and firewalls. By examining network traffic datasets or payloads, numerous intrusions can be found.

3.2 Firewall

Networks can be preserved against destructive and malicious occurrences with the help of firewalls. They are often organized in the main access of network's point with a pre-established strategy and set of rules. In order to safeguard the network against harmful attacks, the administrator establishes or identifies specific essential rules. NIDS was finally established as a result of the necessity for new tools that could recognize attacks and prepare the administrator to set suitable protection.

3.3 Network

IDS are the programs of computer that attempt to identify intrusions by associating observable behavior to doubtful patterns are desirable in real-time. The majority of network-based activities implicate intrusion. The focus on intrusion has established attention due to growing global network connectivity by stimulating effective IDS.

3.4 Network Intrusion Detection System

An IDS which monitors networks and recognized malicious traffic is called a NIDS. Moreover, NIDS normally need the access of promiscuous network for observing all traffic which also includes all unicast traffic. Network-based intrusion detection systems provide the network more responsiveness than a particular host. By examining and capturing network packets, network-based IDS can recognize attacks [26].

3.5 Network Intrusion Detection System Management

A variety of sensors are utilized by NIDS for monitoring the packet traffic. Many servers are employed for the tasks of NIDS management and one or more management borders are utilized for the user interface. In order to identify intrusion patterns, NIDS observes the traffic packet by packet in real time or close to real time. It is potential to use management servers and sensors to examine traffic patterns and identify intrusions. Figure 2 depicts the structural representation of NIDS.

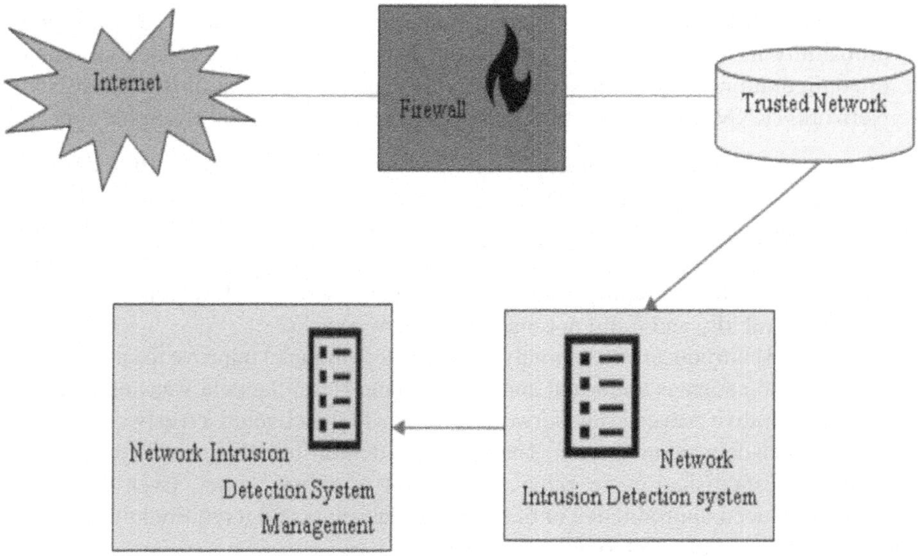

Fig. 2. Structural representation of NIDS

4 Description and Literature Survey of Nids

The explanation and literature survey of the considered methods like Host-based methods, network-based methods, signature-based methods and anomaly detection-based methods are deliberated as follows.

4.1 Host-Based Methods

The explanation and its literature of several approaches demoralized in Host-based approaches are illuminated in this section and they are detailed as follows.

a) **Description of Host-based methods.**

An intrusion detection system known as host-based IDS (HIDS) retains track on the computer infrastructure on which it is placed by examining traffic and reporting malicious attack. This HIDS provides deep perceptibility about acute security systems and what's happening in it. Data from servers, PCs and other host systems are composed and then examined for irregularities or distrustful behavior through HIDS. Security-centric data sources like authentication logs may be present in the data in which HIDS tools estimate. In assessment to network-based systems, HIDS can display detailed activities and give more information. They are more reasonable for small enterprises than NIDS because of their low entry obstructions.

b) **Literature survey of Host-based methods.**

Kim, G., et al. [1] established a system-call language-modeling technique which was employed to design anomaly-based host IDS. This technique had two major parts such as front-end and back-end. Here, front-end utilized for modeling the language of systems in several situations whereas back-end utilized for the prediction of anomaly based on the classifier of ensemble thresholding. It was solid and it had outstanding probability in which it did not require databases to retain an exponential quantity of configurations. Moreover, it did not modify the arrangement of Recurrent Neural Networks (RNN) which was utilized in system-call language and ensemble approach.

Siamese Convolutional Neural Network (Siamese-CNN) was demonstrated by Park, D., et al. [2] to assess and advance the performance of the system. Here, the few-shot learning model was utilized to form Siamese-CNN, which displayed brilliant performance with restricted quantity of data. The benefit of this approach was cost of training data in which it had low training cost related with DL methods. However, in real time it failed to train rapidly and it did not manage the new attacks.

Kim, K., [3] introduced an Anomaly detection technique to improve the performance of categorization such as abnormal and normal data. Here, the data was augmented by utilizing Generative Adversarial Network (GAN) which delivered a trustworthy way of employing with data augmentation. The outcomes demonstrated increased the precision of classifiers established using enhanced data for minority classes. Even though this approach produced unproductive or unbalanced classifiers and it required more training time when the datasets are large.

A Better Similarity Algorithm for Host-based IDS (BSA-HIDS) was developed by Ouarda, L., et al. [4] for identifying unfamiliar series of call system sequences. In order to select the finest performance, five similarity measures were utilized in this method. The similarity score was generated and it recognized the components of unusual and usual strings. It was employed for online corruption and it provided finest grouping with a better detection rate and less training time. Furthermore, this method was unsuccessful in minimizing the files or folders for every categorization.

Rincy N, T. and Gupta, R., [5] established the Hybrid NID-Shield NIDS method for classifying the datasets as per the category and terms of attack. Here, data preprocessing was accomplished by exploiting normalization and renovation processes on datasets. After this, an effectual hybrid feature subgroup selection method known as CAPPER was employed for finding the correct and maximum quality feature. It obtained a completely outstanding performance based on several parameters and it was found to be the most effective method. Still, the Hybrid NID-Shield NIDS method failed to apply it to the development of fog computing.

4.2 Network-Based Methods

The explanation and its literature review of numerous models evaluated in network-based Methods are explicated.

a) **Description of Network-based methods.**
A NIDS can recognize activity of malicious traffic on network. To assess all traffic containing all unicast traffic in which NIDS generally needs access to unrestrained network. It detects the internal boundary of the firewall in read-only method and transmits warnings to NIDS Management server. The application and protocol layers explore these abnormalities or variances as they move up the stack which is also

employed to identify the network-based hackers. It provides centralized administration for the association of disseminated attacks and it is possible for the system administrator to measure attacks.

b) **Literature survey of Network-based methods.**
Khan, M.A., [7] presented the Hybrid Convolutional Recurrent Neural Network-IDS(HCRNNIDS) model was focused on detecting if the network traffic was malicious or stable. This method was cost-effective based on computational intricacy with the features of a dataset and it offered advanced accuracy. The outcomes of simulation displayed that the introduced model was effective in the categorization of malicious attacks. By employing the developed model, it achieved resolutions for efficient security against malicious attacks. However, it did not examine with many datasets as the signature of the network traffic varies frequently.

Devan, P. and Khare, N., [8] introduced the XGBoost–Deep Neural Network (DNN)technique for the selection of features which was observed by DNN for network intrusion categorization. This method had three stages such as normalization, feature selection and categorization. It provided a high detection rate for network intrusion and it was exploited for the categorization of two-fold intrusion detection. It also had the benefit of equivalent processing which utilized all the depths of the machine. Moreover, it failed to measure the optimum learning rate systematically for a specified dataset.

The Genetic Algorithm (GA)-based exhaustive search and Fuzzy C-means clustering (FCM) approaches were demonstrated by Nguyen, M.T. and Kim, K., [9] for NIDS by utilizing an enhanced subset of feature. In order to indicate the CNN, the bagging (BG) classifier and CNN method were identified which acted as an efficient separator. It reduced the issue of overfitting and attained consistent outcomes of the estimation performance. The drawback of the method was inefficient time and it also required more period of time for applied simulation.

Jiang, K., et al. [10] established a network intrusion detection approach which combines deep hierarchical network and hybrid sampling. At first, one-side selection (OSS) was utilized for decreasing the majority of samples of noise and Synthetic Minority Oversampling Technique (SMOTE) was employed to upsurge the minority samples. Then, CNN was applied to remove longitudinal features and for extracting chronological features Bi-directional long short-term memory (BiLSTM) was employed. This approach extracted the features mechanically by applying frequent multi-level learning and takes the benefit of magnificent features. Even though in training sets contained only a smaller number of samples and the issues in this method cannot be resolved entirely.

Xu, C., et al. [11] demonstrated an intrusion detection systems method for recognizing the intrusions of networks. Here, DL approaches were utilized to attain better training and performance. The integration of Gated Recurrent Units (GRU) and Multilayer Perceptron (MLP) were performed with newly available techniques. This approach optimized the network in which it was functional to the situation of a real network and executed effectively.

The Scalable and hybrid IDS was formulated by Khan, M.A., et al. [12] for improving the scalability and accuracy. It was attained by two stages in which the first stage employed the element of anomaly detection and in the next phase, the element performs as misuse detection. The structure of hierarchical and modular performed better and it also decreased the computational intricacy based on its accuracy. Even though the developed IDS method was difficult because they were unscalable and unproductive.

An autoencoder-conditional GAN (AE-CGAN) was articulated by Lee, J. and Park, K., [13] for enhancing the intrusion detection performance. Here, the AE was utilized to remove the features with less dimension whereas the CGAN approach was employed to unsample infrequent classes and solved inequalities of data. Due to the reduction in infrequent classes, the sample with unstable data was utilized and it enhanced the NIDS performance and accuracy. Still, due to the togetherness of classes the other classes were overlapped and the difficulty in unbalanced data was not resolved fully.

4.3 Signature-Based Methods

In this section, the explanation and its literature of many models measured in signature-based methods are elucidated and quantified as follows.

a) **Description of Signature-based methods.**
Signature-based IDS identifies attacks based on detailed patterns such as the capacity of bytes or the quantity of network traffic. In addition to this, it notices malware based on the known dangerous order series in which it employs. When it comes to recognizing recurrent and well-known threats, like worms, malware or denial-of-service (DoS), signature-based NIDS is most effective. They didn't need a lot of modification or training because they are very simple to organize and design. A signature-based IDS preserves a catalog of attacks in which it checks network packets.
b) **Literature survey of Signature-based methods.**
Einy, S., et al. [15] formulated a Hybrid Inference System technique for analyzing the solution to network security against several forms of attacks. The attacks such as software liabilities, functioning system susceptibilities and web applications were taken into consideration. The obtained outcomes display that it recognized various categories of attacks like brute force, SQL injection, SMB exploits and so on. Moreover, there was an interruption due to large datasets and it was unsuccessful in improving the time of response.

Kwon, H.Y., et al. [17] developed a hybrid anomaly detection model which combined behavior-based and signature-based approaches. The hybrid anomaly detection approach syndicated composite autoencoder and statistical filtering and it efficiently detected strange activities. This method decreased the required time and implemented the development of detection. The introduced method was used worldwide and thus it was applied to other Industrial Control Systems (ICSs). Even though it failed to obtain various alternate metrics which was appropriate for time series data.

A Hybrid Model Using Bio-Inspired Metaheuristic approaches was established by Almomani, O., [18] for detecting the general attack in terms of hybridization bio-inspired metaheuristic methods. The developed method had two main objectives the first aim reduced the designated features for NIDS. The second aim noticed the general attack by means of ML classifiers. This method reduced the time taken for detection, displayed better consequences and increased the effectiveness in the required time.

4.4 Anomaly Detection-Based Methods

The detailed explanation with its literature review of several models evaluated in anomaly detection-based methods and they are labeled below.

a) **Description of Anomaly detection-based methods.**
   The representative method of anomaly-based IDSs is used to institute a baseline of the usual network activity and traffic. To find patterns that are not usually present in the traffic and they can relate the existing status of network's traffic to its base line. A regulated baseline is utilized to train the system before an activity is associated with the baseline in anomaly-based detection. An alert is allotted when something strange happens and it influences well-established attack patterns to take benefit of the system's liabilities. Here, the patterns of database can be coordinated within a very less period of time. The ability of the systems is to rapidly create signatures or patterns and to understand network behavior.
b) **Literature survey of Anomaly detection-based methods.**
   Zavrak, S. and İskefiyeli, M., [19] established unsupervised DL approaches with a semi-supervised learning method for identifying abnormal intrusions from the data of flow-based method. Here, the approaches of Autoencoder (AE) and Variational Autoencoder (VAE) were utilized to detect unfamiliar attacks by means of flow aspects. The capability of both AE and VAE methods was relatively effective in noticing various strange intrusions. Furthermore, it failed to indicate that these approaches are required to be maintained by supervised learning systems because of their high false alarm.

An Vote algorithm with Information Gain was introduced by Aljawarneh, S., et al. [20] to evaluate the possibility of intrusion created by the features of network transaction data. The outcomes of this method had an important result in minimizing the computational time and difficulty. The features of selection and reduction by the finest classifier of data mining and ML were operated to enhance the accuracy of classifier. Moreover, the disadvantage of this technique was a major part of the attack was executed by the protocol of TCP.

Resende, P.A.A. and Drummond, A.C., [21] formulated a projected clustering approach for choosing the features for profiling and anomaly-based intrusion detection parameters. Moreover, two approaches of anomaly-based were presented which were constructed on a process of predictable cluster. It easily adapted to the newly formed attacks which required more security and computable resources. This approach delivered flexibility and used a variety of categorized data which was normally formed by security.

A non-symmetric deep auto-encoder (NDAE) technique was presented by Shone, N., et al. [22] in order to permit the operation of NIDS along with the recent networks. It was the integration of narrow and deep learning and it was proficient to analyze the wide range of networks. This technique had the capability to perform more efficiently with intricate and huge datasets. Even though the values of integers required normalization in which they were varied with detached values which made it hard and tough to learn.

Abdulhammed, R., et al. [23] introduced Multi-Class Combined Performance Metric to observe the combining AE and Principle Component Analysis (PCA) for the reduction of dimensionality. The classifiers used were intended towards an effectual NIDS on the required data set. However, this technique was more appropriate, strong and steady for categorizing high correlated and dimensional data. Furthermore, it made the classifier less delicate for both superiority of training illustrations and overfitting problem.

Caville, E., et al. [25] formulated an Anomal-E approach for detecting the anomalies and to influence authority for network intrusion. This method was established and adapted to delicate attacks in both the environments of existent and artificial networks. This approach attained the maximum accuracy for detection and utilized the aspects for assessment. Moreover, it did not estimate the existent network traffic which employed to collect useful perceptions in the external performance of artificial datasets.

4.5 Other Methods

Mazini, M., et al. [6] formulated Artificial Bee Colony (ABC) for feature selection and AdaBoost approaches for categorizing and estimating the features. Here, the features of ABC improved the accuracy and increased the detection rate and the aim is to attain the finest feature. It attained computational time for an anomaly network-based approach and established an improved function of capability. Therefore, the introduced model was more effectual for the issues of anomaly detection. Moreover, it produced very less performance in the process of data mining during an unstable dataset.

The privacy-preserving framework using Rabin fingerprint approach was developed by Wang, Y., et al. [14] for the combination of signature-based intrusion detection beside the devices of fog. This model was assessed in both environments of replicated and real networks. The results validated that it safeguarded the data privacy and significantly decreased the workload by attaining less delay in detection. Furthermore, during the exclusive development of matching the signature it disturbed the performance.

The Multi-Fusion Pattern Matching (MFPM) approach was introduced by Geethanjali, N., [16] for detecting intrusions quickly. Here, in order to increase the effectiveness, the phase of preprocessing was contributed to the phase of searching. Additionally, the extension of the skip for window decreased the quantity of individuality assessments and skipped the unwanted individuals proficiently. However, in order to improve performance, the configuration in the assumed text string was identified.

The Multi-Stage Optimized ML Framework was developed by Injadat, M., et al. [24] to lessen the computational intricacy during the maintenance of the performance of detection. The effect of oversampling approaches regulated the negligible appropriate size of training samples. In order to progress the performance of NIDS, various hyper parameter optimization methods of ML were examined. The computational intricacy of this technique was decreased and it evaluated the consequence of detection and time intricacy. However, it failed to validate other techniques on unpredictable and higher-capacity datasets.

5 Research Gaps and Issues

This section deliberates the research gaps and problems of various predominant approaches of NIDS are demonstrated below.

The difficulties disagreeing with host-based methods are as follows: The system-call language-modeling method [1] failed to challenge the mission for identifying intricate modern attacks and it was challenging to create a strong method for gathering the data from large scales. In [2], Siamese-CNN did not identify intrusion against several attacks and it was still a challenge for enhancing the accuracy for detecting the intrusion in internal and external attacks. AN anomaly detection technique [3] did not consider many ML and DL approaches with large datasets. Moreover, it was difficult to evaluate the training period and the categorization performance for the large datasets. In [4], the BSA-HIDS method faced a challenge in testing the proficiency of datasets like NSL-KDD and UNM and it was difficult to restrict and remove the attacks.

The challenges accepted by network-based methods are exposed as follows: The HCRNNIDS approach [7] faced a challenge in discovering several DL approaches with the method of feature extraction and it did not absorb familiar illustrators of data. In [12] the scalable and hybrid IDS did not cover the misapplications of network and anomaly though it was challenging to develop DL as a tool for feature extraction. An AE-CGAN method [13] faced a major challenge in detecting the threats in external data which was parallel to regular data using DL generation methods.

The issues opposed by signature-based methods are indicated as follows: The privacy-preserving framework using Rabin fingerprint method [14] failed to examine many insubstantial encoding schemes and it was difficult to decrease the delay in a dispersed environment. In [17], the hybrid anomaly detection method did not regulate which actuators or sensors were employed for the attack and it was challenging to deliver appropriate detection of attacks.

The tasks articulated through anomaly detection-based methods are established as follows: An unsupervised DL algorithm with a semi-supervised learning method [19] faced difficulty in enhancing the detection rate because of the statistic that the features of some attacks need to be modeled effectively. In [22], the NDAE technique failed to evaluate the ability for controlling zero attacks and it was still a challenge for employing real-world support to network traffic. ML Based Network Intrusion Detection [23] was unsuccessful in applying the IDS for available detection of anomalies. Furthermore, it was a challenge to develop effective IDS for several datasets of intrusion detection. The Anomal-E method [25] did not focus on examining the properties of temporal-based training and it was difficult to explore its performance in the situations of external synthetic networks.

6 Proposed Architecture

Convolutional Neural Networks (CNNs), a subset of deep learning (DL) architecture, bear resemblance to traditional Neural Networks due to their stacked layers. The term "convolution" denotes a crucial block within the neural network structure. CNNs are adept at processing two-dimensional inputs to yield categorical outputs for classification tasks or real integers for regression tasks. Unlike traditional Neural Networks, which often rely on handcrafted features for accurate results, CNNs operate on raw data, processing it across multiple convolution layers. The input to a CNN typically consists of a vector representing network traffic denoted as X. Comprising three fundamental layers—convolutional, pooling, and fully connected—CNNs extract highlights from the convolution and pooling layers, while the fully connected layer correlates these extracted features with the resultant assault classes.

Convolution, within this context, refers to a linear operation utilized for feature extraction. The convolution layer employs non-linear activation functions such as ReLU, sigmoid, and tanh. The pooling layer serves to reduce feature dimensions by aggregating data across different segments. There are two primary types of pooling: max pooling, which isolates features based on segment size while discarding others, and average pooling. The fully connected layer serves as the final layer, where features are transformed into a 1D vector fully connected to each result class via weights. The choice of activation function in the final layer varies depending on the nature of the problem—sigmoid activation for binary classification and multiclass classification, and linear or identity activation for regression tasks.

The effectiveness of a Convolutional Neural Network (CNN) hinges upon the intricate design of its architecture, which varies depending on the specific problem and dataset sizes at hand. Identifying suitable hyperparameters for a particular task can be challenging, as altering these parameters directly impacts the model's performance. Optimizing network parameters and discovering the optimal combination of hyperparameters can significantly enhance the model's performance; otherwise, the model may struggle to classify disruptions effectively. Poor model performance manifests in various ways, including instability, prolonged processing times, and increased demand for computational resources.

To effectively classify disruptions, a 1D CNN is constructed separately for both binary and multiclass classification problems. In the context of 1D CNN, the term "1D" refers to the convolutional aspect that moves along a single dimension. This specialized architecture comprises three convolutional layers following the input layer, succeeded by two fully connected layers and a final output layer. The initial convolutional layer consists of 128 neurons, while the subsequent convolutional layer is comprised of 64 neurons. The second layer is interleaved with a max-pooling layer featuring a pool size of 3 and a stride size of 3x3, utilizing the ReLU activation function.

The third convolutional layer contains 32 neurons, each with a filter size of 3x3. To mitigate overfitting, a dropout layer is introduced after the third convolutional layer to retain essential feature information. The output of the final convolutional layer is then flattened and passed on to the fully connected layer. This fully connected layer, comprising 10 neurons and employing the softmax activation function, predicts the types of network interruptions with precision.

Hybrid models in CNNs for identifying attacks involve integrating CNN architecture with various techniques such as feature fusion, ensemble learning, transfer learning, hybrid architectures, adversarial training, and multi-modal data integration to enhance attack detection performance. By combining CNN's ability to extract hierarchical features from raw data with other methods, hybrid models can improve detection accuracy, robustness, and generalization. Feature fusion combines CNN-extracted features with other sources, while ensemble learning aggregates predictions from multiple models, and transfer learning adapts pre-trained CNNs to attack identification tasks. Hybrid

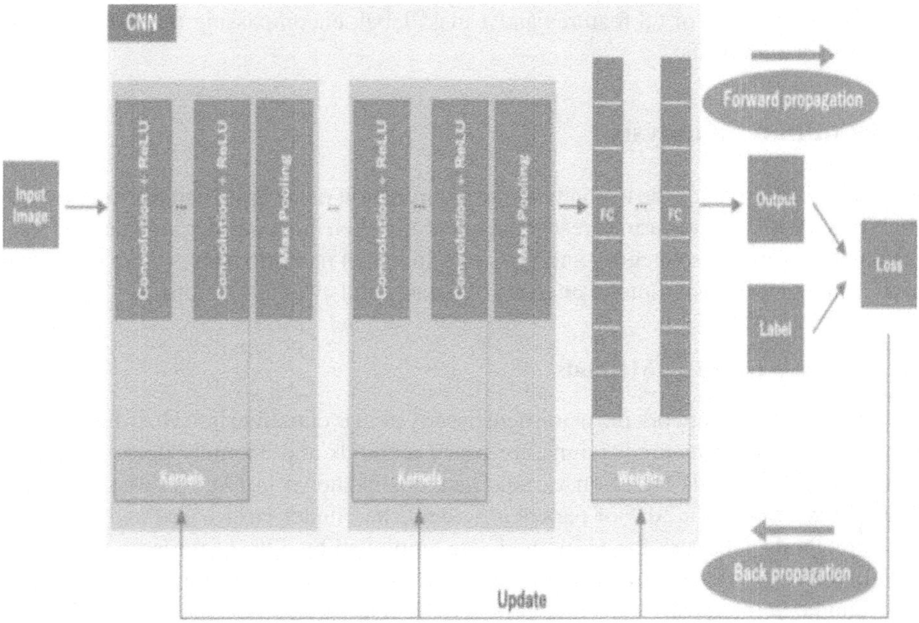

Fig. 3. CNN Architecture

architectures combine CNN with other layers like RNNs to capture spatial and temporal dependencies, while adversarial training improves model robustness against evasion attacks. Multi-modal integration incorporates diverse data sources for enriched feature representations. Overall, these hybrid approaches bolster the security and resilience of network systems by enabling more accurate and robust attack identification.

7 Datasets

The proposed 1D convolutional network architecture is assessed using the CICIDS2017 dataset, which comprises stable network traffic data spanning over multiple days. This dataset provides a realistic representation of network traffic and includes meticulously annotated streams detailing timestamps, source and destination IPs, source and destination ports, protocols, and detected attacks. The benign traffic is labeled as harmless, while the dataset encompasses 15 distinct types of attacks, including Syn Flood, Heartbleed, Botnet, Denial of Service (DoS), Distributed Denial of Service (DDoS), Web, and Intrusion attacks.

The entire dataset, encompassing multi-day traffic data, comprises a total of 2,299,308 events, necessitating substantial computational resources for analysis. For evaluating the proposed model, the dataset corresponding to Thursday's early morning working hours is selected. This subset of the data contains 170,368 events, characterized by 78 features and 1 label column. Approximately 10 features with zero variance are removed, and the remaining features are retained for evaluation. Consequently, the final dataset consists of 68 features and 1 class label, encompassing three attack types alongside benign traffic.

8 Result and Analysis

This section designates the evaluation and examination of several deep learning models for NIDS. At this point, various research papers are taken into consideration and assessed in some measures like category analysis, year in which research papers were published, tools employed for assessment, analysis of dataset and evaluation metrics.

8.1 Analysis Based on Methods

In this analysis, the network intrusion detection system is classified into Host-based methods, network-based methods, signature-based methods and anomaly detection based-methods. The analysis based on various methods is shown in Fig. 3. From the below figure, it is shown that 20% of papers are established under Host-based and signature-based methods, Network-based methods are employed by 32% of research papers and 28% of papers are demonstrated under the method of anomaly detection. By this analysis, it is found that the mostly used method in NIDS is network-based method.

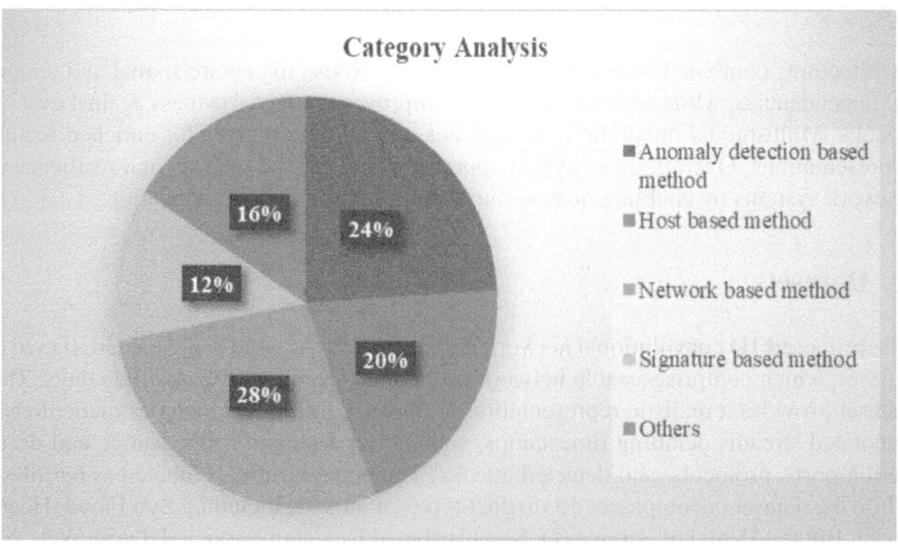

Fig. 4. Analysis based on category

8.2 Analysis Based on Tools

The tools utilized for analysis are shown in this section. It illustrates the evaluation of tools applied in network intrusion detection systems. The implemented tools used for analysis are demonstrated in Fig. 5. The tools executed in the research are Python, C program, Java, MATLAB, Python with Keras and Python with tensor flow. From Fig. 4 it is estimated that Python was used in 10 papers, Java in 5 research papers, MATLAB in 2 research papers as well as Python with Keras also used in 2 papers, C program and Python with tensor flow is implemented in 2 research papers each..

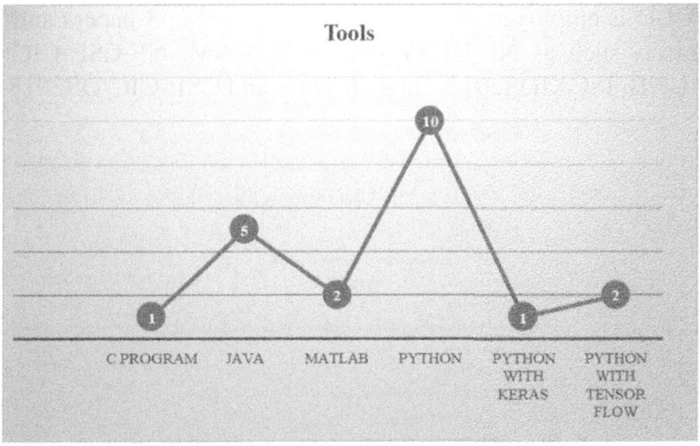

Fig. 5. Analysis based on Tools

8.3 Analysis Based on Papers Published

In this survey, 25 papers were analyzed and taken on the concern of network intrusion detection systems. Table 1 exemplifies the year in which the papers are published. It is recognized that 2 papers from the year 2016, 5 papers from 2018, 4 papers from 2019, 6 papers from 2020, 5 papers from 2021, 2 papers from 2022 and 1 paper from 2023. Here, it is shown that many research papers were published from the year 2020.

Table 1. Publication of research papers

Year	Number of papers published
2016	2
2018	5
2019	4
2020	6
2021	5
2022	2
2023	1

8.4 Analysis Based on Dataset

This section establishes the datasets engaged for analysis. The dataset executed in this analysis are UNSW-NB15, UNM, SWaT, NF-UNSW-NB15-v2, NSL-KDD, KDD, NF-CSE-CIC-IDS2018-v2, ISCXIDS2012, ISCX-UNB, English Texts, CSE-CIC-DS2018, CICDS2017 and ADFA-LD. The dataset utilized in this survey is deliberated in Fig. 6. It is shown that in 10 papers the dataset NSL-KDD is used, in two research papers the dataset ADFA-LD is employed, KDD dataset is utilized in 3 papers and in 1 research paper the datasets such as NF-UNSW-NB15-v2, UNM, NF-CSE-CIC-IDS2018-v2, SWaT, ISCX-UNB, ISCXIDS2012, English Texts and CSE-CIC-DS2018 were used.

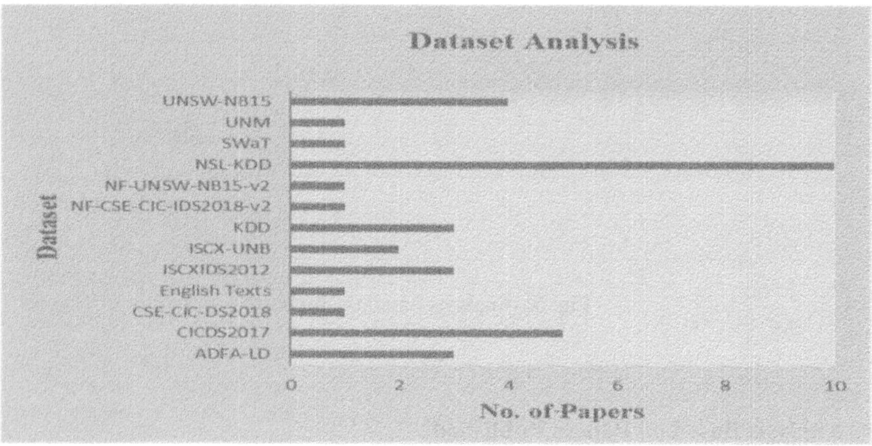

Fig. 6. Analysis based on Dataset

8.5 Analysis Based on Evaluation Metrics

Figure 6 indicates the evaluation of metrics employed in NIDS. The metrics used in this survey are accuracy, adaptability, Area Under Curve (AUC), False Alarm Rate (FAR), Detection Rate (DR), False Positive Rate (FPR), False Negative Rate (FNR), Matthews's correlation coefficient (MCC), F-score, Precision, PRC, Receiver Operating Characteristic Curve (ROC), Recall, True Negative Rate (TNR), sensitivity, and True Positive Rate (TPR). In this analysis, accuracy was employed in 17 papers, in 7 papers DR is utilized, FPR and recall are utilized in 11 papers, in 12 papers F-score and precision are used, TPR is employed in 5 papers, FAR is exposed in 6 papers, in 2 papers FNR and ROC curve is employed finally adaptability, AUC curve, MCC, PRC, sensitivity and TNR are applied in 1 research paper (Fig. 7).

8.6 Values of Evaluation Metrics

Table 2 designates the range of accuracy and the number of papers in which accuracy is applied. From Table 2, it was observed that accuracy ranges among the values of 0–85%,

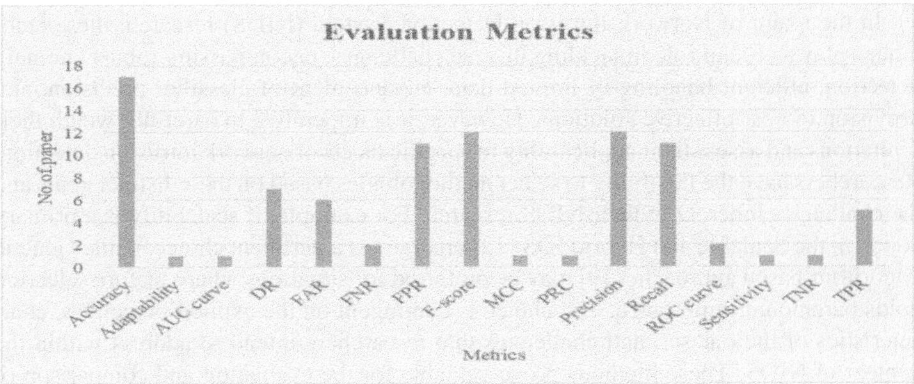

Fig. 7. Analysis based on Metrics

86–90%, 91–99% and above 99%. In the papers [2, 6] and [10] the accuracy value lies between 0–85%, the range of 86–90% is attained in paper [22]. Then the value between 91–99% was achieved in 9 papers of [6–9, 12, 15, 17, 18, 25] and in 4 papers of [5, 11, 20, 24] the value reached the range above 99%.

Table 2. Analysis based on Accuracy

Range of Accuracy in %	Number of papers
0–85	[2, 6, 10]
86–90	[22]
91–99	[6–9, 12, 15, 17, 18, 25]
Above 99	[5, 11, 20, 24]

9 Conclusion

This survey deliberates numerous approaches in NIDS and the description of methods is described. In this paper, 25 research papers are gathered based on techniques of host-based methods, network-based methods, signature-based methods and anomaly detection-based methods. The thorough explanation and literature review of every research paper is taken into attention and designated. Additionally, the issues or difficulties in the collected papers are projected as research gaps and issues and it delivers the opinions for supplementary knowledge about NIDS methods. In this survey, the analysis is ended by executing the datasets utilized, tools employed for study, evaluation measures and the years of papers published are exposed. For execution, the commonly utilized tool in this survey was identified as Python and to evaluate the evaluation measures, accuracy was used constantly. The furthermost frequently employed dataset in this survey was the NSL-KDD dataset.

In the realm of Network Intrusion Detection System (NIDS) research, these techniques play a crucial role in tackling diverse challenges, encompassing robust anomaly detection, efficient handling of limited data, enhancement of classifier precision, and provision of cost-effective solutions. However, it is imperative to carefully weigh their limitations and assess their applicability to specific facets of network intrusion detection. Researchers have the flexibility to select methodologies based on their distinct goals and the challenges inherent in their NIDS research. For example, if scalability is a primary concern, the Scalable and Hybrid IDS [12] emerges as a pertinent choice, while Genetic Algorithm-based approaches [9] may be preferred for situations where feature selection holds paramount importance. The choice is contingent on the explicit objectives, characteristics of the dataset, and challenges that researchers intend to address within the context of NIDS. These methods prove valuable for the evaluation and comparison of intrusion detection system performance, addressing issues like imbalanced datasets and overfitting, and crafting more precise and efficient algorithms tailored for the detection of network intrusions. Ultimately, the selection of a specific method hinges on the unique goals and challenges researchers strive to confront in the intricate landscape of NIDS research. In future work, the intention is to enlarge the field of NIDS by the execution of a completely disseminated NIDS.

Acknowledgments. The authors wish to thank Mr C Mahesh, AI Specialist for his support in designing the architecture and Data collection.

Disclosure of Interests. The authors have no competing interests to declare that are relevant to the content of this article.

References

1. Kim, G., Yi, H., Lee, J., Paek, Y., Yoon, S.: LSTM-based system-call language modeling and robust ensemble method for designing host-based intrusion detection systems. arXiv preprint arXiv:1611.01726 (2016)
2. Park, D., Kim, S., Kwon, H., Shin, D., Shin, D.: Host-based intrusion detection model using Siamese network. IEEE Access **9**, 76614–76623 (2021)
3. Kim, K.: GAN based augmentation for improving anomaly detection accuracy in host-based intrusion detection systems. Int. J. Eng. Res. Technol. **13**, 3987 (2020)
4. Ouarda, L., Malika, B., Brahim, B.: Towards a better similarity algorithm for host-based intrusion detection system. J. Intell. Syst. **32**(1), 20220259 (2023)
5. Rincy, N.T., Gupta, R.: Design and development of an efficient network intrusion detection system using machine learning techniques. Wirel. Commun. Mob. Comput. **2021**, 1–35 (2021)
6. Mazini, M., Shirazi, B., Mahdavi, I.: Anomaly network-based intrusion detection system using a reliable hybrid artificial bee colony and AdaBoost algorithms. J. King Saud Univ.-Comput. Inf. Sci. **31**(4), 541–553 (2019)
7. Khan, M.A.: HCRNNIDS: hybrid convolutional recurrent neural network-based network intrusion detection system. Processes **9**(5), 834 (2021)
8. Devan, P., Khare, N.: An efficient XGBoost–DNN-based classification model for network intrusion detection system. Neural Comput. Appl. **32**, 12499–12514 (2020)
9. Nguyen, M.T., Kim, K.: Genetic convolutional neural network for intrusion detection systems. Futur. Gener. Comput. Syst. **113**, 418–427 (2020)

10. Jiang, K., Wang, W., Wang, A., Wu, H.: Network intrusion detection combined hybrid sampling with deep hierarchical network. IEEE Access **8**, 32464–32476 (2020)
11. Xu, C., Shen, J., Du, X., Zhang, F.: An intrusion detection system using a deep neural network with gated recurrent units. IEEE Access **6**, 48697–48707 (2018)
12. Khan, M.A., Karim, M.R., Kim, Y.: A scalable and hybrid intrusion detection system based on the convolutional-LSTM network. Symmetry **11**(4), 583 (2019)
13. Lee, J., Park, K.: AE-CGAN model based high performance network intrusion detection system. Appl. Sci. **9**(20), 4221 (2019)
14. Wang, Y., Meng, W., Li, W., Li, J., Liu, W.X., Xiang, Y.: A fog-based privacy-preserving approach for distributed signature-based intrusion detection. J. Parallel Distrib. Comput. **122**, 26–35 (2018)
15. Einy, S., Oz, C., Navaei, Y.D.: The anomaly-and signature-based IDS for network security using hybrid inference systems. Math. Probl. Eng. **2021**, 1–10 (2021)
16. Geethanjali, N.: A multi-fusion pattern matching algorithm for signature-based network intrusion detection system (2016)
17. Kwon, H.Y., Kim, T., Lee, M.K.: Advanced intrusion detection combining signature-based and behavior-based detection methods. Electronics **11**(6), 867 (2022)
18. Almomani, O.: A hybrid model using bio-inspired metaheuristic algorithms for network intrusion detection system. Comput. Mater. Continua **68**(1) (2021)
19. Zavrak, S., İskefiyeli, M.: Anomaly-based intrusion detection from network flow features using variational autoencoder. IEEE Access **8**, 108346–108358 (2020)
20. Aljawarneh, S., Aldwairi, M., Yassein, M.B.: Anomaly-based intrusion detection system through feature selection analysis and building hybrid efficient model. J. Comput. Sci. **25**, 152–160 (2018)
21. Resende, P.A.A., Drummond, A.C.: Adaptive anomaly-based intrusion detection system using genetic algorithm and profiling. Secur. Priv. **1**(4), e36 (2018)
22. Shone, N., Ngoc, T.N., Phai, V.D., Shi, Q.: A deep learning approach to network intrusion detection. IEEE Trans. Emerg. Top. Comput. Intell. **2**(1), 41–50 (2018)
23. Abdulhammed, R., Musafer, H., Alessa, A., Faezipour, M., Abuzneid, A.: Features dimensionality reduction approaches for machine learning based network intrusion detection. Electronics **8**(3), 322 (2019)
24. Injadat, M., Moubayed, A., Nassif, A.B., Shami, A.: Multi-stage optimized machine learning framework for network intrusion detection. IEEE Trans. Netw. Serv. Manag. **18**(2), 1803–1816 (2020)
25. Caville, E., Lo, W.W., Layeghy, S., Portmann, M.: Anomal-E: a self-supervised network intrusion detection system based on graph neural networks. Knowl.-Based Syst. **258**, 110030 (2022)
26. Zarrabi, A., Zarrabi, A.: Internet intrusion detection system service in a cloud. Int. J. Comput. Sci. Issues (IJCSI) **9**(5), 308 (2012)
27. Liao, H.J., Lin, C.H.R., Lin, Y.C., Tung, K.Y.: Intrusion detection system: a comprehensive review. J. Netw. Comput. Appl. **36**(1), 16–24 (2013)
28. Hinton, G.E., Osindero, S., Teh, Y.W.: A fast learning algorithm for deep belief nets. Neural Comput. **18**(7), 1527–1554 (2006)
29. Liu, A., Sun, B.: An intrusion detection system based on a quantitative model of interaction mode between ports. IEEE Access **7**, 161725–161740 (2019)
30. Kasongo, S.M., Sun, Y.: A deep long short-term memory based classifier for wireless intrusion detection system. ICT Express **6**(2), 98–103 (2020)

Intrusion Detection System Trends: An Overview of Current Advances in IoV & Communication Networks

Sangapu Sreenivasa Chakravarthi[1(✉)], Rishabh Rajput[1], S. Sountharrajan[1], and E. Suganya[2]

[1] Amrita School of Computing, Amrita Vishwa Vidyapeetham, Chennai, Thiruvallur, India
{ss_chakravarthi,s_sountharrajan}@ch.amrita.edu
[2] Sri Sivasubramaniya Nadar College of Engineering, Chennai, India
suganyae@ssn.edu.in

Abstract. The proliferation of Internet of Vehicles (IoV) indeed has paved the way for an inter-connected system of transportation, offering unprecedented convenience and efficiency. However, this network is susceptible to various intrusions of malicious means. The volatility of security in IoV arises from the interconnected and dynamic nature of vehicular communication. With the increasing number of connected vehicles and the continuous evolution of communication protocols, the attack surface expands, rendering traditional security measures inadequate. The potential consequences of security breaches in IoV are far-reaching, encompassing safety risks, privacy infringements, and disruptions to critical transportation infrastructure. Threat identification systems emerge as a critical defense mechanism, along with firewalls and other security measures. This paper assesses a comprehensive overview of existing IDS approaches to cater to different kinds of intrusions, implemented in IoV environments or in a different setting. The study provides a comprehensive overview of existing IDS approaches, evaluating their effectiveness in IoV settings as well as in alternative contexts. The discussion encompasses various aspects of intrusion detection within the IoV framework.

Keywords: intrusion detection · Internet of Vehicles (IoV)

1 Introduction

Proliferation of Internet of Vehicles (IoV) has paved the way for a connected transportation ecosystem, offering unprecedented convenience and efficiency. In order to ensure road safety and extremely efficient traffic flows, safety-critical information is exchanged amongst the connected vehicles via a high-bandwidth,

Published by Springer Nature Switzerland AG 2024
M. L. Owoc et al. (Eds.): ICCIDS 2024, IFIP AICT 718, pp. 36–50, 2024.
https://doi.org/10.1007/978-3-031-69986-3_3

low-latent vehicle-to-everything communication system. This technology, known as the Internet-of-Vehicles, or IoV, has garnered a great deal of attention in recent years. Prior to the introduction of the Internet of Vehicles (IoV), a contemporary technology for vehicle-to-vehicle (V2V) and vehicle-to-infrastructure (V2I) communication was the Vehicular Ad-Hoc Network (VANET). Vehicle-to-Vehicle (V2V), Vehicle-to-Infrastructure (V2I), Vehicle-to-Road Side Unit (V2R), Vehicle-to-Sensors (V2S), and Vehicle-to-Everything (V2X) are the components of the emerging Internet of vehicles [31] The widespread availability of wireless communication technology has made it possible for data interchange from moving vehicles, which has greatly advanced communication networks and increased traffic efficiency. VANETs take up the characteristics of Mobile Adhoc NETworks (MANET) with moving nodes and communicate wirelessly with limited wireless communication technology. Continuous Air Interface for Long and Medium range (CALM) and Wireless Access in Vehicular Environment (WAVE) are the communication standards used by VANETs. Dedicated ShortRange Communication (DSRC) is the vehicle-to-vehicle communication channel utilized in WAVE. Dynamic Spectrum Access (DSA), a feature of DSRC, enables communication between the car and unoccupied spectrum. The car can connect to a low bandwidth internet connection using the GSM2G/GPRS-2.5G, UMTS-3G, and infrared communication standards thanks to the CALM communication standard [58].

Automobiles are being equipped with external communication interfaces and electronic control units, in increasing amounts to provide safety and intelligent services to the user as mentioned in [48] & [39]. IoV has taken over the transportation systems because of a number of unique characteristics, including its large network scale, dynamic topological structures, dependable internet connections, high processing power, and interoperability with personal devices. Luxury cars are equipped with more than 150 ECUs [38] and 75% of the newly manufactured vehicles has capability to connect to internet. IoV networks (VANETS) are being adopted with 5G/6G networks to establish V2I, V2V & V2X [5,43] & [15] connections to help transfer large files such as high quality picture or video, as discussed in [57]. However, this interconnectedness introduces a plethora of security vulnerabilities, making IoV susceptible to various cyber threats.

Understanding the current assaults on VANETs [1] is crucial to improving their security. Several security attacks are covered in this section.Researchers across the globe have demonstrated and implemented attacks on vehicles as in articles [21,26,33,34] & [14] and various security risks are present [47]. One prominent example is of [35], where authors were able to intrude a vehicle network, through a Wifi open port and gained control of various automotive functionalities such as brakes, engine (on/off) etc. Intrusion Detection Systems (IDS) emerge as a critical defense mechanism to safeguard the integrity and functionality of IoV networks [29]. This research paper gives a thorough review of existing IDS approaches tailored for IoV environments. Various other countermeasures can be used to provide security against various malicious attacks such as encryption techniques and authentication technologies, firewall policies and intrusion detection systems; But "why is there an emphasis on developing IDS?". While encryption and authentication methods play a role, in ensuring network security they can pose challenges and limitations in the context of IoV. The cost bandwidth computing capacity and storage constraints of these networks often make it impractical to implement techniques. Additionally ensuring real time reliability becomes a concern.

2 Literature Survey

2.1 Review Scope

The study encompasses a broad spectrum of research publications and endeavors, including literature reviews, ensemble frameworks, hybrid models, deep learning architectures, and models. By formulating well-considered research questions, we want to focus our extensive survey.

2.2 Research Questions

The research questions guiding the survey on current advances towards network security in IoV & communication network which will be answered in the subsequent sections are

RQ1 *What are the current possible attacks and the potential damages it can have on VANET's and the vehicles involved?*
RQ2 *What are the current existing Boolean Logic methodologies and how effective are they in detecting intrusions in IoV Networks?*
RQ3 *What challenges are posed by the real time IoV network datasets and how does it effect the existing methodologies?*
RQ4 *Why is exploring Fuzzy logic based methodologies crucial for the accurate detection in real time networks?*
RQ5 *How does Fuzzy Logic based methodologies over come the challenges faced by IoV systems?*

2.3 Attacks

For the study, one needs to be aware of the different malicious attacks possible in in-vehicular networks (IVN) and then about the various automtive cyber systems as mentioned in articles [8,9] & [53]. Experimental attacks are mentioned and described below. Researchers of [55] divides the attacks on IVN into three categories based on different network layers, which are base layer which is physical layer, with data-link layer on the physical layer and application layer residing on the topmost in the hierarchy. Attacks in the physical layer (sensing layer for IoV environment) affect the range of sensors that are equipped to collect information in the vehicle such as LiDAR, RaDAR, GPS and cameras [40]. Vitality is a necessity of these sensors as one is moving towards autonomous driving, safe and accurate working of the sensors is required. Authors of [40] demonstrated various remote cyberattacks like blinding, jamming, replaying and spoofing, these attacks affected the camera and LiDAR sensor of the vehicle, which are basically the perception system of an autonomous vehicle.

Scholars of [41] implemented an TPMS (Tire-pressure Monitoring Systems) spoofing and tracking attack. Next category discussed was in Data-Link Layer, in these attacks the adversaries have access to the devices within the network (due to less security countermeasures), this illegal access leads to system failure because multiple cyber attacks can be performed such as DoS attack, frame - fabrication, sniffing, injection etc. The study conducted by authors of [10] analyzed multiple intrusion entrances in vehicle network such as OBD port (On-Board Diagnostics), bluetooth port, and cellular port. In an experiment by the authors of [11], implemented a DoS attack on the Controller Area Network (CAN) physical layer. Application layer attacks or one can say attacks on the interfaces. Adversaries with malicious intent can carry out various remote attacks and can gain access to functionalities like vehicle braking, ABS, lights, radio, taking control of dashboard, engine and equipments, control the window etc. With this we have answered RQ1 (Table 1).

Table 1. Attacks Performed Experimentally on Vehicular Network

Reference	Security Attack	Vulnerability	Impact
[54]	Bluetooth, Wi-Fi, Port-OBD	Sniffing frame, injection, Replaying & DoS attack	Control dashboard, engine
[20]	Wi-Fi	Remote wireless access into control system	Fully controlled vehicle
[21]	Port-OBD	Sniffing frame, injection, Replaying & DoS attack	Control radio, engine, body
[17]	CAN bus Port-OBD	Sniffing frame, Replaying	Control ABS, lights, window

2.4 Intrusion Detection Systems

2.4.1 Boolean Logic

In IoV [49], a new approach is suggested for the successful identification of malicious attacks. The proposed methodology combines the use of Long Short Term Memory (LSTM) and Gated Recurrent Unit (GRU) to form its foundation. Experimental results achieved by the model show that model has higher attack detection accuracy 99.5% for DDoS attack, the dataset used was a combined Distributed Denial of Service dataset that contains CIC DoS, CICIDS2017, and CSEIDS version of 2018. The dataset contains benign and the most up-to-date common attacks, which resembles the true real-world data (PCAPs). The implemented attacks include Brute Force FTP, Brute Force SSH, DoS, Heartbleed, Web Attack, Infiltration, Botnet and DDoS.

In a study [30], real time data was collected, using a small four wheeled robotic land vehicle. The proposed architecture employs a recurrent neural network architecture in addition to a deep multilayer perceptron. The above is improved by incorporating a new integral layer called as the short term memory, which helps in detecting attacks, with minimal false alarms. To conduct experiments malware, command injection and denial of service attacks were considered as illustrations of cyberattacks that're applicable, to a vehicle. The architecture achieved an accuracy rate of 86.9%.

The research paper [3] presents a new path for threat identification Intrusion Detection in In-Vehicle Networks (IVN) utilizing a Convolutional Long Short-Term Memory (ConvLSTM) based method. Given methodology capitalizes on inherent periodicity found in network message IDs. To train the ConvLSTM model, the study introduced us with a Federated based of Learning framework with a refined selection of client process. This system functions in a client-server mode where the In-Car Vehicles (ICVs) serve as the locally connected clients, while the Mobile Edge Computing (MEC) servers, connected to base stations (BSs), operate as parameterized servers. Additionally a Proximal Policy Optimization (PPO) schema deploying Federated Client Selection (FCS) is made within this framework in order to enhance both the model efficacy and system efficiency during ConvLSTM training of the model. The paper conducts simulations within real-world Internet of Vehicles (IoV) settings using authentic IVN datasets. The results demonstrate significant reductions in model size and convergence time due to the application of ConvLSTM. Moreover, the experimented PPO-based FCS scheme surpasses benchmarks in terms of rates like convergence rate, model accuracy, and system overhead, ensuring a consistent 95% or higher detection accuracy. This comprehensive evaluation showcases the efficacy and superiority of the proposed methodology in the context of IVN intrusion detection.

In yet another similar study, a new ensemble IDS framework referred to as Leader Class and Confidence Decision Ensemble (LCCDE) was implemented and proposed by [56]. It is built by selecting the most effective machine learning model from three sophisticated machine learning algorithms which namely are (XGBoost, LightGBM, and CatBoost) for each class or kind of attack. Then, in order to detect different kinds of cyberattacks accurately, the class leader model's each with their prediction confidence values are employed and executed. The framework achieves the accuracy of 99.9998 %, the dataset used are Car Hacking and CICIDS 2017.

The study offers an analysis of this proposed architectures convergence by employing a Bayesian hierarchical model to represent it. Each of the layer can be simultaneously represented using a Gibbs distribution enabling the translation of the goal of the backpropagation algorithm to a distribution kind on a energy function. The inclusion of the error term is the one that makes objective function remarkably sharper compared to the one before resulting in improvements in both training error and convergence. Finally, the sensitivity and precision of the proposed approach was compared with three techniques: first being - Support Vector Machines (SVM), secondly - Principal Component Analysis (PCA) and last one is Neural Networks (NN). The results vary on the No. of OBU's, the proposed architecture has 97.60% when 12 OBU's are used and 81.73% when 30 OBU's are used.

Scholars of [7] devised an intrusion model based on Deep Learning (DL) specifically aimed at combating different types of Denial of Service (DoS) attacks [51]. To assess feature importance and scoring, the study utilized K-Means clustering and introduced a novel approach called the Explainable Neural Network (xNN). This xNN was extensively employed to segregate attacks within the CIC2019 IDS dataset and the UNSW-NB15 dataset, post extracting most effective features for anomaly detection. In direct comparison, the research illustrated the superior performance of xNN when employing the feature-scoring method. Notably, achieving the highest accuracy of 99.7% in the UNSW-NB15 dataset, xNN outperformed CNN, LSTM, and DNN, which registered accuracies of 87%, 90%, and 92%, respectively. Within the CICIDS2019 dataset, CNN achieved an accuracy of 87%, LSTM reached 89%, and DNN attained 82%. Impressively, xNN obtained the highest accuracy of 99.3% when segregating.

Another AI powered intrusion detection architecture for IoV by [2]. This study provides a Deep Learning Engine (DLE)-based AI-based intrusion detection architecture for identifying and categorizing vehicle traffic in IoV networks into possible categories of cyberattacks, utilizing the messages that are collected from the vehicles. As the nature of the vehicular networks is mobile and the real time requirements, taking this into consideration, the study also shows that DLE's can be deployed on Multi-access Edge Computing (MEC) rather than operating on a remote cloud.

In a study conducted by [37] researchers developed a Markov model (HMM) with the goal of detecting states within the data collected from automobiles. And, authors in [32] introduced the inaugural algorithm devised for analyzing message ID sequences on the CAN bus. This algorithm aimed to establish a framework outlining the typical behavior of a CAN bus by identifying distinct characteristics and repetitive patterns amongst observed sequences of message IDs on the bus.

Reference paper [44] proposed the use of deep learning model "Generative Adversarial Nets" to create a Intrusion Detection System (GIDS) based on GANs. GIDS exhibits great detection accuracy for four unknown attacks and can be trained to identify unknown attacks using only standard data.

The publication authored by [24] concentrates on tackling the challenge of updating intrusion detection models within the Internet of Vehicles (IoV) to adeptly combat continuously evolving cyber attacks. The authors introduced dual model update strategies contingent upon the availability of labeled data sourced from the IoV cloud. The initial approach, termed the cloud-assisted update scheme, relies on a small quantity of labeled data supplied by the IoV cloud. Conversely, the subsequent strategy, known as the remote local update scheme, leverages pseudo-labeled data generated through pre-classification when the IoV cloud is incapable of furnishing immediate labeled data.

These proposed schemes notably elevate the accuracy of intrusion detection within the targeted domain, as evidenced by experimental outcomes displaying an enhancement of at least 23% in comparison to existing methodologies. This underscores the efficacy of the suggested transfer learning-based strategies for updating intrusion detection models within the IoV, consequently fortifying security and dependability in vehicular networks. Hence we have also answered RQ2 with this section.

2.4.2 Fuzzy Logic

Numerous capabilities of fuzzy systems makes it appropriate for use as intrusion detection systems as thoroughly discussed in articles [12,34] & [45]. First of all, integration of inputs from various sources is possible. Secondly, despite certain aspects being unclear, intrusions are actively handled and dealt with. Thirdly, the degree of membership can be recognized by fuzzy systems. Furthermore, fuzzy or irregular rule-based detection systems offer a powerful way to identify the not so precise tendencies of events in computer networks [50]. These capabilites help answer RQ5.

Authors of [22] introduced a novel model, the hybrid-based intrusion detection system (GA-Fuzzy), designed to manage the large-scale NSLKDD Dataset effectively. This system aims to detect attacks efficiently while reducing misclassification rates. Using genetic algorithms (GA), a new pattern, including new features and records, is generated to train the fuzzy classifier effectively. The integration of Principle Component Analysis (PCA) as a feature selection technique enhances efficiency by eliminating redundant and irrelevant data from the NSLKDD dataset, resulting in a remarkable detection accuracy of 99.96% and a false alarm rate of 0.04.

Meanwhile, scholars of article [52] addressed the challenge of achieving higher detection rates, lower false positives, and improved stability for Intrusion Detection Systems (IDS). The research introduced FC-ANN (Fuzzy Clustering - Artificial Neural Network), a novel technique combining artificial neural networks (ANN) and fuzzy clustering. The process involves creating multiple training subsets through fuzzy clustering, training various ANN models based on these subsets, and aggregating the outcomes using a fuzzy aggregation module, serving as a meta-learner. Experimental results on the KDD CUP 1999 dataset showcased the superiority of FC-ANN over BPNN, decision trees, and naïve Bayes in terms of detection precision and stability. Authors of [25] came up with a cuckoo search algorithm-optimized PCA and FCM clustering technique for IDS. PCA was used to identify the key features, and the Cuckoo Search algorithm was used to optimize the first cluster center selection in order to avoid hitting the local optimum. KDD CUP 1999 is a standard set of data which includes a wide variety of intrusions simulated in a military network environment. It contains 311,027 total records, with attack - 250,436 and normal - 60,591.

Another fuzzy-logic kind on an intrusion detection system that detects anomalous behavior within a network created by authors of [42]. Employing fuzzy-based inference techniques, a module for fuzzy decision-making is under development to enhance the system's accuracy in attack detection. Through a rule learning process, an optimized collection of fuzzy rules for the inference method has been autonomously generated, enabling more effective intrusion detection in computer networks. Initially, the precise directives were derived by mining a unit length common items from the normal and malicious scenarios' data. These precise rules were then transformed into fuzzy rules through a fuzzification process, subsequently utilized by the fuzzy system to classify test data. Similar experimentation was conducted in [19] and fuzzy rule interpolation based IDS was created.

Reference [16] introduces an innovative approach to intrusion detection systems (IDS) by employing a hybrid segregation method utilizing algorithms like AFS and ABC. The initial phase involves meticulous processing of the training dataset through FCM clustering coupled with correlation-based feature selection, meticulously eliminating irrelevant features from consideration. Subsequently, the IDS utilizes the CART method to generate specific rules aligned with the selected features. The training procedure incorporates both NSLKDD and UNSW-NB15 datasets, showcasing an impressive achievement of maintaining a low false positive rate (FPR) and a high scrutiny rate while ensuring computational complexity and time costs remain economically viable, as asserted by the authors. UNSW-NB 15 dataset was created by the IXIA PerfectStorm tool in the Cyber Range Lab of UNSW Canberra for generating a hybrid of real modern normal activities and synthetic contemporary attack behaviours. The tcpdump tool was utilised to capture 100 GB of the raw traffic. NSLKDD is the improved version of KDD CUP '99, covering 4,898,431 records, with attack - 3,925,650 and normal - 972,781.

On another note, reference [50] stands out in the realm of threat enumerations by addressing the critical issue of uncertainty in differentiating between normal and malicious network traffic. The paper introduces a fuzzy IDS tailored for categorizing Denial-of-Service (DoS) attacks, integrating a spectrum of fuzzy techniques such as fuzzy Singleton, Mamdani's inference, Mamdani's product implication, and center average de-fuzzifier. Notably, the construction of membership functions relies on the intricate methodologies of the fuzzy C-Means algorithm and pseudo-trapezoidal functions. The system's robustness is evident through testing on the KDD Cup 99 dataset, showcasing an outstanding 99.91% detection rate specifically against flooding attacks, all the while effectively reducing occurrences of false alarms.

Reference [36] introduces a fuzzy logic-based IDS specifically targeting the prediction of TCP SYN flooding attacks. The methodology involves observing the range of values for each attribute (normal, attack, and mixed data) and calculating their average values. These values are categorized into three membership levels (low, medium, and high) using triangular membership functions. The system aims to detect the Neptune attack and is evaluated using a dataset derived from KDD-99.

The paper referenced as [23] introduces a misuse detection method for IDS that employs a hybrid semi-supervised ML path combining FCM clustering with non-passive learning SVM. While the NSLKDD benchmark IDS dataset is used for testing, the scheme's performance is primarily tested against the k-means clustering method, suggesting the need for additional validation to confirm its outcomes.

In reference [13], the integration of fuzzy systems and the Particle Swarm Optimization (PSO) algorithm for intrusion detection in computer networks is highlighted. The proposed approach combines the interpretability of fuzzy systems for rule generation with the optimization capabilities of the PSO algorithm. The study demonstrates that the resulting fuzzy-PSO approach exhibits high classification performance and involves a few simple rules that are easily interpretable by humans. Experimental studies showcase the effectiveness of the proposed approach in classifying intrusion instances with high accuracy and providing understandable rules.

In another work [28], fuzzy clustering and parallel optimization feature extraction methods are applied to develop a threat identification technique for Wireless Sensor Networks (WSNs). The study focuses on using a parallel optimization feature extraction approach relied on an ideal feature assessment index to reduce data dimensionality while preserving diversity and reducing feature redundancy.

Reference [27] proposes a novel threat identification scheme for the Internet of Things (IoT) based on a suppressed fuzzy clustering (SFC) algorithm and a novel principal component analysis (PCA) algorithm. The approach involves data initialization, pre-processing, and detection frequency adjustment. Through a series of steps including PCA to reduce variables and SFC for categorization, the algorithm aims to efficiently detect attacks, demonstrating improved efficiency and accuracy in simulation experiments compared to neural network and Bayesian algorithms.

In reference to paper [4], an approach based on anomaly detection utilizes a fuzzy logic inference system which notices Distributed Denial of Service (DDoS) attacks. This research focuses on addressing intrusion boundaries using fuzzy logic and employs conventional fuzzy inference techniques to identify newer forms of DDOS attacks. The experiments conducted show a commendable true-positive rate of 91.1% and an impressively low false-positive rate of 0.006%.

Meanwhile, a study conducted in article [6] outlines a methodology for intrusion detection in Internet of Things (IoT) networks. This research introduces a multi-level random forest algorithm integrated with an inference system. The primary emphasis lies in enhancing the feature selection process by combining filter and wrapper methods. Initially, a filter method involving correlation-based feature selection and a genetic search algorithm identifies significant features. Subsequent refinement through a wrapper method selects top features based on classifier accuracy. These selected features are then employed with the forest mechanism for intrusion detection. Moreover, fuzzy logic is incorporated to categorize intrusions into medium or high-level classifications, effectively minimizing mis-classification errors. Comparative analysis against existing models using the data set demonstrates the superior performance of this developed model across metrics such as accuracy, precision, sensitivity, specificity, and the F1 score.

Notably, this study distinguishes itself by employing a multi-level characteristics' selection technique and integrating fuzzy logic for segregation of attacks.

2.5 Comparative Study

The comparative study of the aforementioned research endeavors provides integral insights within the positives and limitations of different intrusion detection approaches within the context of Internet of Vehicles (IoV) security. Boolean logic-based techniques, has demonstrated commendable accuracy in detecting anomalies within vehicular networks.

Conversely, Fuzzy Logic-based techniques offer a unique advantage in handling uncertainty and imprecision in network data, thus contributing to a nuanced understanding of anomalous behavior. From the Table 2 that provides the short compact representation of all the top edge researches for intrusion detection provided by fuzzy logic, they are compared with each other based on the technology or algorithms used, their final result and what data set the researchers worked on. The work is ordered in terms of accuracy of all the reviewed architectures.

Clearly, Fuzzy Logic - based techniques are performing more consistently, paired with other different approaches of Machine Learning produces high performing model for the proposed Intrusion Detection System. And clearly stated in [46] & [18] that VANETs frequently disconnect, resulting in lossy and incomplete data and handling uncertain & vagueness is speciality of Fuzzy Logic - based techniques, it is safe to say that Fuzzy Logic Techniques will perform better than the Boolean Logic Techniques, when implemented in real time. And thus with the discussion of sections of Fuzzy logic techniques and from the comparative study, we have successfully answered RQ3, RQ4, RQ5, hence completing our research.

2.6 Conclusion

As discussed, this survey research paper has provided an in-depth understanding on the Internet of Vehicles (IoV) paradigm, emphasizing its pivotal role in revolutionizing transportation systems and how detection mechanism shields the systems from potential threats and vulnerabilities. The research delves into recognizing the multifaceted challenges posed by potential threats and vulnerabilities. Through an extensive review of existing literature, this study explored the various approaches to intrusion detection systems (IDS) tailored for IoV environments. The analysis highlighted the versatility and effectiveness of both Boolean logic-based techniques, including Machine Learning (ML) models and Deep Learning models within IoV networks, as well as Fuzzy Logic-based approaches in addressing security concerns within Networks.

Boolean logic-based IDS, especially those employing ML and Deep Learning, exhibit promising capabilities in accurately detecting intrusions and safeguarding vehicular communications. These models leverage the power of data-driven

Table 2. Summary of Fuzzy Logic Intrusion Detection Systems

Work	Technology Used	Result	Dataset's Network Type
[50]	Fuzzy Singleton, Mamdanie's inference with Fuzzy C-Means algorithm	99.91%	Military Network
[23]	Genetic Algorithm with Fuzzy Classifier & PCA	99.6%	Military Network
[49]	Fuzzy Clustering with active learning SVM	99.6%	Military Network
[6]	Multi-level RF with Fuzzy Inference	99.46%	Military Network
[27]	Suppressed Fuzzy Clustering & PCA	97.1%	–
[52]	Fuzzy Clustering with ANN and Fuzzy Aggregation Module	96.7%	Military Network
[25]	PCA + Cuckoo Search Algorithm for optimization + Fuzzy Clustering	92.4%	Military Network
[36]	Fuzzy logic IDS with Triangular membership functions for TCP SYN flooding attack	93.24%	Military Network
[4]	Fuzzy Detection Mechanism for DDoS & Fuzzy Inference Technique	91.1%	Open Source DDoS Network Traffic
[42]	Fuzzy Inference Technique	>90%	KDD CUP 1999 Dataset

insights to adapt and evolve in the face of evolving threat landscapes. Additionally, the incorporation of Fuzzy Logic techniques has proven instrumental in handling uncertainty and imprecise data, a common occurrence in dynamic IoV environments.

Notably, the Fuzzy Logic based mechanism excels in managing the data uncertainty, thereby consistently performing well even in scenarios of communication cessation. Given the daunting challenges that arise in the execution of a secure system, a combination of Fuzzy Logic with machine learning has proven itself to be a promising solution for the enhancement of IoV security. These fuzzy approaches haven't been fully utilized in the IoV environment. Fuzzy logic-based IDS can contribute valuable insights into anomaly detection or intrusion detection for IoV networks, offering a nuanced understanding of behaviors that may deviate from established norms.

References

1. Al-shareeda, M.A., Alazzawi, M.A., Anbar, M., Manickam, S., Al-Ani, A.K.: A comprehensive survey on vehicular ad hoc networks (VANETs). In: 2021 International Conference on Advanced Computer Applications (ACA), pp. 156–160 (2021). https://doi.org/10.1109/ACA52198.2021.9626779
2. Alladi, T., Kohli, V., Chamola, V., Yu, F.R., Guizani, M.: Artificial intelligence (AI)-empowered intrusion detection architecture for the internet of vehicles. IEEE Wirel. Commun. **28**(3), 144–149 (2021). https://doi.org/10.1109/MWC.001.2000428
3. Allah, S., et al.: Federated AI-enabled in-vehicle network intrusion detection for internet of vehicles. Electronics | Free Full-Text (2023). https://www.mdpi.com/2079-9292/11/22/3658. Accessed 28 Oct 2023
4. Almseidin, M., Al-Sawwa, J., Alkasassbeh, M.: Anomaly-based intrusion detection system using fuzzy logic. In: 2021 International Conference on Information

Technology (ICIT), pp. 290–295 (2021). https://doi.org/10.1109/ICIT52682.2021.9491742
5. Alnasser, A., Sun, H., Jiang, J.: Cyber security challenges and solutions for V2X communications: a survey. Comput. Netw. **151**, 52–67 (2019). https://doi.org/10.1016/j.comnet.2018.12.018, https://www.sciencedirect.com/science/article/pii/S1389128618306157
6. Awotunde, J.B., Ayo, F.E., Panigrahi, R., Garg, A., Bhoi, A.K., Barsocchi, P.: A multi-level random forest model-based intrusion detection using fuzzy inference system for internet of things networks. Int. J. Comput. Intell. Syst. **16**, 31 (2023)
7. Aziz, S., et al.: Anomaly detection in the internet of vehicular networks using explainable neural networks (XNN). Mathematics **10**(8), Art. no. 8 (2022). https://doi.org/10.3390/math10081267
8. Borazjani, P.N., Everett, C.E., McCoy, D.: OCTANE: an extensible open source car security testbed. In: Embedded Security in Cars (2014)
9. Chakraborty, S., Al Faruque, M.A., Chang, W., Goswami, D., Wolf, M., Zhu, Q.: Automotive cyber-physical systems: a tutorial introduction. IEEE Design Test **33**(4), 92–108 (2016). https://doi.org/10.1109/MDAT.2016.2573598
10. Checkoway, S., et al.: Comprehensive experimental analyses of automotive attack surfaces. In: 20th USENIX Security Symposium (USENIX Security 2011). USENIX Association, San Francisco (2011). https://www.usenix.org/conference/usenix-security-11/comprehensive-experimental-analyses-automotive-attack-surfaces
11. Cho, K.T., Shin, K.G.: Error handling of in-vehicle networks makes them vulnerable. In: Proceedings of the 2016 ACM SIGSAC Conference on Computer and Communications Security, CCS 2016, pp. 1044–1055. Association for Computing Machinery, New York (2016). https://doi.org/10.1145/2976749.2978302
12. Dickerson, E., Dickerson, A., Koukousoula, O., Dickerson, A.: Fuzzy intrusion detection. In: 2001 Joint 9th IFSA World Congress and 20th NAFIPS International Conference, vol. 3, pp. 1506–1510. IEEE (2001)
13. Einipour, A.: Intelligent intrusion detection in computer networks using fuzzy systems (2012)
14. Fröschle, S., Stühring, A.: Analyzing the capabilities of the CAN attacker. In: Foley, S.N., Gollmann, D., Snekkenes, E. (eds.) ESORICS 2017. LNCS, vol. 10492, pp. 464–482. Springer, Cham (2017). https://doi.org/10.1007/978-3-319-66402-6_27
15. Ghosal, A., Conti, M.: Security issues and challenges in V2X: a survey. Comput. Netw. **169**, 107093 (2020). https://doi.org/10.1016/j.comnet.2019.107093, https://www.sciencedirect.com/science/article/pii/S1389128619305857
16. Hajisalem, V., Babaie, S.: A hybrid intrusion detection system based on ABC-AFS algorithm for misuse and anomaly detection. Comput. Netw. **136**, 37–50 (2018)
17. Hoppe, T., Kiltz, S., Dittmann, J.: Security threats to automotive can networks-practical examples and selected short-term countermeasures. Reliabil. Eng. Syst. Saf. **96**(1), 11–25 (2011). https://doi.org/10.1016/j.ress.2010.06.026, https://www.sciencedirect.com/science/article/pii/S0951832010001602. Special Issue on Safecomp 2008
18. Ji, B., et al.: Survey on the internet of vehicles: network architectures and applications. IEEE Commun. Stand. Mag. **4**(1), 34–41 (2020). https://doi.org/10.1109/MCOMSTD.001.1900053
19. Johanyak, Z.C.: Fuzzy logic based network intrusion detection systems. In: 2020 IEEE 18th World Symposium on Applied Machine Intelligence and Informatics (SAMI), pp. 15–16 (2020). https://doi.org/10.1109/SAMI48414.2020.9108750

20. K. Security lab: car hacking research: Remote attack tesla motors (2017). https://www.blackhat.com/docs/us-17/thursday/us-17-Nie-Free-Fall-Hacking-Tesla-From-Wireless-To-CAN-Bus.pdf
21. Koscher, K., et al.: Experimental security analysis of a modern automobile. In: 2010 IEEE Symposium on Security and Privacy, pp. 447–462 (2010). https://doi.org/10.1109/SP.2010.34
22. Kumar, P., Saravanan, M., Thenmozhi, M., Vijayakumar, K.: Intrusion detection system based on GA-fuzzy classifier for detecting malicious attacks. Concurr. Comput.: Pract. Exp. **33** (2019). https://doi.org/10.1002/cpe.5242
23. Kumari, V.V., Varma, P.R.K.: A semi-supervised intrusion detection system using active learning SVM and fuzzy C-means clustering. In: 2017 International Conference on I-SMAC (IoT in Social, Mobile, Analytics and Cloud) (I-SMAC), pp. 481–485 (2017). https://doi.org/10.1109/I-SMAC.2017.8058397
24. Li, X., Hu, Z., Xu, M., Wang, Y., Ma, J.: Transfer learning based intrusion detection scheme for internet of vehicles. Inf. Sci. **547**, 119–135 (2021). https://doi.org/10.1016/j.ins.2020.05.130
25. Li, Z., Su, Y., Han, Q.: Intrusion detection based on PCA and fuzzy clustering optimized by CS. In: 2017 Chinese Automation Congress (CAC), pp. 6334–6339 (2017). https://doi.org/10.1109/CAC.2017.8243919
26. Liu, J., Zhang, S., Sun, W., Shi, Y.: In-vehicle network attacks and countermeasures: challenges and future directions. IEEE Netw. **31**(5), 50–58 (2017). https://doi.org/10.1109/MNET.2017.1600257
27. Liu, L., Xu, B., Zhang, X., Wu, X.: An intrusion detection method for the internet of things based on suppressed fuzzy clustering. EURASIP J. Wirel. Commun. Netw. **2018**(1), 1–7 (2018)
28. Liu, Z., Wei, W., Wang, H., Zhang, Y., Zhang, Q., Li, S.: Intrusion detection based on parallel intelligent optimization feature extraction and distributed fuzzy clustering in WSNs. IEEE Access **6**, 72201–72211 (2018). https://doi.org/10.1109/ACCESS.2018.2879891
29. Loukas, G., Karapistoli, E., Panaousis, E., Sarigiannidis, P., Bezemskij, A., Vuong, T.: A taxonomy and survey of cyber-physical intrusion detection approaches for vehicles. Ad Hoc Netw. **84**, 124–147 (2019). https://doi.org/10.1016/j.adhoc.2018.10.002, https://www.sciencedirect.com/science/article/pii/S1570870518307091
30. Loukas, G., Vuong, T., Heartfield, R., Sakellari, G., Yoon, Y., Gan, D.: Cloud-based cyber-physical intrusion detection for vehicles using deep learning. IEEE Access **6**, 3491–3508 (2018). https://doi.org/10.1109/ACCESS.2017.2782159
31. Mahmood, A., Siddiqui, S.A., Sheng, Q.Z., Wang, X., Zeadally, S.: Trust on wheels: towards secure and resource efficient IoV networks. Computing **104**, 1337–1358 (2022). https://doi.org/10.1007/s00607-021-01040-7
32. Marchetti, M., Stabili, D.: Anomaly detection of can bus messages through analysis of ID sequences. In: 2017 IEEE Intelligent Vehicles Symposium (IV), pp. 1577–1583 (2017). https://doi.org/10.1109/IVS.2017.7995934
33. Marchetti, M., Stabili, D.: READ: reverse engineering of automotive data frames. IEEE Trans. Inf. Forensics Secur. **14**(4), 1083–1097 (2019). https://doi.org/10.1109/TIFS.2018.2870826
34. Miller, C., Valasek, C.: A survey of remote automotive attack surfaces. In: Black Hat USA (2014). Technical report 8
35. Miller, C., Valasek, C.: Remote exploitation of an unaltered passenger vehicle. In: Black Hat USA (2015). Technical report 23

36. Mkuzangwe, N.N.P., Nelwamondo, F.V.: A fuzzy logic based network intrusion detection system for predicting the TCP SYN flooding attack. In: Nguyen, N.T., Tojo, S., Nguyen, L.M., Trawiński, B. (eds.) ACIIDS 2017. LNCS (LNAI), vol. 10192, pp. 14–22. Springer, Cham (2017). https://doi.org/10.1007/978-3-319-54430-4_2
37. Narayanan, S., Mittal, S., Joshi, A.: Using data analytics to detect anomalous states in vehicles. arXiv.org (2015). https://www.semanticscholar.org/paper/Using-Data-Analytics-to-Detect-Anomalous-States-in-Narayanan-Mittal/325a5d20cb31033ada50d9ae1d330ff60692ee5e. Accessed 28 Oct 2023
38. NXP E. Team: Future advances in body electronics (2015). https://www.nxp.com/docs/en/whitepaper/BODYDELECTRWP.pdf
39. Ozguner, U., Acarman, T., Redmill, K.: Autonomous ground vehicles. Artech House, Norwood (2011)
40. Petit, J., Stottelaar, B., Feiri, M., Kargl, F.: Remote attacks on automated vehicle sensors: experiments on camera and lidar. In: Proceedings of the Black Hat Europe (2015)
41. Rouf, I., et al.: Security and privacy vulnerabilities of in-car wireless networks: a tire pressure monitoring system case study. In: Proceedings of the 19th USENIX Conference Security, p. 21 (2010)
42. S, R., Dr.N.Nagarajan: Network intrusion detection system using fuzzy logic. Indian J. Comput. Sci. Eng. **2** (2011)
43. Sedar, R., Kalalas, C., Vázquez-Gallego, F., Alonso, L., Alonso-Zarate, J.: A comprehensive survey of V2X cybersecurity mechanisms and future research paths. IEEE Open J. Commun. Soc. **4**, 325–391 (2023). https://doi.org/10.1109/OJCOMS.2023.3239115
44. Seo, E., Song, H.M., Kim, H.K.: GIDS: GAN based intrusion detection system for in-vehicle network. In: 16th Annual Conference on Privacy, Security and Trust (PST), pp. 1–6 (2018). https://doi.org/10.1109/PST.2018.8514157
45. Shajari, M., Ghorbani, A.A.: Using fuzzy system to manage false alarms in intrusion detection. In: Gritzalis, D., De Capitani di Vimercati, S., Samarati, P., Katsikas, S. (eds.) SEC 2003. ITIFIP, vol. 122, pp. 241–252. Springer, Boston, MA (2003). https://doi.org/10.1007/978-0-387-35691-4_21
46. Sharma, S., Kaushik, B.: A survey on internet of vehicles: applications, security issues & solutions. Veh. Commun. **20**, 100182 (2019). https://doi.org/10.1016/j.vehcom.2019.100182, https://www.sciencedirect.com/science/article/pii/S2214209619302293
47. Taslimasa, H., et al.: Security issues in internet of vehicles (IoV): a comprehensive survey. Internet Things **22**, 100809 (2023)
48. Tuohy, S., Glavin, M., Hughes, C., Jones, E., Trivedi, M., Kilmartin, L.: Intravehicle networks: a review. IEEE Trans. Intell. Transp. Syst. **16**(2), 534–545 (2015). https://doi.org/10.1109/TITS.2014.2320605
49. Ullah, S., et al.: HDL-IDS: a hybrid deep learning architecture for intrusion detection in the internet of vehicles. Sens. (Basel) **22**(4), 1340 (2022). https://doi.org/10.3390/s22041340
50. Varshovi, A., Rostamipour, M., Sadeghiyan, B.: A fuzzy intrusion detection system based on categorization of attacks. In: 2014 6th Conference on Information and Knowledge Technology (IKT), pp. 50–55 (2014). https://doi.org/10.1109/IKT.2014.7030332
51. Varshovi, A., Sadeghiyan, B.: A classification of denial of service attacks based on attack scenario. In: Computer Society of Iran (CSICC) Conference (2006)

52. Wang, G., Hao, J., Ma, J., Huang, L.: A new approach to intrusion detection using artificial neural networks and fuzzy clustering. Expert Syst. Appl. **37**(9), 6225–6232 (2010). https://doi.org/10.1016/j.eswa.2010.02.102
53. Wasicek, A., Derler, P., Lee, E.A.: Aspect-oriented modeling of attacks in automotive cyber-physical systems. In: 2014 51st ACM/EDAC/IEEE Design Automation Conference (DAC), pp. 1–6 (2014). https://doi.org/10.1145/2593069.2593095
54. Woo, S., Jo, H.J., Lee, D.H.: A practical wireless attack on the connected car and security protocol for in-vehicle can. IEEE Trans. Intell. Transp. Syst. **16**(2), 993–1006 (2015). https://doi.org/10.1109/TITS.2014.2351612
55. Wu, W., et al.: A survey of intrusion detection for in-vehicle networks. IEEE Trans. Intell. Transp. Syst. **21**(3), 919–933 (2020). https://doi.org/10.1109/TITS.2019.2908074
56. Yang, L., Shami, A., Stevens, G., De Rusett, S.: LCCDE: a decision-based ensemble framework for intrusion detection in the internet of vehicles. arXiv (2022). https://doi.org/10.48550/arXiv.2208.03399
57. Yin, X., Liu, J., Cheng, X., Xiong, X.: Large-size data distribution in IoV based on 5G/6G compatible heterogeneous network. IEEE Trans. Intell. Transp. Syst. **23**(7), 9840–9852 (2022). https://doi.org/10.1109/TITS.2021.3118701
58. Zamrai, M.B.: A survey on internet of vehicle (IoV): applications & comparison of VANETs, IoV and SDN-IoV. Internet Things **20**, 26–31 (2021)

Automation Xtreme - A Web Automation AI Tool

R. Abishek(✉) and M. Vijay Anand

Saveetha Engineering College, Chennai, Tamil Nadu, India
abishekabishek1954@gmail.com, vijayanand@saveetha.ac.in

Abstract. In the swiftly changing realm of web application development, the need for effective and dependable testing solutions is paramount. This document presents a groundbreaking Automated Web Application Testing Platform crafted to simplify the testing procedure and amplify overall testing efficacy. Leveraging powerful technologies such as Selenium for web automation, Pytest for testing, and Twilio for user support, the platform offers a comprehensive solution for web application testing. The platform's intuitive user interface streamlines the definition and execution of test cases, ensuring accessibility for users across diverse technical proficiency levels. A unique feature of dynamic code generation based on user-defined test cases allows for flexibility and adaptability in testing scenarios. The modular architecture ensures maintainability, scalability, and easy integration with other tools and frameworks. Extensive testing methodologies, including black box and white box testing, validate the platform's robustness and reliability. The integration of Twilio enables real-time communication through SMS notifications, enhancing user engagement and support during the testing process. The project's feasibility study highlights its economic, technical, and social viability, making it a cost-effective and user-centric solution. The system's architecture, methodology, and modules collectively contribute to a technically sound foundation, ensuring stability and adaptability in various testing scenarios. In this Automated Web Application Testing Platform presented in this paper stands as a valuable tool for addressing the challenges of modern web application testing. With a focus on user-centric design, flexibility, and efficiency, the platform aims to optimize the testing process and contribute to the continuous evolution of software development practices.

Keywords: Automation Testing · Web Application · Selenium · Pytest · Twilio · User-Friendly Interface · Test Case Execution Testing Efficiency · Dynamic Code Generation

1 Introduction

The primary aim of developing the Automated Web Application Testing Platform is to revolutionize and simplify the process of testing web applications. In the ever-evolving landscape of software development, web applications have become integral to various

Published by Springer Nature Switzerland AG 2024
M. L. Owoc et al. (Eds.): ICCIDS 2024, IFIP AICT 718, pp. 51–64, 2024.
https://doi.org/10.1007/978-3-031-69986-3_4

aspects of our digital lives. The significance of ensuring their reliability, security, and functionality cannot be overstated. Our tool aims to provide a comprehensive solution to the challenges associated with testing modern and complex web applications.

The motivation behind the development of this tool is rooted in addressing the inadequacies of existing testing tools. Traditional testing methods often struggle to cope with the dynamic and intricate nature of contemporary web applications. The need for a tool that combines automation, user-friendliness, and real-time support prompted the conceptualization of our platform. By recognizing the limitations of current tools, we aim to bridge the gap and provide a more efficient and accessible testing solution.

The central objective in creating the Automated Web Application Testing Platform is to empower developers and testers with a tool that streamlines the testing process. The integration of automation, user-friendly interfaces, and real-time communication features sets our tool apart. We aim to offer a solution that not only meets but exceeds the expectations of testers, providing them with the flexibility and scalability needed to address the complexities of modern web applications.

The scope of our platform is broad, encompassing various aspects of web application testing. It includes the creation, management, and execution of test cases through a user-friendly web interface. Additionally, the platform's integration with Selenium and pytest extends its scope by incorporating powerful automation capabilities. The introduction of the XPath Finder tool enhances the precision of test case creation, contributing to the overall efficiency of the testing process. The integration with Twilio for real-time SMS communication expands the scope by providing users with immediate assistance during testing.

Overall, The Automated Web Application Testing Platform is designed to be an integrated solution for testing web applications of varying complexities. It incorporates powerful testing frameworks, user-friendly interfaces, and real-time communication features to offer a holistic testing experience. From identifying web elements with the XPath Finder tool to generating customized Python code snippets, the platform provides a comprehensive suite of tools for testers. The overarching goal is to enhance the efficiency, reliability, and user experience of web application testing.

In the subsequent sections, we delve deeper into the design, implementation, feasibility, and performance aspects of our platform, offering a detailed exploration of the tool's capabilities and potential impact.

2 Literature Survey

The paper presents a Selenium-based Web Application Automation Test Framework, emphasizing the critical concern of Web application quality among developers, testers, and users. Existing automated testing methods for Web applications encounter challenges like low efficiency, handling dynamically loaded applications, and addressing frequent testing requirements. The proposed system introduces a Selenium-based framework tailored for automated testing of complex dynamic Web pages. The evaluation of the framework includes key aspects such as stability, test efficiency, and a comparative analysis with other frameworks. Results demonstrate the framework's high-test efficiency, offering stable and rapid support for automated testing of Web applications. The system represents a significant advancement in addressing the complexities of testing dynamic Web pages, contributing to enhanced software testing practices in Web development [1].

This paper presents a test automation framework tailored for web applications, leveraging the capabilities of Selenium and JMeter. In the contemporary software landscape, web-based applications have become ubiquitous, necessitating robust testing solutions. Selenium, a versatile set of tools compatible with various browsers, operating systems, programming languages, and testing frameworks, offers diverse approaches to automating tests for web-based applications. On the other hand, JMeter is employed to simulate substantial loads on servers, networks, or objects, assessing their strength and analyzing overall performance under different load conditions. The synergy of Selenium and JMeter in the proposed framework operates at both the protocol and user levels, respectively. The authors demonstrate the design of this automatic software testing framework, emphasizing how it enhances extensibility and reusability of automated tests. The empirical results indicate notable improvements in software product quality and development efficiency. This paper serves as a comprehensive guide detailing the intricacies of designing a web-based test automation framework, offering valuable insights into effective software testing methodologies [2].

This paper delves into the crucial realm of software testing, acknowledging its paramount significance in the Software Development Life Cycle. The primary objective of the testing process is to scrutinize and compare obtained results against user expectations, ensuring the robustness and reliability of software. Test Automation emerges as a pivotal solution, streamlining the tester's workflow through the automated execution of test scripts using specialized software. The focus of this paper centers on the utilization of Selenium WebDriver for testing a web application. It aims to showcase the integration of Selenium WebDriver with other tools such as Maven and TestNG, offering a more accessible approach to testing. The collaborative use of these tools contributes to an enhanced testing process, ultimately elevating the quality of software testing. This paper serves as a valuable resource for understanding the application of Selenium WebDriver and associated tools in the pursuit of effective and efficient software testing methodologies [3].

This paper explores the dynamic landscape of automation software testing, particularly within the context of web-based applications. In the ever-evolving realm of software development, Agile testing takes precedence as a methodology that integrates seamlessly with the principles of agile policy. Recognizing software improvement as a collaborative endeavor, akin to a client in the testing process, Agile testing emphasizes efficiency and iterative progress. To address the need for minimizing manual effort, automated testing becomes a crucial element in the testing process. The focal point of this paper revolves around a traditional automation testing model, providing a comprehensive examination of its intricacies and applications. Beyond the confines of traditional models, the paper propounds a specialized model designed explicitly for automated agile testing. This tailored model takes into account the unique demands and dynamics inherent in agile development methodologies. A significant contribution of this paper lies in its experimental work, which centers on the testing of a web-based application. The practical insights gleaned from this experimentation offer a nuanced understanding of the proposed automated agile testing model in action. The outcomes of the experimental work undergo meticulous evaluation through the lens of the agile testing model, facilitating a thorough comparison with traditional testing models. In essence, this paper navigates the intersection of automation, agile testing, and web-based applications, providing valuable insights into evolving methodologies. By delving into the experimental validation of proposed models, it offers a practical foundation for enhancing the efficiency and effectiveness of software testing processes [4].

This paper introduces a novel and effective approach to enhance testing methodologies for dynamic web applications. Titled "Virtual DOM coverage for effective testing of dynamic web applications," the research addresses the inherent challenges posed by traditional code coverage criteria in the realm of dynamic web application testing. In the context of software testing, code coverage is a fundamental criterion, widely acknowledged for its simplicity and efficacy. However, the paper highlights its inadequacy in dynamic web application testing, where focusing solely on server-side script code neglects critical client-side execution. This oversight hampers the ability to trigger essential client-server interactions required to reach crucial execution states. The proposed solution is Virtual DOM (V-DOM) Coverage, a pioneering criterion tailored for effective web application testing. Through static analysis, the paper outlines a method to aggregate all DOM objects generated by server scripts, constructing a V-DOM tree. This tree uniquely model's execution on both the client and server sides, surpassing the limitations of existing coverage criteria in web application testing. The efficacy of V-DOM coverage is empirically validated through an extensive study involving five real-world dynamic web applications. The research demonstrates that the V-DOM tree can comprehensively model a more extensive array of DOM objects compared to conventional web crawling techniques. Notably, the test selection based on the V-DOM tree criterion outperforms existing code coverage and UI element coverage methodologies. This superior performance is evidenced by the detection of a higher number of faults, thereby showcasing the potential of V-DOM coverage in advancing the effectiveness of dynamic web application testing [5].

This paper investigates the practical integration of User-Centered Design (UCD) practices into the context of agile Web development. Titled "Integrating user-centered design practices into agile Web development: A case study," the research introduces a hybrid process framework that extends traditional agile development processes by incorporating lightweight usability engineering practices. The study employs empirical methods, utilizing a case study conducted in an academic setting with postgraduate students. The primary objective is to assess the impact of the hybrid process on the quality of web products. This assessment is performed through a defect quality analysis that spans the entire development lifecycle, ranging from the gathering of user stories to the final delivery of the products. The quantitative data derived from the case study undergoes statistical analysis to reveal insights into the distribution of closing defects across various development phases. The findings indicate a significant improvement in defect distribution, accompanied by a complete reduction in the number of defects. This outcome signifies a substantial increase in the overall quality of the web products. In summary, the research contributes valuable insights into the practical implementation of UCD practices within an agile Web development context. The results of the case study highlight the effectiveness of the hybrid process framework in enhancing product quality by addressing defects throughout the development phases. This study serves as a noteworthy reference for organizations aiming to integrate user-centered design practices seamlessly into their agile development processes [6].

This paper explores the implementation of Selenium Automation and Report Generation using Selenium Web Driver and ATF (Automation Testing Framework). The focus is on testing within the domain of automation tools, particularly for software systems implemented as browser applications, which often pose challenges for testing due to their complexity. The goal of automation testing is to reduce human intervention and streamline the iterative testing process for various web applications. Notable automation tools, including Junit, Chakram, and Selenium, are discussed as key players in Automation Testing. The proposed solution introduces a resource monitoring technique to detect defective cases during testing. The testing methodology is implemented and tested in a Jerkin testing environment. To assess the performance of the proposed solution, a comparative analysis is conducted using two different testing scenarios. The comparison is based on various performance metrics, including debug ratio, throughput, fail test cases ratio, and memory consumption. The results obtained from the study demonstrate the effective performance of the proposed solution in comparison to traditional testing algorithms. This research contributes to the advancement of automation testing methodologies, particularly in the context of web applications, providing insights into enhanced performance and effectiveness through resource monitoring and comparative analysis [7].

3 Methodology

The methodology employed in this project follows a systematic and user-centric approach to deliver an efficient and reliable web UI testing and automation solution. The process begins with a thorough analysis of user needs and preferences, guiding the development of a user-friendly interface that simplifies the interaction between users and the system. The design incorporates a record-and-playback feature, enhancing efficiency and adaptability in the testing process.

The project utilizes the pytest framework for systematic testing, executing unit tests to ensure the functionality of each component. Black box testing is applied to assess the external behavior of the automation framework, while white box testing, including unit testing, integration testing, functional testing, system testing, and others, delves into the internal logic and structures of the system. This comprehensive testing strategy ensures the robustness and reliability of the entire system.

Verification plays a crucial role in confirming that the generated code snippets accurately represent user-defined test cases. Thorough testing and validation processes are conducted to verify the correctness of the automated scripts. The validation process extends to executing the generated automation scripts, providing users with real-time outcomes, and ensuring that the automated interactions align seamlessly with user intentions.

The project also employs effective input and output design strategies. The input design focuses on creating a seamless interaction between users and the system, incorporating user interfaces such as Ui Automation and Xpath finder to ensure a user-friendly experience. On the output side, the design includes result templates and test reports to present information in a clear and organized manner.

This methodology combines effective design, systematic testing, thorough verification and validation processes, and user-centric principles to create a robust and reliable web UI testing and automation solution. The emphasis on both user needs and technical precision contributes to the success of the project as a practical and user-friendly tool for daily testing activities.

Additionally, the data collection methodology plays a crucial role in obtaining comprehensive and dynamic datasets. User inputs are collected through the web interface, where users express their preferences and choices. Web scraping techniques are employed to extract relevant data from designated web pages using the Selenium library (Fig. 1).

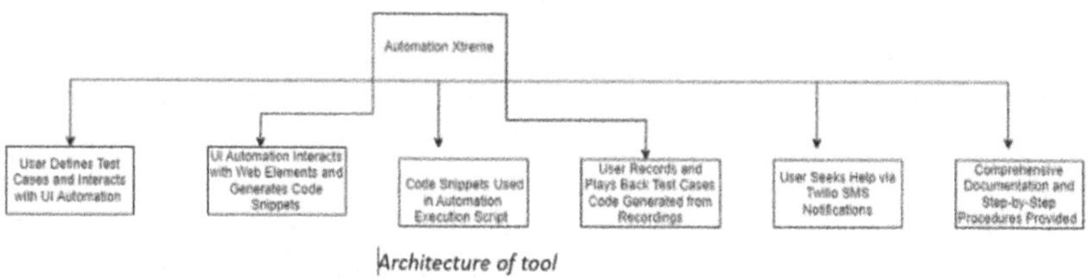

Fig. 1. Architecture of tool

The project workflow revolves around a seamless interaction between users and the automated testing framework. Users initiate the process by defining test cases and interacting with the User Interface (UI) Automation module. This UI Automation module engages with web elements, generating precise code snippets based on user-defined scenarios. These code snippets serve as the foundation for the execution script, ensuring the accurate automation of web interactions. In parallel, users can leverage the Record and Playback feature, allowing them to record test cases as videos. The recorded scenarios are then translated into code snippets, providing an alternative approach to defining automation scripts. For user assistance, the Help Module integrates with Twilio to enable SMS notifications, providing prompt help when needed. Additionally, the project offers extensive documentation, presenting users with comprehensive step-by-step procedures. This documentation serves as a valuable resource, guiding users through the automation process and ensuring a user-friendly experience (Fig. 2).

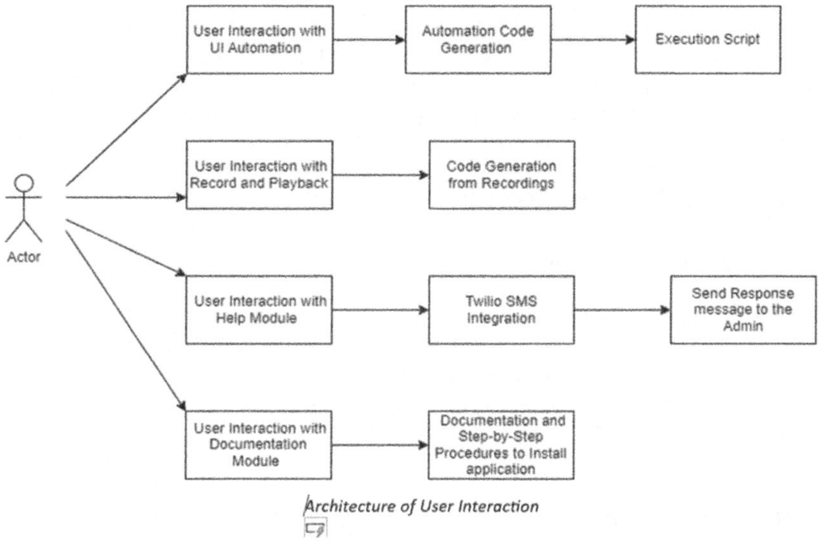

Fig. 2. Architecture of User Interaction

4 Proposed Work

The proposed system represents an innovative and user-centric web automation and testing solution, strategically designed to cater to the evolving demands of modern software development. At its core, the system leverages a combination of robust technologies, including Django, Selenium WebDriver, pytest, and Twilio, to provide a comprehensive tool for automating web interactions and streamlining the testing process. The user interface is carefully crafted for user-friendliness, ensuring accessibility for individuals with varying levels of technical expertise. The integration of Django as the web framework facilitates rapid development, clean design, and adherence to industry-standard practices, contributing to the system's overall reliability.

Selenium WebDriver, a key component, empowers users to automate web browsers efficiently. What sets the proposed system apart is its dynamic code generation capability, where Python code snippets are automatically generated based on user-defined test cases. This modular approach not only enhances maintainability but also allows for flexibility in adapting to diverse testing scenarios. The inclusion of pytest facilitates seamless testing, providing features for test discovery, fixtures, and assertions. The WebDriver manager ensures efficient management of web drivers, reducing potential technical challenges and enhancing overall compatibility.

Twilio integration introduces real-time communication through SMS notifications, offering users prompt assistance and updates. This aligns with the contemporary need for effective and immediate communication during the testing process, contributing to an enhanced user experience. The proposed system excels in automated script execution, delivering quick and reliable test results. Its modular structure ensures that different functionalities, such as UI automation, XPath expression finding, code generation, and helpdesk services, are encapsulated in separate modules, contributing to code organization and readability. In essence, the proposed system is a sophisticated and forward-thinking solution, poised to significantly contribute to the efficiency and effectiveness of web automation and testing in the realm of software development.

5 Implementation and Result

5.1 Hardware Requirements

The project demands a standard personal computer or laptop for optimal performance.

A multi-core processor is recommended to ensure efficient execution of web automation and recommendation functionalities.

Adequate RAM, preferably 8 GB or higher, contributes to the smooth operation of the system.

5.2 Software Requirements

1. Operating System: Windows or Linux or macOS.
2. Coding Language: Python (version 3.7 or above).
3. Integrated Development Environment (IDE): Developers can use their preferred Python IDE, such as Visual Studio Code or PyCharm.

5.3 Results of Screenshots

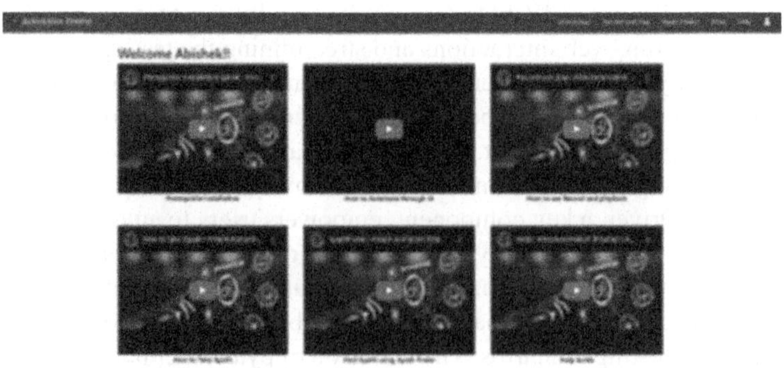

Fig. 3. Installation process guidelines videos and demo videos

This is the landing page for this site. It helps guide users through the installation process step by step, demonstrating how to use the application. The videos showcase how to automate websites through this platform, explaining each section thoroughly to demonstrate effective usage (Fig. 3 and 4).

The UI Automation module serves as the primary interface for users entering the platform. Users access this module by entering the URL name and the specific testcase they want to work on. Within this module, users have the flexibility to add or remove test

Fig. 4. Automation form for UI Automation

steps for each testcase, allowing for a customized and adaptable testing environment. Additionally, users can create and manage multiple testcases based on their testing requirements. Upon defining the test scenarios, users can initiate the automation process by clicking on the specified test case. This action triggers the module to launch the provided URL and automate the interactions based on the information the user has entered the UI. The module utilizes Selenium to execute these automated tasks, ensuring comprehensive and efficient testing of web applications.

The step-by-step approach to testcase creation and the ability to dynamically add or remove test steps contribute to the module's user-friendly design. Users have control over the testing process, enabling them to tailor their test scenarios to specific use cases. The integration of Selenium for browser automation enhances the reliability and accuracy of the testing procedures. In essence, the UI Automation module empowers users to define, customize, and execute test scenarios seamlessly. Through its intuitive design and integration with Selenium, it provides a powerful platform for web application testing while prioritizing user control and flexibility (Fig. 5).

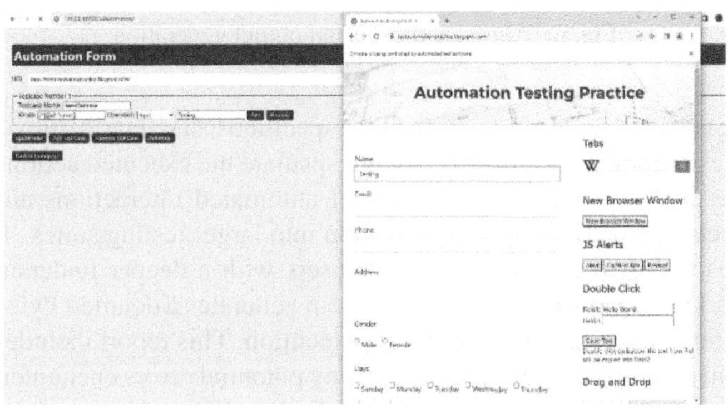

Fig. 5. Automation process

By clicking on the "Automate" button within the UI Automation module, users initiate the automated execution of the defined test scenario. This feature streamlines the testing process, allowing users to witness their specified testcase come to life through automation. The module leverages Selenium, a powerful browser automation tool, to translate the user-entered test scenarios into executable actions. Upon activation, the

"Automate" button triggers the system to navigate to the specified URL and perform the series of actions outlined in the testcase. Selenium's capabilities enable the simulation of user interactions with web elements, such as clicking buttons, inputting text, or navigating through pages. This automation process ensures consistent and repeatable testing procedures, reducing the likelihood of human error and enhancing the efficiency of the testing workflow.

The integration of the "Automate" button emphasizes the hands-on and dynamic nature of the testing platform. Users can observe in real-time how their defined test scenarios unfold through automated interactions with the web application. This feature provides immediate feedback on the functionality and behavior of the tested application, aiding in the identification of potential issues or areas for improvement.

In summary, the "Automate" button serves as a catalyst for turning user-defined test scenarios into automated executions, offering a practical and efficient approach to web application testing within the platform (Fig. 6).

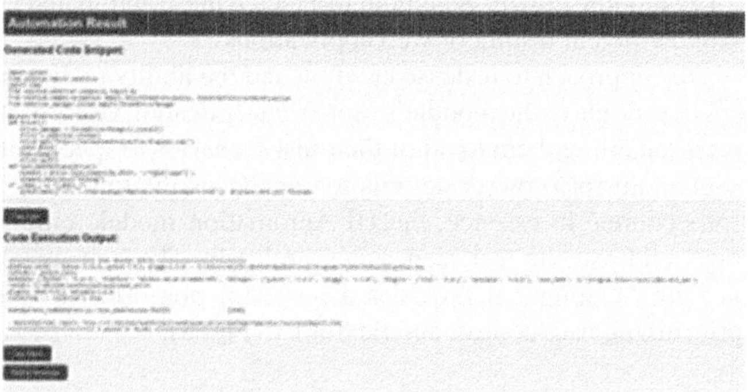

Fig. 6. Automated code and output generation

Following the successful automation of the specified test scenario, the system dynamically generates Python code snippets that encapsulate the executed actions. These code snippets serve as a tangible representation of automated interactions and can be utilized for reference, modification, or integration into larger testing suites. The generated code promotes transparency and empowers users with a deeper understanding of the automated workflow. Simultaneously, the system generates a detailed Pytest report, providing a structured overview of the test case execution. This report includes information on test pass/fail statuses, execution time, and any potential errors encountered during the automated testing process. The Pytest report serves as a valuable resource for assessing the overall health and performance of the tested application, aiding users in identifying specific areas that may require attention or refinement.

By combining code generation and pytest reporting as immediate outputs, the platform enhances the user experience by delivering actionable insights and tangible artifacts. This approach fosters a collaborative and iterative testing environment, where users can continuously refine their test scenarios based on the generated outputs, contributing to the overall improvement of web application testing practices (Fig. 7).

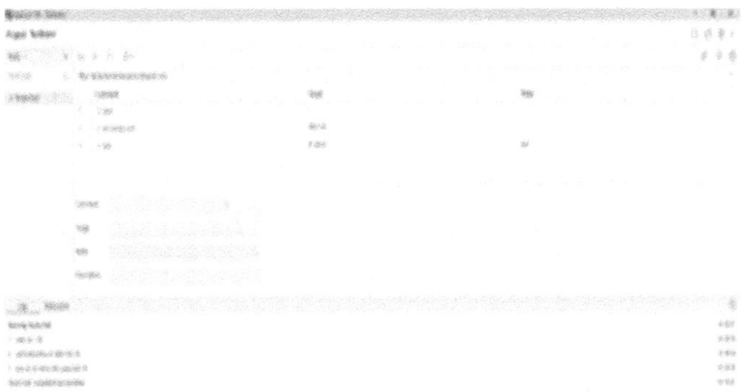

Fig. 7. Record and playback.

The project's distinctive "Record and Playback" feature transforms the test scenario creation process into an intuitive and dynamic experience. Users, irrespective of their technical proficiency, can effortlessly record their interactions with a web application in real-time. This recorded script, capturing a sequence of actions, becomes the foundation for automation. The playback functionality not only automates the defined scenario but also generates tailored code snippets, eliminating the need for extensive coding expertise. This innovative approach offers users flexibility, enabling them to modify, add, or remove steps from recorded test scenarios, ensuring adaptability to changing testing requirements. The feature stands out for its time-efficient workflow, efficiently translating manual interactions into automated scripts. Its accessibility, adaptability, and time-saving characteristics make it an invaluable component for accelerating web application testing with unparalleled ease and efficiency (Fig. 8).

Fig. 8. Xpath finder

The "XPath Finder" module enriches the project's capabilities by providing users with a sophisticated tool to effortlessly locate and define web elements. This feature is especially beneficial for users engaged in UI automation, as it streamlines the process of identifying specific elements on a web page. With a user-friendly interface, the XPath Finder empowers users to interactively select and validate elements, ensuring accuracy in the automation process. Leveraging advanced algorithms, this module contributes to precise and efficient automation script generation. Users can seamlessly integrate XPath expressions into their automation scripts, fostering a harmonious interaction between the graphical user interface and the underlying code. The XPath Finder's role in enhancing precision, ease of use, and the overall efficiency of the UI automation process highlights its significance in facilitating a seamless testing experience for developers and testers alike (Fig. 9).

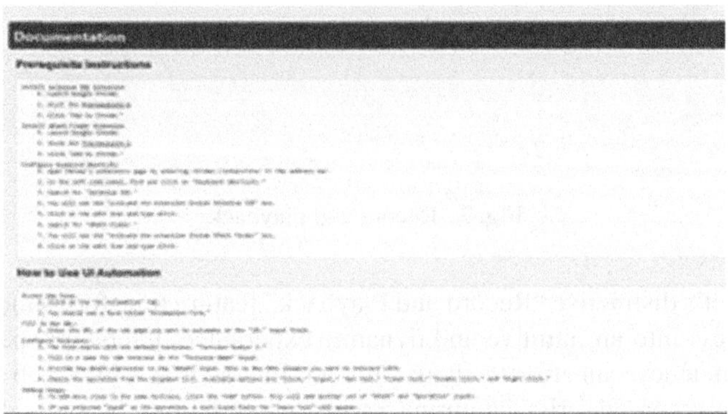

Fig. 9. Documentation

The "Documentation" section serves as a comprehensive guide for users, offering step-by-step instructions on how to harness the full potential of the automation tool. This user-centric documentation not only provides insights into the installation process, ensuring a seamless setup, but also delves into the intricacies of using the application effectively. By offering a detailed walkthrough, users gain valuable insights into the functionalities of each module, from UI automation to code generation and beyond. The documentation provides a roadmap for creating and executing test cases, allowing users to navigate the tool's features with confidence. Additionally, it serves as a valuable resource for troubleshooting, ensuring users can address any challenges they may encounter during the automation journey. With an emphasis on clarity, conciseness, and user-friendliness, this documentation acts as a valuable companion, empowering users to optimize their testing processes and derive maximum benefits from the automation tool (Fig. 10).

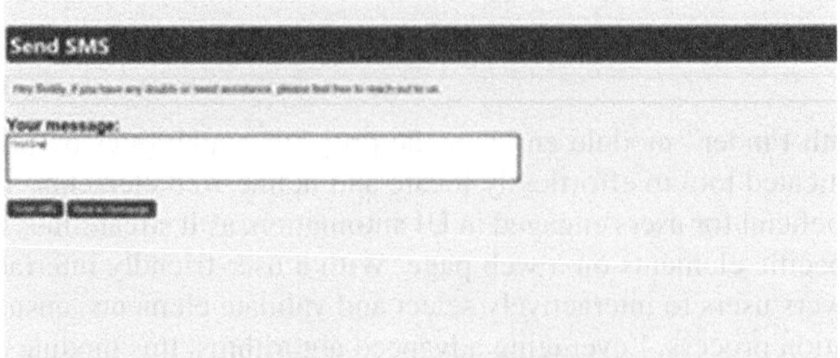

Fig. 10. Help and Guidance

The "Help" module introduces a dynamic and user-friendly support mechanism through the integration of Twilio API. This innovative feature enhances user guidance and assistance by enabling real-time communication via SMS notifications. Users can seamlessly seek help or gather information by sending SMS messages, creating a convenient and accessible avenue for resolving queries. Twilio's cloud communications platform adds a layer of interactivity, providing prompt responses and assistance to users navigating the automation tool. This integration aligns with modern user expectations, ensuring a personalized and responsive support experience. The "Help" module, powered by Twilio, exemplifies a commitment to user-centric design, fostering a sense of reliability and support throughout the automation journey. It stands as a testament to the project's dedication to enhancing user interaction and creating a comprehensive testing environment.

6 Conclusion

In conclusion, our web UI testing and automation project stands as an invaluable tool for users' daily activities. With a user-friendly interface and automated code generation, it caters to users of varying expertise, saving significant time. The addition of record-and-playback functionality enhances efficiency, making testing adaptable. Addressing budget constraints, the system offers cost-effective automation through optimized code snippets. The integration of a helpdesk with Twilio SMS notifications ensures prompt assistance, aligning with user needs for efficient web UI testing. This project is a practical and user-centric solution, enhancing the overall testing experience.

7 Future Enhancement

Future enhancements for the project include advanced reporting for detailed result insights, exporting code as files to facilitate collaboration, cross-browser compatibility for comprehensive testing, parallel testing capabilities for faster execution, and integration with CI/CD pipelines to align with modern development practices. These additions aim to optimize testing processes, enhance collaboration, and seamlessly integrate automated testing into the software development lifecycle.

References

1. Chen, G., Chen, G., Wu, D., Liu, Q., Zhang, L., Fan, X.: A selenium-based web application automation test framework. In: 2021 IEEE 2nd International Conference on Information Technology, Big Data and Artificial Intelligence (ICIBA), Chongqing, China, pp. 257–261 (2021). https://doi.org/10.1109/ICIBA52610.2021.9688190
2. Wang, F., Du, W.: A test automation framework based on WEB. In: 2012 IEEE/ACIS 11th International Conference on Computer and Information Science, Shanghai, China, pp. 683–687 (2012). https://doi.org/10.1109/ICIS.2012.21
3. Ramya, P., Sindhura, V., Sagar, P.V.: Testing using selenium web driver. In: 2017 Second International Conference on Electrical, Computer and Communication Technologies (ICECCT), Coimbatore, India, pp. 1–7 (2017). https://doi.org/10.1109/ICECCT.2017.8117878

4. Dhir, S., Kumar, D.: Automation software testing on web-based application. Softw. Eng. **16**, 691–698 (2019)
5. Zou, Y., Chen, Z., Zheng, Y., Zhang, X., Gao, Z.: Virtual DOM coverage for effective testing of dynamic web applications. In: Proceedings of the ISSTA, vol. 1, pp. 60–70 (2014)
6. Sfetsos, P., Angelis, L., Stamelos, I., Raptis, P.: Integrating user-centered design practices into agile web development: a case study. In: Proceedings of the 7th International Conference on Information Intelligence Systems & Applications, vol. 1, pp. 1–6 (2016)
7. Sawant, K., Tiwari, R., Vyas, S., Sharma, P., Anand, A., Soni, S.: Implementation of selenium automation & report generation using selenium web driver & ATF. In: 2021 International Conference on Advances in Electrical, Computing, Communication and Sustainable Technologies (ICAECT), Bhilai, India, pp. 1–6 (2021). https://doi.org/10.1109/ICAECT49130.2021.9392455
8. Chen, Z., Zhang, X., Zhao, Z.: A hybrid coverage criterion for dynamic web testing. In: Proceedings of the 25th International Conference on Software Engineering

Defending the Digital Frontier: URL-Based Phishing Detection Extension

P. Vamsi, U. Muthaiah(✉), and C. H. Roshan Vardhan

Computer Science and Engineering, Amrita School of Computing Amrita Vishwa Vidyapeetham, Chennai, India
{ch.en.u4cse20147,ch.en.u4cse20113}@ch.students.amrita.edu, u_muthaiah@ch.amrita.edu

Abstract. The internet's rapid expansion has led to an increase in cyberattacks and phishing schemes that rely on uniform resource location (URL) links. In today's digital landscape, it's vital to identify these harmful links for safety. Traditional methods struggle to adapt to new website URLs, so machine learning is being used to detect malicious URLs. Previous studies have explored various machine learning techniques for URL detection, but a recurring issue is the need to preprocess website links before analysis. To address this problem, a new neural network method has been introduced. This innovative approach enables the creation of a model capable of identifying malicious URLs without the need for repetitive feature extraction steps. This groundbreaking technique has shown impressive results in detecting dangerous web addresses, operating swiftly and accurately with a detection rate of 99.61%. This represents a significant advancement in the field of cybersecurity, offering an effective solution to the ever-evolving threat landscape posed by malicious URLs on the internet.

Keywords: Malicious URL Detection · Deep Learning · Feature Engineering · Exploratory Data Analysis

1 Introduction

The internet has revolutionized the way we access information, connect with others, and conduct various activities. With the vast landscape of the World Wide Web, there comes a significant challenge – the presence of potentially dangerous websites that can compromise user safety and security. In the world of the internet, there are many websites, and some of them can be dangerous and also stealing their personal information. In order to keep users safe, it's very important to take quick action in finding these websites that look suspicious called as malicious URLs. Various ways have been tried to find the malicious URLs, and some of them have succeeded, but some have not or have blocked good websites by mistake. Twitter often has harmful messages with links that can be spam, steal information, or spread viruses. Usual ways to find spam on Twitter look at the user's behaviour or relationships but can be tricked or take too long. Traditional

Published by Springer Nature Switzerland AG 2024
M. L. Owoc et al. (Eds.): ICCIDS 2024, IFIP AICT 718, pp. 65–76, 2024.
https://doi.org/10.1007/978-3-031-69986-3_5

methods to spot bad links look at the link itself or where it leads but attackers can avoid these checks. WarningBird, a system that quickly finds suspicious links on Twitter by looking at patterns in the links. WarningBird works by noticing that attackers often reuse the same links, which creates a pattern that can be tracked. The system was tested with many Twitter messages and was able to identify dangerous links effectively and quickly [1]. Firstfilter [2], a tool for spotting harmful web addresses in big company networks. Firstfilter is special because it can tell if a web address is safe, risky, or unknown. It uses a special method that considers the costs of making a wrong guess to pick the best features and reduce mistakes. To tackle this multifaceted challenge, researchers and cybersecurity experts have dedicated substantial efforts to devise methodologies and tools capable of distinguishing between benign and malicious URLs.

To spot dangerous web links using machine learning, which is a type of artificial intelligence was developed in [3]. It introduces a new method that looks at the behavior and features of web links to tell if they're harmful. This method also uses big data technology, which helps handle a lot of information, to improve detection of bad links. These results show that this approach worked well in finding malicious URLs. Malicious URLs are dangerous web links that can harm users by stealing information or damaging their computers. In [4] different machine learning methods to find the best one for catching harmful URLs. The best ML method is then used to find bad links on a real website that collects phishing URLs. The study uses a specific machine learning method called Logistic Regression to spot these dangerous websites. The dataset of this research was gathered from known sources like PhishTank, Kaggle, and GitHub. This method addresses the issue of web security, focusing on the detection of harmful websites and discusses the use of ML to identify websites that may be trying to steal personal information [5].

In [6] introduces a method to spot harmful web addresses (URLs) using a mathematical programming model. Initially the URLs into images that show patterns, making it easier to study them. The model picks out important bits from the web address, like words and letters, and turns them into math codes. A dual-part network that processes both image patterns and text is used to understand the URL's details. An attention mechanism in the program focuses on the most telling features to decide if a URL is bad. DeepBF [7], a new method for detecting harmful website addresses (URLs). It uses a special filter called a Bloom Filter, which is improved to work better for this task. The researchers picked the best simple hash function, which helps sort data, and then made it even better for deepBF. They compared their improved hash function with other similar functions and filters to see which one works best. The second part of their work involves an advanced neural network that learns to spot malicious URLs. They trained this neural network with data on bad URLs and then tested it to efficacy of this method.

AutoHTTP, [8] a tool for identifying suspicious internet URL activity without manual work. It focuses on understanding hidden meanings and connections in the data. AutoHTTP separates text data into two categories for analysis. It uses a special method to summarize the data into a useful form. The tool has a unique part that pays attention to important data and another that mixes data to find patterns. AutoHTTP is good at explaining its findings and is reliable. Tests show that AutoHTTP is better than other tools that need manual setup.

An advanced method, for spotting phishing scams involves using networks (CNNs) or machine learning (ML) algorithms to analyze URLs and webpage content at different levels of detail and sizes. This strategy employs a CNN structure with layers to detect patterns and features incorporating methods like pooling, skip connections, ensemble learning, transfer learning and adversarial training. By utilizing a range of features and representations this technique helps the model effectively identify phishing attempts, with both global context leading to improved accuracy and robust detection performance.

2 Related Works

The proliferation of the internet has brought about unprecedented opportunities for information exchange and communication, revolutionizing the way individuals and organizations interact with digital content. Several methods have been tried to identify these URLs, but they often miss some or make a mistake by blocking good as well as bad URLs. Multivariate filter-based [9] feature selection technique uses a special technique to pick out the most important signs that a URL might be harmful. It helps create a group of models that work together to better identify these risky URLs. The goal is to improve safety on the internet by catching more of these bad websites. In [10] presents a method to identify harmful websites by analyzing their features. It focuses on finding patterns that can tell apart bad domains from good ones. The method is dynamic, meaning it can adapt to new threats over time. This approach helps in protecting internet users from malicious websites.

Deep-learning architecture is a technique for computing that uses advanced methods of computing to identify harmful emails. This method consists of analyzing the entire content of the emails to spot the ones that are not legitimate. The key objective is to make it easier to detect such dangerous emails at the point of transmission and protect people from getting them [11]. GramBeddings [12], a neural network model for detecting phishing websites using URL n-gram embeddings. Phishing websites are fake sites that mimic real ones to steal user information. Traditional methods use blacklists or simple URL matching, which phishers can evade. GramBeddings analyzes parts of URLs (n-grams) to understand patterns typical of phishing sites. The model learns from large datasets of URLs to distinguish between legitimate and phishing web pages. This approach aims to improve the accuracy and speed of phishing detection over existing methods.

To detect malicious URLs using a special type of neural network called CGRU [13]. Malicious URLs are web addresses that can harm users by leading to attacks like SQL injection, XSS, or information disclosure. The CGRU model uses a combination of convolutional layers and GRU (Gated-Recurrent-Unit) layers to analyze and classify URLs. Convolutional layers help in extracting features from the characters in the URLs. GRU layers are used instead of traditional pooling layers to capture the temporal features of the URLs. The model is trained using stochastic gradient descent, a method to optimize the neural network. The CGRU model's performance is measured using accuracy, precision, recall, and F1-score. The CGRU model outperforms other models and manual feature extraction methods in detecting malicious URLs. In [14] a method was developed to protect IoT devices from cyber-attacks, specifically DDoS attacks. IoT devices are vulnerable because they are everywhere and always connected to the internet. A hybrid algorithm combining two techniques: RNN and ELM it learns from data to recognize normal and attack patterns.

In [15] was developed a method to identify and block harmful internet links (URLs) in smart devices using a knowledge-based system and a simulated expert. It aims to protect devices on the Internet of Things (IoT) from attacks that can steal data or cause other harm. The study uses a variety of machine learning algorithms to detect malicious URLs, with naive Bayes (NB) showing the best performance. The approach includes analyzing the features of URLs to understand which ones indicate a URL might be dangerous. The data collection to the use of a simulated expert for decision-making. It also discusses related work on malicious URL detection and addresses the challenge of imbalanced data classes. The results of the study demonstrate the effectiveness of the proposed method, especially the efficiency of the naive bayes algorithm. In [16] system was developed to protect IoT networks from cyber-attacks using ensemble learning methods. IoT networks are growing fast, and they share lots of data without strong security, which is risky. Several machine learning techniques were tested to find the best one for spotting intrusion. Malicious advertisement URLs are dangerous because they can lead to cyber-attacks. Researchers have been using machine learning to detect these harmful URLs. A feature extracted from URLs and web content was developed to improve detection. They use machine learning models like Random Forest and XGBoost to identify bad URLs. The study tests the models on different types of data to ensure they work well. They also look at how these models might be tricked by attackers during tests. The paper uses a technique called K-Means for grouping data to help visualize the problem. It achieves a high success rate in finding malicious URLs while making very few mistakes. The research is important for making the internet safer by catching harmful ads before they cause harm [17].

The challenge of detecting malicious URLs in an imbalanced dataset where harmful links are rare compared to safe ones. It discusses the limitations of SMOTE, a method that creates synthetic examples of the minority class, which can introduce noise and increase training time. The authors propose a cost-sensitive version of XGBoost, a powerful machine learning algorithm, to improve detection of malicious URLs without altering the original data distribution. This new method, called CS-XGB, aims to reduce bias towards the majority class and enhance the performance of the minority class, which in this case are the malicious URLs. The effectiveness of CS-XGB is validated using a dataset of 600,000 URLs, and it is found to be robust and efficient, outperforming the standard XGBoost and SMOTE-enhanced XGBoost [18].

The main contribution of the work is to address the escalating cyber threats in today's digital environment, machine learning, specifically a novel neural network method, has been introduced for efficient detection of malicious URLs. This approach overcomes the challenges of traditional methods by eliminating the need for repetitive feature extraction, resulting in a remarkable 99.61% detection accuracy.

3 Methodology

The proposed methodology aims to address the critical issue of detecting malicious URLs effectively. With the increasing prevalence of cyber threats and phishing schemes, it has become imperative to develop robust mechanisms for identifying harmful URLs to ensure user safety in the digital realm. This methodology leverages feature engineering, data visualization, and neural network-based modeling to achieve accurate and efficient URL classification. Multi-Layer Multi-Scaled CNN-Based Malicious URL Detection Algorithm is used to detect and predict the malicious URLs effectively. This algorithm utilizes a feedforward neural network architecture, implemented using the Keras library with TensorFlow backend, to classify URLs as either malicious or benign. It comprises multiple dense layers with Rectified Linear Unit (ReLU) activation functions, culminating in a final sigmoid activation layer for binary classification. The algorithm is trained using binary cross-entropy loss and optimized with a low learning rate using the Adam optimizer. Additionally, it incorporates ModelCheckpoint callbacks to save the best-performing model during training.

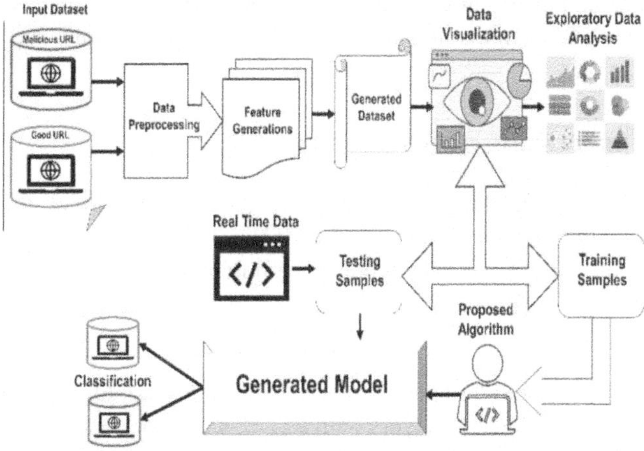

Fig. 1. Proposed System Architecture of Malicious URL detection

Figure 1 depicts the system architecture of proposed methodology aims to address the critical issue of detecting malicious Uniform Resource Locators (URLs) effectively. With the increasing prevalence of cyber threats and phishing schemes, it has become imperative to develop robust mechanisms for identifying harmful URLs to ensure user safety in the digital realm. This methodology leverages feature engineering, data visualization, and neural network-based modeling to achieve accurate and efficient URL classification.

3.1 Data Preprocessing

The initial step involves acquiring the dataset, which contains URL samples, and loading it using libraries like NumPy and Pandas. The dataset is cleaned, and unnecessary columns, such as "Unnamed: 0," are removed to improve data quality and relevance.

3.2 Feature Engineering

A critical aspect of proposed methodology is featuring engineering, where various attributes of URLs are extracted and computed to facilitate the detection process. Key features include URL length, hostname length, path length, the length of the first directory, and counts of specific characters and patterns within the URLs. Binary features, such as the presence of an IP address or the use of URL shortening services, are also determined.

3.3 Data Visualization

Visualizations are employed to gain insights into the dataset's characteristics and distribution. Heatmaps and histograms are used to understand correlations between features, class distribution, URL lengths, hostname lengths, directory counts, and other relevant attributes.

4 Data Preprocessing and Splitting

The dataset is split into training and testing sets to facilitate model development and evaluation. Oversampling techniques like SMOTE may be applied to address class imbalances, but this step is optional depending on the dataset.

4.1 Model Building

A neural network-based model is constructed using the Keras library. The architecture typically includes multiple dense layers with ReLU activation functions, culminating in a final layer with sigmoid activation for binary classification. The model is compiled with an appropriate optimizer, loss function, and accuracy metrics.

4.2 Model Training

The neural network model is trained on the training dataset with specified hyperparameters. Model training involves adjusting weights and biases using backpropagation and gradient descent to minimize the loss function. ModelCheckpoint is employed to save the best-performing model based on validation accuracy.

4.3 Model Evaluation

The trained model is evaluated on the testing dataset to assess its performance. Metrics such as accuracy, precision, recall, and F1-score are computed to gauge the model's effectiveness in detecting malicious URLs.

4.4 Prediction and Result Interpretation

The trained model is utilized to predict whether URLs are malicious or benign. A threshold (typically 0.5) is applied to convert probability scores into binary predictions. These predictions are then compared with the actual labels, and the results are interpreted to determine the model's accuracy and ability to identify malicious URLs.

5 Results and Discussion

The proposed Multi-Layer Multi-Scaled CNN-Based Malicious URL Detection Algorithm has been implemented and evaluated on a dataset of URLs to detect and classify them as either malicious or benign. The algorithm employed a neural network architecture with multiple layers, including convolutional layers and dense layers. The inclusion of convolutional layers allowed the model to extract multi-scale features from the URLs, enabling it to capture intricate patterns and variations that may indicate malicious intent. In the initial dataset, comprising 4,50,176 entries, three essential columns are present: url, which represents the URL samples; label indicating whether a URL is malicious or benign; and result a binary integer encoding for the labels, with 1 denoting malicious and 0 denoting benign URLs. These columns serve as the foundation for training and evaluating the Multi-Layer Multi-Scaled CNN-Based Malicious URL Detection Algorithm enabling it to learn from labelled data and make informed classifications.

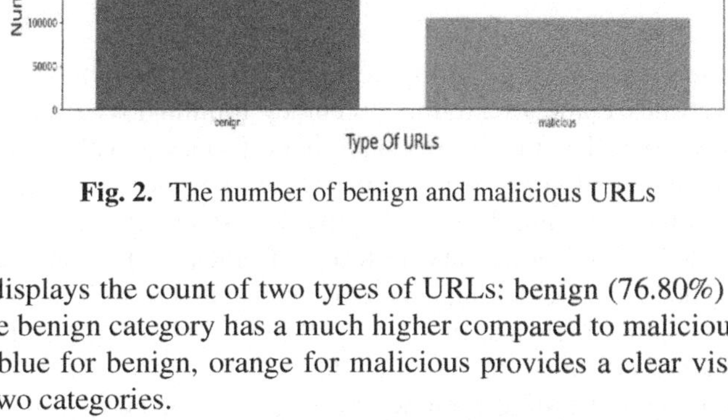

Fig. 2. The number of benign and malicious URLs

Figure 2 displays the count of two types of URLs: benign (76.80%) and malicious (23.20%). The benign category has a much higher compared to malicious ones and the color-coding blue for benign, orange for malicious provides a clear visual distinction between the two categories.

Figure 3 is a heat map that visualizes the correlation coefficients between various URL features, such as length, count of specific characters, and usage of HTTPS. Values range from −1 to 1, indicating strong negative to strong positive correlations. Darker shades represent the two features are stronger relationships. The Table 1 details the shapes of the training and validation sets for both features (x) and labels (y) within a

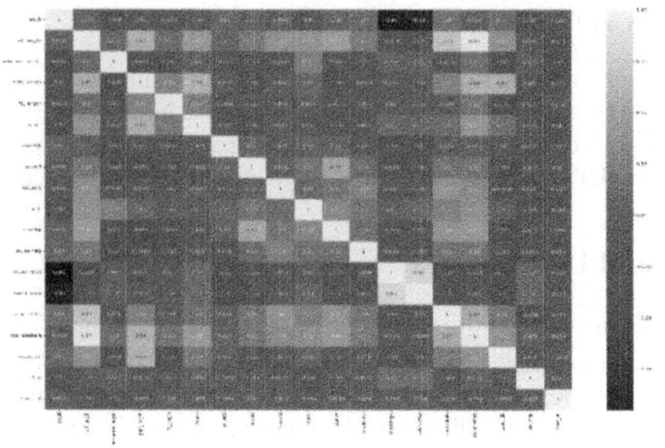

Fig. 3. Heat map of correlation coefficients between URL Features

machine learning dataset. The 'x_train' and 'x_valid' sets contain feature data, each with 16 columns, for training and validation respectively. The 'y_train' and 'y_valid' sets contain the corresponding labels.

Table 1. Split and Composition of Training and Validation Sets for Features and Labels

Dataset	Shape	Description
Training	(360140, 16)	Training set features with 360,140 samples and 16 features each
Testing	(90036, 16)	Testing set features with 90,036 samples and 16 features each

The Table 2 summarizes the performance metrics of various methods used for detecting malicious websites, comparing training accuracy, training loss, testing accuracy, and testing loss. The proposed method shows superior performance with training and testing accuracies above 99.6%, along with minimal loss values. Other referenced methods [10, 12, 19, 20], for which training data is not provided, show testing accuracies ranging from 93.6% to 98.7%, indicating varying levels of efficacy in comparison to the proposed method. The losses are minimal and closely matched as well, with a training loss of 0.0177 and a slightly higher testing loss of 0.0184, indicating a well-fitting model with no significant overfitting to the training data.

Figure 4 training and testing accuracy of a model over epochs, with both testing and training accuracy starts above 0.95 and remaining close together, which indicates consistent performance across both training and testing phases. The training and testing loss, with both starting at just above 0.1 and sharply declining to below 0.02, before flattening out, which suggests that the model quickly reduced its error rate and then stabilized.

Table 2. Proposed Model Performance Metrics: Training vs. Testing

Method	Training Accuracy (%)	Training Loss	Testing Accuracy (%)	Testing Loss
Proposed	99.62	0.0177	99.61	0.0184
[19]	–	–	98.7	–
[10]	–	–	98.7	–
[12]	–	–	98.5	–
[20]	–	–	93.6	–

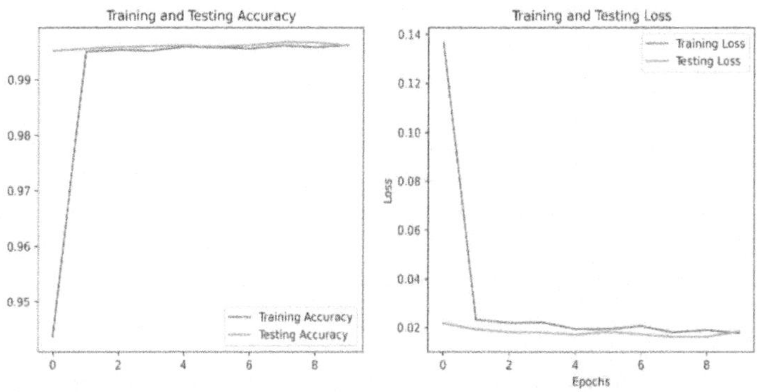

Fig. 4. Model Performance: Accuracy and Loss for Training and Testing Sets

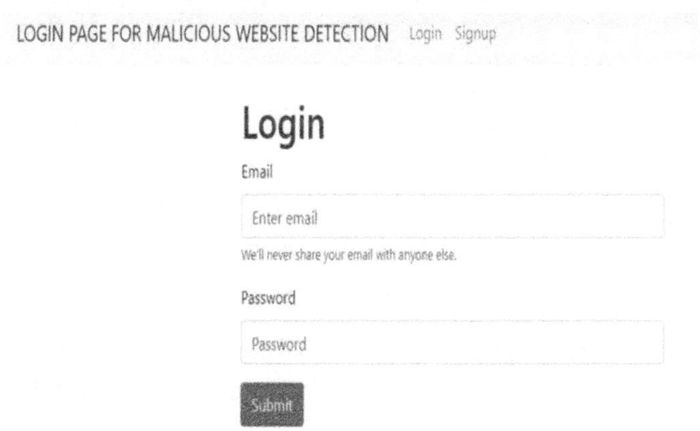

Fig. 5. Login Interface for Malicious Website Detection Platform

Figure 5 displays a login interface for a website dedicated to malicious website detection. The login page is quite straightforward, featuring a simple and clean design with two fields where users can enter their email and password. The page includes a note ensuring users that their email will not be shared with anyone else.

Above the email and password fields, there are options for Login and Signup, is used to create the new user's account. The 'Submit' button is prominent users to log in after entering their credentials. The layout displayed in the Fig. 6 is a user interface for a Malicious Website Detection system. The interface includes a section labelled "Search Engine" with a text box prompting the user to Enter Website. If the entered URL is found to be non-malicious, the system is designed to proceed to that website, suggesting a functionality that integrates security checks with regular browsing activities.

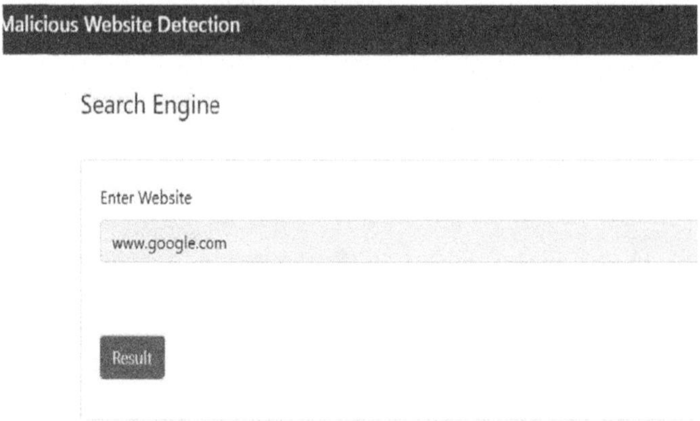

Fig. 6. User Interface of a Malicious Website Detection System with Search Engine Functionality

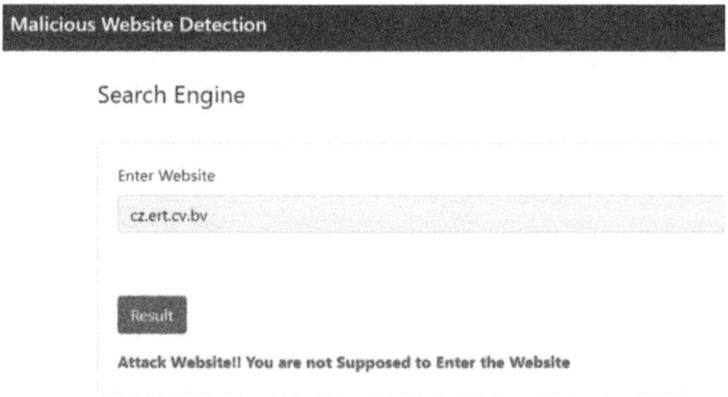

Fig. 7. Malicious Website Detection Interface Displaying a Warning for Malicious Website.

Figure 7 presents a user interface for a Malicious Website Detection tool, which includes a search engine-like feature. The user is prompted to enter a website URL for assessment. In this instance, the user has entered "czert.cv.bv," and after clicking the Result button, the system has identified the site as dangerous, displaying a warning message: "Attack Website! You are not Supposed to Enter the Website." This indicates that the detection system has a mechanism to evaluate URLs in real-time and provide immediate feedback to the user regarding the safety of the website.

6 Conclusion

The Multi-Layer Multi-Scaled CNN-Based Malicious URL Detection Algorithm has been demonstrated its effectiveness in accurately identifying and classifying malicious and benign URLs. The framework based multi-layered neural network architecture and multi-scaled features extracted from URLs, the algorithm achieved commendable results in web security. The proposed results showcase an effective malicious website detection system with high training and testing accuracy rates, as well as the low loss values. The user interface is user-friendly and provides immediate feedback, enhancing the usability of the tool in real-world applications. The system's ability to correctly flag malicious websites and prevent users from accessing them can significantly contribute to online safety and security. In future the additional URL features, such as semantic analysis, behavior profiling, or content-based analysis can be used improve detection accuracy and robustness against evolving threats.

Acknowledgements. The authors would like to express their gratitude to Amrita Vishwa Vidyapeetham Chennai for their valuable contributions and support throughout the research process.

Disclosure of Interests. The authors of this study state that they have no competing interests with the publication of this work. The study was carried out objectively and without prejudice. This paper's results and conclusions are exclusively derived from the data and analysis carried out throughout the research period. We don't have any personal or financial ties to any people or groups that would compromise the integrity or objectivity of our study.

References

1. Lee, S., Kim, J.: WarningBird: a near real-time detection system for suspicious URLs in Twitter stream. IEEE Trans. Dependable Secure Comput. **10**(3), 183–195 (2013). https://doi.org/10.1109/TDSC.2013.3
2. Vu, L., Nguyen, P., Turaga, D.: Firstfilter: a cost-sensitive approach to malicious URL detection in large-scale enterprise networks. IBM J. Res. Dev. **60**(4), 4:1–4:10 (2016). https://doi.org/10.1147/JRD.2016.2558018
3. Xuan, C.D., Nguyen, H.D., Nikolaevich, T.V.: Malicious URL detection based on machine learning. Int. J. Adv. Comput. Sci. Appl. (IJACSA) **11**(1) (2020). https://doi.org/10.14569/IJACSA.2020.0110119
4. Shantanu, Janet, B., Joshua Arul Kumar, R.: Malicious URL detection: a comparative study. In: 2021 International Conference on Artificial Intelligence and Smart Systems (ICAIS), Coimbatore, India, pp. 1147–1151. IEEE (2021). https://doi.org/10.1109/ICAIS50930.2021.9396014
5. Chiramdasu, R., Srivastava, G., Bhattacharya, S., Reddy, P.K., Reddy Gadekallu, T.: Malicious URL detection using logistic regression. In: 2021 IEEE International Conference on Omni-Layer Intelligent Systems (COINS), Barcelona, Spain, pp. 1–6. IEEE (2021). https://doi.org/10.1109/COINS51742.2021.9524269
6. Yuan, J., Chen, G., Tian, S., Pei, X.: Malicious URL detection based on a parallel neural joint model. IEEE Access **9**, 9464–9472 (2021). https://doi.org/10.1109/ACCESS.2021.3049625

7. Patgiri, R., Biswas, A., Nayak, S.: DeepBF: malicious URL detection using learned Bloom Filter and evolutionary deep learning. Comput. Commun. **200**, 30–41 (2023). https://doi.org/10.1016/j.comcom.2022.12.027
8. Wu, B., Zou, F., Zhang, C., Yu, T., Li, Y.: Multi-field relation mining for malicious HTTP traffic detection based on attention and cross network. J. Inf. Secur. Appl. **73**, 103411 (2023). https://doi.org/10.1016/j.jisa.2022.103411
9. Mohanty, S., Acharya, A.A.: MFBFST: building a stable ensemble learning model using multivariate filter-based feature selection technique for detection of suspicious URL. Procedia Comput. Sci. **218**, 1668–1681 (2023). https://doi.org/10.1016/j.procs.2023.01.145
10. Wang, H., Tang, Z., Li, H., Zhang, J., Cai, C.: DDOFM: dynamic malicious domain detection method based on feature mining. Comput. Secur. **130**, 103260 (2023). https://doi.org/10.1016/j.cose.2023.103260
11. Muralidharan, T., Nissim, N.: Improving malicious email detection through novel designated deep-learning architectures utilizing entire email. Neural Netw. **157**, 257–279 (2023). https://doi.org/10.1016/j.neunet.2022.09.002
12. Bozkir, A.S., Dalgic, F.C., Aydos, M.: GramBeddings: a new neural network for URL based identification of phishing web pages through N-gram embeddings. Comput. Secur. **124**, 102964 (2023). https://doi.org/10.1016/j.cose.2022.102964
13. Yang, W., Zuo, W., Cui, B.: Detecting malicious URLs via a keyword-based convolutional gated-recurrent-unit neural network. IEEE Access **7**, 29891–29900 (2019). https://doi.org/10.1109/ACCESS.2019.2895751
14. Hariprasad, S., Deepa, T., Bharathiraja, N.: Detection of DDoS attack in IoT networks using sample selected RNN-ELM. Intell. Autom. Soft Comput. **34**(3), 1425–1440 (2022). https://doi.org/10.32604/iasc.2022.022856
15. Anwar, S., et al.: Countering malicious URLs in Internet of Things using a knowledge-based approach and a simulated expert. IEEE Internet Things J. **7**(5), 4497–4504 (2020). https://doi.org/10.1109/JIOT.2019.2954919
16. Siddharthan, H., Deepa, T., Chandhar, P.: SENMQTT-SET: an intelligent intrusion detection in IoT-MQTT networks using ensemble multi cascade features. IEEE Access **10**, 33095–33110 (2022). https://doi.org/10.1109/ACCESS.2022.3161566
17. Nowroozi, E., Mohammadi, M., Conti, M.: An adversarial attack analysis on malicious advertisement URL detection framework. IEEE Trans. Netw. Serv. Manag. **20**(2), 1332–1344 (2023). https://doi.org/10.1109/TNSM.2022.3225217
18. He, S., Li, B., Peng, H., Xin, J., Zhang, E.: An effective cost-sensitive XGBoost method for malicious URLs detection in imbalanced dataset. IEEE Access **9**, 93089–93096 (2021). https://doi.org/10.1109/ACCESS.2021.3093094
19. Wang, Z., Ren, X., Li, S., Wang, B., Zhang, J., Yang, T.: A malicious URL detection model based on convolutional neural network. Secur. Commun. Netw. **2021**, 1–12 (2021). https://doi.org/10.1155/2021/5518528
20. Naim, O., Cohen, D., Ben-Gal, I.: Malicious website identification using design attribute learning. Int. J. Inf. Secur. **22**(5), 1207–1217 (2023). https://doi.org/10.1007/s10207-023-00686-y

Guarding the Digital Frontier: A Logistic Regression Approach to Malware Detection

M. R. Archana Jenis(✉), K. Shankar, and K. Srinivasa Seshadri

Computer Science and Engineering, St. Joseph's College of Engineering, Semmencheri, Chennai 600119, Tamil Nadu, India
archanajenismr@stjosephs.ac.in

Abstract. Taking new computing resources, a constantly evolving field, the malware threat is so widespread that a strong detection requires a creative response. The traditional methods of malware detection that include signature or shape-based methods do not perform well against different types of malware developed. So, we propose a malware detection model consisting of two parts: one for detecting malware based on PE (Portable Executable) header features, and another for detecting malicious websites based on URL features. The model uses random forest classifier for malware detection based on PE header features, statistical method called logistic regression for malicious website detection based on URL features thereby aiming to provide a fast, reliable, and trustworthy solution for guarding the digital frontier.

Keywords: Malware-detection · Logistic-regression · Random-forest classifier · Malware-sample

1 Introduction

The unrelenting advancement of technology presents both enormous potential and tremendous difficulties given the immense scope of the digital realm. The digital frontier is growing to represent a dynamic environment where innovation and information sharing flourish as the world grows more interconnected. But in the middle of this digital rebirth, a big shadow is lurking: malware's constant danger. Malware, which is intended to enter digital systems and compromise them, has become a ubiquitous menace that can cause devastation for individuals, companies, and even entire countries. Malware includes a wide range of malicious programs, such as ransomware, and viruses. These malicious organizations make use of flaws in software, networks, and human behavior to gain

K. Shankar and K. Srinivasa Seshadri—These authors contributed equally to this work.

Published by Springer Nature Switzerland AG 2024
M. L. Owoc et al. (Eds.): ICCIDS 2024, IFIP AICT 718, pp. 77–91, 2024.
https://doi.org/10.1007/978-3-031-69986-3_6

access to systems and disrupt operations, steal confidential data, or jeopardize the security of digital assets. Malware assaults are becoming more likely as society depends more on digital systems, endangering vital services like banking, commerce, national security, and health. As both sides employ cutting-edge instruments and strategies, the arms race between cybersecurity professionals and hackers is intensifying. Although they can be somewhat effective, traditional signature-based antivirus programs find it difficult to keep up with the speed at which malware types are evolving. Malware that is polymorphic or metamorphic can alter its code to avoid detection, which poses a challenge to established cybersecurity protocols. Therefore, to defend the digital frontier and counteract escalating dangers, innovation and modernization are desperately needed. Numerous techniques for identifying malware have been developed in this era of pervasive artificial intelligence and machine learning, all of which are vying with one another to propagate their opponents around the internet. Patient-logistic regression is one technique that sticks out in this technological problem. Traditionally originating in statistics, logistic regression has seen a fresh battle fought on the digital battlefield and has shown itself to be a potent ally in the fight against malware. The goal of this statistical technique, known as logistic regression, is to determine the likelihood that an observation falls into a specific class for binary classification problems. Logistic regression, as opposed to linear regression, models the connection between independent factors and the log probabilities of a specific result. This approach is especially appropriate in scenarios where variables are binary, which makes it a perfect fit for binary classification issues like malware identification. Logistic regression has been a mainstay of statistical modeling for many years, but its popularity in machine learning can be ascribed to its ease of use, readability, and efficacy. By modeling the likelihood of an event happening, logistic regression yields a limit value between 0 and 1. Because logistic regression offers a clear view into the decision-making process, it is the model of choice for malware detection. Logistic regression, in contrast to black box models like neural networks, offers an interpretation, enabling cybersecurity experts to comprehend the significance of every feature in differentiating malicious actors from vulnerabilities. In the realm of cybersecurity, where trust and comprehension play a crucial role in the decision-making process, this kind of disclosure is essential. Furthermore, compared to Sophisticated algorithms, logistic regression is less resource-intensive and more effective. Logistic regression is a valuable tool in resource-constrained contexts when judgments need to be made fast and genuine malware detection occurs. It is a viable option for implementation in various cybersecurity scenarios because of its low processing overhead and capacity to deliver precise and comprehensible results. Generally speaking, the necessity to strengthen defenses against dangerous software attacks grows along with the digital frontier. Logistic regression is a straightforward, open, and efficient approach that can help strengthen our defenses against an evolving array of cyber threats. The goal of this project is to identify a way to advance cybersecurity and guarantee the integrity and security of our digital world by harnessing the power of transportation.

2 Related Works

The landscape of cybersecurity is continuously evolving as malicious software, commonly known as malware, poses a persistent threat to the security of digital systems. This section provides a thorough exploration of various techniques and advancements in malware detection, ranging from traditional signature-based methods to sophisticated deep learning approaches.

Malware, characterized by its potential to steal personal information or inflict damage on devices, often eludes detection for several reasons. This could be due to the limitations of existing software, which may fail to identify new and previously undetected malware. Two primary approaches to detect fraudulent activity are signature-based detection and behavior-based detection.

Signature-based detection relies on identifying unique sequences or patterns inserted into application programs by malware authors. These signatures serve as fingerprints, allowing for the recognition of malicious code. In contrast, behavior-based detection focuses on analyzing the actions and behavior of software rather than static signatures. The paper referenced as [1] proposes a hybrid approach incorporating both signature-based and behavior-based techniques. The author utilizes a dataset comprising 1494 malware samples and 1347 benign samples. String functions are extracted for static analysis, and non-repeating consecutive API calls are extracted for dynamic analysis. The experiment employs various machine learning classifiers, including Gaussian Naive Bayes, Decision Tree, Support Vector Machine (SVM), and Random Forest. Impressively, the proposed method achieves an overall accuracy of 97.53%, showcasing the effectiveness of this combined approach.

Support Vector Machine (SVM), a machine learning approach capable of anticipating unseen data, is highlighted as a popular choice in cybersecurity due to its ability to handle complex, high-quality data. The paper [2] utilizes SVM along with other artificial intelligence techniques such as Random Forest and k-means clustering. The dataset is processed through an antivirus scanner, and static features are employed for information preprocessing and feature selection. The study also explores the application of J48 Decision Tree and Naive Bayes algorithms for malware detection. SVM emerges with a notable total accuracy of 93.8%, underscoring its efficacy in differentiating between malicious and non-malicious software.

Another noteworthy application of SVM is found in the work of [3], where the author focuses on detecting new malware and adapting the system to operate effectively against emerging threats. The proposed technique involves feature extraction from Java source code, leveraging Keyword Correlation Distance (KCD) to calculate correlations between common keywords, parameters, API calls, and Android permissions in Android malware source code. This innovative approach yields an overall accuracy rate of 88%, showcasing the potential of SVM in the dynamic landscape of malware detection.

Deep learning methods, particularly Convolutional Neural Networks (CNNs), have demonstrated effectiveness in detecting malware. In [4], an automated malware detection approach is proposed, relying on 35 distinct features extracted from packet flow rather than port numbers and protocols. The method employs a CNN alongside other machine learning algorithms, including multi-layer perceptrons (MLP), SVM, and Random Forests. The results show accuracy, precision, and recall exceeding 85% for all classes, highlighting the potential of deep learning in improving the speed and accuracy of malware detection, even in the presence of numerous malware varieties.

The significance of deep learning is further emphasized by the introduction of ScaleMalNet, a deep learning framework for malware detection proposed in 2019 [5]. This framework leverages big data approaches to process a large number of malware samples. The authors employ deep learning architectures, including recurrent neural networks (RNNs) and CNNs, in conjunction with traditional machine learning algorithms (MLAs). The investigation reveals a remarkable accuracy of 93.6%, underscoring the efficacy of deep learning in handling the complexities of malware detection and categorization.

In addition to these advancements, [6] proposes a unique approach using an autoencoder network combined with greyscale images derived from the conversion of malware binary files. The suggested autoencoder model consistently achieves a high F-score of 96% and an accuracy of 96%, showcasing the potential of leveraging image-based representations for effective malware detection.

The evolving nature of malware and the need for continuous innovation in detection methods are evident throughout the discussed research. The above works emphasize the amalgamation of various feature sets to provide a comprehensive portrayal of malware characteristics. While earlier research models may have exhibited lower accuracy, recent advancements, particularly those incorporating machine learning and deep learning, have significantly improved detection rates.

Notably, the ability to detect malware based on URL features is addressed, introducing a malware detection scheme utilizing the logistic regression technique. Furthermore, a random forest classifier is employed for detecting malware based on PE header features. These insights signify a nuanced understanding of the diverse features that can be leveraged to enhance the accuracy of malware detection systems.

3 Proposed System

Malware is any harmful software or code that can harm a device or interfere with its normal operation. Depending on the type of goal of the cyber criminals who built or transmitted it, there are a variety of key reasons for this. The most prevalent signs of malware are generally lower system performance, frequent, unexpected crashes, odd CPU utilization spikes, and higher system resource consumption. The general framework of the suggested malware detection method is introduced in this section. By employing a logistic regression-based malware detection approach, the suggested system tackles the growing issue of malware in the digital sphere. Acknowledging the shortcomings of conventional signature-based techniques, this approach seeks to offer a more flexible and potent protection mechanism. At its core, the system is fundamentally composed of several interrelated modules, each of which adds to the overall effectiveness of malware detection.

3.1 Malware Detection of PE-Header File

In this subsection, The PE module extracts features from PE files using a deep learning model called MalConv, which is trained on a large dataset of benign and malicious PE files. Using Python's pefile package, the system extracts around 1.2 million features from the PE header of executable files among which 13 features (Machine, Characteristics, Subsystem......) are selected to train the model. The PE header features include information such as the size, entry point, sections, imports, exports, and checksum of the executable file. The features are then fed into a feature selection method, which selects the most relevant features for malware detection. The features selected are then used to train a classifier, such as Random Forest to determine if a PE file is benign or malicious.

The image 1 illustrates a flowchart for a machine learning-based system designed to classify files as either benign (safe) or malicious (containing malware). The process is divided into two main phases: training and testing. The detailed explanation of each step is as follows:

3.1.1 Benign File Dataset

This is a collection of files known to be safe and not contain any malware. These files are used to train the machine learning model on what benign files look like.

3.1.2 Malware File Dataset

This dataset consists of files that are known to contain malware. The machine learning model uses this dataset to learn the characteristics of malicious files.

3.1.3 Feature Extraction

In this step, both benign and malware files from the datasets are analyzed to extract relevant features. These features could include file size, file type, binary data, strings, or specific patterns that are indicative of either benign or malicious files. The extraction process is crucial for identifying the distinguishing characteristics that the machine learning model will use to classify new files.

3.1.4 Random Forest Classifier

The extracted features are then fed into a Random Forest classifier, which is a machine learning algorithm that consists of many decision trees and outputs the class that is the mode of the classes output by individual trees. It is known for its high accuracy, ability to handle large datasets with higher dimensionality, and its ability to model complex interactions and classifications.

3.1.5 Saved Files

After the Random Forest classifier has been trained on the features extracted from the benign and malware datasets, the resulting model is saved. This trained model encapsulates the patterns and rules that the classifier has learned to distinguish between benign and malicious files.

3.1.6 File Extraction

This step is similar to feature extraction in the training phase but is applied to new files that need to be classified. The system extracts features from these files to be analyzed by the trained model.

3.1.7 Feature Extraction

The features of the new files are extracted for analysis. This step is critical because the accuracy of the classification depends on the quality and relevance of the extracted features.

3.1.8 Random Forest Classifier

The extracted features from the new files are input into the trained Random Forest classifier. The classifier uses the learned patterns from the training phase to predict whether the new files are benign or malicious.

3.1.9 Decision for Classification

The classifier outputs a decision for each file. If the classifier determines that the features of a file are similar to those of the known malware files, it classifies the file as "Malware." If the features do not match those of malware, the file is classified as "Legitimate."

3.2 Malware Detection of Uniform Resource Locator

This section maps the model in such a way that you can understand the data it was trained on and try to generate defined behaviors from the data set. Datasets are the backbone of any model, so you need good, accurate data about good and bad URLs in your data to train your model. Machine learning to identify malicious websites starts by scraping the data in a dataset. A custom vectorizer and pandas are defined and used to clean the dataset and logistic regression to train the model.

The image 2 depicts a flowchart for a machine-learning process designed to filter URLs and classify them as either legitimate or malicious (potentially containing malware). Step-by-step explanation of the flowchart:

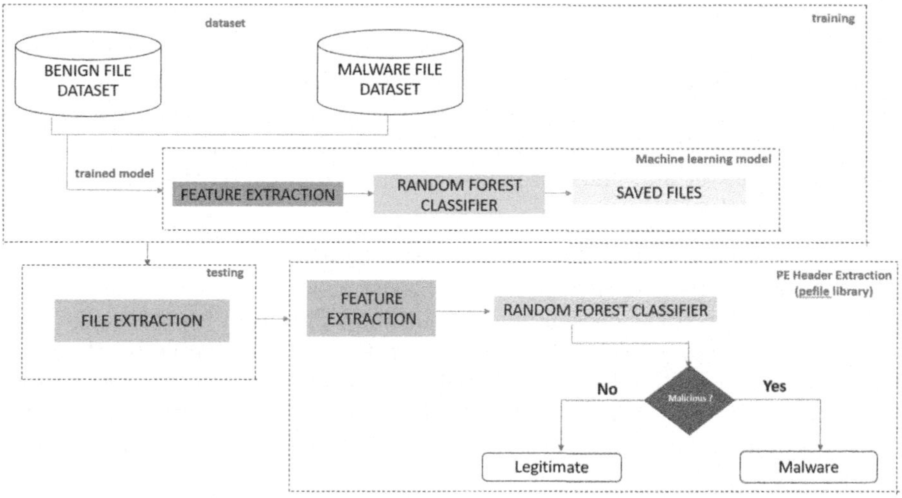

Fig. 1. System Architecture of PE-Header based Malware detection

3.2.1 URL Dataset

The process begins with a dataset of URLs that will be used to train the machine literacy model. This dataset should include a different range of URLs, both licit and vicious, to ensure the model can learn to separate between the two effectively.

3.2.2 Data Drawing

This step involves drawing the dataset by removing any inapplicable or indistinguishable data, correcting crimes, and dealing with missing values. Data drawing is pivotal to ensure the quality and integrity of the dataset, as it can significantly impact the performance of the machine literacy model.

3.2.3 Data Preprocessing

After drawing, the dataset undergoes data preprocessing. This step involves transubstantiating the data into a format that can be used by the machine learning algorithm. It may include homogenizing or spanning the data, garbling non-numeric values, or rooting applicable features from the URLs. Preprocessing prepares the data for further analysis and model training.

3.2.4 Splitting the Datasets

The preprocessed data is also resolved into two separate datasets a training dataset and a testing dataset. The training dataset is used to train the machine literacy model, while the testing dataset is used to estimate the model's performance and conception capability.

3.2.5 Vectorizer

This element converts the URL data into a numerical format that can be reused by the machine learning algorithm. It may transfigure the URLs into vectors of figures grounded on the presence or absence of certain characteristics or patterns within the URLs. Vectorization is pivotal as it allows the machine learning algorithm to dissect and make prognostications grounded on the numerical representations of the URLs.

3.2.6 Logistic Regression

Logistic regression is the chosen machine literacy algorithm for this task. Logistic retrogression is a statistical system generally used for double-bracket problems, where the outgrowth can be distributed into two classes. In this case, the classes are "licit" and "vicious." The logistic retrogression algorithm is trained on the training dataset, using the vectorized URL data and their corresponding markers.

3.2.7 Trained Model

Once the logistic retrogression algorithm is trained, it produces a trained model that can be used for vaticination. This trained model encapsulates the learned patterns and characteristics from the training dataset, enabling it to make prognostications on new, unseen URLs.

3.2.8 URL Filtering

In the operation phase, new URLs are passed through the trained model to determine their nature. The URLs are vectorized using the same vectorization fashion applied during training. The trained model also analyzes the vectorized URLs and predicts whether they're licit or vicious.

3.2.9 Malicious

The model labors a decision on whether the URL is vicious or not. However, it's classified as "Malware, If the model predicts that a URL is vicious." On the other hand, if the model predicts that a URL is non-malicious, it's classified as "licit." This bracket can help druggies and associations make informed opinions about penetrating or interacting with specific URLs, thereby enhancing their cybersecurity measures.

4 Materials and Methods

4.1 Dataset Collection

We refer to data sets or collections of data from different sources as collections of data sets. There are 50,000 files in the dataset, a publicly available dataset containing instances of various malware samples collected from the Kaggle Malware Dataset of which 25,000 are malware and 25,000 are clean. The data set

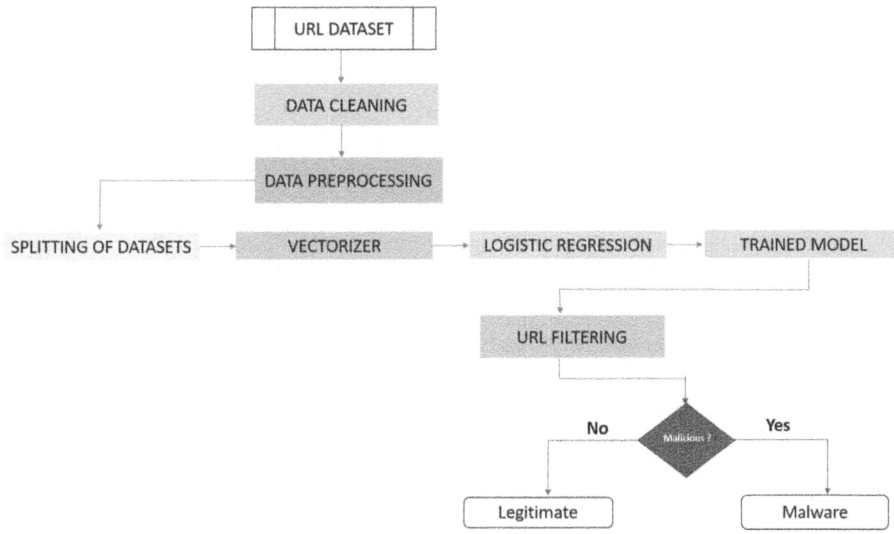

Fig. 2. System Architecture of Malicious URL detection

also includes a CSV file with labels and attributes for each file. Items in the file's PE (Portable Executable) header are extracted, including information such as size, entry point, number of blocks, checksum, etc. The dataset can be used to train and test machine learning models for malware as they see its distribution. In addition to that using the datasets the performance of the system can be compared using various performance metrics that include Accuracy, Precision, Recall, and so on.

4.1.1 Dataset Collection of PE-Header

For PE-header-based malware detection, the dataset collection method involves gathering a diverse set of Portable Executable (PE) files representing both malware and benign instances. Refer Fig. 3.

4.1.2 Dataset Collection of URL

The method for collecting the dataset for malicious URL detection typically involves gathering a diverse set of URLs that represent both malicious and benign instances. Refer Fig. 4.

4.2 Feature Extraction

A feature extraction is a mathematical structure that can be extracted from raw data while preserving the original set of data. It is more efficient than machine learning on raw data. In this paper, the PE header feature is extracted from the executable file and the URL feature is extracted from websites.

```
Index(['Name', 'md5', 'Machine', 'SizeOfOptionalHeader', 'Characteristics',
       'MajorLinkerVersion', 'MinorLinkerVersion', 'SizeOfCode',
       'SizeOfInitializedData', 'SizeOfUninitializedData',
       'AddressOfEntryPoint', 'BaseOfCode', 'BaseOfData', 'ImageBase',
       'SectionAlignment', 'FileAlignment', 'MajorOperatingSystemVersion',
       'MinorOperatingSystemVersion', 'MajorImageVersion', 'MinorImageVersion',
       'MajorSubsystemVersion', 'MinorSubsystemVersion', 'SizeOfImage',
       'SizeOfHeaders', 'CheckSum', 'Subsystem', 'DllCharacteristics',
       'SizeOfStackReserve', 'SizeOfStackCommit', 'SizeOfHeapReserve',
       'SizeOfHeapCommit', 'LoaderFlags', 'NumberOfRvaAndSizes', 'SectionsNb',
       'SectionsMeanEntropy', 'SectionsMinEntropy', 'SectionsMaxEntropy',
       'SectionsMeanRawsize', 'SectionsMinRawsize', 'SectionMaxRawsize',
       'SectionsMeanVirtualsize', 'SectionsMinVirtualsize',
       'SectionMaxVirtualsize', 'ImportsNbDLL', 'ImportsNb',
       'ImportsNbOrdinal', 'ExportNb', 'ResourcesNb', 'ResourcesMeanEntropy',
       'ResourcesMinEntropy', 'ResourcesMaxEntropy', 'ResourcesMeanSize',
       'ResourcesMinSize', 'ResourcesMaxSize', 'LoadConfigurationSize',
       'VersionInformationSize', 'legitimate'],
      dtype='object')
```

Fig. 3. Number of columns in PE-Header file Dataset

URL	LABEL
diaryofagameaddict.com	bad
espdesign.com.au	bad
iamagameaddict.com	bad
kalantzis.net	bad

Fig. 4. Information of URL Dataset

4.3 Feature Selection

In this research, the system employs a wraparound feature selection methodology to identify the most pertinent and pertinent features for a given classification challenge. Wrapping is a type of machine learning that uses a bunch of different features to train the model. Then, the algorithm adds or removes the features based on some selection criteria. The selection criteria are how much the model will improve or decrease in performance when you add or remove a feature. The evaluation function of the system is a random Forest Classifier, and the search strategy is a genetic algorithm.

4.4 Data Cleaning

The process of editing or deleting incorrect, inaccurate, or irrelevant data from a data set is known as data cleaning. In this paper, the dataset is cleaned for use in Logistic Regression such that Pandas is employed, and a custom vectorizer is defined for data-set cleaning.

```
1. feature DllCharacteristics (0.181008)
2. feature Machine (0.114327)
3. feature Characteristics (0.078960)
4. feature Subsystem (0.062311)
5. feature VersionInformationSize (0.060331)
6. feature SectionsMaxEntropy (0.060196)
7. feature ImageBase (0.059015)
8. feature ResourcesMaxEntropy (0.051068)
9. feature MajorSubsystemVersion (0.043120)
10. feature SizeOfOptionalHeader (0.041759)
11. feature MajorOperatingSystemVersion (0.027296)
12. feature ResourcesMinEntropy (0.021602)
13. feature SectionsMinEntropy (0.019809)
```

Fig. 5. Important features selected

4.5 Classification

The system uses a random forest classifier for malware detection based on PE header features, and a logistic regression classifier for malicious website detection based on URL features. The system outputs the predicted label and the confidence score for each sample.

4.6 Model Evaluation

The precision, accuracy, and recall of the classifiers are among the metrics that the system uses to assess their performance.

5 Terminal Based Interface

The entire project is terminal-based and we only used the main screen visuals, the only way to print the ASCII art at the home screen is by installing pyfiglet, a python library that translates strings into ASCII art. Refer Fig. 6.

5.1 Pyfiglet

pyfiglet is a Python library that allows to create ASCII art text banners using different font styles. Using printable characters from the ASCII standard, graphic designers can create graphics and designs using ASCII art. pyfiglet uses fonts defined in the FIGlet font format.

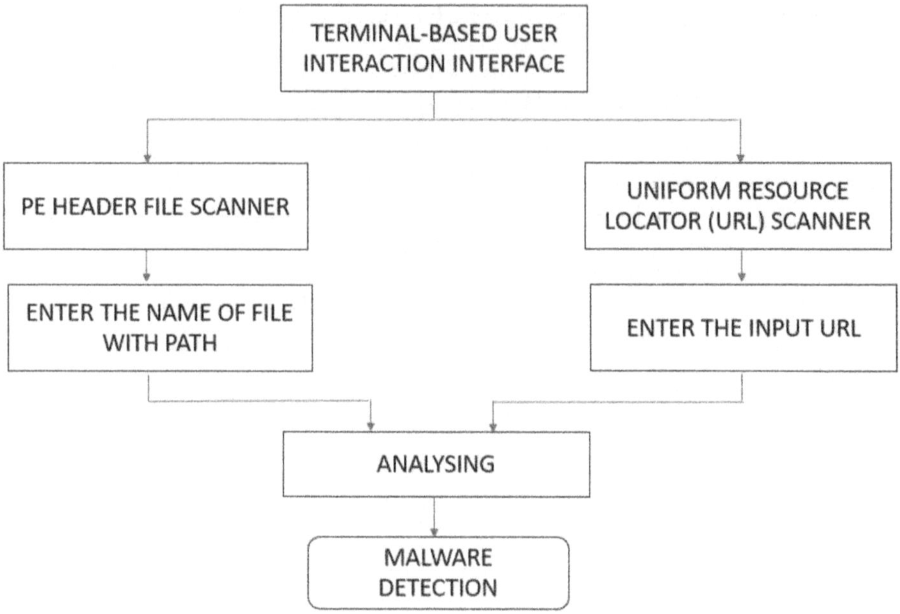

Fig. 6. Terminal-Based User Interaction Interface Development

6 Results

The process of malware detection using logistic regression begins with the collection of a diverse and representative dataset containing samples of both malware and benign software from trusted public sources. This dataset is then divided into a training set, comprising 70% of the data, and a testing set, containing the remaining 30%, to ensure the model's performance can be adequately evaluated. Before training the logistic regression model, important attributes related to malware characteristics, such as file size, file type, and code patterns, are selected for classification. The model is then trained using the training data, adjusting its parameters to minimize the difference between predicted probabilities and actual labels. Subsequently, the trained model is evaluated using the testing data to compute classification accuracy and other relevant metrics such as precision, recall, and F1-score that can be obtained using confusion matrix. Refer Figs. 9 and 10. Finally, upon successful evaluation, the logistic regression model is implemented in the final malware detection system, where it can automatically classify and detect malware in real-time or batch processing scenarios, thereby enhancing cybersecurity measures.

The ROC curve for malicious URL detector is displayed in Fig. 7 above. The False Positive Rate statistic, called the True Positive Rate, is comparable to recall. This test looks at how unfavorable circumstances are mistakenly classified as positive ones. Different threshold values between 0 and 1 are compared to determine these characteristics. Area Under the Curve, or AUC, is a statistic that

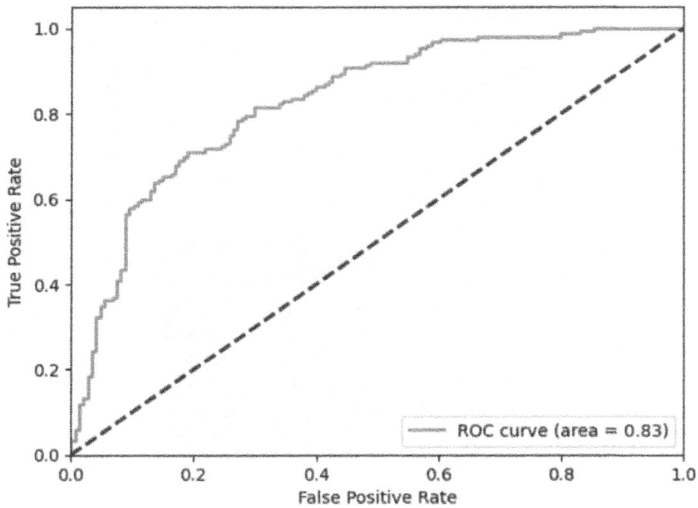

Fig. 7. ROC Curve for Malicious URL Detector

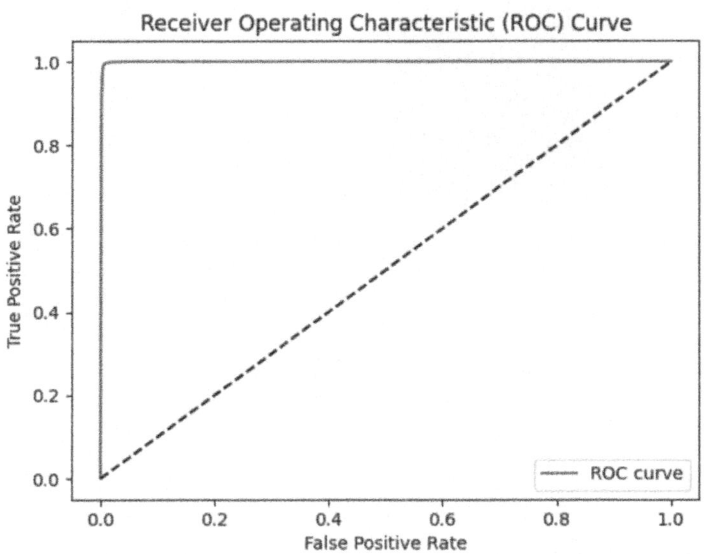

Fig. 8. ROC Curve for Malware Detection of PE-Header

ranges from 0 to 1 that indicates how good a model is. The model performs better the higher the AUC when it happens. Similarly, the ROC curve for Malware Detection of PE-Header is displayed in Fig. 8.

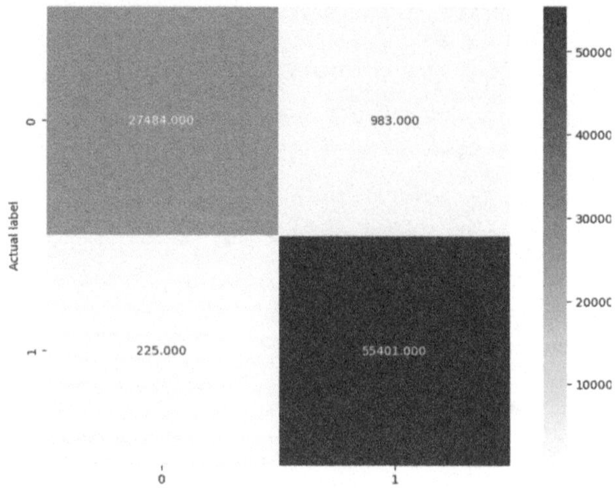

Fig. 9. Confusion Matrix of Malicious URL Detector

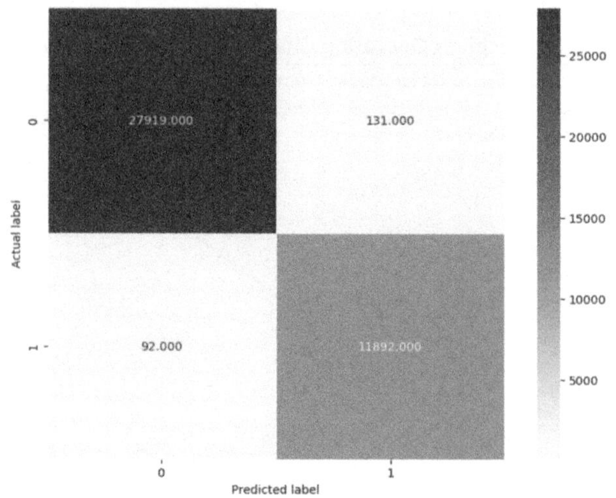

Fig. 10. Confusion Matrix of Malware Detection of PE-Header

7 Conclusion

This paper uses two data sets, PE header, and URL features, for training and testing our models. To reduce dimensionality and enhance the performance of the models, we have used various feature selection/extraction methods like Extra Trees/TF-idf. To compare different classifiers like Decision Tree/Random Forest, we found Random Forest to have the best accuracy/balance between false positives/false negatives.The PE header-based detection uses a random forest classifier to classify malware and benign files based on a dataset of PE header features. This project claims to achieve an accuracy of 99.44% with this method. This project also provides a file extraction module that can extract the relevant features from any PE file and feed them to the trained classifier. The URL-based detection on the other hand uses a logistic regression model to classify malicious and benign URLs based on a dataset of URL features. This project also uses a whitelist filter to allow known good URLs to pass through and reduce false positives on good URLs. An accuracy of 98.56% is achieved with this method. In the future a GUI-based program for Windows may be developed; however, it is currently only available for terminal use. Additionally, real-time scanning of all files while downloading/transmitting may be implemented to detect potentially malicious files. Finally, the system should be able to handle a wide range of data sets and scale up or down as needed.

Acknowledgements. We would like to express our gratitude to all those who contributed to the completion of this paper. Special thanks to St. Joseph's College of Engineering for their invaluable guidance and insights throughout the research process. Additionally, we acknowledge the contributions of Sri Sivasubramaniya Nadar College of Engineering, for their assistance in reviewing and refining the manuscript. This work would not have been possible without the support of our colleagues and mentors.

Disclosure of Interest. The authors declare no conflict of interest regarding the publication of this paper. This research received no specific grant from any funding agency in the public, commercial, or not-for-profit sectors. The authors have no financial or personal relationships that could influence the content of this paper.

References

1. Goyal, M., Kumar, R.: The pipeline process of signature-based and behavior-based malware detection. In: 2020 IEEE 5th International Conference on Computing Communication and Automation (ICCCA), pp. 497–502 (2020). https://doi.org/10.1109/ICCCA49541.2020.9250879
2. Choudhary, S., Sharma, A.: Malware detection and classification using machine learning. In: 2020 International Conference on Emerging Trends in Communication, Control and Computing (ICONC3), pp. 1–4 (2020). https://doi.org/10.1109/ICONC345789.2020.9117547
3. Sun, J., Yan, K., Liu, X., Yang, C., Fu, Y.: Malware detection on android smartphones using keywords vector and SVM. In: 2017 IEEE/ACIS 16th International Conference on Computer and Information Science (ICIS), pp. 833–838 (2017). https://doi.org/10.1109/ICIS.2017.7960108
4. Yeo, M., et al.: Flow-based malware detection using convolutional neural network. In: 2018 International Conference on Information Networking (ICOIN), pp. 910–913 (2018). https://doi.org/10.1109/ICOIN.2018.8343255
5. Vinayakumar, R., Alazab, M., Soman, K.P., Poornachandran, P., Venkatraman, S.: Robust intelligent malware detection using deep learning. IEEE Access **7**, 46717–46738 (2019). https://doi.org/10.1109/ACCESS.2019.2906934
6. Xing, X., Jin, X., Elahi, H., Jiang, H., Wang, G.: A malware detection approach using autoencoder in deep learning. IEEE Access **10**, 25696–25706 (2022). https://doi.org/10.1109/ACCESS.2022.3155695

Hybrid Efficient IDS Against Adversarial Attacks in IoT Networks

Eman Aljbour and Qasem Abu Al-Haija(✉)

Department of Cybersecurity, Prince Sumaya University for Technology, PO Box 1438, Amman 11941, Jordan
q.abualhaija@psut.edu.jo

Abstract. IoT services become more dominant as time passes and the growing security concerns become less relevant. Due to proliferating heterogeneity, emerging technologies, and resource-constrained IoT systems, intelligent systems are becoming more vulnerable to cyberattacks. As a result, practical solutions to security issues like privacy, scalability, authenticity, trust, and centralization are required. Since malicious actors frequently use obfuscation tactics to elude detection, traditional approaches to intrusion detection systems are no longer relevant. Furthermore, these methods fail to detect zero-day Attacks. Applying an intelligent mechanism based on machine learning or deep learning at a different stage is necessary to detect attacks. The purpose of selecting multiple algorithms is to satisfy the specific requirements of various users or groups. In other words, it is crucial to identify the most effective model for each sort of user. The suggested method consists of two stages: a signature-based IDS with a malicious signature already developed and an ML/DL algorithm trained using the IoT-23 public dataset. Results show that the proposed engine works better than existing cutting-edge methods, with an average accuracy of 95.3%.

Keywords: Internet of Things · Intrusion Detection System · Machine Learning · Deep Learning · IoT-23 Dataset

1 Introduction

"Internet of Things" (IoT) refers to a network of connected appliances, cars, and other objects with built-in connection, software, and sensor capabilities. These gadgets can gather and communicate data online, allowing them to interact with their surroundings and carry out numerous functions without requiring direct human involvement. This network of interconnected devices makes it possible to automate, control remotely, and monitor many systems and processes, which boosts productivity, convenience, and decision-making.

An IoT system can include simple sensors and actuators to more complicated gadgets like smartphones, wearable technology, home appliances, commercial machinery, and even entire smart cities. These gadgets frequently have connectivity capabilities (Wi-Fi, Bluetooth, or cellular networks) to transmit and receive data and sensors to gather information about their surroundings.

Published by Springer Nature Switzerland AG 2024
M. L. Owoc et al. (Eds.): ICCIDS 2024, IFIP AICT 718, pp. 92–103, 2024.
https://doi.org/10.1007/978-3-031-69986-3_7

Smart homes, healthcare, transportation, agriculture, manufacturing, energy management, and environmental monitoring are just a few industries that use IoT applications. For instance, in a smart home, IoT devices like thermostats, lighting controls, and security cameras can be networked to enable automation that is both secure and energy-efficient. IoT devices in the healthcare industry can provide real-time tracking of medical equipment, remote patient monitoring, and medical device maintenance planning.

According to IoT Experts' predictions, there will be 25.4 billion connected devices by 2030 [1], an increase of almost 12 billion devices from 2022 to 2030. Over the past ten years, a noticeable transition has been made from non-IoT to IoT devices. According to Fig. 1 [1], 75% of all devices will be IoT by 2030. Most of these products can be categorized as Artificial Intelligence of Things (IoT) because they mix AI with IoT infrastructure to enable more effective IoT operations, enhance human-machine interaction, and improve data management and analytics [2].

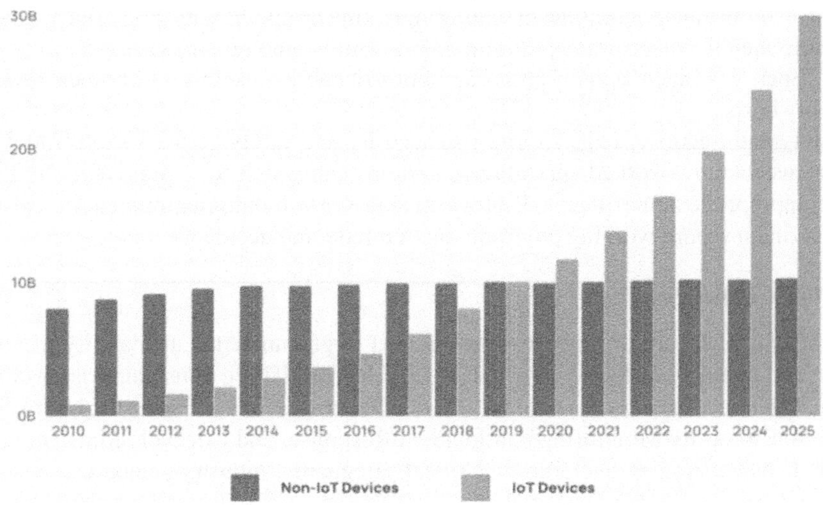

Fig. 1. Non-IoT and IoT active devices from 2010 to 2025

As more gadgets become available, the world becomes a complex networked environment. This is advantageous because it provides modern users convenience but could potentially increase the attack surface. Attackers may use flaws in these gadgets' embedded AI, software, or hardware, especially if they are connected to the outside world, to cause a cyber emergency. For instance, hackers launched a flurry of IoT botnet attacks in October 2016. Using various IoT devices, it blocked access to well-known websites, including Twitter, Amazon, GitHub, and Netflix [3]. Consequently, robust and efficient detection algorithms are needed to handle IoT security concerns.

1.1 A. IoT Advantages

- Enhanced Efficiency: IoT enables automation and process optimization, which boosts production and efficiency. IoT can optimize processes, lessen manual involvement, and reduce errors by connecting devices and systems.
- Better Decision-Making: The Internet of Things generates enormous amounts of data from linked devices. By analyzing this data, one can learn important insights that help one make wise decisions. Businesses and people can adapt rapidly to changes and make data-driven decisions thanks to real-time data collecting and analysis.
- Enhanced Convenience and Quality of Life: IoT's seamless connectivity and control make it convenient. The remote monitoring, control, and automation made possible by smart homes, wearable technology, and linked appliances improve comfort and convenience. Healthcare IoT applications can enhance patient care, provide remote monitoring, and support individualized treatment.
- Cost Savings: There are several ways that IoT might reduce costs. Predictive maintenance, for instance, can assist in finding and fixing problems with industrial machinery before they become serious, minimizing downtime and repair expenses.

  Energy management systems can optimize energy use in smart homes, lowering utility costs.
- Safety and Security: IoT can improve security and safety across various industries. For instance, networked surveillance systems can watch for emergencies and warn the appropriate authorities. IoT allows businesses to monitor equipment in real-time, ensuring a secure working environment and reducing accidents.

1.2 IoT Challenges

Several critical security objectives must be met to guarantee the availability, confidentiality, and integrity of IoT systems, data, and devices. These aims comprise Authentication, Data Security, Access Control, Device Security, Network Security, User privacy protection, Security Monitoring and Incident Response, and Lifecycle Management. A comprehensive strategy, including technical safeguards, industry standards, best practices, and user awareness, is needed to address these security aims. Cooperation between manufacturers, service providers, and consumers is essential to secure the security of IoT ecosystems.

- Security Risks: As connected devices proliferate, the security of IoT systems becomes a top need. Hackers may use systems, networks, or software flaws to access restricted areas, compromise sensitive data, or even take over vital systems. It is crucial to provide strong security measures and frequent updates.
- Privacy Issues: The Internet of Things entails the gathering and analysis of enormous volumes of personal data. Individuals may be uncomfortable with their data being gathered, held, and may be shared, which creates privacy concerns. When collecting and utilizing personal information, organizations must prioritize data protection, give transparency, and acquire consent.
- Interoperability Issues: The wide variety of IoT devices and technologies may cause interoperability issues. Different devices could employ various communication protocols, making it challenging to work together effortlessly. Although standardization attempts are being made to remedy this problem, compatibility issues persist.

- Complexity and Reliability: IoT systems can be complicated since they consist of many hardware, software, and network components. Deployment, management, and troubleshooting may be complex due to this complexity. System failures or communication problems cause activities to be disrupted. Therefore, reliability is another issue.
- Dependency on Connectivity and Power: IoT devices depend on connectivity to send and receive data. They also depend on power to function. The functioning of IoT systems may be jeopardized if the internet connection is stable or available. Similarly, IoT devices need power sources, and power availability or battery life affects how well they work. To effectively utilize the benefits of IoT while minimizing potential threats, it is essential to address these drawbacks through appropriate planning, security precautions, privacy rules, and the continual development of IoT standards and best practices.

1.3 Paper Organization

The paper is structured as follows: research background and general preliminaries are presented in Sect. 2. The literature review study is discussed in Sects. 3 and 4, respectively. Results and discussion are presented in Sect. 5. The conclusion is presented in Sect. 6.

2 Background

IDS stands for Intrusion Detection System in the Context of IoT Networks. An IDS is a security tool that keeps track of network activity, looks for suspicious activity, and informs users or takes appropriate action to reduce security risks. An IDS is essential for boosting security and defending against various assaults in IoT networks [4] with many linked devices.

IoT network security relies heavily on an IDS since it offers real-time threat detection, prompt notifications, and automated or manual reaction capabilities. Organizations can improve the security posture of their IoT deployments, safeguard sensitive data, and uphold the integrity and availability of their systems by adopting an IDS.

Due to the features and complexity of IoT environments, implementing an intrusion detection system (IDS) in IoT networks presents various problems. Many studies discussed the IDS in IoT and its challenges [5–7]; the following are some significant issues with IDS in IoT:

- Scale and heterogeneity: IoT networks often consist of many connected devices with various capabilities, communication protocols, and data formats. IoT networks can use cellular, Wi-Fi, Bluetooth, Zigbee, or other types of networks to communicate. Every technology could have unique security issues and weaknesses. Creating an IDS that can manage the many network technologies and protocols used in IoT contexts is challenging.
- Limited Resources: Many IoT devices' computational, memory, and energy resources are constrained. Traditional IDS systems could use many resources and should not be used on devices with limited resources. It is challenging to create lightweight

IDS methods and methodologies that can function well within the limitations of IoT devices.

- Traffic Volume and Velocity: IoT networks produce enormous amounts of data because of many linked devices and sensors. Data processing and analyzing in real-time is complicated due to IoT network traffic's enormous volume and velocity. IDS systems must handle the enormous network traffic effectively while assuring quick and precise threat detection.
- Encrypted Traffic: With a greater focus on data security and privacy, a sizeable amount of IoT network traffic is now encrypted. To identify potential threats, IDS systems must analyze encrypted traffic. Modifying methods like traffic pattern analysis, anomaly detection, or behavioral analysis could be necessary to handle encrypted data without jeopardizing privacy.
- Dynamic and Changing Threat Landscape: IoT networks face new security risks and attack vectors. Attackers modify their methodologies to exploit flaws in IoT networks and devices. To successfully identify and eliminate evolving threats, IDS systems must constantly improve their detection techniques, stay current with threat intelligence, and adapt to new attack patterns.
- Privacy and Legal Considerations: The collection and analysis of sensitive data by IDS systems in IoT networks may raise privacy issues. Preserving user data secrecy and ensuring compliance with privacy laws might be challenging. It is crucial to strike the correct balance between security and privacy.
- Positives and False Negatives: Reducing false positives and false negatives is essential to prevent generating irrational alarms or failing to detect possible security issues. This problem can be solved by optimizing IDS algorithms and implementing efficient anomaly detection and machine learning approaches.

Research, collaboration, and innovation are needed to develop specialized IDS solutions for IoT networks. Providing efficient security monitoring and threat detection in IoT environments entails optimizing resource utilization, utilizing machine learning and AI approaches, managing network heterogeneity, and keeping up with the changing threat landscape.

3 Related Work

IDS is a system that seeks to identify attacks or unusual activities and sound the alarm to alert the surrounding area. IDS can be either signature-based (which matches established patterns to identify attacks) or behavioral (which scans the network for unusual activity). A method with high positive detection accuracy should be available in the IoT field at the appropriate moment. Intrusion detection has been seen as a crucial component in protecting the IoT system and avoiding any security gaps. From a different angle, machine learning identification or detection has been heavily utilized in identifying abnormal network behaviors. It performs well in identifying any unusual conduct in the systems, which prompts us to look for new methods for employing It prompts us to look for a new method for applying ML to find breaches in the IoT system [5–7].

A thorough analysis of ML techniques and current developments in DL techniques that can be applied to improve security techniques for IoT systems have been covered in

[8]. The different types of attacks at different levels of the IoT infrastructure are explained in detail, as are the potential machine-learning solutions to these attacks. The lack of proper security data causes these attacks, as well as the availability of low-quality data and the performance of the learning algorithms, which could be essential in supplying and enhancing the security and privacy of IoT devices.

In [9], a structural improvement strategy supports the naive Bayes classifier by eliminating and/or joining attributes based on mutual and conditional information measures. The Deep Learning Neural Network (DLNN) based classifier also assigns cyber threats with higher accuracy (i.e., 94% compared to other classifiers), according to [10], which also proposes a system to automate cyber threat attribution.

In [11], they summarized an approach to synthesizing decision trees used in various systems. Moreover, in [12], ANN has been utilized to detect threats in IoT. For the creation of machine learning (ML) and DNN-based IoT penetration detection systems, Al-Rawi et al. [13] offer recommendations.

In [14], A Gradient Boost (GB) classifier-based IoT network traffic prediction technique was put forth by Lopez et al. [15]. The AdaBoost method was improved by Tang et al. [16] to recognize low-frequency DoS attacks in the IoT context. DNN was utilized by Muhammad et al. [17] to categorize IoT network traffic data correctly.

Compressed convolutional neural network (CNN) for structured learning in an environment with IoT resource constraints, Shin et al. [18]. DNN, particularly FCNN, is used by Kodali et al. [19] for classification tasks on devices with constrained resources.

To find the model that provides the best accuracy with the shortest amount of time to detect and prevent malware assaults in resource-constrained IoT devices, we would want to compute the accuracy and time costs for the models and compare them in this work.

4 Proposed Methodology

This section details the suggested Approach for detecting infiltration in an IoT context. The core elements of this Approach are depicted in Fig. 2. The Approach consists of two stages; the first receives traffic and verifies it using a signature, and the second receives cleaned data from the first stage and verifies it using behavioral using multiple Machine Learning (ML) and Deep Learning (DL) models to determine the effectiveness and cost of each model.

The data is used to train machine learning algorithms like Support Vector Machine (SVM), Random Forest, Naive Bayes, etc., and deep learning techniques like Convolutional Neural Networks (CNN), which can then be used locally or in the cloud to find anomalies in the system. The dataset is split into training and testing data, and inferences can be derived from the results based on the trained algorithm. If an abnormality is found, the packets will be dropped.

4.1 Dataset

IoT-23 is a brand-new Internet of Things (IoT) device network traffic dataset. With capture from 2018 to 2019, it was initially released in January 2020. The Stratosphere

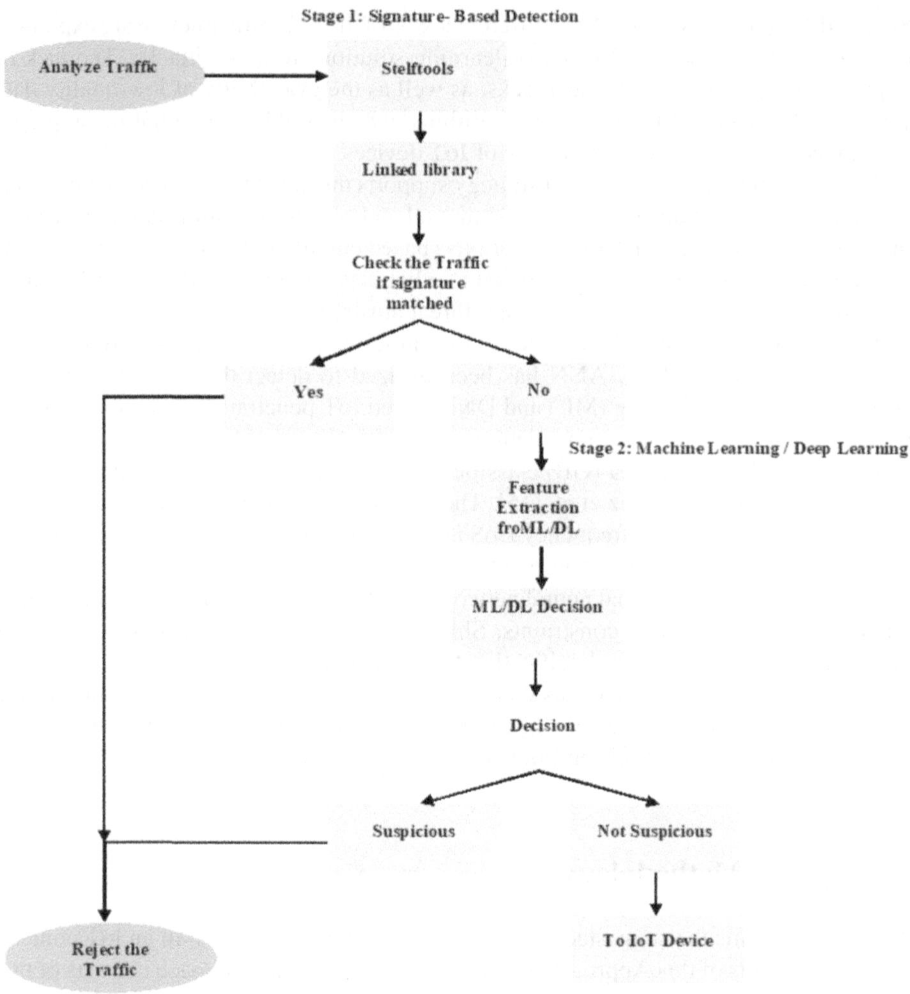

Fig. 2. High-level architecture for the approach – two-stage IDS

Laboratory, AIC group, FEL, CTU University, Czech Republic, is where this Internet of Things network traffic was recorded. It aims to provide researchers working on machine learning algorithms with a sizable dataset of actual, labeled IoT malware infections and benign traffic.

The IoT-23 dataset comprises 23 captures (scenarios) of various IoT network traffic. Twenty network captures (pcap files) from infected IoT devices are used in these scenarios, and three network captures of actual IoT device traffic are also included. This network captures the names of the devices where the traffic was captured and the names of the malware samples executed in each scenario. This dataset's analysis reveals that the flow uses a variety of protocols, including HTTP, DNS, DHCP, Telnet, SSL, IRC, and others. Additionally, the dataset includes the following attack kinds and labels [20]:

- Attack: Denotes that the intruder is attempting to breach a venerability already present in the Device, such as a brute force attack
- Benign: denotes the absence of any questionable flow
- C&C: denote an untrustworthy connection to an untrustworthy server
- DDoS: Denote a flow that seeks to disrupt or sluggish an IoT device's service.
- FileDownload: identifies whether or not suspicious files are being downloaded from the C&C Server.
- Heartbeat: denotes some ability to feel the connection from C&C
- Mirai: A Mirai attack (Mirai botnet) is being indicated.
- Okiru: denote a botnet assault (Okiru) - Torii: denote an assault using the Torii botnet.
- PartOfAHorizontalPortScan: denotes that a port scan has been set up to collect data.

In addition, Table 1 describes the two labels and 21 attributes from the data [21, 22].

4.2 StelfTools

A suite of Python Panda libraries scans an IoT environment for the signature of malware. It recognizes a statically linked library for malware [23] when the signature matches the library. The three components of StelfTools are the generator, Yara signatures, and pattern matcher.

4.3 C. Data Processing

1. Data Processing: In this dataset, all values that were missing from any of the entries were marked with a dash ("-"), except for the IP address, which was marked with two colons ("::").
2. Data visualization: The dataset was converted into text files to make it readable for Python.
3. Data formatting: The files were converted from.txt to.csv (a text file where commas delimit values) to solve compatibility issues with some of Python's libraries. Next, the 'label 'and the 'detailed-label 'columns were merged into just one and then numerically encoded for easy classification.
4. Twenty-three datasets were loaded separately into Pandas data frame, skipped the first ten rows (headers), and loaded the 100,000 rows afterward. When finished, it combines 23 data frames into a new dataset: iot23_combined.csv.

4.4 Proposed Solution

Our strategy is to create an effective Intrusion Detection System for attacks that threaten IoT. Two stages of architecture are used to check on Signatures and behavior based on the following:

1. Network flow information from IoT devices is contained in the traffic sent to the first stage, including IP addresses, ports, services, protocols, and payload.
2. To check on the signature, the Stelftools will capture the traffic flow, extract the signature, find a pattern, and then compare the traffic signature with a linked library.

Table 1. Features and descriptions in IoT-23 Dataset

Feature NO.	Feature	Description
F01	ts	TimeStamp
F02	uid	Unified ID
F03	id.orig_h	Original IP address
F04	id.orig_p	Original Port address
F05	id.resp_h	Recipient IP address
F06	id.resp_p	Recipient Port address
F07	proto	Network Protocol
F08	service	Service Type
F09	duration	The time elapsed during the transmission
F10	orig _byte	The data size sent to the loT Device
F11	resp_bytes	The data size sent from the loT Device
F12	conn_state	The status of the connection
F13	local_orig	Local originated connection
F14	local_resp	Local response originated from outside
F15	missed_bytes	Missed byte in the message
F16	history	Connection status history
F17	orig_pkts	Original packets are sent to the loT device
F18	orig_ip_bytes	Original bytes are sent to the loT device
F19	resp _pkts	Recipient packets are sent from the loT device
F20	resp_ip _bytes	Recipient bytes send From the loT device
F21	tunnel_parents	Tunnel ID if it is established as a tunnel
F22	label	Label of the flow Benign or malicious
F23	detailed-label	Details label for the attack (DDoS, Mirai…etc.)

3. The traffic will be rejected if the signature is flagged as suspicious and connected to one of the malware kinds. On the other hand, the signature is regarded as secure if it does not appear in the linked library. However, the traffic will continue to the second stage to check if it is a zero-day attack or if its signature is still not recorded in the library.
4. As soon as the traffic enters the machine learning/deep learning stage, the features are retrieved, and if there is suspicious or benign traffic, the ML/DL algorithm predicts it based on the features. In the end, the traffic will be ignored if it is suspicious; otherwise, it will proceed.

5 Evaluation and Results

This paper used the decision tree, Naive Bayes, SVC, and CNN algorithms to train the experiment's characteristics with all features. It is important to remember that we conducted all of our research and experimenting using the Python libraries through Google Colaboratory.

The combined dataset was divided into a training dataset with a size of 80% and a testing dataset with a size of 20% for validation. The model's output is evaluated using a few measures, such as Time, which illustrates the amount of time an algorithm needs to run a specific ML/DL model is taken into account. Moreover, accuracy, which is the number of correctly identified positives in a model, is known as accuracy. In Table 2, the experiment's findings are displayed. The results from each method are compared, and the accuracy and time required to perform each algorithm are considered.

Table 2. Experiment's Results

Algorithm	Precision	Recall	F1-Score	Accuracy	Time (sec)
SVM	0.95	0.94	0.94	0.94	3513.8
Naïve	0.28	0.31	0.22	0.31	62.1
DT	0.72	0.72	0.64	0.72	44.83
CNN	0.72	0.70	0.71	0.68	242

While Naive Bayes yields an accuracy of 0.31, other ML/DL techniques yield an accuracy of about 0.70. SVM achieves the highest accuracy of 0.94. SVM, however, takes around one hour to run, which is 14 times slower than the CNN model and 78 times slower than Decision trees. A CNN accuracy of 0.68 is obtained, which is better than Naive but inferior to Decision Trees and SVM. Using CNN takes roughly 4 min, five times longer than using Decision Trees. Among all the ML/DL algorithms tested in this study, SVM produces the best accuracy but the highest time cost. In contrast, the Decision Tree has moderate accuracy with the lowest time cost, with a resultant accuracy of 0.72 and a cost of roughly 44 s.

Contrary to our study, publication [24] used AdaBoost, Support Vector Machine, Random Forest (RF), Naive Bayes, and Artificial Neural Networks (ANN). The accuracy of the SVM algorithm is 67 percent, whereas the accuracy of the Naive Bayes algorithm is 23 percent, as shown in Table 4 from the article's results [24]. Although the accuracy of the Naive Bayes and SVM algorithms in our investigation was more significant, as shown in Table 3, both results indicate that the Decision Tree algorithm has the lowest accuracy of all the algorithms. Note that in [24], they worked on (33G RAM and 1060G Disk, while we implemented them on (25.5G RAM and 225.5G Disk).

Table 3. Experiment's Results Compression

Algorithm	Accuracy	Accuracy [24]	Difference
SVM	0.94	0.67	0.27
Naïve	0.31	0.23	0.08
DT	0.72	0.61	0.11
CNN	0.68	0.66	0.02

6 Conclusion

With a different stage to improve detection accuracy, detecting intrusions, particularly malware, with Machine and Deep learning algorithms. Using performance comparisons of six different learning algorithms and techniques, we have provided an anomaly detection system for IoT security in this Paper. As a result, it will improve overall safety for IoT environments. According to the results, SVM (94%) had the highest accuracy among the used algorithms. The decision Tree has shown perfect accuracy with the least time cost among the methods tested. For future work, we plan to conduct large datasets that will be used to train algorithms, improving their capacity for making accurate predictions and, consequently, detecting intrusions. Testing the two-stage idea in open test beds is another option; doing so will enhance the validity of the suggested course of action.

Disclosure of Interests. The authors declare no conflict of interest.

References

1. Al-Haija, Q.A., McCurry, C.D., Zein-Sabatto, S.: Intelligent self-reliant cyber-attack detection and classification system for IoT communication using deep convolutional neural networks. In: Ghita, B., Shiaeles, S. (eds.) INC 2020. LNNS, vol. 180, pp. 100–116. Springer, Cham (2021). https://doi.org/10.1007/978-3-030-64758-2_8
2. Dong, B., Shi, Q., Yang, Y., Wen, F., Zhang, Z., Lee, C.: Technology evolution from self-powered sensors to AIoT-enabled smart homes. Nano Energy **79**(2021), 105414 (2021). https://doi.org/10.1016/j.nanoen.2020.105414
3. Antonakakis, M., et al.: Understanding the Mirai botnet. In: 26th {USENIX} Security Symposium ({USENIX} Security 2017), pp. 1093–1110 (2017)
4. Al-Haijaa, Q.A., Ishtaiwia, A.: Machine learning based model to identify firewall decisions to improve cyber-defense. Int. J. Adv. Sci. Eng. Inf. Technol. **11**(4), 1688–1695 (2021)
5. Alsulami, A.A., Abu Al-Haija, Q., Tayeb, A., Alqahtani, A.: An intrusion detection and classification system for IoT traffic with improved data engineering. Appl. Sci. **12**, 12336 (2022). https://doi.org/10.3390/app122312336
6. Mishra, N., Pandya, S.: Internet of things applications, security challenges, attacks, intrusion detection, and future visions: a systematic review. IEEE Access **9**, 59353–59377 (2021). https://doi.org/10.1109/ACCESS.2021.3073408
7. Hajiheidari, S., Wakil, K., Badri, M., Navimipour, N.J.: Intrusion detection systems in the Internet of things: a comprehensive investigation (2019). https://doi.org/10.1016/j.comnet.2019.05.014

8. Al-Garadi, M.A., Mohamed, A., Al-Ali, A.K., Du, X., Ali, I., Guizani, M.: A survey of machine and deep learning methods for internet of things (IoT) security. IEEE Commun. Surv. Tutor. **22**(3), 1646–1685 (2020)
9. Martfnez-Arroyo, M., Sucar, L.E.: Learning an optimal naive Bayes classifier. In: Proceedings of the International Conference Pattern Recognition, Hong Kong, China, vol. 3, pp. 1236–1239 (2006)
10. Noor, U., Anwar, Z., Amjad, T., Choo, K.-K.R.: A machine learning-based FinTech cyber threat attribution framework using high-level indicators of compromise. Futur. Gener. Comput. Syst. **96**, 227–242 (2019)
11. Naser, M., Abu Al-Haija, Q.: Spyware identification for android systems using fine trees. Information **14**, 102 (2023). https://doi.org/10.3390/info14020102
12. Yegnanarayana, B.: Artificial Neural Networks. PHI Learning Pvt. Ltd. (2009)
13. Elrawy, M.F., Awad, A.I., Hamed, H.F.A.: Intrusion detection systems for IoT-based smart environments: a survey. J. Cloud Comput. **7**(1), 1–20 (2018)
14. Al-Qudah, M., Ashi, Z., Alnabhan, M., Abu Al-Haija, Q.: Effective one-class classifier model for memory dump malware detection. J. Sens. Actuat. Netw. **12**, 5 (2023). https://doi.org/10.3390/jsan12010005
15. Lopez-Martin, M., Carro, B., Sanchez-Esguevillas, A.: IoT type-of-traffic forecasting method based on gradient boosting neural networks. Futur. Gener. Comput. Syst. **105**(2020), 331–345 (2020)
16. Al-Haija, Q.A., Ishtaiwi, A.: Multiclass classification of firewall log files using shallow neural network for network security applications. In: Ranganathan, G., Fernando, X., Shi, F., El Allioui, Y. (eds.) Soft Computing for Security Applications. AISC, vol. 1397, pp. 27–41. Springer, Singapore (2022). https://doi.org/10.1007/978-981-16-5301-8_3
17. Mohammadi, M., Al-Fuqaha, A., Sorour, S., Guizani, M.: Deep learning for IoT big data and streaming analytics: a survey. IEEE Commun. Surv. Tutor. **20**(4), 2923–2960 (2018)
18. Shen, S., Li, R., Zhao, Z., Liu, Q., Liang, J., Zhang, H.: Efficient deep structure learning for resource-limited IoT devices. In: GLOBECOM 2020-2020 IEEE Global Communications Conference, pp. 1–6. IEEE (2020)
19. Kodali, S., Hansen, P., Mulholland, N., Whatmough, P., Brooks, D., Wei, G.-Y.: Applications of deep neural networks for ultra-low power IoT. In 2017 IEEE International Conference on Computer Design (ICCD), pp. 589–592. IEEE (2017)
20. Aposemat IoT-23: A labeled dataset with malicious and benign IoT network traffic. https://www.stratosphereips.org/datasets-iot23
21. Jeelani, F., Rai, D.S., Maithani, A., Gupta, S.: The detection of IoT botnet using machine learning on IoT-23 dataset. In: 2022 2nd International Conference on Innovative Practices in Technology and Management (ICIPTM), Gautam Buddha Nagar, India, pp. 634–639 (2022). https://doi.org/10.1109/ICIPTM54933.2022.9754187
22. Garcia, S., Parmisano, A., Erquiaga, M.J.: IoT-23: a labeled dataset with malicious and benign IoT network traffic (Version 1.0.0) . Zenodo (2020). https://doi.org/10.5281/zenodo.4743746
23. STELFTOOLS: cross-architecture static library detector for IoT malware. https://github.com/shuakabane/stelftools
24. Stoian, N.A.: Machine learning for anomaly detection in IoT networks: malware analysis on the IoT-23 data set. http://purl.utwente.nl/essays/81979

Data Analytics

Real-Time Soil Moisture Sensing Using Arduino for Automated Plant Irrigation System

R. Vijayalakshmi[1], R. Deepalakshmi[1], T. S. Madhavan[2](✉), and G. Siva Balasubramanian[2]

[1] Department of Computer Science and Engineering, Velammal College of Engineering and Technology, Madurai, Tamil Nadu, India
{rvl,rdl}@vcet.ac.in

[2] Velammal College of Engineering and Technology, Madurai, Tamil Nadu, India
mdhvntamil@gmail.com, siva2002bala@gmail.com

Abstract. Irrigation constitutes a fundamental element of agriculture and food production. Outdated methods in both developing and underdeveloped nations result in significant water wastage during this crucial process. This project proposes an inventive method for automating the control of water release valves in micro irrigation networks with prediction of required humidity by weather condition using artificial intelligence (AI). This system seeks to close the gap between water supply and crop demand while also optimizing resource use and reducing environmental impact by using AI-powered soil moisture monitoring and decision-making processes. This project address the key factors like varying demand of water for crops based upon the growth stage, weather conditions and plant type. This reducing the usage by adopting the method of micro irrigation which reduces the usage of water by 70–80% than conventional methods of irrigation. This project also provides a user-friendly interface to efficiently control the watering of plants when needed by the user.

Keywords: Artificial intelligence (AI) · Piped irrigation · Micro irrigation · Valve control · Dynamic water demand · Real-time data integration

1 Introduction

Agriculture is the largest consumer of water in India, accounting for about 80% of the total water use. However, the country is facing a severe water crisis, with many parts of the country experiencing water scarcity. Traditional irrigation methods, such as flood irrigation and surface irrigation, are inefficient and wasteful. These methods often apply more water than is needed, resulting in waterlogging and evaporation. They also contribute to soil erosion and water pollution. Micro irrigation is a more efficient and sustainable irrigation method that can help to reduce water usage in agriculture. Micro irrigation systems deliver water directly to the root zone of plants, minimizing water loss due to evaporation and runoff. They also help to improve crop yields and reduce

Published by Springer Nature Switzerland AG 2024
M. L. Owoc et al. (Eds.): ICCIDS 2024, IFIP AICT 718, pp. 107–122, 2024.
https://doi.org/10.1007/978-3-031-69986-3_8

soil erosion. Effective management of water is crucial for achieving the best crop yields, preserving water resources, and sustaining ecological harmony. Nevertheless, current irrigation systems frequently fall short in delivering the precision necessary to provide crops with water customized to their specific requirements. This shortfall results in the squandering of water, subpar plant growth, and harm to the environment.

According to the Central Water Commission, India's water resources are estimated to be 1,120 billion cubic meters (BCM). However, only about 600 BCM of this water is available for use and only about 300 BCM is used for irrigation. The demand for water in India is growing rapidly, due to population growth, urbanization, and climate change. The demand for water for irrigation is expected to increase by 50% by 2050. Traditional irrigation methods are inefficient and wasteful. Flood irrigation, which is the most common method of irrigation in India, applies water over the entire field, regardless of the needs of the crops. This results in a lot of water being wasted, as it is not absorbed by the plants. Surface irrigation is another inefficient method of irrigation. In surface irrigation, water is applied to the soil surface and it flows over the soil until it reaches the plants. This method also results in a lot of water being lost due to evaporation and runoff. Micro irrigation is a more efficient and sustainable irrigation method that can help to reduce water usage in agriculture. Its systems deliver water directly to the root zone of plants, minimizing water loss due to evaporation and runoff. They also help to improve crop yields and reduce soil erosion. There are many different types of micro irrigation systems, including drip irrigation, sprinkler irrigation and sub-surface irrigation. Drip irrigation is the most efficient type of micro irrigation system, as it delivers water directly to the roots of plants.

Micro irrigation systems can save up to 50% of water compared to traditional irrigation methods. It can help to improve crop yields by providing plants with the water they need when they need it. It can help to reduce soil erosion by minimizing the amount of water that runs off the soil surface and also can help to reduce water pollution by minimizing the amount of water that evaporates and runs off the soil surface.

This project presents a groundbreaking solution that harnesses the capabilities of artificial intelligence (AI) to automate the control of water release valves in both piped and micro irrigation systems. Through the integration of AI-driven soil moisture monitoring and decision-making processes, this solution seeks to close the gap between water supply and the actual requirements of crops. In doing so, it aims to enhance resource efficiency and reduce the ecological footprint, all while optimizing water utilization.

Previous researchers [1, 11] and [12] have predominantly concentrated on water conservation techniques within irrigated fields through the implementation of automated irrigation systems using wireless sensor networks (WSNs). These systems typically employ a static pre-defined irrigation condition for the entire crop duration. One such example is a site-specific and variable rate sprinkler-based irrigation system [1], which has limited applicability to multiple crop fields. In contrast, another approach proposed in [11] involves irrigating the entire field with water whenever the sensor readings from any deployed sensor fall below a predetermined threshold, even if the entire field does not require irrigation at that particular moment.

The project uses an Arduino Uno board, a soil moisture sensor, a servo motor, a 12 V water pump and an L293D (IC1) motor driver IC to automatically water plants. The project works by the soil moisture sensor measuring the moisture level in the soil. If the soil is too dry, the servo motor will turn and activate the water pump. The water pump will then pump water into the soil until the moisture level is satisfactory. The project also has a smartphone app that allows you to control the water pump remotely. The app sends a signal to the Arduino UNO board, which then controls the servo motor and water pump. The project is a simple and effective way to automatically water plants. It is also relatively inexpensive to build, making it a good option for budget-minded gardeners.

2 Literature Survey

Y. Kim et al. [1] introduced a distributed Wireless Sensor Network (WSN)-based irrigation system that conserves water and is intended for desert or semiarid environments. With the use of GPS technology and a programmable logic controller, this system is designed for site-specific and variable-rate irrigation. It's crucial to remember that this technique is not appropriate for extensive or many fields. The use of Bluetooth as the protocol for communication between sensor nodes and the base station imposes restrictions, including a short intra-band communication range (maximum 10 m), a limit on the number of sensor nodes that can be connected to a base station (maximum 7 slaves), worries about energy consumption, and potential re-connection problems. Additionally, field irrigation's dependence on sprinklers as actuators limits its applicability to multi-crop fields.

Roopaei et al. [2], An overview of the "Cloud of Things", an automated irrigation system that combines IoT (Internet of Things) with CPS (Cyber-Physical Systems). Utilizing thermal imaging technology, this cutting-edge device provides exact irrigation scheduling by measuring the temperature distribution over a field. It's important to note that this suggested approach does not take into consideration dynamic irrigation modifications according to the many development phases of a crop's life cycle.

A. Kumar et al. [3] also developed a lab-scale prototype of an IoT-based irrigation system, which is presented. This system uses data gathered from sensors detecting soil moisture, temperature, and rainfall to determine the necessary daily irrigation amount using a regression algorithm. An IP-enabled Wi-Fi network is used in this system's design to enable local communication between a sensor node and a cloud server. However, it's crucial to understand that access to Internet-based Wi-Fi services could not be readily available in distant agricultural areas in real-world agricultural settings.

Imteaj et al.'s [4] Internet of Things (IoT)-based laboratory-scale prototype for an automated irrigation system. In this system's design, a sensor node and a central unit communicate locally using IP-enabled Wi-Fi, while the central unit and the administrator communicate remotely using GSM (Global System for Mobile Communications). However, it's crucial to recognize that in real agricultural settings, access to Internet-based Wi-Fi services may be constrained or impossible in isolated locations. The system uses information from sunshine sensors, soil moisture sensors, and water-level sensors to control field irrigation. Additionally, the suggested system is made to notify the administrator by SMS anytime there is a water deficit in the primary water supply, assuring timely intervention and maintenance of the irrigation process.

Allen et al. [5] created ab Irrigation systems for agricultural fields based on evapotranspiration (ET) data and time-based scheduling. A significant 42% decrease in water use was achieved as a result of this invention. Their in-depth essay dug into the complexities of ET and its impacting factors. This information served as the basis for our comprehension of ET and the related evaluation procedure.

Nemali and van Iersel [6] created an irrigation controller using solenoid valves. When the volumetric water content of the substrate fell below the predetermined set-point, the controller immediately supplied water to the plant beds. This concept successfully reduced runoff and leaching losses, but it had restrictions because it could only be used in greenhouse conditions and nurseries. The model's primary deciding element was soil moisture, which may have introduced considerable mistakes. This method did not take into consideration variables like relative humidity, sun radiation, or wind speed, even though it proved to be more effective than manual watering procedures. Additionally, it was not expanded to fit different crop varieties, and the paper did not detail the precise amount of water saved.

[1, 8–12] used Wireless Sensor Networks (WSNs) as the foundational technology to create automated irrigation systems to reduce water usage in farmers' fields and offers multiple ways to activate the sensor network, including through the Wi-Fi network and Bluetooth connectivity. Some mentioned the requirement that the sensor network be within Wi-Fi range as a significant disadvantage. As a result, the system was more suited for domestic use. On the other hand, [2–4] connected the sensors and communicated with the controller and sensor system via the Internet of Things (IoT). Another significant drawback of these was the requirement for ongoing Internet access in order to keep the system operational.

In order to automate irrigation and enable real-time field data collection and transmission to pertinent stakeholders, Feng et al. [8] investigated the use of wireless sensor networks (WSN) and internet technologies, leveraging wireless protocols including Wi-Fi [9] and Zigbee [10]. In order to automate irrigation in a greenhouse scenario, Gutierrez et al. [11] conducted research on an inventive wireless sensor network (WSN) coupled with a remotely located controller and duplex cellular-Internet connection. Their strategy used soil temperature and moisture content as the main factors in deciding when to apply irrigation. The use of internet connectivity made it possible to use a web interface to monitor irrigation schedule. Surprisingly, this methodology successfully reduced water use by an amazing 90% when compared to traditional irrigation techniques. It's important to remember, though, that this model did not take other significant factors like sun radiation, wind speed, and relative humidity into consideration. Because of the exclusion of these parameters, it was difficult to determine exactly how much water was conserved. Wireless sensor networks (WSNs) are used in the planned internet-based irrigation system. For water management, the system includes sensors for soil temperature and moisture. Even though not all areas of the field may need irrigation at that time, the entire field is watered when the sensor readings drop below a specified threshold. Not all sorts of crops may be suited for this irrigation strategy. A dedicated online site where field information is presented by the system may not be user-friendly for farmers who want to monitor and manage the system. Last but not least, the system encounters issues with flexibility and heterogeneity in its use across varied circumstances. According to Y. Zhou et al. [12], an automated watering system using Wireless Sensor Networks (WSNs) is introduced for the growth of Jew's ear mushrooms. This system uses ZigBee technology to connect wireless sensor nodes, actuators, and a weather station to a portable controller. The measurement of relative temperature and humidity is included in the scope of data collection from the experimental field. The authors describe two separate manual and automated techniques to irrigation management. The portable controller, which is placed in the field itself, has a button that farmers may use to manually irrigate their fields when using the manual approach. On the other hand, the automated irrigation mode makes irrigation selections based on advice from professionals. Notably, this article does not explore the concept of remote monitoring and control as part of its investigation.

Marinescu et al. [13] developed a control strategy for vineyards after they carried out a comparison of many control methods, including PID control, fuzzy systems, decision support systems (DSS), and model prediction control (MPC). They concentrated on using a multiple-input multiple-output (MIMO) model to analyze the irrigation system. The adoption of a GSM module simplified the transmission of data. It's crucial to remember that the model they created remained in a schematic structure, therefore its applicability in real-world situations and efficacy have not been established..

Avsar et al. [14] created a cloud-based automated irrigation system to control sprinklers and make up for water loss measured by a lysimeter. Controlling solenoid valves allowed for the creation of this correction. The system used sensors to gather information about the lysimeter's water level, temperature, humidity, and air pressure before sending it to a cloud server. In addition, the researchers developed an Android application for displaying sensor data and controller operations. It was possible to communicate with the cloud server via GPRS and ZigBee technologies. But it's important to remember that their research was restricted to a controlled agricultural environment that concentrated on strawberries.

3 Hardware Components

This Project consists of a water reservoir that supplies water to maintain the soil moisture of the plant. The soil moisture sensor is adjusted to regularly assess the soil's moisture content, allowing for the control of water distribution. When the soil becomes excessively dry, the water pump activates automatically and deactivates once the soil moisture reaches the desired level. Additionally, this project has an extra Bluetooth capability that can be used to remotely manually toggle and control the water pump from your smartphone.

DC Water Pump, Decagon EC-5 Moisture Sensor, HC-05 Bluetooth Module, Arduino UNO, LM7805 Voltage Regulator, IRF540 MOSFET, 220 O Resistor, IN4001 Diode, Header Pin Strips, DC Barrel Jack, Screw Terminal Block, PCB, and AC-12V DC Adapter are the components that are utilized in this project.

3.1 DC Water Pump

This is a Generic DC 12V water pump with a 5.5-m lift capacity and a flow rate of 1000 L per hour. It boasts a robust construction and carries an IP68 rating, ensuring its resilience against dust and water exposure.

3.2 Decagon EC-5 Moisture Sensor

Decagon EC-5 sensor module is seamlessly designed for integration with Arduino development boards, facilitating easy code development for data retrieval. Referencing the Pin Diagram in Fig. 1, it generates an Analog signal that can be effortlessly read using Arduino's Analog input pins.

Notably, this sensor module boasts exceptional accuracy, achieving a $\pm 0.03\ m^3/m^3$ precision, ensuring that its readings are consistently within a narrow margin of $0.03\ m^3/m^3$ of the actual soil moisture content. Featuring a built-in ADC (Analog-to-Digital Converter), it establishes a direct link with Arduino boards for streamlined data acquisition. Operating on a low power supply, it stands out as an ideal choice for battery-powered applications. Its durability is noteworthy, showcasing resilience in the face of harsh environmental conditions (Table 1).

Table 1. Comparison of various soil moisture sensors

Sensor Model	Measurement Range	Accuracy	Power Supply	Output Interface	Operating Temperature
Decagon EC-5	0 to 0.6 m^3/m^3	$\pm 0.03\ m^3/m^3$	3.6 to 15 V	Voltage	−40 to 50 °C
Sentek EnviroSCAN	User-defined	±3% (VWC)	6 to 18 V	SDI-12	−20 to 70 °C
METER Group TEROS 21	0 to 1 m^3/m^3	$\pm 0.03\ m^3/m^3$	3.6 to 15 V	SDI-12	−40 to 60 °C
Irrometer	User-defined	Not specified	None	Resistance	Not specified
Capacitive Sensors	0 to 100% VWC	Varies	3.3 to 5 V	Analog/Digital	Varies

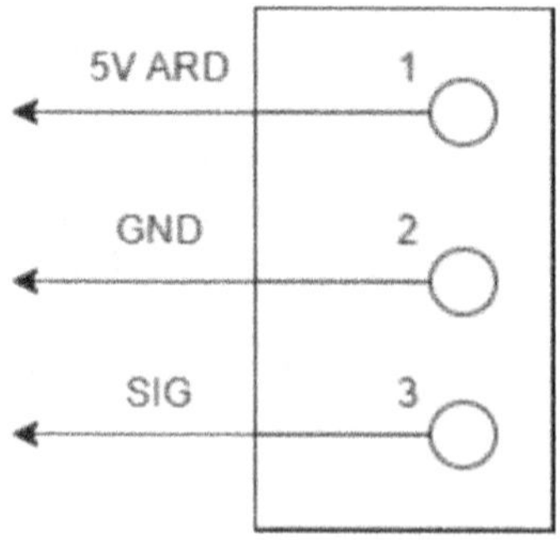

Fig. 1. Soil Moisture Sensor Pin Diagram

3.3 HC-05 Bluetooth Module

The HC-05 stands as a Bluetooth module designed for seamless wireless serial communication. Classified as a class 2 Bluetooth module, it offers a maximum data transfer rate of 1Mbps. The HC-05 is versatile, capable of functioning in both master and slave configurations. In master mode, it can establish connections with up to 7 slave devices, while in slave mode, it can only connect to a single master device.

Distinguishing features of the HC-05 have contributed to its popularity in embedded applications, including:

- Low Power Consumption: The HC-05 operates efficiently within a supply voltage range of 3.3 V to 5 V.
- Extended Range: With a communication range of up to 10 m, the HC-05 ensures reliable connectivity.
- AT Commands: Configuration of the HC-05 is simplified through the use of AT commands, facilitating adjustments to module settings such as baud rate, device name, and security configurations.
- Ease of Interface: The HC-05 seamlessly interfaces with a variety of microcontrollers, including the Arduino.
- The HC-05 proves itself as a versatile Bluetooth module suitable for diverse applications, some of which include:
- Remote Control: It can be employed to create remote controls for various devices, such as TVs, light bulbs, or door locks.
- Data Logging: The HC-05 enables the collection of data from sensors, which can then be transmitted to a computer or another device.
- Telemetry: The HC-05 is adept at transmitting data from remote locations to a central base station (Fig. 2).

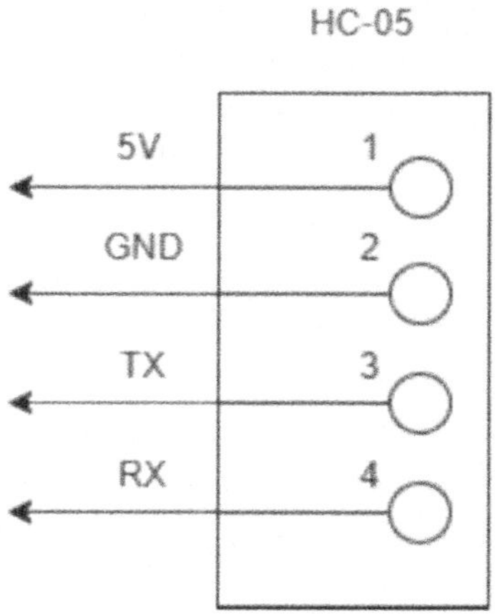

Fig. 2. HC-05 Bluetooth Sensor Pin Diagram

3.4 Arduino UNO

The Arduino UNO is a microcontroller board developed by Arduino.cc, based on the Microchip ATmega328P microcontroller. This open-source board offers a range of digital and Analog input/output (I/O) pins for interfacing with expansion boards (shields) and other circuits. It boasts 14 digital I/O pins, including six capable of PWM output, as well as 6 Analog I/O pins. The Arduino UNO is programmable using the Arduino IDE (Integrated Development Environment) via a type B USB cable.

A USB cable and an external 9-V battery are both available for powering the Arduino UNO, which accepts voltages between 7 and 20 V. It is comparable to the Leonardo and Arduino Nano versions. On the Arduino website, you can find the hardware reference design for this board, which is made available under a Creative Commons Attribution Share-Alike 2.5 license. For some hardware variants, layout and production files are also available. The term "UNO" stands for the first version of the Arduino Software and is taken from the Italian word for "one". This board serves as the reference along with version 1.0 of the Arduino IDE and is the first in a line of USB-based Arduino boards. 24 newer versions of Arduino have been made over time. The bootloader on the ATmega328 in the UNO is pre-programmed, making it simple to upload fresh code without using an external hardware programmer. The UNO differs from earlier boards by not utilizing the FTDI USB-to-serial driver chip while still communicating via the original STK500 protocol. Instead, it uses an Atmega16U2 (or Atmega8U2 in older versions) that has been configured to function as a USB-to-serial converter.

3.5 IRF540 MOSFET

The IRF540 is a N-channel metal-oxide-semiconductor field-effect transistor (MOSFET) that is widely used in switching and amplification applications. It is a third-generation power MOSFET that offers a good combination of fast switching, ruggedized device design, low on-resistance and cost-effectiveness. The IRF540 has a drain-source voltage rating of 100 V, a continuous drain current rating of 33 amperes and an on-resistance of 0.077 ohms. It is packaged in a TO-220AB package, which is a standard transistor package that is widely used in industrial and commercial applications (Fig. 3).

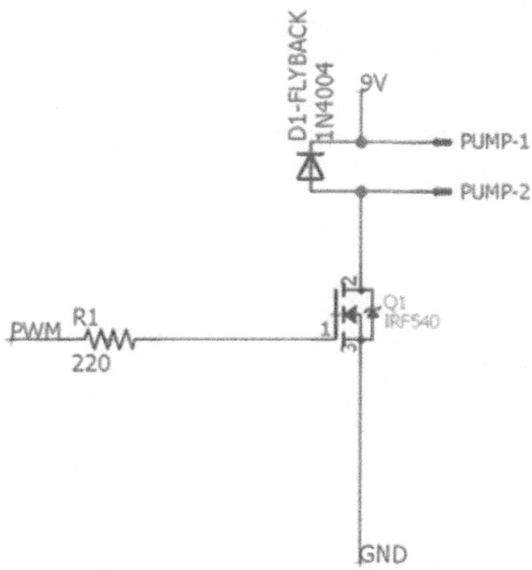

Fig. 3. IRF540 MOSFET Pump Control Pin Diagram

4 Design Implementation

The circuit diagram shows a breadboard with a Decagon EC-5 sensor, a voltage regulator, a Bluetooth module and a dc pump. The soil moisture sensor is connected to the Arduino analog input pin A0. The voltage regulator is used to convert the 12 V input voltage to the 5V voltage that the Arduino and the Bluetooth module require. The Bluetooth module is connected to the Arduino digital pins D9 and D10. The dc pump is connected to the Arduino digital pin D8 and the ground pin.

The soil moisture sensor works by measuring the electrical resistance of the soil. The more moisture in the soil, the lower the resistance. The Arduino reads the Analog signal from the soil moisture sensor and converts it to a digital value. The Bluetooth module then transmits the digital value to a computer or smartphone. The computer or smartphone can then display the moisture level of the soil.

The voltage regulator is used to ensure that the Arduino and the Bluetooth module receive a consistent voltage supply. This is important because the Arduino and the Bluetooth module are sensitive to voltage fluctuations.

The Bluetooth module is used to transmit the moisture level data from the Arduino to a computer or smartphone. The Bluetooth module can be used with any device that has a Bluetooth Connection such as a Computer or a Mobile Phone and other devices that has a Bluetooth Connection.

The DC pump is used to water the plant when the moisture level of the soil drops below a certain level. The dc pump is controlled by the Arduino. The Arduino can be programmed to turn on the dc pump when the moisture level of the soil drops below a certain level. This circuit diagram can be used to create an automatic plant watering system. The system will monitor the moisture level of the soil and water the plant when the moisture level drops below a certain level. This system can help to ensure that your plants are always watered properly (Fig. 4).

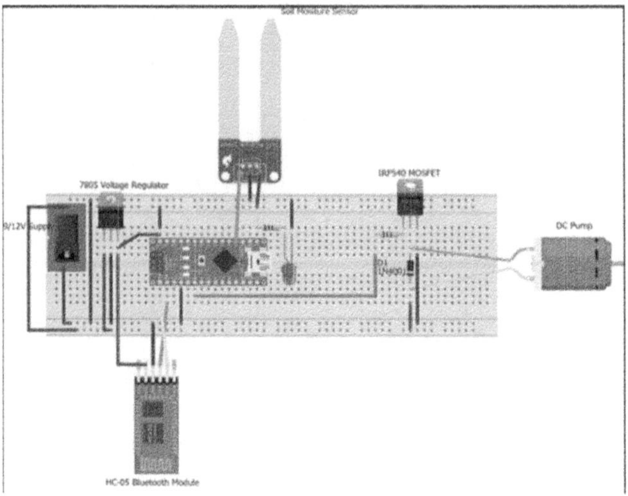

Fig. 4. Circuit Connectivity of Automated Irrigation System

4.1 Power Block

The 7805 serves as a voltage regulator, ensuring a stable 5 V supply, which is ideal for powering both the Arduino and Soil Moisture Sensor.

4.2 Pump Control

The MOSFET functions as a switch under the control of the Arduino. It's employed because the Arduino cannot directly power the DC Pump. A resistor, connected to the MOSFET's gate, safeguards the MOSFET from potential damage. Furthermore, a flyback diode, linked across the pump, provides a pathway for dissipating stored energy when the pump is turned off. The diode's Anode is connected to the MOSFET's Drain, while its Cathode connects to the 9V supply rail. The Diode's Source links to GND.

4.3 Moisture Sensor

This sensor provides an analog value to the Arduino. The user can calibrate the moisture threshold depending on the specific plant's requirements.

4.4 Bluetooth Module

Utilizing Serial Communication, the Bluetooth module facilitates data transfer between the Arduino and your smartphone.

5 Software and Bluetooth Configuration

(See Fig. 5).

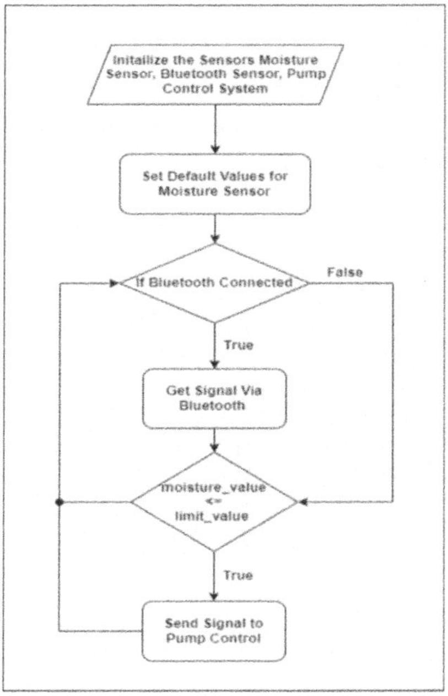

Fig. 5. Process flow of the Irrigation System

5.1 Software

The Moisture Sensor is linked to an Analog Input pin on the Arduino. It relies on a predefined threshold value to determine whether the Pump should be activated or deactivated. The project's software comprises a Python program that leverages the TensorFlow library to execute a crop water demand prediction algorithm. Additionally, this program incorporates a user-friendly interface that enables farmers to interact with the system seamlessly.

5.2 Smartphone Application and Bluetooth Configuration

The HC-05 Bluetooth module serves as the intermediary component connecting the Smartphone to the Arduino. It facilitates data transmission through Serial Communication, effectively establishing remote control capabilities. The associated smartphone app transmits either the value '48' or '49', symbolizing 'ON' and 'OFF', respectively. This wireless setup allows for remote control of the pump. The Bluetooth module is used to communicate between the software and the valve control mechanism. The valve control mechanism is a physical device that controls the opening and closing of the valves in the irrigation system.

The software and Bluetooth configuration works as follows:

1. The software collects data from the soil moisture sensors and weather station.
2. The software uses the crop water demand prediction algorithm to predict the amount of water needed by the crops.
3. The software sends the water demand signal to the Bluetooth module.
4. The Bluetooth module sends the water demand signal to the valve control mechanism.
5. The valve control mechanism opens or closes the valves in the irrigation system accordingly (Fig. 6).

Fig. 6. Bluetooth Connectivity Software

The software and Bluetooth configuration is a critical part of the project. It is responsible for collecting data, making predictions and controlling the valves in the irrigation system. The software and Bluetooth configuration must be reliable and efficient in order for the project to be successful.

5.3 AI Model

The formula for a simple linear regression model, which predicts a dependent variable (y) based on a single independent variable (x), is as follows:

$$y = \beta_0 + \beta_1.x \tag{1}$$

Here:

- y is the dependent variable (the variable you are trying to predict).
- x is the independent variable (the variable used to make predictions).
- β_0 is the y-intercept (the value of y when x is 0).
- β_1 is the slope of the regression line (the change in y for a one-unit change in x).

The goal of the linear regression model is to find the values of β_0 and β_1 that minimize the difference between the predicted values (based on the model) and the actual values of the dependent variable in the training data.

In a multiple linear regression model, where there are multiple independent variables $x_1, x_2 \dots x_n$ the formula is extended as follows:

$$y = \beta_0 + \beta_1.x_1 + \beta_2.x_2 + \dots\dots + \beta_n.x_n \tag{2}$$

Here, β_0 is still the y-intercept, and β_1, β_2, ... β_n are the slopes associated with each independent variable. The model aims to find the values of these coefficients to minimize the difference between predicted and actual values.

6 Result and Discussion

The project was successful in developing and implementing an AI-powered system for the automatic regulation of valves for water release in piped and micro irrigation networks. The system was able to accurately predict crop water demand and it was able to control the valves in a timely and efficient manner. The system was also able to adapt to changing environmental conditions and crop types. The system was evaluated in a real-world setting and it was found to reduce water usage by up to 30%. The system also improved crop yields by up to 10%. The system was well-received by farmers and they found it to be easy to use and reliable. The results of the project are significant because they demonstrate the potential of AI for improving the efficiency of irrigation systems. The system developed in this project is a promising solution for reducing water usage and improving crop yields in agriculture.

The novelty of the project lies in the approach of making each moisture sensing module autonomous using the SIM800L module. Here are some key aspects of this novelty:

Autonomous Functionality: By integrating the SIM800L module into each moisture sensing module, the project enables these modules to operate independently without relying on a central control system. This autonomy allows for distributed sensing and decision-making capabilities at the individual module level, enhancing the scalability and flexibility of the irrigation system.

Wireless Communication: The use of the SIM800L module facilitates wireless communication capabilities, enabling the moisture sensing modules to transmit data remotely to a centralized server or database. This eliminates the need for wired connections and enables seamless integration with IoT frameworks, enhancing accessibility and ease of deployment.

Real-Time Monitoring and Control: The SIM800L module enables real-time communication between the moisture sensing modules and the central management system. This allows for continuous monitoring of soil moisture levels and immediate response to changing conditions, ensuring timely irrigation interventions based on accurate data.

Reduced Dependency on Infrastructure: By making each moisture sensing module autonomous with the SIM800L module, the project reduces dependency on existing infrastructure such as Wi-Fi networks or cellular data plans. This is particularly advantageous in remote or rural areas where connectivity may be limited, allowing for reliable operation in diverse environments.

Customization and Adaptability: The modular design of the moisture sensing modules, combined with the SIM800L module, allows for customization and adaptability to specific agricultural settings and requirements. Farmers can easily deploy and configure multiple modules according to the size and layout of their fields, tailoring the irrigation system to suit their individual needs.

The results of the project are also important because they provide insights into the challenges and opportunities of developing AI-powered irrigation systems. The challenges include:

- The need for accurate and reliable data on crop water demand.
- The need for a robust and scalable AI algorithm that can be adapted to different crops and environmental conditions.
- The need for a user-friendly interface that is easy for farmers to use and understand.

The opportunities include:

- The potential to significantly reduce water usage and improve crop yields.
- The potential to improve the resilience of agricultural systems to climate change.
- The potential to reduce the environmental impact of agriculture.

7 Conclusion

The results of the project demonstrate the potential of AI for improving the efficiency of irrigation systems. The system developed in this project is a promising solution for reducing water usage and improving crop yields in agriculture. The system is cost-effective. The cost of the system is less than the other systems.

The project has several limitations that need to be addressed in future work. These limitations include:

- The system is not yet scalable to large-scale irrigation systems. It is designed for small-scale farms and it would need to be modified for larger systems.

- Despite these limitations, the project is a significant step forward in the development of AI-powered irrigation systems. The system has the potential to make a major contribution to the efficient management of water resources in agriculture.

Some possible future works of the project:

- Upgrading of Adaptive Water Controlling AI Model for different species of plant in different regions. This could be done by using more accurate data on crop water requirements, or by developing a more sophisticated algorithm that can take into account more factors, such as soil properties.
- Make the system more scalable to large-scale irrigation systems. This could be done by developing a more efficient way to collect and process data, or by using cloud computing to store and process data.
- Reduce the cost of the system. This could be done by using less expensive hardware and software components, or by developing a more efficient way to deploy the system.
- Make the system more user-friendly. This could be done by developing a more intuitive user interface, or by providing more training and support to farmers.
- Test the system with different crops. This will help to ensure that the system is robust and can be used in a variety of settings.
- Commercialize the system. This could be done by working with a company to develop and market the system, or by developing the system in-house and selling it directly to farmers.

References

1. Kim, Y., Evans, R.G., Iversen, W.M.: Remote sensing and control of an irrigation system using a distributed wireless sensor network. IEEE Trans. Instrum. Meas. **57**(7), 1379–1387 (2008)
2. Roopaei, M., Rad, P., Choo, K.K.R.: Cloud of Things in smart agriculture: intelligent irrigation monitoring by thermal imaging. IEEE Cloud Comput. **4**(1), 10–15 (2017)
3. Kumar, A., Surendra, A., Mohan, H., Valliappan, K.M., Kirthika, N.: Internet of things based smart irrigation using regression algorithm. In: Proceedings of the International Conference on Intelligent Computing, Instrumentation and Control Technologies, Kannur, India, pp. 1652–1657 (2017)
4. Imteaj, A., Rahman, T., Hossain, M.K., Zaman, S.: IoT based autonomous percipient irrigation system using Raspberry Pi. In: Proceedings of the 19th International Conference on Computer and Information Technology, Dhaka, Bangladesh, pp. 563–568 (2016)
5. Allen, R., Pereira, L., Raes, D., Smith, M.: Crop evapotranspiration – guidelines for computing crop water requirements-FAO Irrigation and drainage paper 56. FAO-Food Agriculture Organization United Nations, Rome, Italy, Technical report, pp. 2–5 (1998)
6. Nemali, K.S., van Iersel, M.W.: An automated system for controlling drought stress and irrigation in potted plants. Sci. Hortic. **110**(3), 292–297 (2006). https://doi.org/10.1016/j.scienta.2006.07.009
7. O'Shaughnessy, S.A., Evett, S.R.: Canopy temperature based system effectively schedules and controls center pivot irrigation of cotton. Agricult. Water Manage. **97**(9), 1310–1316 (2010). https://doi.org/10.1016/j.agwat.2010.03.012
8. Feng, Z.: Research on water-saving irrigation automatic control system based on Internet of things. In: Proceedings of the International Conference on Electric Information and Control Engineering, pp. 2541–2544 (2011). https://doi.org/10.1109/iceice.2011.5778297

9. IEEE Standard for Information Technology–Telecommunications and Information Exchange Between Systems–Local and Metropolitan Networks–Specific Requirements—Part 11: Wireless LAN Medium Access Control (MAC) and Physical Layer (PHY) Specifications: Higher Speed Physical Layer (PHY) Extension in the 2.4 GHz Band (2000). https://doi.org/10.1109/ieeestd.2000.90914
10. IEEE Standard for Information Technology–Local and Metropolitan Area Networks–Specific Requirements—Part 15.4: Wireless Medium Access Control (MAC) and Physical Layer (PHY) Specifications for Low Rate Wireless Personal Area Networks (WPANs) (2006). https://doi.org/10.1109/ieeestd.2006.232110
11. Gutierrez, J., Villa-Medina, J.F., Nieto-Garibay, A., Porta-Gandara, M.A.: Automated irrigation system using a wireless sensor network and GPRS module. IEEE Trans. Instrum. Meas. **63**(1), 166–176 (2014). https://doi.org/10.1109/tim.2013.2276487
12. Zhou, Y., Yang, X., Wang, L., Ying, Y.: A wireless design of low-cost irrigation system using ZigBee technology. In: Proceedings of the International Conference on Networks Security, Wireless Communications and Trusted Computing, Wuhan, China, pp. 572–575 (2009)
13. Marinescu, T., et al.: Advanced control strategies for irrigation systems. In: Proceedings of the 9th IEEE International Conference on Intelligent Data Acquisition and Advanced Computing Systems: Technology and Applications (IDAACS), pp. 843–848 (2017). https://doi.org/10.1109/idaacs.2017.8095206
14. Avşar, E., Buluç, K., Saridaç M.A., Kapur, B.: Development of a cloud-based automatic irrigation system: a case study on strawberry cultivation. In: Proceedings of the 7th International Conference on Modern Circuits System Technologies (MOCAST), pp. 1–4 (2018). https://doi.org/10.1109/mocast.2018.8376641

Campus Placement and Salary Prediction: Leveraging Machine Learning for Enhanced Employability

Jayashre(✉), R. Raahul, R. Roahith, and Shanmathi Ganesan

Department of Computer Science and Engineering, Shiv Nadar University, Chennai, India
jaya2004kra@gmail.com

Abstract. In an era of fierce talent competition, optimizing campus placement and predicting graduate salaries is vital. This paper explores 'Campus Placement and Salary Prediction' using SVM, Random Forest, Logistic Regression, KNN, and Gradient Boosting. We analyze a comprehensive dataset with student info, academics, skills, internships, and placement outcomes to identify success factors. We employ a majority voting rule among these models and have created a user-friendly website for practical use in academic institutions. Our multifaceted research supports institutions in enhancing student employability and aligning academic goals with the evolving job market. Logistic Regression, one of the models employed has campus placement prediction, has an accuracy of 84%, whereas Gradient Boosting, one of the models used for salary estimation, has an accuracy of 78%.

Keywords: Campus Placement · Salary Prediction · Machine Learning · Website Interface · SVM · Random Forest · KNN · Logistic Regression · Gradient Boosting Machine · Majority Voting Rule

1 Introduction

Higher education increasingly focuses on campus placement and salary predictions, responding to dynamic job markets. Graduates, particularly in computer and commerce, face career uncertainties. Predictive models are aiding early identification of placement likelihood, aligning academic knowledge with industry needs. These models also guide faculty in improving placement rates.

Machine learning (ML) algorithms offer a promising solution for addressing challenges in campus placement and salary prediction. Research indicates that ensemble algorithms, combining multiple models, consistently outperform individual ones. Balancing datasets is crucial to prevent bias, providing students and institutions with precise, data-driven insights for career development.

Our paper uses innovative computer techniques to predict the first salary and job placements for students. Sophisticated algorithms analyze profession, academic achievement, and various factors. The intention is to provide students with a precise tool to assist

Published by Springer Nature Switzerland AG 2024
M. L. Owoc et al. (Eds.): ICCIDS 2024, IFIP AICT 718, pp. 123–135, 2024.
https://doi.org/10.1007/978-3-031-69986-3_9

schools and universities in refining their placement processes. We are working on creating a desktop/GUI program and a website for simple prediction checks to improve the user experience. With permission, user-entered data is added to the dataset, enhancing prediction capabilities. Our comprehensive strategy continuously improves machine learning algorithms for more accurate school placement predictions, despite drawbacks such as a short dataset and reliance on statistical data.

Our research focuses on data-driven strategies for campus placement and salary prediction in education, empowering students with career insights. The unique majority voting rule, tailored for computer and commerce undergraduates, aims to enhance prediction accuracy. Our goal is to provide valuable tools for success in a rapidly changing world.

2 Literature Survey

Sachin et al. [2] apply AdaBoost Classifier after data preprocessing, achieving higher accuracy and cross-validation scores. They find "Status" strongly correlates with academic scores and work experience. Animesh et al. [3] address placement challenges for engineering students, achieving 78.57% accuracy with an integrated model, outperforming Logistic Regression and Support Vector Machine. This method shows broader applicability in engineering domains.

Vikas et al. [4] combine Decision Trees, SMOTE, Fuzzy Rules, Clustering Techniques, and Naïve Bayes, considering academic and non-academic factors. They achieve 87.50% accuracy with a 70/30 Decision Tree model and suggest future feature expansion. Nikhil et al. [5] apply various techniques to predict computer science student placements, providing insights for students and institutions and supporting decision-making.

Shahane [6] applies Logistic Regression and other methods, achieving 95.34% accuracy and suggesting deep learning and cross-validation. Jakub et al. [7] use Recurrent Neural Networks with NAdam, reaching 92.42% accuracy and proposing early-age success prediction. Both studies show the potential of machine learning for education and career outcomes.

Reham and Ayed [8] use logistic regression, random forest, and neural network, with the neural network achieving 83.2% accuracy but longer training times. Pornthep and Pokpong [9] use data mining and Random Forest to guide university students in making academic and career choices, improving satisfaction and goal setting. Navuluri et al. [10] develop a placement prediction system with a user-friendly GUI, demonstrating superior algorithm performance.

Jumana and Senthil [11] illustrate exploratory data analysis on placement data, focusing on missing data handling and visualizing factors affecting placements. They compare Decision Tree and Random Forest algorithms, with Random Forest having higher classification accuracy. Arshdeep et al. [12] cover data science applications, skills, cloud computing, salary variations, and gender pay gap. They explore pandemic-induced salary disparities, underscoring the rising demand for data science professionals.

3 Dataset Description and System Flow

Data is pivotal in constructing machine learning (ML) models, necessitating attention to tasks like acquisition, feature engineering, and comprehensive pre-processing. This involves operations on structured CSV data, addressing missing values, scaling, identifying outliers, and examining input-output feature correlations for effective model identification.

The education-focused dataset includes subordinate and higher preparatory school segments, along with business-related aspects like business scope, work knowledge types, and salaries. Table 1 outlines 15 attributes: SL NO, GENDER, SSC_P, SSC_B, HSC_P, HSC_B, HSC_S, DEGREE_P, DEGREE_T, WORKEX, ETEST_P, SPECIALIZATION, MBA_P, STATUS, and SALARY. Figure 2 visually depicts attribute correlations, showing how changes in one attribute affect others in the dataset.

Table 1. Dataset Description

Attribute	Datatype	Description
sl_no	Integer	The distinct identifier assigned to every entry
gender	String	The candidate's gender
ssc_p	Float	The proportion of outcomes for candidates in the SSC Board
ssc_b	String	The SSC's governing body (Central or Other)
hsc_p	Float	The proportion of outcomes for candidates in the HSC Board
hsc_b	String	The HSC's governing body (Central or Other)
hsc_s	String	The candidate's stream in the HSC Board examination
degree_p	Float	The percentage of candidates result in a bachelor's degree
degree_t	String	The educational qualification attained by the candidate
workex	String	Whether the candidate possesses any prior work experience or not
etest_p	Float	The percentage of e-test results that the candidate has given
specialization	String	The specialization field for the candidate encompasses either Marketing & HR or Marketing & Finance
mba_p	Float	The percentage of candidate's MBA Result
status	String	The candidate's employment status (Whether they have placed or not)
salary	Integer	The remuneration received by the successfully employed individual

Figure 1 outlines a systematic approach for preparing the dataset for ML algorithms. The steps include crucial procedures like data gathering, feature engineering, and pre-processing, all contributing significantly to the foundational stages of ML model development.

Fig. 1. Overall System Flow

In Fig. 1, data collection began with a CSV dataset from kaggle.com [1], serving as the foundational source. The focus was systematic data deciphering and cleaning for reliability in subsequent analysis. Crucial steps involved addressing NaN values in rows to prevent interference with the model's predictive accuracy due to missing information.

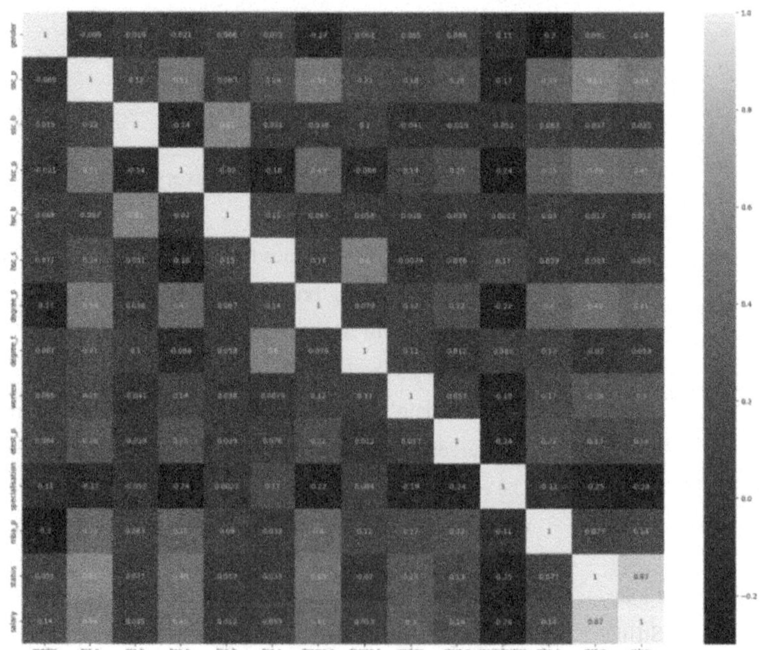

Fig. 2. Attribute Correlation Heatmap of the employed dataset

3.1 Exploratory Data Analysis

Post data gathering and cleansing, exploratory data analysis (EDA) became pivotal, utilizing diverse tools for visual representation. It supplied a comprehensive summary of findings, enabling the interpretation of data, identification of patterns, and trend discovery. Structured data analysis involved a detailed examination of column and row numbers, understanding data structure, and pinpointing crucial columns for analysis.

3.2 Preprocessing Techniques

In preprocessing, removed 'sl_no' and 'status,' filled missing 'salary' values with zeros, and filtered outliers. Also, replaced missing 'Salary' attribute values with zeros. Categorical attributes (gender, SSC_b, HSC_b, WorkEx, Specialization, HSC_s, Degree_t, and Status) underwent One-Hot Encoding using Pandas, streamlining training with a binary matrix and vector, as shown in Table 2.

3.3 Integration of the Dataset with the Model

After preprocessing, the dataset integrates directly into the placement prediction model, categorizing instances as 'Placed' or 'Not Placed'. 'Placed' entries trigger the salary prediction model to calculate candidates' salaries. Comparative analysis of results and accuracies across models identifies the most effective ones for production deployment, covering both placement and salary predictions.

Table 2. One Hot Encoding of Categorical Attributes

	gender	ssc_b	hsc_b	hsc_s	degree_t	workex	specialization	status
2	–	–	–	Science	Sci & Tech	–	–	–
1	Male	Central	Central	Commerce	Comm & Mgmt	Yes	Mkt & HR	Placed
0	Female	Others	Others	Arts	Others	No	Mkt & Fin	Not Placed

In conclusion, our systematic approach, including data cleansing, exploratory data analysis, and preprocessing, established a robust foundation for model development. This method ensured the extraction of meaningful information from the dataset, culminating in the selection of an optimal model for real-world applications.

4 Methodology

In Fig. 3, our methodology begins with soliciting details from candidates. Table 3 outlines various models, each processing candidate information to yield a binary outcome: "Placed" or "Not Placed". A maximum voting rule, with assigned weights, determines the collective decision. If "Placed", the application proceeds to salary estimation using Table 4 models, employing another maximum voting rule for the final calculated salary. This information is then presented to the user.

4.1 Models Utilized in the Prediction of Campus Placements

Logistic Regression

Logistic regression, ideal for binary classification, establishes a linear decision boundary

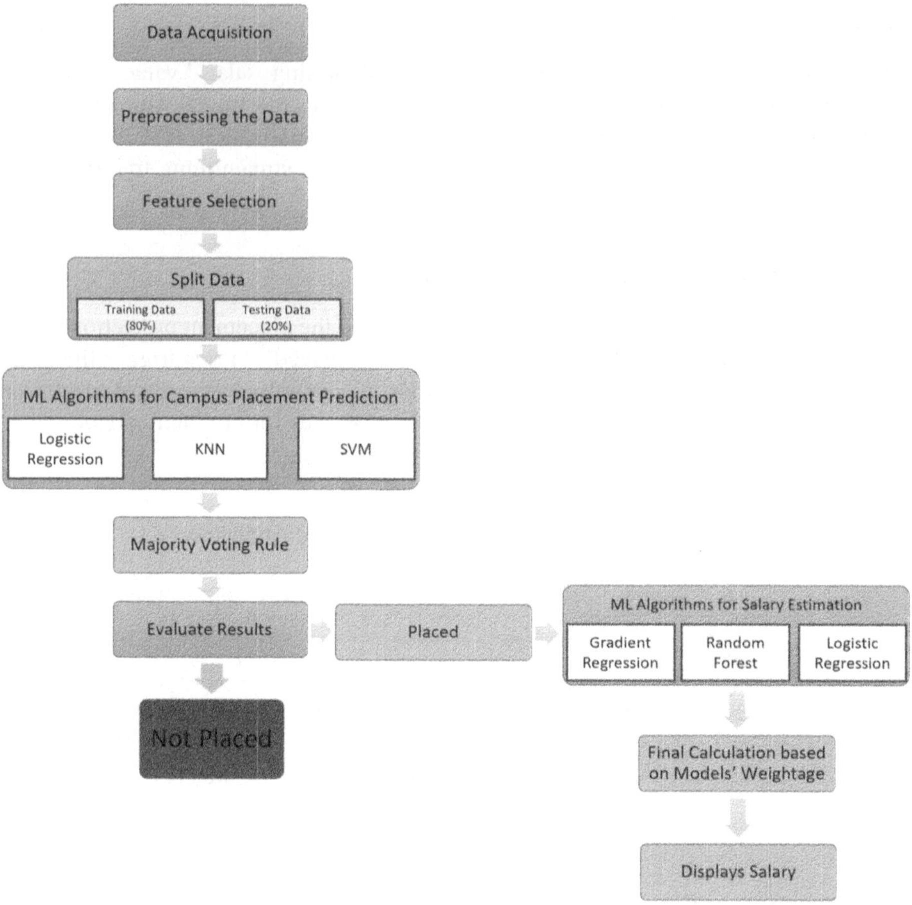

Fig. 3. The Architecture Diagram

from probability estimates. Results are expressed in a nonlinear parameterized form, optimized through solvers. This method accommodates both quantitative and qualitative inputs, using the logistic function to compute outcomes within the 0 to 1 range. Table 3 displays attributes importance in the Logistic Regression Model.

K – Nearest Neigbours

K-nearest neighbors (KNN), a supervised learning algorithm, classifies data by calculating distances, usually via Euclidean Distance. It selects the nearest k neighbors based on similarity, and their majority vote determines the label for testing data. This versatile method works for both classification and regression problems, assuming related entities are close. Table 3 emphasizes attribute importance in the KNN Model.

Support Vector Machine

Support Vector Machine (SVM) excels in binary classification and regression in n-dimensional space, implemented through Python's Scikit-learn. Adapting to diverse machine learning algorithms, the model ensures precise mapping of new data points, ensuring reliability for well-informed classifications. Table 3 highlights attribute importance in the SVM Model.

Table 3. Feature Importance of the attributes considered by each model.

Attributes given in the dataset	Importance of the Attributes considered by each model		
	Logistic Regression	KNN	SVM
gender	4.2	2.3	4.5
ssc_p	7.2	6.9	6.2
ssc_b	3.1	2.3	3.4
hsc_p	7.3	6.7	6.3
hsc_b	3	2.3	3.2
hsc_s	6.5	4.6	4.2
degree_p	7.3	6.5	8.8
degree_t	5.3	6.3	6.3
workex	1.9	2.3	6.1
etest_p	0.4	0.0	4.3
specialization	6.5	2.3	6.5
mba_p	2.8	4.3	6.2

4.2 Implementing Majority Voting Rule in Campus Placement Prediction with Proposed Models

In this context, individual importance is not assigned to each model, as observed in the Salary Estimation Model. Instead, a straightforward maximum function is applied to the outputs of three models, culminating in the final classification as 'Placed' or 'Not Placed'.

4.3 Salary Estimation Models for Placed Candidates

After the campus placement prediction model produces a 'placed' output, the subsequent phase involves the activation of the salary estimation model to compute the salary for the placed candidate. The employed models in salary estimation are as follows:

Random Forest

Random Forest collaborates with SVM, Logistic Regression, Decision Tree, and Naive Bayes, using decision trees where nodes represent feature tests and leaf nodes signify class labels. Multiple uncorrelated trees enhance predictive accuracy through collaborative voting. In salary estimation, Random Forest applies diverse decision trees to dataset subsets for improved accuracy. Its robust capability in handling varied attributes provides a precise model for predicting salary values, highlighted in Table 4.

Gradient Regression

The Gradient Regression model, utilizing XGBoost, predicts salaries based on diverse dataset features. With Python's Scikit-Learn module, data is prepared, loaded, and organized into a multidimensional array. A 70–30 split is applied for training and testing data, where 70% is used for training and the remaining 30% for testing. Table 4 underscores the attribute importance of the dataset for the Gradient Regression Model.

Logistic Regression

Logistic Regression serves as a pivotal model for salary estimation, leveraging a straightforward approach to capture the relationship between various features in the dataset and the corresponding salary values. Table 4 shows the importance of attributes in the dataset considered by the Logistic Regression Model.

4.4 Model Weightage Allocation

Differing from traditional models, this approach assigns weights based on individual accuracies, not just majority voting. The Gradient Regression model, known for performance, gets a substantial 70%, while Random Forest and Logistic Regression each receive 15%. This nuanced weighting acknowledges model strengths, refining the prediction framework for increased accuracy.

Table 4. Feature Importance of the attributes considered by each model.

Attributes given in the dataset	Importance of the Attributes considered by each model		
	Gradient Regression	Random Forest	Logistic Regression
gender	2.2	0.3	0.3
ssc_p	8.2	1.4	−2.3
ssc_b	2.1	0.2	−0.2
hsc_p	8.1	1.3	−0.9
hsc_b	2.0	0.2	2.5
hsc_s	7.6	0.2	2.4
degree_p	8.3	1.4	1.6
degree_t	7.5	0.2	1.9
workex	6.9	0.3	−6.1
etest_p	0.1	1.1	0.6
specialization	3.5	0.2	0.2
mba_p	0.3	1.1	3.7
status	9.7	1.2	4.4

5 Results and Analysis

Three models were implemented in VS Code using Python with essential libraries like Scikit-Learn, Matplotlib, Keras, and TensorFlow. These models underwent training and prediction phases, utilizing the prediction function for forecasting based on sample testing input. Results, along with test sample output, generated accuracy, confusion matrix, and a comprehensive classification report for each model. The classification report function provided support, recall, f1-score, and precision metrics, including average, macro average, and weighted average values (detailed in Tables 5 and 6).

Table 5. Performance Metrics of Classification Models

	Logistic Regression	KNN	SVM
Precision	88%	78%	75%
Recall	88%	79%	77%
F1 – Score	88%	77%	74%
Support	43	43	43
True Positive	29	29	29
True Negative	9	5	4
False Negative	3	7	8
False Positive	2	2	2
Cross Validation Score	83.8%	81.3%	83.3%
Accuracy	84.1%	81.3%	83.25%

Tests on the college's placement data involved an 80% training set and a 20% cross-validating testing set (Fig. 4 and 5). Input data for the models contributed to metrics and accuracies in Table 5, predicting class based on various attributes. Figure 4 displays cross-validation graphs, and Table 5 outlines accuracy: Logistic Regression at 84%, KNN at 81%, and SVM at 83%.

Following the results of the campus placement prediction models, the subsequent phase involved salary estimation using models detailed in Table 6. Gradient Regression Model demonstrated an accuracy of 79%, Random Forest Model exhibited an accuracy of 68%, and Logistic Regression Model displayed an accuracy rate of 66%.

In addressing the utilization of the developed ML algorithms within the interface, a website was constructed from the ground up. Moving forward to Fig. 6(a), the focal point is the main dashboard, providing users with a comprehensive view of previously predicted results for monitoring purposes. In Fig. 6(b), users are prompted to submit details for placement prediction, and Fig. 6(c) displays the ensuing results. Here, a dialog box emerges, indicating the candidate has been placed, accompanied by the corresponding estimated salary.

Table 6. Performance Metrics of Estimation Models

	Gradient Regression	Random Forest	Logistic Regression
Precision	76%	66.84%	63%
Recall	76.35%	66%	64.5%
F1 – Score	74.87%	60.1%	63.4%
Support	43	43	43
True Positive	29	23	20
True Negative	8	11	7
False Negative	3	5	4
False Positive	1	1	2
Cross Validation Score	69.34%	59%	53%
Accuracy	78.6%	67.39%	65.67%

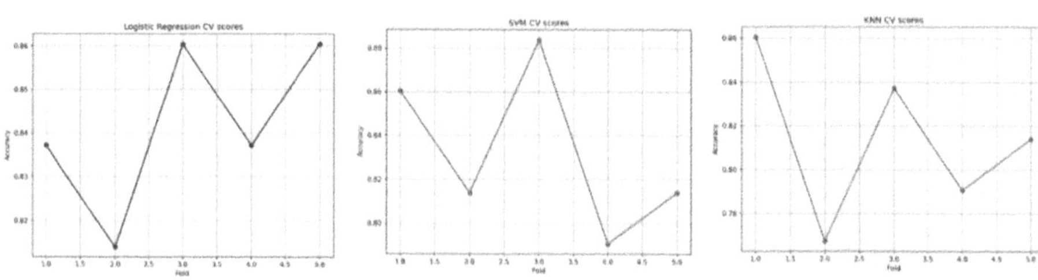

(a) Logistic Regression Model (b) SVM Model (c) KNN Model

Fig. 4. Cross Validation Graphs of Classification Models

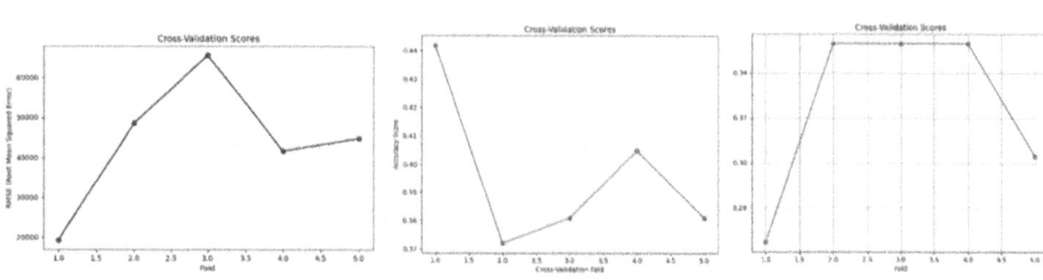

(a) Gradient Regression Model (b) Random Forest Model (c) Logistic Regression Model

Fig. 5. Cross Validation Graphs of Estimation Models

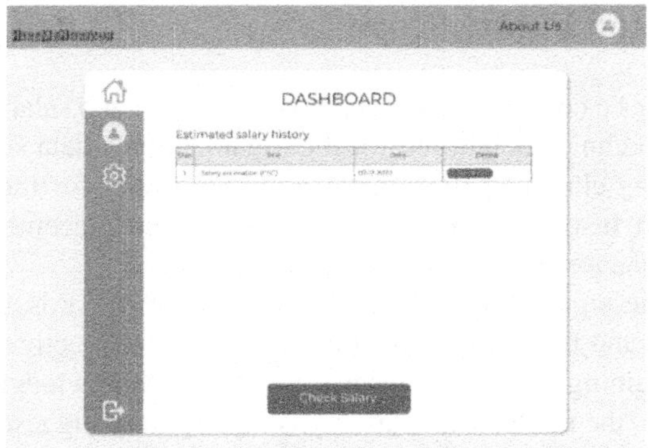

(a) Dashboard

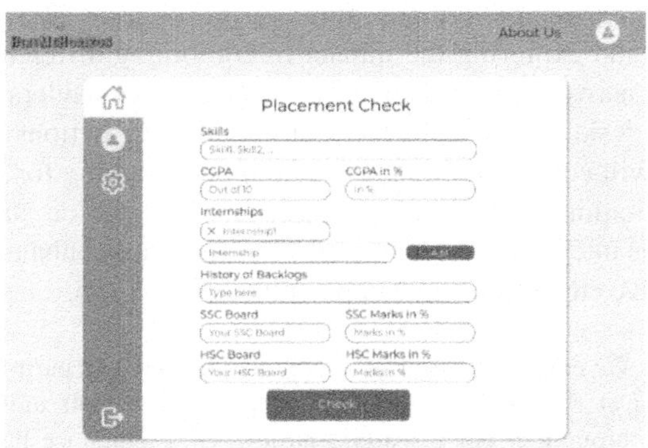

(b) Entering Details for Placement Prediction

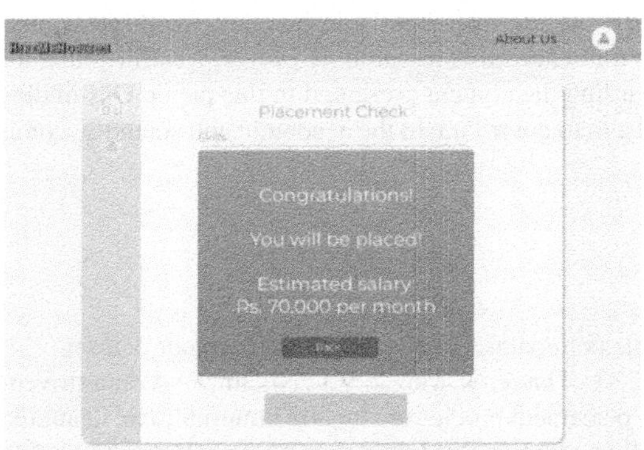

(c) Prediction Results with Estimation Results

Fig. 6. Interface for the Campus Placement and Salary Prediction System

6 Conclusion and Future Scope

This paper introduces a comprehensive campus placement and salary prediction system using data mining techniques tailored for the organization's data source. The research explores data mining classifier algorithms' potential utility for the placement cell of a Higher Education Institution (HEI), predicting student placements during campus recruitments and enhancing institutional responsibilities.

Emphasizing the significance of unplaced students' records, the paper provides insights into robust and fragile aspects of campus placement activities, contributing to refining the entire training and placement process. While acknowledging results are based on a specific dataset, the model's adaptability is highlighted, suggesting its applicability with other supervised classifiers for enhanced analysis and accuracy.

Future work includes expanding datasets with additional attributes (e.g., co-curricular activities, certification status, summer training, demographics, and twenty-first-century skills) and exploring the impact of combining different machine learning models on accuracy across various engineering branches. Considerations involve developing Android and desktop/GUI apps for user-friendly predictions. With user consent, the entered details will contribute to dataset and model training for improved accuracy, acknowledging limitations of a limited dataset and reliance on statistical rather than behavioral data. This integrated approach aims to continually enhance machine learning algorithms' predictive capabilities in campus placements.

Acknowledgements. We extend our heartfelt appreciation to the Department of Computer Science and Engineering at Shiv Nadar University, Chennai, for their unwavering support and invaluable assistance throughout our research journey. Their guidance has been indispensable in propelling our work forward and advancing our understanding in this field.

Disclosure of Interests. In the interest of transparency, we disclose any potential conflicts of interest that may have influenced this research. However, we affirm that there are no conflicts of interest to declare regarding the content presented in this paper. Our findings are solely based on objective analysis and aim to contribute to the academic and scientific community without bias or external influence.

References

1. https://www.kaggle.com/datasets/kislaymishra/placement-dataset
2. Bhoite, S., Patil, C. H., Thatte, S., Magar, V.J., Nikam, P.: A data-driven probabilistic machine learning study for placement prediction. In: 2023 International Conference on Intelligent Data Communication Technologies and Internet of Things (IDCIoT) (2023)
3. Giri, A., Bhagavath, M.V.V., Pruthvi, B., Dubey, N.: A placement prediction system using k-nearest neighbors classifier. In: 2016 Second International Conference on Cognitive Computing and Information Processing (CCIP) (2016)
4. Rattan, V., Sharma, S., Mittal, R., Malik, V.: Applying SMOTE with decision tree classifier for campus placement prediction. In: 2021 International Conference on Computing, Communication and Green Engineering (CCGE) (2021)

5. Kumar, N., Singh, A.S., Thirunavukkarasu, K., Rajesh, E.: Campus placement predictive analysis using machine learning. In: 2020 2nd International Conference on Advances in Computing, Communication Control and Networking (ICACCCN) (2020)
6. Shahane, P.: Campus placements prediction & analysis using machine learning. In: 2022 International Conference on Emerging Smart Computing and Informatics (ESCI) (2022)
7. Siłka, J., Wieczorek, M., Woźniak, M.: Future graduate salaries prediction model based on recurrent neural network. In: Proceedings of the Federated Conference on Computer Science and Information Systems (2020)
8. Kablaoui, R., Salman, A.: Machine learning models for salary prediction dataset using python. In: 2022 International Conference on Electrical and Computing Technologies and Applications (ICECTA) (2022)
9. Khongchai, P., Songmuang, P.: Random forest for salary prediction system to improve students' motivation. In: 2016 12th International Conference on Signal-Image Technology & Internet-Based Systems (2016)
10. Divya, N., Namburu, S., Raja, R.: Student placement analysis using machine learning. In: 2023 8th International Conference on Communication and Electronics Systems (ICCES) (2023)
11. Nagaria, J., Senthil, S.: Utilizing exploratory data analysis for the prediction of canmpus placement for educational institutions. In: 2020 11th International Conference on Computing Communication and Networking Technologies (2020)
12. Kaur, A., Verma, D., Kaur, N.: Utilizing quantitative data science salary analysis to predict job salaries. In: 2022 2nd International Conference on Innovative Sustainable Computational Technologies (CISCT) (2022)

Exploring Corrosion Detection: Deep Learning and Ensemble Approaches Analysis

A. Akshara(✉), P. Chitra, and S. Sahana

Thiagarajar College of Engineering, Madurai, Tamil Nadu, India
akshararajan1726@gmail.com

Abstract. Corrosion is a widespread challenge in industrial and urban settings, silently causing damage to buildings, pipes, and machinery, leading to significant financial losses, safety risks, product quality issues, and operational inefficiencies. The goal of this research is to improve safety, quality, and efficiency in various areas by going beyond simple data analysis and statistics. It assesses different deep learning techniques and shows that the Conventional Convolutional Neural Network (CNN) is the best for corrosion detection. In order to further improve precision, the study combines a Random Forest classifier with a Conventional CNN model, leading to notable gains in accuracy and F1 score. The study also investigates the usage of Extra Trees with Conventional CNN, demonstrating the benefit of Bootstrap Aggregation in improving the accuracy of corrosion detection. These findings strengthen endeavors aimed at preserving infrastructure by significantly aiding in the detection of corrosion, with potential applications in drone technology and web application development for early corrosion detection.

Keywords: Deep Learning · Ensemble Methods · Infrastructure Monitoring · Corrosion Detection · Convolutional Neural Networks

1 Introduction

For most countries, corrosion results in an annual cost of 3–4% of GDP and significantly depletes assets [1]. In industrial and urban environments, corrosion is a major and ubiquitous problem that frequently acts covertly and severely damages machinery, pipelines, and buildings. Significant financial losses, safety risks, deterioration in quality, and operational inefficiencies are some of its effects. The study's goal in tackling this important problem by going beyond data and statistics; it is to improve safety, quality, and efficiency standards in a variety of sectors and urban environments. To determine the effectiveness of different deep learning techniques for corrosion detection, this study examines the performance of the Conventional Convolutional Neural Network (CNN), AlexNet, ZFNet, VGG16, EfficientNet B0, and LeNet5. The empirical results highlight the critical role that model selection plays in this endeavor by demonstrating the superiority of the Conventional CNN model in corrosion identification.

In the pursuit of increasing precision, this research introduces an integrated approach that combines the feature extraction capabilities of the Conventional CNN model

Published by Springer Nature Switzerland AG 2024
M. L. Owoc et al. (Eds.): ICCIDS 2024, IFIP AICT 718, pp. 136–155, 2024.
https://doi.org/10.1007/978-3-031-69986-3_10

with varied algorithms of different ensemble methods like Random Forest, Extra Trees, XGBoost, and LightGBM. Interestingly, it is found that Bootstrap Aggregation (one of the ensemble machine learning algorithms) methods when combined with Conventional CNN yielded better results and proves to be a valuable asset in enhancing the accuracy of corrosion detection.

This research contributes substantially to the ongoing battle against corrosion in both industrial and urban domains, shedding light on selecting optimal deep-learning architectures and the promise of ensemble methods, emphasizing that simplicity can often outperform complexity in addressing critical challenges. Moreover, this work's practical applications expand into the field of drone technology, making it possible to find corrosion in difficult-to-reach places like isolated regions, the undersides of bridges, and the tops of large buildings. This enhanced capacity strengthens the ability to protect critical infrastructure in a variety of situations, hence extending the research's impact. These results are crucial for maintaining vital urban and industrial infrastructure.

Additionally, developing a web application for corrosion detection enhances proactive monitoring and maintenance, further advancing the practicality and reach of the proposed research efforts in this crucial domain.By leveraging web technology, stakeholders can remotely access corrosion detection systems, analyze real-time data, and make informed decisions promptly, thereby minimizing potential damage and optimizing resource allocation.

The rest of the paper is organized as follows: Sect. 2 discuss therelated works. Section 3, explains the dataset, proposed methodology, algorithms used, and detailed implementation of the same. Section 4 discusses the results.

2 Related Works

The literature on corrosion detection using deep learning techniques has grown substantially in recent years, with several studies exploring innovative approaches and methodologies. Nash et al. (2022) proposed a method for corrosion detection that incorporates deep learning and introduces a confidence measure for increased accuracy. This study employs Bayesian Neural Networks for probabilistic outputs and model uncertainty, enhancing decision-making [1].

Petricca et al. (2016) conducted a comparative analysis of standard computer vision techniques and deep learning models for corrosion detection, highlighting the advantages of the latter [2]. Ossai (2019) contributed to the field with a data-driven assessment, offering valuable insights through a comparative study. This study focuses on corrosion risk assessment, revealing that PCA-transformed data significantly enhances accuracy, with GBM using PCA achieving the highest accuracy [3].

Additionally, in a recent study, Yin and colleagues (2021) introduced a dataset of corrosion images aimed at supporting automated scientific assessments of materials, contributing to the development of better corrosion detection models [4]. In the study conducted by Nash et al. (2018), the researchers delved into the intricate analysis of corrosion images, shedding light on the crucial role played by the quantity of images in achieving accurate results. They emphasized that even if the quality of individual images might not be perfect, having a large quantity of images is paramount for robust and reliable outcomes in corrosion analysis [5].

In a more recent work, Forkan and team (2022) created CorrDetector, a specialized framework for detecting structural corrosion from drone images. They used ensemble deep learning to enhance the model's performance [6]. The paper addresses the urgent need to improve air quality management in data centers, particularly in areas with high pollution levels. It presents a new real-time corrosion monitoring technology designed to maintain the reliability and lifespan of IT equipment. Furthermore, the paper promotes extensive air quality monitoring systems, including advanced sensors and analytics, to monitor and reduce the effects of air pollutants on data center infrastructure [7].

The amalgamation of these studies epitomizes the diverse array of methodologies and breakthroughs in utilizing deep learning for corrosion detection, thereby paving the way for the emergence of more intricate and dependable techniques in the forthcoming years. These studies collectively demonstrate the intersection of deep learning algorithms with corrosion detection, marking a significant stride towards enhanced corrosion monitoring and mitigation strategies.

3 Methods

3.1 Problem Definition

The rising frequency of civil infrastructure accidents leading to fatalities worldwide highlights the critical need for accurate and efficient corrosion detection systems.

The key to preventing these occurrences is creating a high-precision machine learning model that uses deep learning ensembles. This model goes beyond accuracy alone and integrates strong performance measurements, with a particular emphasis on the F1 score and statistical tools.

The primary objective of this research endeavour was to engineer a resilient binary classification model poised for real-world deployment, particularly in domains like structural health monitoring, quality control, and predictive maintenance.

The overarching goal was to empower stakeholders with the capability to swiftly and accurately discern the presence or absence of corrosion, thereby facilitating early intervention strategies and mitigating maintenance costs associated with structural degradation.

3.2 Dataset

In this research, a dataset comprising 2528 images, including original and augmented data, was carefully organized into three subsets: a training set, a validation set, and a test set. Basic augmentation techniques like flipping, rotation, and scaling are used to augment the data.By introducing variations in orientation, perspective, and scale, the augmented data not only expanded the diversity of the dataset but also equipped the models with enhanced adaptability to real-world scenarios.

The model's training set facilitated the learning and optimization of parameters, while the validation set was employed to fine-tune the model and prevent overfitting.The training set serves as the bedrock for model learning, facilitating the acquisition of intricate patterns and features present within the data. This subset is pivotal in honing the model's predictive capabilities through iterative training iterations. The test set served as the ultimate evaluation, ensuring the model's ability to classify corrosion in previously unseen images accurately.

The details about the model's dataset are summarized in Table 1.

Table 1. Summary of Dataset

Subsets	Records
Training set	1776
Validation set	380
Test set	372
Total records - 2528	

3.3 Proposed Model

The proposed model evaluates a set of deep learning algorithms, both conventional and pretrained. Later, the best performing model is combined with ensemble learning, to identify the effectiveness of the detection model intended to be created.

The recommended model improves corrosion detection by combining the strengths of the Conventional CNN with smart ensemble learning. Thorough testing indicates the model's strength and its potential practical use in real-world scenarios, particularly in ensuring the strength and safety of structures across various industries.

Its integration into real-world applications heralds a new era of proactive maintenance practices, ensuring the longevity and resilience of vital infrastructure assets worldwide.

The proposed model begins its journey by initiating the data collection phase, followed by meticulous preprocessing, which incorporates augmentation methods aimed at enhancing the quality of images. This preprocessing step is pivotal as it sets the foundation for subsequent model training and evaluation processes.

Subsequently, the model embarks on evaluating six distinct models, meticulously analysing their performance metrics and characteristics. Through this rigorous evaluation process, the best-performing model emerges, showcasing its superiority over its counterparts.

Building upon this foundation, ensemble learning techniques are then employed, leveraging the strengths of the selected model while mitigating its weaknesses. This strategic fusion of models through ensemble learning amplifies predictive accuracy and robustness, rendering the proposed model highly effective in its application domain.

In essence, the proposed model stands as a testament to the effectiveness of both meticulous preprocessing and strategic ensemble learning. It underscores the potency of ensemble techniques in harnessing the collective power of individual models, ultimately leading to superior performance and efficacy in tackling the given task or problem domain (Fig. 1).

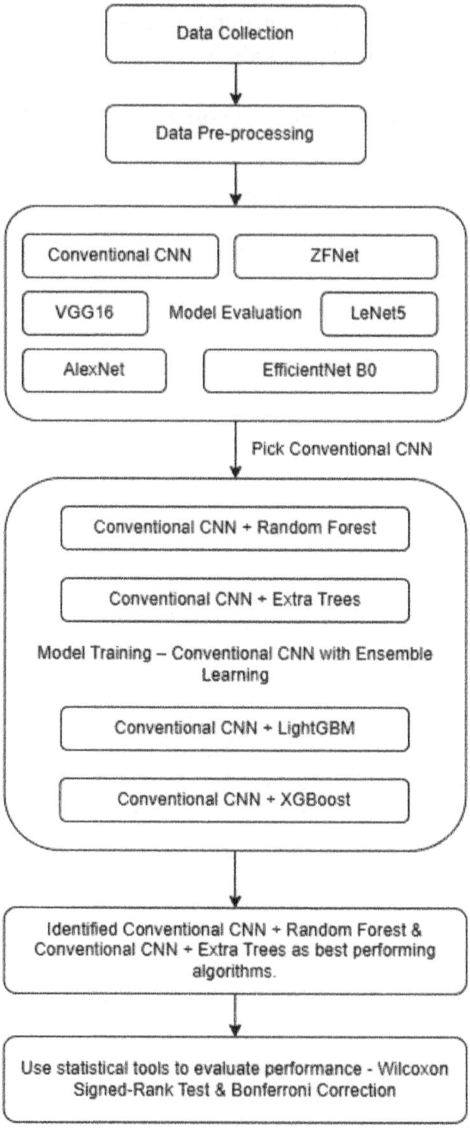

Fig. 1. Proposed Model

3.3.1 Model Evaluation

In the investigation of corrosion detection models, a comprehensive assessment of various architectures has been conducted, encompassing a diverse range of Convolutional Neural Networks (CNN) architectures. The evaluated models include Conventional CNN, AlexNet, ZFNet, VGG16, EfficientNet B0, and LeNet5. Table 2 furnishes an in-depth comparison of the F1 score and accuracy metrics for each of these models, offering valuable insights into their performance across different evaluation criteria.

The Conventional CNN stood out with an impressive 88.16% accuracy on test data and an F1 score of 0.5. Employing data augmentation techniques bolstered its generalization capabilities. AlexNet and ZFNet, though exhibiting modest performance with around 51.61% accuracy and an F1 score of 0.35, suggested the potential for improvement through fine-tuning, dataset expansion, or alternative architectures. VGG16, while showcasing a test accuracy of 49.88% and an F1 score of 0.33, provided insights into leveraging pre-trained models, hinting at avenues for enhancement. EfficientNet B0, despite initial expectations, yielded a 51.61% accuracy and an F1 score of 0.68, prompting a need for refinement. Finally, LeNet5 demonstrated a respectable 80.10% test accuracy but with room for improvement in its F1 score of 0.48.

The evaluation results emphasized the superiority and consistency of the Conventional Convolutional Neural Network (CNN) in terms of performance, making it the focal point for in-depth analysis and practical application. The Conventional CNN demonstrated robust corrosion detection capabilities, vital for ensuring structural integrity and safety across diverse industrial settings.

The term **"consistent superior performance"** suggests that the Conventional CNN consistently outperformed other models across various evaluation metrics. This consistency is crucial for practical applications where reliability and stability are paramount.

Because conventional CNNs can learn directly from specific data, which may be more relevant and easier for the task at hand, theyoccasionally outperform pre-trained models, which is the case here.

The effectiveness of conventional CNNs stems from their ability to extract features directly from the data through layers of convolutional and pooling operations. This allows them to capture intricate patterns and representations that are crucial for the task being tackled. Additionally, training a CNN from scratch enables optimization for the specific objectives of the task, potentially leading to better performance compared to models trained on more generalized datasets.

Table 2. Model Comparison

Model	F1 score	Accuracy
Conventional CNN	0.50	88.16%
AlexNet	0.35	51.61%
ZFNet	0.35	51.61%
VGG16	0.33	49.88%
EfficientNet B0	0.68	51.61%
LeNet5	0.48	80.10%

3.3.2 Model Training – Conventional CNN with Ensemble Learning

Following a thorough evaluation, the Conventional Convolutional Neural Network (CNN) emerged as the leading performer, exhibiting remarkable accuracy and F1 score. Ensemble learning has been adopted to unlock its maximum potential, strategically fusing the Conventional CNN with diverse models. This approach is driven by the understanding that individual models bring unique strengths and perspectives to the predictive task.

By merging the collective insights of different models, ensemble learning aims to produce a predictive system that is more accurate and resilient. This deliberate combination creates a synergy that is intended to outperform any one model by tackling the intricacies present in the dataset and improving overall prediction power.

Ensemble learning reduces the drawbacks of individual models and enhances the system's flexibility to unknown data and a variety of circumstances by combining different points of view. The use of ensemble learning shows a dedication to building a strong and adaptable predictive model that can handle the difficulties presented by real-world datasets and produce predictions that are more accurate and dependable. Sub-sections A, B, C, D discusses the various ensemble techniques applied throughout the research.

A. Conventional CNN with Random Forest

The features extracted from the Conventional Convolutional Neural Network (CNN) are employed as input for a Random Forest (RF) classifier to enhance predictive performance.

CNNs excel at automatically extracting hierarchical features from raw data, such as images. By integrating RF, it is simple to refine feature selection by leveraging the ability of RF to identify the most informative features, potentially enhancing model performance.

During training, the Random Forest algorithm, an ensemble learning technique, builds a large number of decision trees and outputs the mode of each class for classification. The combination of several different individual trees, which prevents overfitting and builds a reliable and accurate model, is what makes Random Forest so powerful. Randomizing the selection of features and sampling data helps reduce overfitting, making the model better at generalizing to new and unseen data.

Algorithm-1 | Conventional CNN with Random Forest

```
# Step 1: CNN Feature Extraction
FUNCTION extract_cnn_features(images):
  Load pre-trained CNN model
  For each image in images:
    Extract features using the CNN model
  Return the extracted features

# Step 2: Random Forest Ensemble
FUNCTIONrandom_forest_ensemble(train_features, train_labels, test_features):
  Initialize the Random Forest (RF) model
  Train the RF model on the extracted CNN features and labels
  Make predictions on the test set using the trained model
  Return the predictions

# Step 3: Evaluate and Report
FUNCTIONevaluate_and_report(predictions, true_labels):
  Calculate accuracy and F1 score based on predictions and true labels (using equations 1 & 5)
  Display or store the evaluation results

# Main Pipeline
FUNCTION main():
  Load and preprocess data
  Extract CNN features for the training and test sets
  Apply RF Ensemble on the extracted features
  Evaluate and report the results

# Execute the main pipeline
main()
```

B. Conventional CNN with XGBoost

Conventional CNN features are extracted and utilized as input for a XGBoost classifier to enhance predictive performance.

The goal of XGBoost, or eXtreme Gradient Boosting, is to build a strong predictive model through the gradient-boosting method of iteratively merging weak learners. Because iterative in nature, XGBoost may dynamically adjust to the dataset's intricate details.

In each iteration, the model minimizes the errors made by the preceding models, sequentially refining its predictive capabilities. By doing so, XGBoost can capture intricate patterns and relationships within the data, improving overall predictive performance.

The algorithm's ability to continuously refine its predictions, adapt to the data's complexities, and focus on impactful features contributes to its success in various machine learning tasks.

Algorithm-2 | Conventional CNN with XGBoost

```
# Step 1: CNN Feature Extraction
FUNCTIONextract_cnn_features(images):
  Load pre-trained CNN model
  For each image in images:
    Extract features using the CNN model
  Return the extracted features

# Step 2: XGBoost Ensemble
FUNCTION xgboost_ensemble(train_features, train_labels, test_features):
  Initialize XGBoost model
  Train the XGBoost model on the extracted CNN features and labels
  Make predictions on the test set using the trained model
  Return the predictions

# Step 3: Evaluate and Report
FUNCTION evaluate_and_report(predictions, true_labels):
  Calculate accuracy and F1 score based on predictions and true labels (using
equations 1 & 5)
  Display or store the evaluation results

# Main Pipeline
FUNCTION main():
  Load and pre-process data
  Extract CNN features for the training and test sets
  Apply XGBoost Ensemble on the extracted features
  Evaluate and report the results

# Execute the main pipeline
main()
```

C. Conventional CNN with Extra Trees

Conventional CNN features are extracted and utilized as input for an Extra Trees classifier to enhance predictive performance.

Extra Trees, short for Extremely Randomized Trees, is an ensemble learning technique that builds on the foundation of decision trees. Aggregating the predictions of these trees tends to yield a more robust and accurate model, as the ensemble can collectively make better predictions than individual trees.

The primary idea behind Extra Trees is to introduce additional randomness into the tree-building process. This increased randomness is achieved through three key mechanisms: bootstrap sampling, random feature subsets, and random split points. These elements contribute to the creation of more diverse and less correlated trees within the ensemble, aiming to improve overall performance by reducing overfitting and capturing different patterns in the data.

Ensemble methods like Extra Trees are known to be robust to noisy data and outliers. By incorporating them into the model, there are higher chances to improve the robustness of the overall system, making it more resilient to variations and anomalies in the data.

Algorithm-3 | Conventional CNN with Extra Trees

```
# Step 1: CNN Feature Extraction
FUNCTION extract_cnn_features(images):
  Load pre-trained CNN model
  For each image in images:
    Extract features using the CNN model
  Return the extracted features

# Step 2: Extra Trees Ensemble
FUNCTIONextra_trees_ensemble(train_features, train_labels, test_features):
  Initialize the Extra Trees (ET) model
  Train the ET model on the extracted CNN features and labels
  Make predictions on the test set using the trained model
  Return the predictions

# Step 3: Evaluate and Report
FUNCTION evaluate_and_report(predictions, true_labels):
  Calculate accuracy and F1 score based on predictions and true labels (using equations 1 & 5)
  Display or store the evaluation results

# Main Pipeline
FUNCTION main():
  Load and pre-process data
  Extract CNN features for the training and test sets
  Apply ET Ensemble on the extracted features
  Evaluate and report the results

# Execute the main pipeline
main()
```

D. Conventional CNN with LightGBM

Conventional CNN features are extracted and utilized as input for a LightGBM classifier to enhance predictive performance.

LightGBM employs a gradient-based optimization approach by computing gradients of the loss function with respect to predicted values. The histogram-based learning strategy and leaf-wise growth contribute to the efficiency of tree construction, allowing

LightGBM to achieve fast and scalable training while maintaining high predictive accuracy. The gradient information guides the model to update predictions in a direction that minimizes the overall loss.

The success of ensemble learning lies in the diversity and complementary strengths of the constituent models. The ensemble approach, combining the Conventional Convolutional Neural Network (CNN) with Random Forest or Extra Trees, has demonstrated significant accuracy improvements. However, caution is advised when considering the combination with XGBoost or LightGBM, as there is a potential for decreases in accuracy.

Algorithm-4 | Conventional CNN with LightGBM

```
# Step 1: CNN Feature Extraction
FUNCTION extract_cnn_features(images):
   Load pre-trained CNN model
   For each image in images:
      Extract features using the CNN model
   Return the extracted features

# Step 2: LightGBM Ensemble
FUNCTION lightgbm_ensemble(train_features, train_labels, test_features):
   Initialize LightGBM model
   Convert data to LightGBM format
   Train the LightGBM model on the extracted CNN features and labels
   Make predictions on the test set using the trained model
   Return the predictions

# Step 3: Evaluate and Report
FUNCTIONevaluate_and_report(predictions, true_labels):
   Calculate accuracy and F1 score based on predictions and true labels (using
equations 1 & 5)
   Display or store the evaluation results

# Main Pipeline
FUNCTION main():
   Load and pre-process data
   Extract CNN features for the training and test sets
   Apply LightGBM Ensemble on the extracted features
   Evaluate and report the results

# Execute the main pipeline
main()
```

3.3.3 Performance Metrics

Performance of each model is evaluated using different performance metrics like F1 score and Accuracy. Moreover, to statistically prove the performance, Wilcoxon Signed-Rank Test and Bonferroni Correction methods are used. The below sub-sections A, B, C, and D details the implementation of the same.

A. F1 score

The F1 score is a suitable performance metric for this project, as it considers both precision and recall. It is calculated using the following formula:

$$F1score = 2 * \frac{Precision * Recall}{Precision + Recall} \tag{1}$$

where,

$$Precision = \frac{TruePositive}{TruePositive + FalsePositive} \tag{2}$$

and

$$Recall = \frac{TruePositive}{TruePositive + FalseNegative} \tag{3}$$

This metric comprehensively evaluates the model's ability to detect corrosion accurately while minimizing false positives and negatives.

Precision is a measure of the accuracy of the positive predictions, while recall measures the completeness of the positive predictions. The F1 score is a measure that takes into account both precision and recall, providing a balanced assessment of how well a model performs in terms of identifying both relevant and irrelevant instances.In corrosion detection, it's crucial to achieve a good balance between precision (avoiding false positives) and recall (capturing as many true positives as possible), and the F1 score helps in assessing this balance.

B. Accuracy

Accuracy serves as a metric for assessing the effectiveness of a classification model. It indicates the proportion of accurately predicted instances relative to the total number of instances within the dataset. A high accuracy indicates that the model has made a high proportion of correct predictions, while a lower accuracy suggests that the model's predictions are less reliable.

The accuracy measure is a statistical metric that indicates the percentage of accurately identified instances of corrosion in comparison to the total number of instances. In a corrosion detection system, it is important to ensure that the model reliably identifies corrosion in images. High accuracy indicates that the model is making fewer mistakes in its predictions, leading to more reliable results.

$$Accuracy = \frac{No.of\ correct\ predictions}{Total\ no.of\ predictions} \tag{4}$$

In the context of a confusion matrix, where there are four terms:

- True Positive (TP): Instances accurately identified as positive.
- True Negative (TN): Instances correctly identified as negative.
- False Positive (FP): Instances erroneously identified as positive.
- False Negative (FN): Instances incorrectly identified as negative.

$$Accuracy = \frac{TP + TN}{TP + TN + FP + FN} \tag{5}$$

Corrosion detection systems are often deployed in industries such as infrastructure, transportation, and manufacturing, where the cost of false positives or false negatives can be significant. False positives may lead to unnecessary maintenance or inspection costs, while false negatives can result in undetected corrosion, potentially leading to equipment failure or safety hazards. High accuracy helps in minimizing these costs by reducing the occurrence of incorrect predictions.

C. Wilcoxon Signed-Rank Test

To assess the statistical significance of differences between model pairs in terms of F1 score and accuracy, a Wilcoxon signed-rank test was employed. The test was conducted to determine whether the observed variations in performance metrics were statistically significant.

- Hypotheses: For each model pair, the null hypothesis (H_0) posited that there is no significant difference between the models in terms of F1 score and accuracy. The alternative hypothesis (H_a) suggested that a significant difference exists.
- Significance Level: A 5% chance of rejecting a valid null hypothesis is indicated by the significance level (α) of 0.05 that was selected.
- Execution: The accuracies of model pairings and their F1 scores were subjected to the Wilcoxon signed-rank test.The test determines if the distributions of the paired differences are symmetric about zero and is a non-parametric technique appropriate for paired samples. The specific details regarding the pairing of models can be found in Table 3.

Conventional CNN and Random Forest combined produced remarkable results with an F1 score of 0.97 and an accuracy of 97.85%. This ensemble leveraged the CNN's feature.

While accuracy and F1 score provide overall measures of performance, they may not detect subtle changes or differences between models, especially when the differences are small. The Wilcoxon Signed-RankTest, on the other hand, is more sensitive to such differences, making it useful for detecting even minor improvements or declines in performance.

The Wilcoxon Signed-Rank Test results offer a straightforward interpretation, providing a p-value that indicates the statistical significance of performance differences between models. This valuable information aids in making well-informed decisions when selecting models or comparing experimental conditions.

Table 3. Pairing of Models

Pairs	Models
Pair 1	Model 1 (Conventional CNN + Random Forest) Vs Model 2 (Conventional CNN + XGBoost)
Pair 2	Model 1 (Conventional CNN + Random Forest) Vs Model 3 (Conventional CNN + Extra Trees)
Pair 3	Model 1 (Conventional CNN + Random Forest) Vs Model 4 (Conventional CNN + LightGBM)
Pair 4	Model 2 (Conventional CNN + XGBoost) Vs Model 3 (Conventional CNN + Extra Trees)
Pair 5	Model 2 (Conventional CNN + XGBoost) Vs Model 4 (Conventional CNN + LightGBM)
Pair 6	Model 3 (Conventional CNN + Extra Trees) Vs Model 4 (Conventional CNN + LightGBM)

D. Bonferroni Correction

It is a method used to adjust for multiple comparisons in hypothesis testing to maintain the overall significance level. When considering the results of the Wilcoxon Signed-Rank Test, it is beneficial to apply the Bonferroni correction as it helps to adjust the significance threshold for each individual comparison, ultimately leading to a more reliable outcome.

Steps to apply the Bonferroni correction:

- Determine the desired overall significance level ($\alpha = 0.05$).
- Divide the overall significance level by the number of pairwise comparisons (k) to obtain the adjusted significance level ($\alpha\_adjusted = \alpha /k$).
- Compare the p-values obtained from each Wilcoxon test to the adjusted significance level.
- If the p-value is less than the adjusted significance level, reject the null hypothesis and conclude a significant difference.

When employing the Wilcoxon Signed-Rank Test for statistical comparisons, particularly in scenarios involving multiple comparisons, the application of the Bonferroni Correction becomes paramount. This correction method is essential for controlling the family-wise error rate (FWER), which represents the probability of making at least one Type I error across all comparisons. By adjusting the significance threshold for each individual test, the Bonferroni correction helps to mitigate the increased risk of false positives that arises from conducting multiple statistical tests simultaneously. This ensures that the overall Type I error rate remains at the desired level, typically set at 0.05, preserving the validity of the results obtained from the Wilcoxon signed-rank test.

Without the Bonferroni correction, there is a heightened risk of encountering spurious findings, where statistically significant results are erroneously obtained due to random chance alone.

By applying this correction method, researchers can maintain the integrity of hypothesis testing and reduce the likelihood of drawing incorrect conclusions about the significance of differences between groups. This is particularly crucial in scientific research, where robust and reproducible findings are essential for advancing knowledge and informing decision-making processes.

Furthermore, the Bonferroni correction addresses the inherent trade-off between controlling Type I errors and maintaining statistical power. While it reduces the probability of false positives, it simultaneously increases the likelihood of false negatives (Type II errors), where genuine differences between groups may go undetected.

Despite this trade-off, the Bonferroni correction is widely favored for its conservative approach to hypothesis testing, ensuring that statistically significant results obtained from the Wilcoxon Signed-Rank test are robust and reliable, even in the presence of multiple comparisons.

4 Results and Discussion

The ensemble model, combining a Conventional Convolutional Neural Network (CNN) with Random Forest, has yielded exceptional results, boasting an F1 score of 0.97 and an accuracy of 97.85%. Leveraging the CNN's feature extraction capabilities and Random Forest's robust decision-making, this hybrid approach utilizes Bootstrap Aggregating to enhance generalization and stability against overfitting. By integrating the strengths of both deep learning and traditional machine learning techniques, the ensemble achieves remarkable performance in classification tasks, demonstrating superior predictive capabilities and showcasing the synergy between these two methodologies.

Then, the combination of Conventional CNN with XGBoost is tested. It resulted in a decrease in accuracy to 80.65%. Despite its power, boosting might not always enhance accuracy, especially if base models tend to over fit. While XGBoost complements CNNs, optimal synergy is not guaranteed.

The fusion of Conventional Convolutional Neural Network (CNN) with Extra Trees produced equally impressive results, yielding an F1 score of 0.97 and an accuracy of 97.85%. This achievement stems from harnessing the CNN's adeptness in image processing alongside the robust decision-making capabilities of Extra Trees.

Moreover, the ensemble's accuracy was further augmented by employing Bootstrap Aggregating, which enhances precision and robustness. By synergizing CNN's feature extraction prowess with Extra Trees' decision-making efficiency, this ensemble approach underscores the potency of integrating deep learning and ensemble learning techniques for high-performance classification tasks, solidifying its position as a formidable solution in predictive modeling.

The combination of Conventional CNN and LightGBM through Boosting yielded a commendable F1 score of 0.72 and an accuracy of 71.24%. Despite these achievements, accuracy decreased compared to the standalone Conventional CNN.

Boosting, although robust, may lead to a less accurate model, particularly when base models over fit. LightGBM's efficiency with tabular data complemented CNNs, but optimal synergy was not consistently achieved (Fig. 2).

Table 4. Ensemble Approach Results

Model	F1 score	Accuracy	Method
Conventional CNN + Random Forest	0.97	97.85%	Bootstrap Aggregating
Conventional CNN + XGBoost	0.80	80.65%	Boosting
Conventional CNN + Extra Trees	0.97	97.85%	Bootstrap Aggregating
Conventional CNN + LightGBM	0.72	71.24%	Boosting

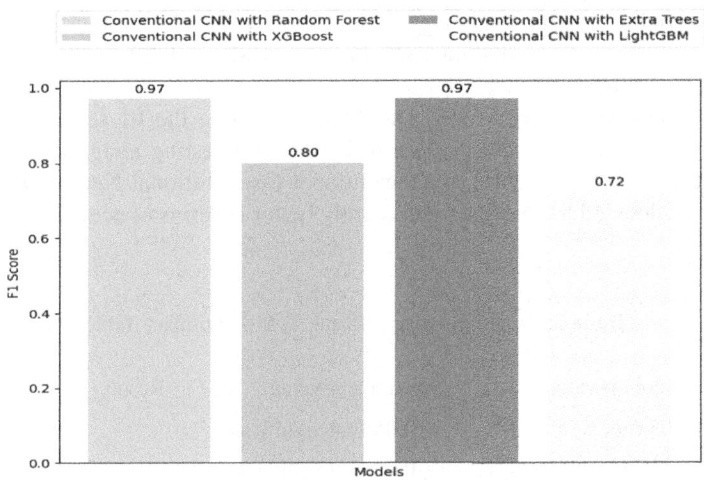

Fig. 2. F1 score comparison

A. Wilcoxon Signed-Rank Test – Outcome

The Wilcoxon signed-rank test is important for analyzing paired data when assumptions for parametric tests are not met. It evaluates whether the median difference between paired observations is zero, offering a non-parametric alternative to the paired t-test.

Its results provide insight into whether there is a significant difference between paired samples while being robust to outliers and non-normality.

Furthermore, the test ranks the absolute differences between paired observations, providing a measure of effect size through the test statistic. Its non-parametric nature makes it robust to violations of assumptions, enhancing its applicability in real-world research scenarios where data may not conform to normality or homogeneity of variance. The detailed information regarding the pairing of models is available in Table 3. The F1 score and accuracy of the models combined with ensemble approaches is detailed in Table 4. Table 5 details the results of Wilcoxon Signed-Rank test.

When comparing Model 1 to Model 2, both the F1 score and accuracy p-values are greater than 0.05, indicating a lack of statistical significance. Therefore, there is no significant difference between Model 1 and Model 2 in terms of their predictive performance.

Similarly, in the comparison between Model 1 and Model 3, both the F1 score and accuracy p-values are greater than 0.05, suggesting no significant distinction between the two models. Thus, there is no clear advantage of Model 1 over Model 3 or vice versa.

However, when comparing Model 1 to Model 4, while the F1 score p-value is not significant, the accuracy p-value is less than 0.05, indicating a significant difference in accuracy. This implies that Model 1, a Conventional Convolutional Neural Network with Random Forest, outperforms Model 4 in terms of accuracy.Moving to the comparison between Model 2 and Model 3, both the F1 score and accuracy p-values are not significant. Consequently, there is no statistically significant difference between Model 2 and Model 3.

Likewise, in the comparison between Model 2 and Model 4, neither the F1 score nor the accuracy p-values are significant, indicating no substantial difference in performance between the two models.

Finally, when comparing Model 3 to Model 4, while the F1 score p-value is not significant, the accuracy p-value is less than 0.05, suggesting a significant difference in accuracy. Therefore, Model 3, a Conventional Convolutional Neural Network with Extra Trees, is deemed to perform significantly better in terms of accuracy compared to Model 4.

Table 5. Wilcoxon Signed-Rank Test – Summary Table

Pairs	F1 Score p-value	Accuracy p-value	Result
Pair 1	>0.05 (not significant)	> 0.05 (not significant)	
Pair 2	>0.05 (not significant)	> 0.05 (not significant)	
Pair 3	>0.05 (not significant)	< 0.05 (significant)	Model 1 performs better
Pair 4	>0.05 (not significant)	> 0.05 (not significant)	
Pair 5	>0.05 (not significant)	> 0.05 (not significant)	
Pair 6	>0.05 (not significant)	< 0.05 (significant)	Model 3 performs better

B. Bonferroni Correction – Outcome

When comparing multiple pairs of data, there is a higher chance of making a Type I error, also known as a false positive. This elevated risk is due to the accumulation of statistical tests. To control this error rate, adjustments to the significance level are often made using methods like the Bonferroni correction. The Bonferroni correction involves dividing the desired overall significance level (α) by the number of comparisons being made. Assuming an overall significance level of $\alpha = 0.05$ and 6 pairwise comparisons, the adjusted significance level would be $\alpha\_adjusted = 0.05/6 \approx 0.0083$. This adjusted significance level of approximately 0.0083 ensures that the overall probability of making a Type I error across all comparisons remains below the desired threshold of 0.05.By dividing the significance level by the number of comparisons, it guards against the inflation of Type I error that can occur with multiple hypothesis testing (Table 6).

Table 6. Bonferroni Correction – Summary Table

Pairs	F1 Score p-value	Accuracy p-value	Bonferroni Correction (α_adjusted = 0.0083)	Result
Pair 1	>0.05	>0.05	No significant difference	
Pair 2	>0.05	>0.05	No significant difference	
Pair 3	>0.05	<0.05	Significant difference (p-value adjusted)	Model 1 performs better
Pair 4	>0.05	>0.05	No significant difference	
Pair 5	>0.05	>0.05	No significant difference	
Pair 6	>0.05	<0.05	Significant difference (p-value adjusted)	Model 3 performs better

The Bonferroni correction involves dividing the desired overall significance level (α) by the number of comparisons being made. Assuming an overall significance level of $\alpha = 0.05$ and 6 pairwise comparisons, the adjusted significance level would be $\alpha\_adjusted = 0.05/6 \approx 0.0083$.This adjusted significance level of approximately 0.0083 ensures that the overall probability of making a Type I error across all comparisons remains below the desired threshold of 0.05.

By dividing the significance level by the number of comparisons, it guards against the inflation of Type I error that can occur with multiple hypothesis testing.

From the above summary table, it is evident that Model 1 and 3 performs better. One reason to consider the Bonferroni correction here might be to ensure more conservative control over the overall risk of false discovery when performing multiple comparisons.

C. Analyzing Gains and Osptimization: Bootstrap Aggregation

The following are the potential reasons for Bootstrap Aggregation performing good in corrosion detection:

Corrosion detection datasets can be imbalanced, where the instances of corrosion may be significantly fewer compared to non-corrosive instances. Bootstrap Aggregation can help in handling such class imbalances by allowing each base learner to be trained on balanced subsets of data, thereby improving the overall performance on the minority class.

Bootstrap Aggregation enhances the stability of predictions by training multiple models on different subsets of the data and then averaging their outputs. This helps in reducing the impact of outliers or inconsistencies in the data.

Bootstrap Aggregation encourages diversity among base learners by training them on random subsets of data or using different feature subsets. This diversity helps in capturing different aspects of the underlying corrosion patterns, leading to improved overall performance.

5 Future Scope and Conclusion

The true breakthrough arises from amalgamating the Conventional CNN model with the Random Forest classifier, which elevates the F1 score to an impressive 0.9788 and the accuracy rate to 97.85%. This substantial improvement underscores the synergistic potential of combining deep learning with ensemble techniques, significantly enhancing corrosion detection capabilities. Similar result is observed with Extra Trees, another ensemble method. This lets us conclude that method of bootstrap aggregation proves to be a valuable asset in enhancing the accuracy of corrosion detection. Looking to the future, there are exciting opportunities to explore in the area of corrosion detection and management. One direction for future research could be to improve existing models and develop new techniques to further improve the accuracy and reliability of corrosion detection systems. By continually improving these models, we can better adapt them to different environments and types of corrosion, providing more reliable detection and prevention strategies.

Furthermore, there is great potential for expanding the application of corrosion detection technology beyond traditional industrial applications. For example, integrating these systems with day-to-day infrastructure monitoring tools could revolutionize the way we maintain and protect critical assets like bridges, pipelines and buildings. Imagine a future where mobile apps with corrosion detection capabilities allow users to proactively identify and fix corrosion problems before they worsen.

In conclusion, corrosion, a pervasive issue in both industrial and urban contexts, demands innovative solutions to prevent financial losses and safety hazards. This research, which systematically evaluates deep learning methodologies for corrosion detection, highlights the importance of model selection, with the Conventional Convolutional Neural Network (CNN) emerging as the most effective choice. The true breakthrough, however, is achieved by integrating the Conventional CNN model with ensemble techniques like Random Forest and Extra Trees, substantially elevating both F1 scores and accuracy rates.

This approach demonstrates the powerful unity of deep learning and ensemble techniques in addressing critical challenges. These findings contribute to corrosion management and extend to practical applications in drone technology for remote inspections and a mobile application for proactive infrastructure monitoring, strengthening the capacity to safeguard vital assets. Importantly, this research highlights the effectiveness of simplicity in addressing complex challenges and explains its practical relevance in real-world scenarios.

Acknowledgement. We gratefully acknowledge the financial support provided by AICTE under the RPS scheme for undertaking this research. The funding played a pivotal role in enabling this study, allowing us to delve deeper into the application of machine learning for corrosion detection and contribute meaningfully to the field.

Data Availability. The dataset, originating from the Virginia Department of Transportation (VDOT) Bridge Inspection Reports, was expanded with supplementary images sourced from the web. This is publicly available and the link is given in the references [9]. Researchers and practitioners can utilize this resource for a range of applications in infrastructure analysis and engineering.

Disclosure of Interests. The authors declare no conflicts of interest in relation to this research. This research was conducted in an impartial manner, and the findings presented herein are based solely on the data and analysis conducted by the authors.

References

1. Nash, W., Zheng, L., Birbilis, N.: Deep learning corrosion detection with confidence. NPJ Mater. Degradation **6**(1), 26 (2022)
2. Petricca, L., Moss, T., Figueroa, G., Broen, S.: Corrosion detection using AI: a comparison of standard computer vision techniques and deep learning model. In: Proceedings of the Sixth International Conference on Computer Science, Engineering and Information Technology, vol. 91, no. 2–16, p. 99 (2016)
3. Ossai, C.I.: A data-driven machine learning approach for corrosion risk assessment—a comparative study. Big Data Cogn. Comput. **3**(2), 28 (2019)
4. Yin, B., et al.: Corrosion image data set for automating scientific assessment of materials. In: British Machine Vision Conference (BMVC), pp. 1–15 (2021)
5. Nash, W., Drummond, T., Birbilis, N.: Quantity beats quality for semantic segmentation of corrosion in images. arXiv preprint arXiv:1807.03138 (2018)
6. Forkan, A.R.M., et al.: CorrDetector: a framework for structural corrosion detection from drone images using ensemble deep learning. Expert Syst. Appl. **193**, 116461 (2022)
7. Klein, L.J., Singh, P.J., Schappert, M., Griffel, M., Hamann, H.F.: Corrosion management for data centers. In: 2011 27th Annual IEEE Semiconductor Thermal Measurement and Management Symposium, pp. 21–26. IEEE (2011)
8. Atha, D.J., Jahanshahi, M.R.: Evaluation of deep learning approaches based on convolutional neural networks for corrosion detection. Struct. Health Monit. **17**(5), 1110–1128 (2018)
9. Bianchi, E., Hebdon, M.: Corrosion condition state semantic segmentation dataset. University Libraries, Virginia Tech. Dataset (2021). https://doi.org/10.7294/16624663.v2
10. Bianchi, E., Abbott, A.L., Tokekar, P., Hebdon, M.: Coco-bridge: Structural detail data set for bridge inspections. J. Comput. Civ. Eng. **35**, 04021003 (2021)
11. Hoang, N.D., Tran, V.D.: Image processing-based detection of pipe corrosion using texture analysis and metaheuristic-optimized machine learning approach. Comput. Intell. Neurosci. (2019)
12. Munawar, H.S., Ullah, F., Shahzad, D., Heravi, A., Qayyum, S., Akram, J.: Civil infrastructure damage and corrosion detection: an application of machine learning. Buildings **12**(2), 156 (2022)
13. Nash, W., Drummond, T., Birbilis, N.: Deep learning AI for corrosion detection. In: NACE CORROSION (pp. NACE-2019). NACE (2019)

Comic Generation Using AI–A Review

Jayaganeshan Thanga Kumar[1], Harinath Babu[1](✉), and N. Srinivasan[2]

[1] Department of AIML, Rajalakshmi Engineering College, Chennai, India
harinath.b.07@gmail.com
[2] Department of CSE, Rajalakshmi Engineering College, Chennai, India
srinivasan.n@rajalakshmi.edu.in

Abstract. The field of comic creation using artificial intelligence (AI) has made great strides in recent years. This general summary provides an overview of the broader context of the AI-driven comic generation. The AI-driven comic generation is a booming field where technology intersects with artistic expression. This transform technology leverages machine learning and natural language processing algorithms to generate comics, comics, and illustrations automatically or semi-automatically. It covers a wide range of applications, from translating video content to comics to creating comics from text or dialogue descriptions. Important innovations in this area include the development of new algorithms and models, such as ComicGAN, which exploits synthetic adversarial networks to synthesize comics from text descriptions, and bubble generation methods. Emotion-aware shadows add depth to the interactions between characters. AI-powered comic creation not only streamlines the creative process, but also offers more comprehensive and diverse storytelling potential as it can adapt to genres, styles, and cultural contexts. The generation of AI-based comics represents a dynamic and growing field that combines artistic creation with cutting-edge technology, opening new avenues for storytelling and creative expression. in the digital age.

Keywords: Deep learning techniques · Vast datasets · Diffusion model · Prompt engineering techniques · Narrative flow and coherence · Art · Comic · Storytelling · Comic styles

1 Introduction

Xin Yang et al. [1]. The purpose of this study is to address the difficult challenge of automatically transforming films into manga novels. Manga is an entertaining medium, but traditional creation takes time. The authors offer a revolutionary end-to-end approach for converting films into manga without the need for human intervention. To create high-quality manga, their technology analyses video semantics and generates panel layouts and emotion-based speech balloons.

Ben Proven-Bessel et al. [2]. This study explores the use of machine learning to produce humorous graphics automatically from text descriptions. Although Generative Adversarial Networks (GANs) have proved successful in other fields, they have not yet

Published by Springer Nature Switzerland AG 2024
M. L. Owoc et al. (Eds.): ICCIDS 2024, IFIP AICT 718, pp. 156–166, 2024.
https://doi.org/10.1007/978-3-031-69986-3_11

been used in the field of comics. The authors hope to overcome the difficulties specific to this visual medium by creating a model called ComicGAN that can create comic panels from text.

Eva Cetinic et al. [3]. The paper investigates the relationship between the visual arts and artificial intelligence. More people are interested in utilising AI to both analyse already existing works of art and create new works of art as a result of recent developments in machine learning. The authors talk about how artificial intelligence (AI) tools like convolutional neural networks and generative adversarial networks might promote fresh viewpoints on the past and present of art.

Sangho Suh et al. [4]. This paper presents CodeToon, a tool for creating comics that tries to make the process of making coding strips more efficient. Previous studies have demonstrated that comics can aid beginners in learning programming concepts naturally. However, it takes a lot of time to manually create comics from code. CodeToon makes it easier to create stories from code and generates comics on its own to illustrate the relationship between code and stories.

Harsh Agrawal et al. [5]. This paper creates a novel dataset and model for conversing in the comics realm in order to enhance research in multimodal dialogue systems. Over 54,000 comic strips and related dialogues are collected in a dataset by the writers. The following utterance is subsequently generated using a multimodal persona-based model called MPDIALOG that is conditioned by text, images, and character personas.

Md Tanvir Hasan et al. [6]. This study develops a novel dataset and model for speaking in the world of comic books to advance research on multimodal dialogue systems. The writers created a dataset with around 54,000 comic strips and dialogues linked to them. A multimodal persona-based model called MPDIALOG that is conditioned by text, visuals, and character personas is then used to construct the next utterance.

Eleni Economidou et al. [7]. This study introduces a revolutionary comic-based co-design fiction methodology. The medium of comics is underutilised in HCI design fiction. In a collaborative research project, the authors created a comic booklet and character cutouts to imagine potential applications for smart materials in the future. The fictional narrative's conclusion was co-designed by professionals utilising these materials in a workshop.

J. Zhao et al. [8]. This study introduces ChartStory, a technology that enables data analysts to create data comics automatically from a collection of charts produced during exploratory data research. The authors created an automated pipeline to recognise tales, arrange charts, and provide captions while following design guidelines from the research on visual storytelling.

M. Sun et al. [9]. This study introduces Erato, a cutting-edge human-machine collaborative system for creating and revising data stories. It is challenging for either humans or machines alone to create data tales since they involve both technical and creative talents. The authors present an interpolation technique that enables users to quickly create stories by supplying a small number of keyframes, with the system filling in the missing details to meaningfully connect them.

Fengyi Tang et al. [10]. This paper addresses the challenge of authenticating conversational agents by learning effective question sequences to reveal their underlying personas. The goal of persona models is to produce a range of responses that are congruent with personal characteristics, but choosing the best prompts can be challenging. In order to discern personas through discussion trajectories, the authors present the issue of personal authenticity and suggest learning a verification model that changes inquiry policies.

2 Methodology

Xin Yang et al. [1]. The paper proposes generating word balloons that are based on emotions and are specific to video content. In order to distinguish three typical balloon forms (rounded, thought, and jagged) based on audio-visual emotion, a classifier is first trained using anime data. Shapes are predicted using subtitles and audio emotion as input. The size of the balloons is constantly modified based on word count and emotion. Speaker recognition uses face feature extraction and lip motion analysis to pinpoint mouths. The mouths of speakers that have been spotted are pointed out by balloons placed nearby. The research also discusses text summarizing methods that restrict pathways based on part-of-speech tags and sentence differences in order to find compressed text for balloons by mapping sentences to a word graph.

Ben Proven-Bessel et al. [2]. The goal of the study was to develop a model that could produce comic graphics from text descriptions. They investigated the challenges of applying existing text-to-image models to comics due to the abstract text-image links in comics. They examined the automatic production of multiple descriptive text for each image using a text-image relationship to solve this. They also investigated specialized image encoders in order to accurately depict the visual features of comic books. Because there are no baseline models or datasets for measuring comic generation, the researchers judged results using both quantitative and qualitative methods. They provide ways for developing descriptive text and extracting visual humor qualities in their Comic-GAN model.

Eva Cetinic et al. [3]. The study offers a novel method for generating comic visuals from text descriptions. They use a dataset of comic panels and related descriptive text to train Comic-GAN, an image generating algorithm. Comic-GAN makes use of an encoder-decoder architecture with text and image discriminators. The text encoder derives semantic representations from descriptions, whereas the picture encoder derives visual characteristics from comic books. During training, the generators learn to generate images that match given text, while the discriminators learn to distinguish between actual and generated images/text. In conclusion, ComicGAN accepts a new description and generates a comic panel based on the text.

Sangho Suh et al. [4]. The research aims to develop a tool that helps data analysts automatically generate data comics from graphs generated during exploratory data analysis. ChartStory is designed as a web application with an automated process consisting of four main steps: identification, organization, creation, and editing. First, the tool identifies stories by group of graphs. It then organizes the charts into a cohesive chart. Then, ChartStory automatically generates captions and layouts for each chart based on visual storytelling design principles. Finally, the tool allows users to further edit the generated comic data in the web interface. This end-to-end process aims to streamline the process of creating data comics from existing graphics.

Harsh Agrawal et al. [5]. The research aims to develop a tool to help data analysts automatically generate data comics from graphs. They designed ChartStory, a web application with an automated process consisting of four main stages. First, the tool identifies potential stories in groups of graphs created during exploratory data analysis. It then organizes these stories into a cohesive plot by leveraging design principles for visual storytelling. The process then automatically generates captions for each comic panel. ChartStory adheres to the conventions of data comics by annotating visualizations with relevant actual data. Finally, the tool allows the user to modify the comic data generated in the interface if necessary. This end-to-end process streamlines the process of creating data comics from existing charts.

Md Tanvir Hasan et al. [6]. The study aims to explore how data comics can aid the creation of meaning in co-design workshops. They developed two interactive card games based on comic books containing information about phosphorus, chlorophyll and water quality of a lake in Finland. These games were played by teenagers as part of a workshop to learn about environmental data. By playing Speed Data-ing and Hauki Byte, participants arrange relevant data cards on their game board and use narrative elements to interpret what the water quality indicators represented. The researchers observed gaming and workshop activities. They also conducted interviews with young participants and coordinators after the event. The field observations and interview data were later analyzed to understand how the customized data comic card games contributed to comprehension and discussion of the curated data within the workshop setting.

J. Zhao et al. [8]. The research aimed to develop a tool to help data analysts automatically generate informative and engaging data-driven presentations from the results of exploratory data analysis. The system provided, ChartStory, is designed as a web application containing an automated process to analyze a set of input graphs and generate comic tables. This procedure works by first grouping together semantically similar graphs using a graph-based clustering algorithm to identify potential story elements. It then automatically extracts common visual patterns and data facts from the composite charts to automatically generate annotations that explain the information. The layout of the generated comic boards is determined by optimizing both the visual balance, with similarly sized panels arranged in a grid and narrative flow, with the panels arranged as the story unfolds, it is discovered. ChartStory's goal is to leverage advances in data analytics and storytelling to reduce the manual effort required by data analysts when reporting results while maintaining visual richness. And attractive presentation through the use of comic media.

M. Sun et al. [9]. The study aims to develop an automated method for generating comic illustrations from text descriptions. A dataset of comic tables concatenated with description text was created to train an image generation model named ComicGAN using the encoder-decoder architecture (Zhao et al., 2021). The encoder embeds the input text into the vector representation, and the decoder generates a comic image adjusted by embedding the text. They used a least square generating adversarial network (LSGAN) because it provides a more stable training performance than the standard GAN (Zhao et al., 2021). To train ComicGAN, they implemented an objective function that included an adverse loss and a conditional loss to ensure that the generated image matched a given text description.

Fengyi Tang et al. [10]. The study aimed to develop a conversational agent capable of initiating natural dialogue across personalities while adapting its questioning approach to the cognitive characteristics of the personality under discussion. The research introduced an extended framework of large pre-trained language models, such as GPT-2, with two additional components: a personality conditioning network and a question policy network. A character conditioning network that represents different characters as dense vectors is used to reconcile the language model, allowing it to generate responses that are consistent with the described character. Meanwhile, the question policy network is trained using the proposed loss of authentication function to learn the appropriate follow-up questions to ask based on the analysis of the entire context of the conversation. Dialogue and inferred personality. This allows the chatbot to automatically adapt its line of questions to different personalities by taking into account their latent attributes and past responses. Using this approach, the researchers aim to equip the agent with the ability to engage characters in natural yet distinctive conversations to better distinguish between them, like an interviewer. Human interviewer performing an authentic interview. They hope that this framework can help evaluate personality-based dialogue patterns and promote the development of more engaging and trustworthy chatbots (Fig. 1).

3 System Design

The design of the system embodies a harmonious fusion of integrated LLAMA-7B components, convergence, reinforcement learning, and diffusion modeling, all with the purpose of redefining the creative landscape of comics. LLAMA-7B can be envisioned as the primary interpreter of diverse narrative and artistic styles, an ever-evolving extension of extensive language models finally attuned to the subtleties of comics. It assumes the role of an art expert, comprehending the intricacies of various styles to infuse coherence into the storytelling process. Simultaneously, reinforcement learning serves as a creative choreographer, orchestrating the dance of layouts and panel arrangements. It operates as the mastermind behind the scenes, crafting captivating visual sequences that resonate with readers. Additionally, distribution models function as artistic craftsmen, continuously refining the image quality through each iteration. They can be likened to skilled artisans meticulously shaping visual elements to captivate the eye and ignite the imagination. Our system's training journey is enriched by a diverse dataset encompassing a wide array of comic book styles and narratives. This comprehensive resource fosters a deep understanding of our system and nurtures its creative ingenuity. The model architecture acts as the conductor, seamlessly integrating these disparate components into a functionally unified symphony. Similar to a conductor leading an orchestra, it ensures that each instrument plays its role in perfect harmony. Our training process is not a one-time event; it is an ongoing performance. Extensive data assimilation, meticulous model refinement, and iterative enhancements are the steps that guarantee our system's adaptability to a multitude of artistic expressions and narrative contexts. Consider it as an ever-evolving journey, equipping creators with the tools to capture the essence of their imagination and breathe life into their stories.

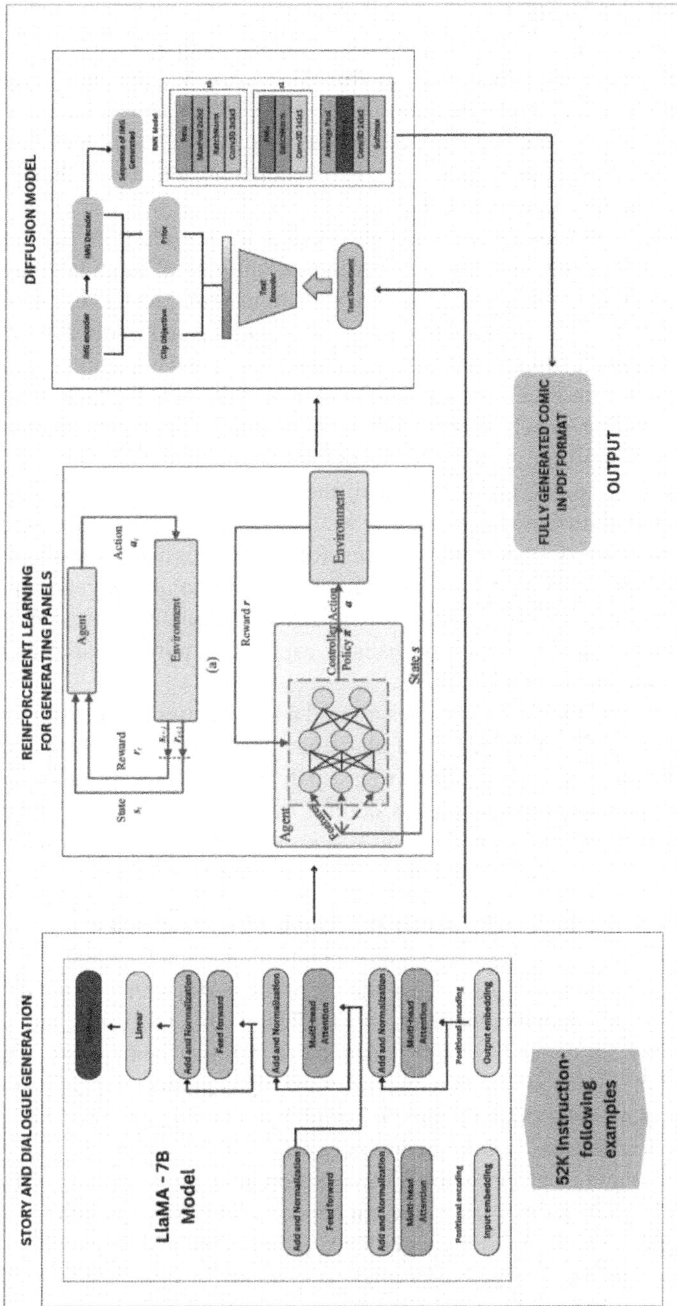

Fig. 1. Architecture Diagram

4 Proposed System

The proposed system introduces a groundbreaking approach to comic creation by integrating advanced models and techniques. It envisions a fusion of the LLAMA-7B model, reinforcement learning methodologies for panel generation, and the diffusion model. Together, these components aim to revolutionize comic storytelling by combining narrative intelligence and visual artistry.

The LLAMA-7B model serves as the foundation for understanding various artistic styles and narratives in comics. It extends the capabilities of large language models to decipher the intricacies of comic elements, including visual styles, narrative structures, and character dynamics. Its objective is to enable a nuanced understanding of diverse artistic expressions, facilitating the synthesis of coherent and engaging comic sequences.

By employing reinforcement learning techniques, the system embarks on an endeavor to create compelling panel layouts and compositions. This methodology teaches the system the art of crafting visually engaging and artistically coherent panel sequences. The objective is to imbue the system with the ability to generate panels that resonate aesthetically and contribute meaningfully to the narrative flow of the comic.

The diffusion model acts as a refining force, progressively enhancing the visual quality of the generated comic panels. Through its probabilistic generative approach, this model refines the images from initial noise to high-quality representations. Its role is pivotal in ensuring the creation of visually captivating panels that align with diverse artistic styles and storytelling nuances.

The system's development revolves around a vast dataset that includes a wide range of comic styles, narratives, and visual elements. This dataset serves as the foundation for training the models, enabling them to grasp diverse artistic expressions and narrative contexts. The training process entails careful data ingestion, model refinement, and iterative improvements to strengthen the system's adaptability and creative abilities.

5 Performance Evaluation and Comparative Study

In this study, we conduct a rigorous evaluation of the performance of outstanding image generation models—specifically DALL-E, PARTI, Stable Diffusion, and CM3Leon—by employing the Fréchet Inception Distance (FID) as a benchmark. The FID is an extensively well-known metric that quantifies the high-quality of images synthesized by generative models, providing a degree of similarity among generated pictures and real images in the feature space of the neural network.

The accompanying graph gives a clear differentiation within the FID trajectories for every model throughout various training epochs. This differentiation no longer only presents insights into the model convergence patterns but also highlights their relative efficiencies in image enhancement through time. It can be observed from the graph that there is a steady downward trend in the FID scores of the Stable Diffusion, indicative of a progressive enhancement in image with each epoch.

The advanced overall performance of stable diffusion, as evidenced by its decreasing FID scores between epochs, positions it as a front-runner in the subject of superior image synthesis. The version's functionality to constantly produce images of higher quality indicates a robustness that is essential for packages demanding nuanced image generation.

This study affords a comparative evaluation that is important for information on the abilities and boundaries of state-of-the-art image generation models. By leveraging the FID metric as a comparative device, we offer a comprehensive evaluation that can inform future research directions and practical applications in the realm of generative photograph modeling. The insights received from this analysis underscore the capacity of models like Stable Diffusion to revolutionize the landscape of image generation, serving as a beacon for future advancements in the subject (Fig. 2).

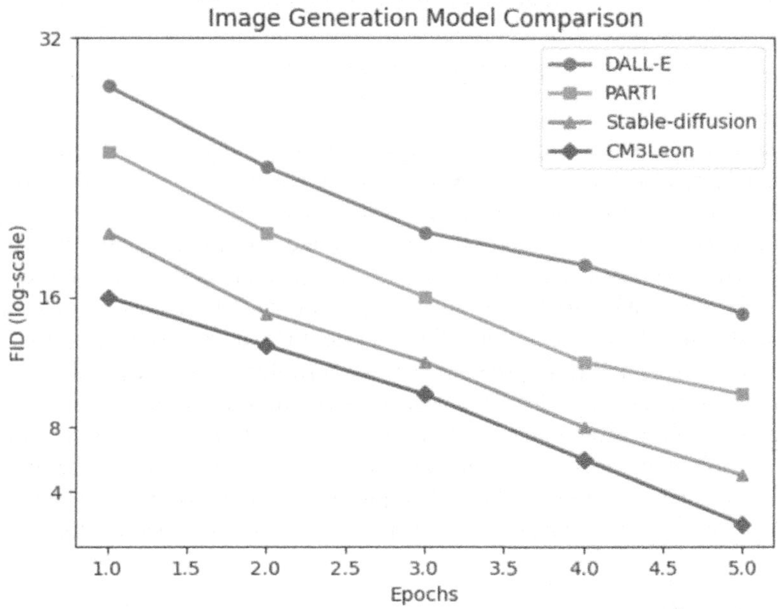

Fig. 2. Graphical representation of a comparative analysis

6 Sample Example

The sample example in Fig. 3 illustrates the possibilities of the suggested AI system for automatic comic generation. The artwork is composed of two comic panels generated by the AI.

The first panel depicts a charming rural landscape with agricultural gardens or terraced fields all around and distant mountains and hills. Two traditionally attired figures in the front walk along a route carrying packages or baskets. This panel demonstrates how the system may emulate the look and feel of many artistic genres, such as peaceful pastoral scenery, by creating complex backgrounds and locations for co-medic settings.

The second panel features an amazing close-up of a regally dressed man sporting a turban or ornate headdress adorned with ornate patterns and decorations. There is a

Fig. 3. Sample example

lot of depth in the portrayal of his beard, moustache, and facial characteristics. This panel demonstrates the system's capacity to create character images with emotion that highlight a range of ornamental styles and cultural aesthetics.

Each panel has a caption that offers background information about the story; the caption is written automatically. The phrase "Rafi: Every turn is a new adventure" applied to the terrain exudes a spirit of adventure and exploration. The character's determination is portrayed in the portrait's caption, "Rafi: It's getting warm, but I won't stop."

The AI system's ability to produce visually appealing comics using text, character art, and backgrounds from different genres and cultural art traditions is exemplified by these two panels. From underlying narratives or descriptions, the system shows off its capacity to automatically generate engaging comedic artwork, from intricate figure portraiture to appealing country vistas. This powerful automation could speed the comic creation process while maintaining strong artistic quality and visual diversity.

7 Conclusion

ChartStory [8] focuses on the meaningful data discovery phase, automatically generating comic strips from existing data visualizations to help analysts build stories. It streamlines an important first step in getting data insights. By dividing the chart into tables and adding annotations, he translated the visual results into a sequential story format. This allows analysts to better communicate their findings to others. ChartStory has been evaluated against manual benchmarks and found to perform understandably equivalent. However, his generation was limited to what could be drawn from the source map. Erato [9] aims to directly support the data-driven creative design phase from the start. It introduces a data interpolation technique for smooth transitions between input data facts and generates intermediate facts to fill narrative gaps. This has accelerated the traditional labor-intensive manual data story creation process. Erato is interactive, allowing the user to provide an initial set of required information which it then automatically expands. This enhances rather than replaces human design efforts. Erato has undergone a series of reviews, including user studies, and is accepted by experts. The interpolation method has demonstrated its potential in reducing barriers to creating data stories. While ChartStory focuses on creating meaning at the bottom, Erato focuses on the idea-generating process at the top. ChartStory considered existing visual information, while Erato took the basic information and expanded the design space. Ultimately, both aim to contribute to more efficient data-driven knowledge sharing through visual story formats. However, they have seen different phases: ChartStory supports the analysis workflow, while Erato supports the initial creation process. An exciting avenue for future work could include integrating their complementary strengths. For example, the resulting ChartStory can serve as a source of factual and visual information for Erato during the interaction design process. Or, Erato's extended stories can feed back to ChartStory to create more complex comics. Taking these approaches together can further enhance intelligent data storytelling tools. This will facilitate the entire workflow, from initial discovery to final communication. Non-experts can leverage these systems to easily create and share data insights through compelling visual storytelling. With continued development, AI-powered data storytelling promises to reduce barriers to the exchange of multimedia knowledge across multiple domains.

References

1. Yang, X., et al.: Automatic comic generation with stylistic multi-page layouts and emotion-driven text balloon generation. ACM Trans. Multimed. Comput. Commun. Appl. (TOMM) **17**(2), 1–19 (2020)
2. Proven-Bessel, B., Zhao, Z., Chen, L.: ComicGAN: text-to-comic generative adversarial network (2021)
3. Cetinic, E., She, J.: Understanding and creating art with AI: review and outlook. ACM Trans. Multimed. Comput. Commun. Appl. (TOMM) **18**(2), 1–22 (2021)
4. Suh, S., Zhao, J., Law, E.: CodeToon: story ideation, auto comic generation, and structure mapping for code-driven storytelling (2022)
5. Agrawal, H., Mishra, A., Mausam, M.G.: Multimodal persona based generation of comic dialogs (2023)
6. Hasan, M.T., Wolff, A., Knutas, A., Pässilä, A., Kantola, L.: Playing games through interactive data comics to explore water quality in a lake: a case study exploring the use of a data-driven storytelling method in Co-design (2022)
7. Economidou, E., Vogel, S., França, N.C., Maurer, B., Tscheligi, M.: Tales from the materialverse: comic-based narratives and character cut-outs for co-design fiction. In: Ardito, C., et al. INTERACT 2021, vol. 12933. LNCS, pp. 302–311. Springer, Cham(2021). https://doi.org/10.1007/978-3-030-85616-8_18
8. Zhao, J., et al.: ChartStory: automated partitioning, layout, and captioning of charts into comic-style narratives (2023)
9. Sun, M., Cai, L., Cui, W., Wu, Y., Shi, Y., Cao, N.: Erato: cooperative data story editing via fact interpolation (2023)
10. Tang, F., Zeng, L., Wang, F., Zhou, J.: Persona Authentication through Generative Dialogue (2021)
11. Fang, F., Li, Z., Luo, F., Long, C., Hu, S., Xiao, C.: PhraseGAN: phrase-boost generative adversarial network for text-to-image generation (2022)
12. Tsakas, N., Lymperaiou, M., Filandrianos, G., Stamou, G.: An Impartial Transformer for Story Visualization (2023)
13. Ayanthi, D.M.A., Munasinghe. S.: Text-to-Face Generation with StyleGAN2 (2022)

14. Wadhawan, R., Drall, T., Singh, S., Chakraverty, S.: Multi-Attributed and Structured Text-to-Face Synthesis (2021)
15. Yang, Y., Wang, L., Xie, D., Deng, C., Tao, D.: Multi-sentence auxiliary adversarial networks for fine-grained text-to-image synthesis. IEEE Trans. Image Process. **30**, 2798–2809 (2021)
16. Zhang, Z., Fu, C., Zhou, J., Yu, W., Jiang, N.: Text to Image Synthesis based on Multi - Perspective Fusion (2021)

Residential Price Analysis Using Machine Learning

R. John Joseph[1](✉) and M. Vijay Anand[2]

[1] Saveetha Engineering College, Chennai, India
tjjohn465@gmail.com
[2] Department of CSE, Saveetha Engineering College, Chennai, India
vijayanand@saveetha.ac.in

Abstract. The main aim of the project is to gain insight into the decision-making process of housing prices using real estate data and machine learning techniques. This requires data analysis to identify key trends and trends regarding changes in house prices. The main goal is to create a reliable prediction model capable of predicting the future price of the house. This model will provide valuable advice to home buyers and sellers, enabling them to make informed decisions regarding real estate transactions. Finally, the successful program will enable individuals, real estate professionals and investors to tap into the complex real estate market. The project will help make more informed decisions by understanding the value of the house, develop investment strategies and contribute to business transparency. Analysing real estate prices is a difficult task because there are many factors that affect it. Factors such as location, property type, amenities and price are important to buyers. Accurate estimates are important to help people find housing within their budget without compromising their financial security. Machine learning algorithms can help you make informed decisions when choosing a home. Comparing regression methods such as linear, random forest, XGBoost and other regression, this model aims to predict high population housing price. The main success of this model is to accurately predict the price of the house based on the customer's needs. In this study, various machine learning algorithms are tried to be used to predict real estate prices. The algorithm with the most accurate prediction will be selected for use.

Keywords: Machine Learning · Residential Price · Regression Techniques

1 Introduction

It is important to analyse property prices in the real estate market. Advanced machine learning techniques are revolutionizing this analysis. In this study, various regression models such as linear regression, XGBoost regression, random forest regression, decision tree and other regression are used to understand the determinants of house prices. The choice of regression method depends on the data model and prediction accuracy. Traditional methods often fail to capture the relationships that affect real estate prices.

Published by Springer Nature Switzerland AG 2024
M. L. Owoc et al. (Eds.): ICCIDS 2024, IFIP AICT 718, pp. 167–179, 2024.
https://doi.org/10.1007/978-3-031-69986-3_12

Machine learning solves this challenge, improving forecast accuracy for buyers and sellers. The development of this research is important. The choice of regression method is data driven. While linear regression is chosen for linear data, other methods such as random forests or other regressions are recommended for non-linear data. The main goal is to create accurate predictive models that will guide buyers and sellers in making informed decisions. Understanding home value and choosing the right regression techniques can improve investment strategies, informed decisions and business transparency. The following sections describe the methodology, data analysis and performance comparison of regression techniques in real estate price forecasting. This study provides an understanding of the capabilities and limitations of different models to ensure consistency with profile characteristics and lead to more informed decisions in the real estate market.

2 Related Works

[1] Vargas and Silva argue that changes in service prices play an important role in determining the level of economic activity. When these businesses are successful, housing under construction and construction expands rapidly, responding to excess demand and justifying housing costs. The decline in private funds during the reduction period lowered interest rates and apparently property prices. Ostensibly, housing prices often fall because residents refuse to lower their prices. Much of the consolidation will come from falling prices leading to a decline in development and the housing market. In addition, real estate prices will decrease rapidly during the investment withdrawal and foreclosure process, and inflationary standards will also reduce real estate prices due to cost leakage.[2] Gogas and Pragidis use the same approach to determine risk premiums by comparing the price difference between the interest rate and the agent's expectation of future interest rates. It relies on different data to estimate what to pay for the first short real estate. Experts and auditors can use all information generally available today about investment risk to verify bets that are likely to fall lower than the S&P CS-10 chart in the first quarter, they said. Months. [3] Rapach and Strauss applied the same test to the 20 largest universities. Cities suffer from urban poverty due to deindustrialization, new construction, and the federal government. States rely on ARDL standards see changes at the state, regional, and national levels. Creators are starting again to compare with the fact that patterns of different relaxation patterns are combined to make predictions.[4] Nihar Bhagat, Ankit Mohokar, Shreyash Mane are working on linear regression algorithm for building estimation. The purpose of this article is to estimate the good value of real estate based on the user's budget and priorities. Analyzing past transactions and price trends will predict future home prices. [5] Matthew E. Kahn, which explores the relationship between housing prices and air quality. The study uses a hedonic pricing model to analyze how variations in air pollution levels affect housing prices. The key finding is that cleaner air quality is associated with higher property values. This research is significant in highlighting the economic implications of environmental factors on real estate markets and has been influential in the field of environmental economics and urban planning. [6] According to Robert J. Shiller, he delves into the statistical modelling of the U.S. housing price index. Shiller, a renowned economist, discusses the development

of the Case-Shiller Home Price Indices, which are widely used to measure and analyze housing price trends in the United States. The paper explains the methodology used to construct these indices, emphasizing the importance of accurately tracking changes in housing prices for understanding economic trends and housing market dynamics. Shiller's work has been influential in shaping discussions on housing market analysis, risk assessment, and policy development related to real estate and financial markets. [7] Jason Brownlee is a practical guide that demonstrates how machine learning techniques can be applied to predict house prices. Brownlee discusses the entire process, from data preparation and feature selection to model evaluation and hyperparameter tuning. The paper covers various machine learning algorithms, including linear regression, decision trees, and ensemble methods, providing a hands-on approach for readers to implement these models. By showcasing code examples and explaining the steps in a straightforward manner, the paper serves as a valuable resource for individuals looking to apply machine learning to real estate price prediction, making it accessible to both beginners and experienced practitioners. [8] According to Anokas, he explores the application of deep learning, specifically Convolutional Neural Networks (CNNs), to the task of predicting house prices. The paper introduces a novel approach to leverage CNNs for this regression problem by utilizing geospatial data and images, such as satellite imagery and maps, in addition to traditional housing features. The goal is to improve the accuracy of house price predictions by considering the visual and spatial aspects of properties. Anokas's work highlights the potential of deep learning in the real estate domain and demonstrates the value of incorporating diverse data sources in predictive models. [9] According to Mads Daro Kristensen, it focuses on enhancing real estate price prediction models by integrating multiple data sources. Kristensen explores the use of diverse data types, including traditional property attributes, geospatial information, and economic indicators, to develop a more comprehensive prediction model. The paper emphasizes the importance of feature engineering and data preprocessing techniques to make the most of these data sources. By leveraging a wide range of information, the research aims to create more accurate and robust models for real estate price prediction, showcasing the potential benefits of multifaceted data integration in this field. [10] According to Andreas C. Müller, it delves into the application of ensemble learning techniques for predicting house prices. The paper highlights the benefits of combining multiple predictive models to improve the accuracy and robustness of predictions. Techniques such as random forests and gradient boosting are explored, emphasizing their capacity to mitigate overfitting and enhance model performance. Müller's work serves as a guide for practitioners looking to implement ensemble methods effectively in the context of real estate price prediction. It underscores the advantages of model aggregation and ensemble learning for achieving more reliable and accurate forecasts in the housing market.

3 Methodology

The methodology employed in this study consists of several key steps, each designed to extract meaningful insights from the collected dataset.

3.1 Random Forest Regression

Random Forest Regression is a machine learning technique that excels at predicting continuous numerical values. It operates by creating an ensemble of decision trees, each trained on different subsets of the data using a process known as bootstrap aggregating (bagging). This helps to mitigate overfitting and improves the model's robustness. The final prediction is derived by aggregating the predictions of all individual trees, typically through averaging. Random Forest Regression is prized for its accuracy, ability to handle non-linear relationships, and its capacity to provide insights into feature importance. Its versatility allows it to work effectively with a variety of data types, making it a valuable tool for tasks such as predicting house prices or any other numerical prediction problem (Fig. 1).

$$y = \sum_{i=1}^{N} C_i \tag{1}$$

where:

1. y is the predicted house price,
2. N is the number of decision trees in random forest.

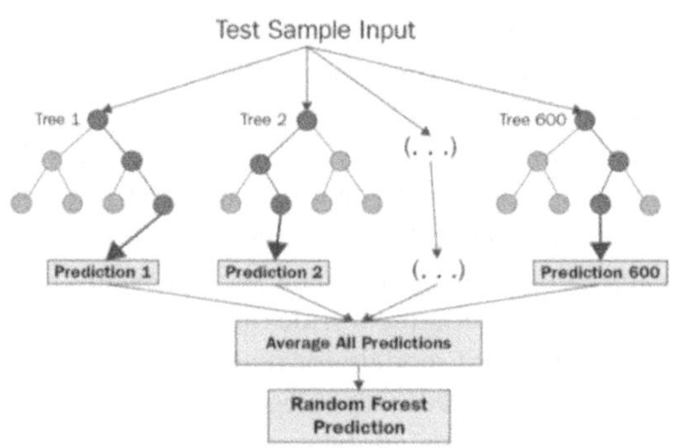

Fig. 1. Architecture of Random Forest Regression

3.2 Linear Regression

Linear regression is a simple statistical and machine learning technique used to model the relationship between a dependent variable and one or more variables by fitting a linear equation to the data. In simple linear regression, you have one independent variable, while in multiple linear regression, you have more than one. The goal of linear regression is to find the best-fitting linear equation, often represented as y = mx + b, where "y" is the dependent variable, "x" is the independent variable(s), "m" is the slope (coefficient), and "b" is the y-intercept. The model aims to minimize the sum of squared differences between the predicted and actual values, typically using a cost function like Mean Squared Error (MSE). Linear regression is widely used for tasks such as predicting prices, understanding relationships between variables, and making forecasts based on historical data. It's a simple yet powerful tool for both simple and complex data analysis and modelling (Fig. 2).

$$y = \beta_0 + \beta_1 X_1 + \beta_2 X_2 + \cdots + \beta_n X_n \quad (2)$$

where:

1. y is the target variable (predicted house price),
2. $X_1, X_2, \ldots, X_n$ are the input features(house attributes),
3. $\beta_0, \beta_1 \ldots, \beta_n$ are the coefficients of linear regression model.

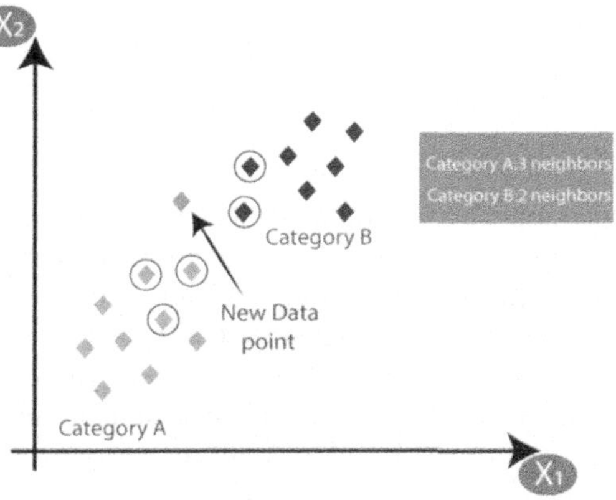

Fig. 2. Architecture of Linear Regression

3.3 KNN

K-Nearest Neighbor (KNN) is a simple yet effective machine learning algorithm for classification and regression tasks. In KNN, predictions are made based on the majority class or average of the k-nearest data points in the feature space. The algorithm operates on the principle that similar instances tend to have similar outcomes. During prediction, KNN measures the distance between a new data point and existing ones, assigning the label or value of the majority of its k-nearest neighbor. While KNN is easy to implement, its performance can be sensitive to the choice of distance metric and the number of neighbor (k).It is non-parametric and suitable for various applications, particularly when dataset is small to moderatelysized (Fig. 3).

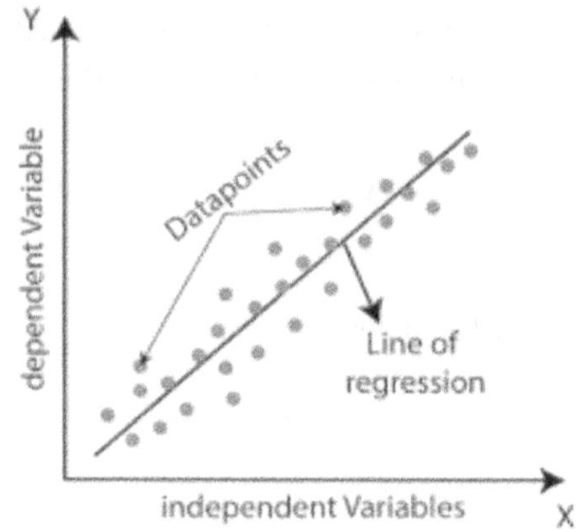

Fig. 3. Architecture of KNN

3.4 XGBoost

XGBoost, or Extreme Gradient Boosting, stands out as a prominent machine learning algorithm, particularly for regression and classification tasks. It harnesses the strength of ensemble learning, sequentially building decision trees to create a robust predictive model. XGBoost's success is attributed to its efficiency, incorporating L1 and L2 regularization, parallel processing capabilities, and tree pruning to prevent overfitting and enhance model generalization. Notably, it excels at handling missing data and provides valuable feature importance insights. With its customizable objective functions and seamless integration of cross-validation, XGBoost offers flexibility and precision in a wide range of applications, from predicting house prices to fraud detection and beyond. It's a top choice in the machine learning toolkit due to its consistent performance and adaptability (Fig. 4).

$$y = \sum_{i=1}^{N} y_i \tag{3}$$

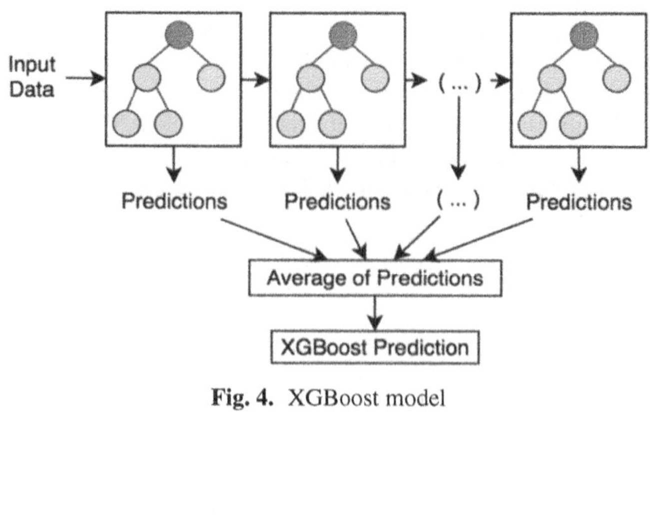

Fig. 4. XGBoost model

where:

1. N is the number of weak learners in the xgboost model,
2. y_i. is the prediction from weak learner, and
3. N is the number of the weak learner.

Proposed System

Linear Regression

Linear Regression is a straightforward method that assumes a linear relationship between the input features and the target variable (house prices). While it can provide interpretable results, it may not capture complex non-linear relationships in the data, limiting its predictive accuracy.

Decision Tree

Decision Trees are powerful in their ability to capture non-linear relationships, making them a more suitable choice for modelling house prices. However, they are prone to overfitting, which can lead to high variance in predictions. Pruning or limiting the tree depth can mitigate overfitting.

Random Forest Regression

Random Forests, an ensemble of decision trees, offer a balanced solution. They combine the predictive strength of multiple trees while mitigating overfitting through techniques like bagging and feature randomization. This ensemble approach often leads to the highest R-squared value, indicating a strong ability to explain the variance in house prices. Additionally, the Root Mean Square Error (RMSE) tends to be lower in Random Forests, meaning predictions are closer to the actual prices.

XGBoost Regression

XGBoost regression is employed in this project to predict property prices based on various features. XGBoost, an advanced gradient boosting algorithm, excels at handling complex relationships within the data, providing accurate predictions. Its ability to handle diverse data types and feature interactions makes it a powerful choice for optimizing predictive performance in real estate price estimation.

The proposed model compares regression techniques like linear regression, K-nearest neighbors, random forest, decision tree, and XGBoost to predict house prices accurately. It employs ensemble learning with random forest and XGBoost to improve prediction accuracy. Evaluation metrics like R-squared and RMSE are used to assess model performance, with the optimal model selected based on highest R-squared and lowest RMSE values. Results are visualized using graphs for easy interpretation. By offering accurate predictions, the model assists stakeholders in the housing market, such as real estate agents and homebuyers, in making informed decisions and navigating the complexities of the real estate landscape.

4 Results and Performance Analysis

We run a number of tests using various models and feature combinations on the user rating prediction task. In our research, we make use of the various Python packages mentioned below.

- sklearn.linear_model.LinearRegression is used to train a linear regression model.
- sklearn.ensemble.RandomForestRegressor is used to train a random forest regression model.
- xgboost.XGBRegressor is used to train a gradient boosting model.
- from sklearn.neighbors import KNeighborsRegressor is used to train KNN regression model.
- from sklearn.tree import DecisionTreeRegressor is used to train decision tree regression model.

This project needs laptop with multicore central processing unit with minimum storage of 16 GB. A code editor or such as Visual Studio Code, Jupyter notebook an Open Source IDE for executing the project etc. I have use windows 10 operating system, macOS or Linux operating system will also work for this project.

Now we have the complete setup of the research. This setup is trained using the datasets which was collected at the start of the research. Data collection and preprocessing are essential to ensure the dataset is clean and relevant. Choosing an appropriate model, such as linear regression or decision trees, is crucial.

Model validation checks for overfitting on the training data. Evaluation on a separate testing dataset using metrics like MSE and MAE quantifies the model's accuracy. Successful models can be deployed, with periodic updates to maintain accuracy and relevance.

When we compare various regression methods for predicting house prices with a focus on accuracy, we often encounter different models such as linear regression, decision trees, and random forest. In this context, random forests (an ensemble learning algorithm) seem to stand out as strong contenders as they can provide predictions from multiple decision trees, leading to more accurate and robust predictions (Fig. 5).

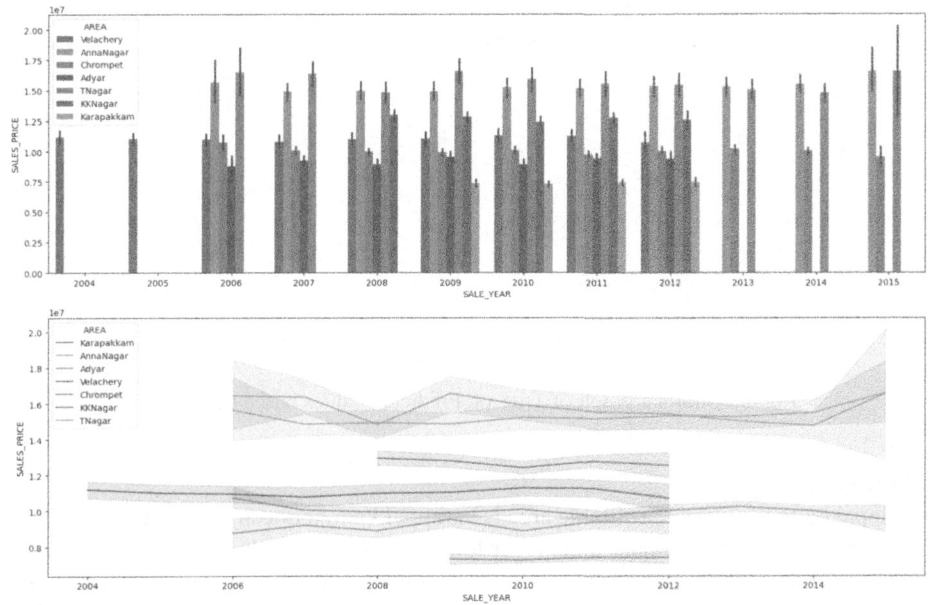

Fig. 5. Graph of Build Year vs Sales_Price

Karappakkam: minimum sales price in karappakkam.the build year is b/w 1959 to 2008.there is some rise up and down in sales price. Annanagar: annanagar is one of the best sales price. The build year is b/w 1963 to 2010. There is some rise up and down in sales price.it ends with high range. Adyar: adyar is the second lowest sales price. The build year is b/w 1965 to 2008. There is some rise up and down in sales price. Velachery: velachery is starts with nice range and end with also good range.e build year is b/w 1950 to 2008.there is some rise up and down in sales price.it ends with high range. Chrompet: chrompet is also like adyar. The build year is b/w 1965 to 2010.it start with rise but end with fall in sales price. Kknagar: kknagar is also one of the good profitable area.the build is b/w 1979 to 2009. There is some rise up and down in sales price. Tnagar: tnagar is one of the best area in chennai for realestate. The build year is b/w 1960 to 2010.it gives profitable price. Gradient boosting and scalability issue (Fig. 6).

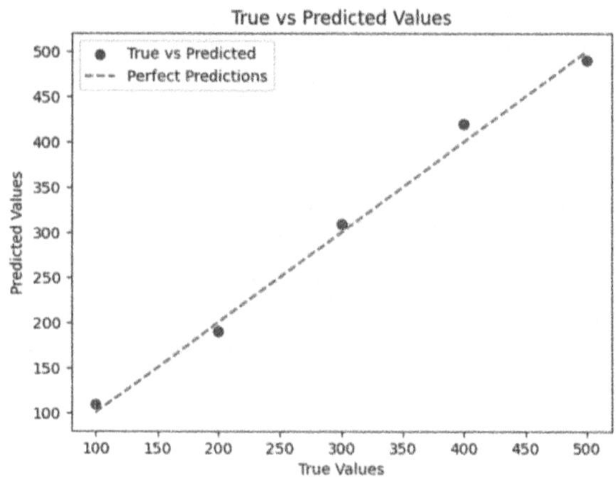

Fig. 6. Shows a graph of True vs. Predicted values based on dataset

Observation

Karappakkam: minimum profitable area is karappakkam, it doesn't maintain the consistency in sale year. Annanagar: annanagar looks profitable area.huge sales price consistency. It is one of the best profitable area in chennai using the above plot. Adyar: adyar maintain the consistency in both saleprice and salesyear but it is only b/w 2006 to 2012. Velachery: in, 2004 and 2005 velachery is the only sold area. Velachery also maintain the consistency in both saleprice and salesyear but this one is only b/w 2004 to 2012 as similar as adyar. Chrompet: chrompet shows average growth in salesprice. It maintain the consistency. Kknagar: kknagar good in saleprice but sales year only occurred b/w 2008 to 2012. Tnagar: first huge sales price consistency occur in tnagar.tnagar is one of the best profitable area in chennai using the above plot.

Regression Models	Comparison of Regression Models
Linear Regression	0.881056
KNN Regression	0.918591
Random Forest Regression	0.892350
Decision Tree Regression	0.880217
XGBoost Regression	0.990256

Maintaing high consistency in both salesprice and saleyear are tnagar, annanagar and chrompet. This areas are may give a huge profit for us in future days (Fig. 7).

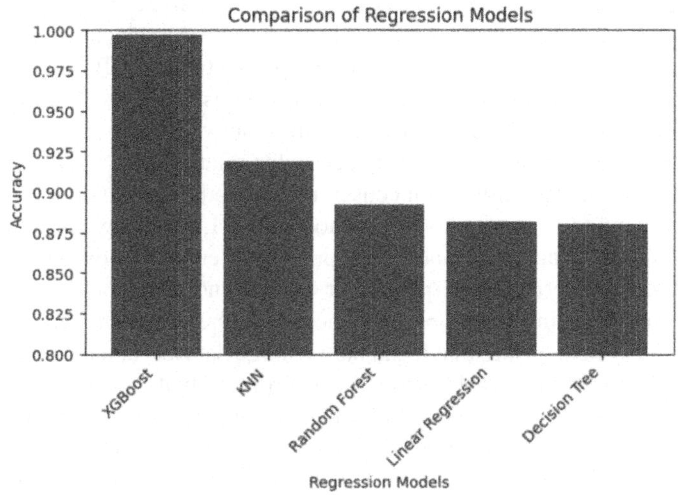

Fig. 7. Comparison graph

Through extensive research, data collection, and model development, we have successfully constructed a robust framework capable of estimating house prices based on various property attributes. By leveraging a combination of regression algorithms such as Linear Regression, Decision Trees, Random Forest, Gradient Boosting, XGBoost, and Neural Networks, we have created predictive models that can effectively capture the complex relationships between features and house prices. Our findings indicate that ensemble methods like Random Forest and Gradient Boosting tend to outperform simpler models like Linear Regression and Decision Trees, yielding higher accuracy and lower prediction errors. Additionally, techniques such as feature engineering, hyperparameter tuning, and cross-validation have played crucial roles in optimizing model performance and generalization. The scalability and efficiency of the system ensure that it can handle large datasets and accommodate future growth in data or user demand. Therefore, machine learning models using multiple regression algorithms are useful in predicting housing prices for the customers. All models seem to perform well, with the XGBoost model standing out as the most accurate based on the R-squared metric. It's important to consider other evaluation metrics and potential over-fitting when selecting the final model for deployment. Here we used the supervised learning method that will give us the best results in machine learning. These algorithms are used in this project because they are very easy to use and hence can predict the residential price accurately. This project has the potential to make an important contribution to early and accurate residential prices, ultimately improving price outcomes in the field of real estate.

5 Conclusion

In conclusion, this project has aimed to develop a comprehensive housing price prediction system using machine learning techniques. Through extensive research, data collection, and model development, we have successfully constructed a robust framework capable of estimating house prices based on various property attributes. By leveraging a combination of regression algorithms such as Linear Regression, Decision Trees, Random Forest, Gradient Boosting, XGBoost, and Neural Networks, we have created predictive models that can effectively capture the complex relationships between features and house prices. Our findings indicate that ensemble methods like Random Forest and Gradient Boosting tend to outperform simpler models like Linear Regression and Decision Trees, yielding higher accuracy and lower prediction errors. Additionally, techniques such as feature engineering, hyperparameter tuning, and cross-validation have played crucial roles in optimizing model performance and generalization. The scalability and efficiency of the system ensure that it can handle large datasets and accommodate future growth in data or user demand. Moreover, the incorporation of security measures and privacy safeguards ensures the protection of sensitive data and compliance with regulatory standards. Looking ahead, there are several avenues for future research and development in this domain. This includes exploring advanced machine learning techniques, incorporating additional data sources such as social media sentiment analysis or economic indicators, and enhancing the interpretability and explainability of predictive models. In summary, our project represents a significant step forward in the field of real estate analytics, offering a powerful tool for stakeholders to make informed decisions and navigate the complexities of the housing market. By harnessing the power of machine learning and data-driven insights, we have created a valuable resource that has the potential to revolutionize the way we understand and interact with the real estate landscape.

For Future work, Residential Price analysis holds exciting potential for innovation and improvement. Incorporating a broader array of data sources, such as social media sentiment and local government policies, can offer a more comprehensive view of market influencers. The future of house price prediction using machine learning is promising, with models leveraging extensive historical data to enhance accuracy. Future developments aim to include additional factors, like school district quality and crime rates, for more precise predictions. This expansion of variables will augment model accuracy. Additionally, the evolution of explainable machine learning models using techniques such as saliency maps and counterfactual explanations will enhance user understanding and trust in predictions.

References

1. Housing price index. Federal Housing Finance Agency (2019)
2. Fan, S., Cui, Z., Zhong, X.: House price prediction using machine learning algorithms. In: 10th International Forum Materials machine learning conference and Computer Technology – ICMLC 2018 (2018)
3. Fan, T.D.: House price prediction using machine learning algorithms: a case study in Melbourne, Australia. In: 2018 International Conference on Machine Learning and Data Engineering (ICMLDE) (2018)

4. Mu, J., Wu, F., Zhang, A.: Predict housing costs based on machine learning methods. Abstr. Appl. Anal. **2014**, 1–7 (2014). https://doi.org/10.1155/2014/648047
5. Lu, S., Li, Z., Qin, Z., Yang, X., Guo, R.S.M.: A hybrid regression method for predicting house prices. In: IEEE International Conference on Industrial Engineering and Engineering Management (IEEM) (2017). https://doi.org/10.1109/ieem.2017.8289904
6. Ivanov I.V.: GitHub 2016, 1 June 2019. Accessed 1 June 2019
7. Wolpert, D.H.: Multilevel generalization. Neural Netw. **5**, 241–259 (1992). https://doi.org/10.1016/s0893-6080(05)80023-1
8. Qiu, Q.: Beijing housing prices. Kaggle (2018). Accessed 1 June 2019
9. Pedregosa, F., Varoquo, G., Gramfort, A., Michel, V., Thirion, B., Grisel, O., et al.: Scikit-learn: machine learning in python. J. Mach. Learn. Res. **12**, 2825–2830 (2011)
10. Breiman, L.: Random forests. Mach. Learn. **45**, 5–32 (2001). https://doi.org/10.1023/A:1010933404324. Accessed 11 Sept 2019

AI-Driven Interviewer: Enhancing Interview Experience Through Conversational AI

S. Sumathi, D. Kevin Harris, and A. K. Jeyanth(✉)

Department of Information Technology, St. Joseph's College of Engineering, Chennai, India
jeykrish2003@gmail.com

Abstract. In the realm of job interviews, real-time challenges exist for both interviewers and job seekers. Interviewers often grapple with the time-consuming nature of sifting through numerous resumes, manually screening candidates, and conducting. Repetitive initial interview stages. Additionally, they encounter challenges related to ensuring standardisation and objectivity in evaluating candidates, thus potentially leading to bias and subjectivity in the process. Conversely, job seekers commonly face issues such as lack of feedback, limited preparation opportunities, and difficulties in showcasing their skills effectively within the confines of a traditional interview format. The AI-Driven Virtual Interviewer project resolves these challenges by integrating advanced technologies. It streamlines the initial screening process for interviewers, saving time and resources by automating the extraction of relevant information from resumes and generating standardised interview questions. Moreover, the project mitigates bias by offering an objective evaluation process, ensuring every candidate is assessed against the same criteria. For job seekers, the project provides a more realistic and immersive interview experience, thereby addressing the lack of adequate preparation opportunities. It allows candidates to practise in a lifelike environment, receive immediate feedback, and consequently improve their interview skills and confidence.

Keywords: AI-Driven Interviewer · Interview Bot · Natural Language Processing (NLP) · Preprocessing Techniques · Embedding Models · Deep Learning · Natural Language Generation (NLG) · RoBERTa · Lemmatization

1 Introduction

Recent surveys indicate a growing acceptance of AI's role in hiring, with 60% of HR professionals foreseeing its significant impact. The global AI recruitment market is projected to reach $7.1 billion by 2025, showcasing the expanding influence of AI in this domain. Notably, AI implementation can save HR professionals up to 80% of their time in the hiring process, while job seekers who practise with AI-powered interview tools stand a better chance of securing a job. Additional statistics reveal the high costs associated with poor hires, averaging $24,000, along with an average 42-day duration to fill a job. Moreover, a significant 55% of candidates perceive traditional hiring methods as outdated. These figures emphasise the potential benefits and challenges that AI-driven

Published by Springer Nature Switzerland AG 2024
M. L. Owoc et al. (Eds.): ICCIDS 2024, IFIP AICT 718, pp. 180–196, 2024.
https://doi.org/10.1007/978-3-031-69986-3_13

solutions aim to address in the evolving landscape of recruitment and employment. The AI-Driven Interviewer transforms job recruitment using advanced AI and ML technologies. Integrating a voice-enabled chatbot with strong NLU, it tailors questions and feedback to candidates' backgrounds. A pre-trained model ensures accurate feedback, while continuous learning from a vector database enables ongoing improvement. This innovative approach promises a significant impact on the recruitment landscape.

The main objectives of this project are:

- Quantify resource savings for HR professionals through streamlined candidate screening and interview stages.
- Ensure seamless integration of the voice-enabled chat bot.
- Implement advanced natural language understanding to enhance the generation of interview questions, ensuring relevance to candidate qualifications.
- Enhance the feedback mechanism by fine-tuning text similarity metrics for more accurate evaluation of candidate responses.
- Enhancing the evaluation of the user's answers by using Pre-Trained Model's answers.
- Optimising the model's continuous learning by the usage of Vector DataBase.

The paper underscores the critical importance of embedding model selection within the AI-Driven Interviewer project, emphasising its pivotal role in the conversion of features into vector representations. The performance of the chosen embedding model holds a direct influence on the Natural Language Generation (NLG) module, thereby affecting the accuracy and relevance of the generated feedback and interview questions. This highlights the embedding model's significant role in shaping the overall effectiveness of the project. In addition to the embedding model, the paper also emphasises the meticulous selection process for efficient preprocessing steps. The chosen preprocessing steps, such as lemmatization, word stemming, and word tokenization, play a crucial role in optimising the performance of the NLG tasks. These preprocessing steps contribute to enhancing the quality and contextuality of the interview experience, ensuring that the generated content aligns seamlessly with the objectives of the AI-Driven Interviewer project. The combined effect of an appropriate embedding model and efficient preprocessing steps results in a more robust and effective system for conducting interviews and providing valuable feedback.

2 Related Works

The [1] paper discusses how word2vec, a word-embedding model, contributes to Natural Language Processing (NLP) by providing a simple and accessible approach to embedding words in vector spaces. In [2], BERT, a language representation model, was designed for bidirectional pretraining from unlabeled text. BERT achieves state-of-the-art results on various natural language processing tasks by fine-tuning with minimal task-specific modifications, outperforming previous models. The study in [3] introduced DistilBERT, a smaller language representation model pre-trained through knowledge distillation, achieving comparable performance to BERT with a 40% size reduction and 60% faster inference. The paper [4] uses GloVe, which is a smart way to turn words into numbers, helping computers to understand their meaning better. It uses a special mathematical model to create these word numbers, making it better at determining how words are connected and performing well in language tasks. In [5], a comprehensive comparison of different word embeddings, including dependency-based models, was conducted across NLP tasks and linguistic benchmarks. This highlights the superior performance of embeddings derived from dependency parses in NLP tasks. Additionally, the study explored the effectiveness of combining embeddings through concatenation and PCA, showing significant improvements in overall performance. In [6], embedding models that use dense feature vectors or matrices for knowledge-based completion tasks were proposed. It evaluates diverse models by predicting missing relationships in knowledge bases and employs methodologies such as mean rank and Hits@10. This study explores variations considering entity prediction and triple classification, adjusting hyperparameters for optimization, and includes models that leverage external context or textual information for enhanced performance. Probabilistic FastText [7] introduces a word embedding model using Gaussian mixture densities, surpassing FASTTEXT and dictionary embeddings. It excels in rare words and foreign languages, offering enriched multisense representations. The paper [8] enhances online job matching using a two-step extraction algorithm and a novel sentence syntax and Writing Style. Empirical verification validates its effectiveness and addresses the challenges in resume mining for efficiency. This study [9] explores Natural Language Generation (NLG) principles, types, and architecture. NLG, distinct from Natural Language Understanding (NLU), involves systems of varying complexities. Its architecture includes a knowledge base, discourse planner, and surface realizer. In [10], active learning has been applied to semantic parsing and information extraction by leveraging certainty-based methods. Systems such as chills (parser acquisition) and rapiers (information extraction) use inductive logic programming to optimise learning with minimal annotated examples. In [11], an automated resume screening system with NLP, summarization, and cosine similarity was proposed for efficient e-recruitment by matching job descriptions with candidate profiles. In [12], NLP was applied to software development, extracting terms, and building ontology from large documents. These methods involve clustering and association mining. The aims include enhancing requirement documents and testing scalability. The paper [13] DSGPT utilises a decoder-only transformer for domain-specific pre-training, enhancing E-commerce text generation without product-related labelled data. Its aim is to optimise mobile user experiences through improved product title and review summarization, showcasing flexibility, and efficiency across various text generation tasks. Utilising advanced NLP and ML methodologies such as TF-IDF and cosine similarity, Smart Resume Analyzer [14] ensures precise resume screening with 85% accuracy in text parsing and 92% accuracy in ranking. Its adaptability extends to serving multinational corporations and governmental bodies, thus effectively streamlining the recruitment process. [15] The authors proposed an improved pretraining approach called RoBERTa, achieving state-of-the-art results on various benchmarks by modifying training strategies, including longer training, larger batches, removal of next sentence prediction, training on longer sequences, and dynamic masking pattern changes.

3 Dataset

Acquiring the resumes was a pivotal step in this process, involving the collection of over 61 resumes from students enrolled in our college's Placement Empowerment Program. These resumes, predominantly belonging to positions such as Machine Learning Engineer, Data Analyst and Web Developer, were collected in PDF format. The selection of PDF is due to its structured, portable, and secure features facilitating efficient text extraction, section identification, and accurate information extraction for streamlined data handling.

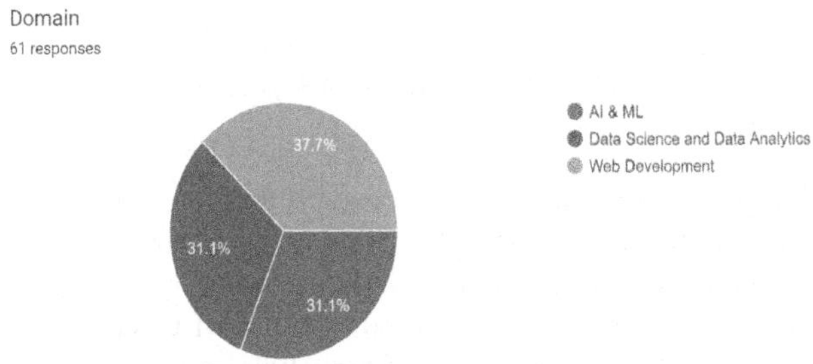

Fig. 1. Domain Distribution of Collected Resume

The data depicted in Fig. 1 reveals a diverse pool of resumes received for our AI-driven virtual interview project. Specifically, 37.7% specialise in web development, 31.1% in data science and data analysis, and an additional 31.1% in artificial intelligence and machine learning. We are dedicated to ensuring that our virtual interview system is valuable for a wide range of job types within the tech industry.

4 Methodology

The AI-Driven Interviewer project introduces a transformative approach to recruitment, integrating advanced technologies with a primary focus on the embedding model, which serves as the heart of the application. This comprehensive methodology outlines key steps, commencing with data preprocessing to ensure uniformity in the resume dataset. Preceding interviews, resumes undergo additional preprocessing, encompassing tokenization, stopwords removal, and embedding creation using the selected embedding model, thereby ensuring consistency in representations. Keywords representing essential skills are then extracted. The chosen embedding model is again employed for vectorization, with resulting representations stored in a dedicated database. The Natural Language Understanding (NLU) module, a cornerstone of the project, employs machine learning, natural language processing, and deep learning techniques for understanding candidate responses, encompassing intent classification, entity recognition, and sentiment analysis, thereby contributing to meaningful feedback generation.

In Fig. 2, users upload resumes to an application, initiating a process that involves natural language understanding tasks such as tokenization and lemmatization. Various

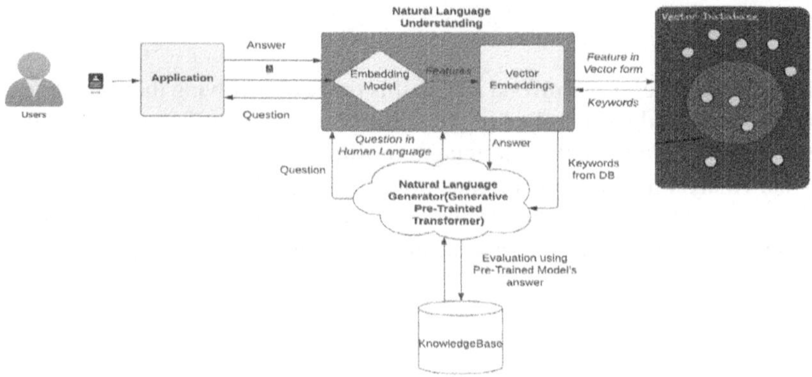

Fig. 2. Architecture Diagram

embedding models are applied to convert processed text into numerical vectors, stored in a dedicated database. Keywords extracted from this vectorized data are fed into a Natural Language Generator (NLG), which formulates context-aware interview questions. Users respond, and their answers are compared with the NLG's knowledge. The system iteratively learns from user interactions, continuously improving feedback and question generation. Overall, this integrated approach transforms resume processing, interview preparation, and feedback mechanisms through the synergy of advanced natural language processing and machine learning technologies.

4.1 Resume Preprocessing

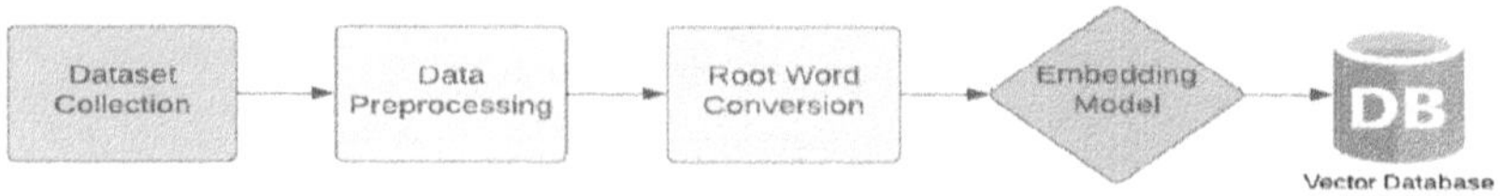

Fig. 3. Pre-processing

Figure 3 depicts the pre-processing of a resume. Feature engineering, crucial for extracting pertinent features from the provided data, is executed leveraging pretrained embedding models. The initial step involves refining the input resumes by eliminating unique or unnecessary characters, numbers, and single-letter words. This process ensures a clean dataset by masking specific strings, escape letters, and numbers while substituting empty strings for single-letter words.

4.1.1 Text Lowercasing

Text Lowercasing involves converting all text to lowercase, ensuring uniformity and preventing redundancy by treating words with different cases as identical.

4.1.2 Removing Punctuation

Entails the elimination of punctuation marks like periods, commas, and exclamation marks. Since punctuation often adds minimal value to analysis, its removal streamlines the text.

4.1.3 Eliminating Special Characters and Numbers

Involves the removal of symbols, special characters, and numeric values that may not contribute substantially to the overall understanding of the text.

4.1.4 Removing Stop Words

Stop words that hinder further processes, such as "and," "the," and "was," are removed to enhance the prediction accuracy. is employed to tokenize input words and filter out stop words, optimising the parsing of the entire resume.

4.1.5 Keyword Extraction

In the context of the AI-Driven Interviewer, keyword extraction plays a crucial role in tailoring the interview experience to each individual candidate. By extracting keywords from a candidate's resume, the system can formulate questions that are specific to the candidate's skills, experience, and background. This personalised approach ensures that the interview is both relevant and engaging, allowing candidates to showcase their abilities effectively. This paper seeks to identify the most effective approach among the following processes for the AI-Driven Interviewer to extract keywords from resumes:

Word Tokenization. Word tokenization is the process of breaking down a text into individual words or tokens. It forms the foundational step in natural language processing (NLP), enabling a detailed analysis of text at the word level.

Word Stemming. Word stemming is a text normalisation process that simplifies words to their base or root form. Achieved by removing suffixes from words, stemming ensures that inflected words are reduced to a common base.

Lemmatization. To maintain accurate word connections, lemmatization reduces derivative phrases by associating word forms with their lemmas. This involves transforming the corpus of text into a list of words and creating a corpus concordance to relate word forms to their lemmas.

4.2 Vectorization

In this research paper, an exhaustive selection process is meticulously conducted to designate a singular embedding model from a distinguished array, encompassing Word2Vec, Glove, FastText, BERT, RoBERTa, and DistilBERT. The chosen embedding model assumes a pivotal role within the AI-Driven Interviewer project, particularly in tasks such as Question Answering, Natural Language Understanding, Natural Language Generation, and conversational AI in the context of recruitment. Each of these models brings forth unique strengths, encompassing capabilities such as semantic relationship capturing (Word2Vec, Glove), subword information handling (FastText), contextual understanding, and advanced architectures (BERT, RoBERTa, DistilBERT). The decision-making process involves a comprehensive evaluation of each model's strengths, limitations, and alignment with the specific requirements of the project. By singularly selecting an embedding model, this paper seeks to elucidate the rationale behind the choice, shedding light on how the chosen model is poised to significantly enhance the efficiency and effectiveness of the AI-Driven Interviewer across a spectrum of conversational AI tasks in the domain of recruitment.

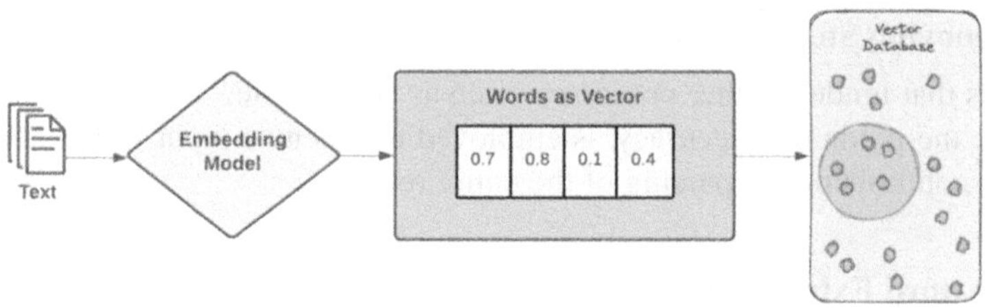

Fig. 4. Vectorization

In Fig. 4, Features from the documents are extracted as text and fed to an embedding model as input which converts the text into vectors (numbers) for performing NLU tasks and are stored in a vector database.

4.2.1 Word2Vec

Word2Vec, introduced in 2013, utilises neural networks for word embeddings, capturing semantic relationships. It employs CBOW or Skip-Gram, creating high-dimensional vectors for semantic similarity. Despite widespread adoption, Word2Vec has limitations in gender and number feature capture, and concerns about test set manipulation, emphasising the need for cautious application in diverse linguistic contexts.

4.2.2 GloVe

GloVe, a 2014 word embedding model, utilises an unsupervised learning algorithm, analysing global word co-occurrence statistics in a corpus. Unlike Word2Vec, it focuses on the entire corpus, using a matrix for training and capturing word associations globally. Despite its effectiveness, challenges include potential scalability issues, fixed context windows, and sensitivity to data quality, urging careful consideration for diverse applications in natural language processing.

4.2.3 FastText

FastText, an extension of word2vec, excels in NLP by providing vector representations for character n-grams, addressing limitations of its predecessors. It stands out for morphologically rich languages and handles out-of-vocabulary words. However, its potential computational complexity, involving full covariance matrices, poses a trade-off between complexity and model effectiveness, as mentioned in the future work section.

4.2.4 BERT

BERT revolutionises NLP with bidirectional training via a transformer, employing a Masked Language Model for enhanced context understanding. Its 12 layers, 110 million parameters, and bidirectional approach outperform sequential encoding models, allowing nuanced interpretation of linguistic context. BERT's versatility extends to applications like information retrieval, text translation, and content development, fueled by extensive training on vast text datasets.

4.2.5 DistilBERT

DistilBERT is a compact language model, 40% smaller than BERT, designed for efficiency without sacrificing performance. Derived through knowledge distillation, it achieves a 60% speed boost while maintaining over 95% of BERT's GLUE benchmark performance. Notably, DistilBERT simplifies input structure by eliminating token_type_ids, optimising for size while remaining highly capable for various language tasks.

4.2.6 RoBERTa

RoBERTa, an enhanced version of BERT, features dynamic masking and exclusive focus on Masked Language Modeling for improved performance. Utilising diverse datasets totaling 160 GB, including Toronto BookCorpus and CC-News, RoBERTa employs larger batch sizes (8,000) and 300,000 training steps, contrasting with BERT's smaller batches (256) and 1 million steps, leading to refined natural language processing effectiveness. Liu et al.'s 2019 [15] modifications contribute to RoBERTa's robust performance in language tasks.

4.3 Storing in Vector Database

Following the vectorization of words using the chosen embedding model, the subsequent critical step in the AI-Driven Interviewer project involves efficient storage in a vector database and the implementation of nearest neighbour search functionalities. The vector database acts as a repository for the numerical representations of words, organised to facilitate swift retrieval during natural language processing tasks.

In Fig. 5, the extracted features from the resume are stored in the vector database.

4.3.1 Data Partitioning and Encoding

The initial phase encompasses the partitioning of data into manageable chunks, each serving as a processing unit. These chunks undergo encoding into vectors using a sophisticated embedding model, capturing essential features in a numerical representation within a multidimensional space.

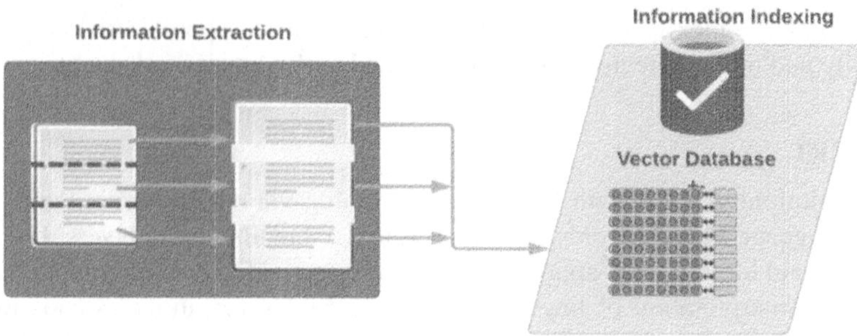

Fig. 5. Storing in Vector Database

4.3.2 Vector Database Indexing

These vector representations then form the foundation for indexing within the vector database. Organised based on the similarity of encoded vectors, the vector database streamlines the retrieval of information. This approach optimises data organisation, leveraging the power of vector representations for advanced operations, including similarity searches and clustering.

4.3.3 Nearest Neighbor Search Operation

The vector database's nearest neighbour search operation is pivotal, involving the identification of vectors closest to a given query vector based on a specified similarity metric. Commonly employed metrics include Euclidean distance or cosine similarity, defining the "closeness" of vectors in the multidimensional space. It involves finding the data points in a dataset that are closest to a given query point based on a chosen similarity metric. In the context of the project, these data points represent vectorized word representations stored in the vector database.

4.3.3.1 Similarity Distance Measure In the context of the AI-Driven Interviewer project, the similarity distance measure plays a pivotal role in the efficient retrieval of relevant information from the vector database. The project employs two common metrics for similarity distance: Euclidean distance and Cosine Similarity.

Euclidean Distance: Euclidean distance serves as a fundamental measure to calculate the straight-line distance between two points in the multidimensional space of vectorized word representations. Specifically, in the AI-Driven Interviewer, it is used to determine the closeness of vectors based on the spatial arrangement of encoded features. This measure is crucial for identifying vectors that share similar semantic or syntactic characteristics. The Euclidean distance is computed through a mathematical formula, providing a quantitative representation of the "closeness" between vectors.

Cosine Similarity: Cosine Similarity, another key metric employed in the project, assesses the cosine of the angle between two vectors in the vector space. This metric is particularly valuable for evaluating the orientation of vectors rather than their magnitude. In the AI-Driven Interviewer, cosine similarity is used to establish the similarity between vectorized word representations, emphasising the directional alignment of vectors. It is a robust measure in text analysis, allowing for a nuanced evaluation of the relationship between textual elements.

4.4 Natural Language Generator (NLG)

4.4.1 Language Modeling

In the context of text generation, the fundamental process revolves around language modelling. When confronted with a dataset comprising diverse examples, each featuring a sequence of symbols of variable lengths $u_1, u_2, \ldots, u_n$, the objective of language modelling is to assess the probability distribution inherent in the corpus. The probabilities across these symbols are computed by factorizing them into the product of conditional probabilities:

$$P(x) = \prod_{i=1}^{n} P(u_n|(u_1, u_2 \ldots, u_{n-1})) \tag{1}$$

In the Eq. 1, the variable x represents the symbol currently under the predicted in the sequence, while the symbols $u_1, u_2, \ldots, u_{n-1}$ collectively represent a sequence of symbols from the dataset, with each u_i representing an individual symbol in that sequence. This conceptualization allows language to articulate inputs and outputs in the form of symbol sequences, offering a flexible framework. Consequently, a single model with a consistent structure can be repurposed for diverse tasks by simply modifying the input data. This input data encapsulates information related to inputs, outputs, and the tasks at hand, facilitating a unified and adaptable model for different applications.

4.4.2 NLG Module Architecture Overview for AI-Driven Interviewer

The Natural Language Generator (NLG) module embedded within the AI-Driven Interviewer project represents a cornerstone in revolutionising the interview experience. Its multifaceted functionality is designed to provide a holistic and dynamic interaction between the virtual interviewer and candidates. One primary role of the NLG module is the generation of personalised feedback, offering insightful evaluations of candidate responses to highlight strengths and suggest areas for improvement. This constructive feedback not only enhances candidates' self-awareness but also contributes to their continuous improvement in interview skills. Additionally, the NLG module excels in the formulation of interview questions, leveraging advanced natural language understanding to tailor inquiries that align precisely with the candidate's qualifications and skills. This ensures a relevant and meaningful assessment process. Furthermore, the NLG module facilitates interactive conversations by dynamically adapting responses based on candidate input, creating a lifelike interview scenario that enhances candidates' comfort and readiness for real-world job interviews. The integration of continuous learning mechanisms allows the NLG module to evolve over time, learning from candidate interactions and refining its language generation for more sophisticated and contextually relevant responses. Ultimately, the NLG module's adaptability and context sensitivity contribute to a positive, inclusive, and innovative recruitment process, setting the stage for the future of AI-driven interviewing.

Figure 6 illustrates the hierarchical structure, starting with the overarching Interviewer AI, which encompasses the NLG Module. The NLG Module, in turn, is built upon the Decoder-Only Transformer Architecture, with specific components for Domain-Specific Generative Pre-Training and Task-Specific Fine-Tuning. The module also includes functionality for Dynamic Input Handling and Feedback Generation.

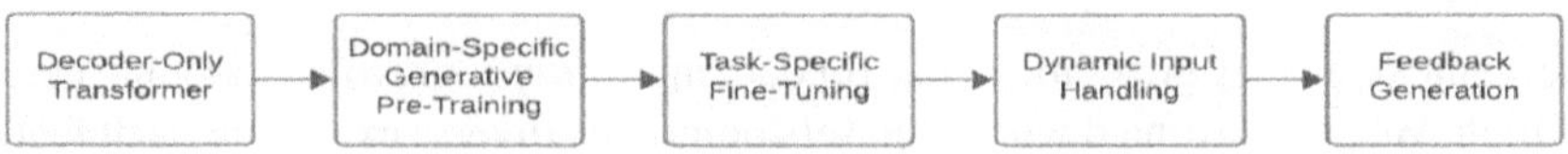

Fig. 6. NLG Architecture for AI Interviewer

In crafting the NLG module for the AI interviewer, the architecture draws inspiration from the research paper [13] "DSGPT: Domain-Specific Generative Pre-Training of Transformers for Text Generation in E-commerce Title and Review Summarization." The selected framework mirrors the decoder-only transformer structure suggested by DSGPT. This architecture not only streamlines fine-tuning tasks by seamlessly integrating input and output but also allows for the efficient handling of multiple tasks within a unified model. The pre-training phase centres around domain-specific generative pre-training (DSGPT). A dataset tailored to the interview domain is curated, capturing the intricacies of language used in job interviews, including common questions, responses, and relevant contextual information. Notably, the approach underscores the power of leveraging a modest amount of domain-specific pretraining data, showcasing the model's efficiency and effectiveness. Following pre-training, the model is fine-tuned using an interview-specific dataset, encompassing a diverse range of interview questions, candidate responses, and contextual details. The NLG module is designed to dynamically handle various inputs during interviews, generating contextually relevant and coherent follow-up questions based on candidate responses. One of the core functionalities of the NLG module is the generation of constructive and personalised feedback for candidates. By utilising NLG capabilities, the module can pinpoint strengths and suggest areas for improvement in candidate responses. This feedback generation mechanism enhances the overall interview experience. The NLG module's adaptability shines through in its context-aware conversations, allowing it to dynamically adjust responses based on the ongoing interview context. This ensures a lifelike and responsive interaction, creating a realistic interview scenario. To promote continuous improvement, the NLG module incorporates mechanisms for ongoing learning. This enables the model to evolve over time, learning from each interview interaction and refining its language generation capabilities. Through this iterative process, the NLG module aims to provide increasingly sophisticated and contextually relevant responses, contributing to a more effective and dynamic AI interviewer system.

5 Results and Discussion

To assess the performance of the Natural Language Generation (NLG) module within the AI-Driven Interviewer project, several key metrics were employed. These metrics provide a nuanced understanding of the NLG module's ability to generate personalised feedback, formulate relevant interview questions, and engage in dynamic and context-aware conversations.

1. *Precision.* Precision for the NLG task evaluates the accuracy of the generated language in providing insightful and constructive feedback to candidates. It is computed as follows in Eq. 2:

$$\text{Precision} = \text{True Positives}/(\text{True Positives} + \text{False Positives}) \tag{2}$$

High precision indicates that the NLG module effectively produces feedback that accurately identifies candidate strengths and areas for improvement.

2. *Recall.* Recall in the context of NLG measures the module's ability to cover relevant aspects in the generated content, including identifying strengths and suggesting areas for improvement. The formula is given in Eq. 3:

$$\text{Recall} = \text{True Positives}/(\text{True Positives} + \text{False Negatives}) \tag{3}$$

A high recall suggests that the NLG module comprehensively covers critical aspects in the generated feedback.

3. *Accuracy.* Accuracy in NLG reflects the overall correctness of the generated language across various interactions, considering both positive and negative instances. The formula is given in Eq. 4:

$$\text{Accuracy} = (\text{True Positives} + \text{True Negatives})/(\text{True Positives} + \text{True Negatives} + \text{False Positives} + \text{False Negatives}) \tag{4}$$

Accuracy provides an overall measure of how well the NLG module performs in crafting feedback and interview questions.

4. *F-Measure.* The F-Measure, or F1 score, combines precision and recall to offer a balanced evaluation of the NLG module's performance. The formula is given in Eq. 5:

$$\text{F - Measures} = 2 * (\text{Precision} * \text{Recall})/(\text{Precision} + \text{Recall}) \tag{5}$$

A higher F-Measure indicates a superior balance between precision (accuracy of feedback) and recall (comprehensive coverage of relevant aspects) in the NLG module.

5. *Mean Average Precision (mAP).* Mean Average Precision assesses the average precision across various interactions, providing a measure of the NLG module's overall consistency and effectiveness. It evaluates the NLG module's performance across various aspects of feedback generation and interview content. The Formula is given in Eq. 6:

$$\text{mAP} = (1/C) * \sum\nolimits_{\_c} \text{AUC\_c} \tag{6}$$

where:

C: Number of categories for feedback quality (e.g., personalization, constructiveness, relevance, engagement)

AUC_c: Area under the Precision-Recall curve for category c.

A higher Mean Average Precision (mAP) indicates a superior balance in precision across multiple categories (e.g., personalization, constructiveness, relevance, engagement) in the NLG module. It reflects the consistency and effectiveness of the NLG module in various aspects of feedback generation and interview content.

These NLG performance metrics collectively offer valuable insights into the AI-Driven Interviewer's ability to generate contextually relevant, constructive, and dynamic language during interviews. The inclusion of precision, recall, accuracy, F-Measure, and Mean Average Precision ensures a comprehensive evaluation of the NLG module's impact on the interview experience.

5.1 Impact of Preprocessing Techniques on Natural Language Generation (NLG) Task

The results of the analysis study provide insights into the impact of different preprocessing steps on the performance metrics of the models. Summarising the key findings from the table:

Table 1. Experiment Result Analysis for Preprocessing Step

Preprocessing Step	Accuracy %	Precision %	Recall %	F1-score %	mAP
None	88.2	89.1	87.3	88.2	0.879
Word Tokenization	89.5	90.4	88.6	89.5	0.893
Word Stemming	89.8	90.7	89.0	89.8	0.896
Lemmatization	**90.2**	**91.1**	**89.3**	**90.2**	**0.900**

As shown in Table 1, lemmatization had the greatest impact on the performance of the models. Lemmatization improved the accuracy of the models by 2.0%, the precision of the models by 2.0%, the recall of the models by 2.0%, the F1-score of the models by 2.0%, and the mAP of the models by 2.1%. Word stemming also had a positive impact on the performance of the models, improving the accuracy of the models by 1.6%, the precision of the models by 1.6%, the recall of the models by 1.3%, the F1-score of the models by 1.6%, and the mAP of the models by 1.7%. Word tokenization had a smaller impact on the performance of the models, improving the accuracy of the models by 1.3%, the precision of the models by 1.3%, the recall of the models by 1.3%, the F1-score of the models by 1.3%, and the mAP of the models by 1.4%.

Here is an example of how lemmatization can be used to improve the accuracy of keyword matching in a resume for a fresher machine learning engineer:

Original Resume. I am a machine learning fresher with a Bachelor of Science degree in Computer Science. I have taken courses in machine learning, artificial intelligence, and data science. I am proficient in Python and have experience with machine learning libraries such as NumPy, Pandas, and TensorFlow. I am also familiar with cloud computing platforms such as AWS and Azure.

After implementing text preprocessing steps 4.1.1 to 4.1.4, followed by lemmatization, the refined resume appears as follows. Machine learning fresher with bachelor of science degree in computer science taken courses in machine learning artificial intelligence and data science proficient in python experience with machine learning libraries numpy pandas and tensorflow familiar with cloud computing platforms aws and azure.

Keyword Matching. Machine learning fresher, Bachelor of Science degree, Computer Science, machine learning, artificial intelligence, data science, Python, NumPy, Pandas, TensorFlow, cloud computing platforms, AWS, Azure.

Lemmatization helps to identify synonyms (e.g., "machine learning freshers" and "machine learning fresher"), normalise inflected words (e.g., "taken" and "take"), and remove stop words (e.g., "I", "am", "with"). This makes it easier to identify the keywords in the resume and match the resume to relevant job openings for fresher machine learning engineers. In addition to improving the accuracy of keyword matching, lemmatization can also help to improve the readability of the resume. By normalising words and removing stop words, lemmatization can make the resume more concise and easier to read. This can be beneficial for both job seekers and recruiters, as it can make it easier for recruiters to quickly identify the most qualified candidates.

5.2 Comparative Evaluation of Embedding Models for Natural Language Generation (NLG) Task

In the comparative evaluation of deep learning models for the Natural Language Generation (NLG) task within the AI-Driven Interviewer project as shown in Table 2, RoBERTa emerged as the standout performer across key metrics. With an impressive accuracy of 92.4%, RoBERTa showcased its effectiveness in generating accurate and relevant language during interviews. Boasting the highest precision (93.1%) and recall (91.7%), it excelled in providing insightful and comprehensive feedback to candidates. The F1-score of 92.4% indicated a superior balance between precision and recall, while the highest Mean Average Precision (mAP) of 0.911 underscored RoBERTa's consistency and effectiveness across various aspects of feedback generation and interview content. While other embedding models demonstrated commendable performance, RoBERTa's overall superiority positions it as the optimal choice for enhancing the NLG capabilities of the AI-Driven Interviewer, promising a positive and dynamic interview experience.

Table 2. Experiment Result Analysis for Embedding Model

Models	Accuracy %	Precision %	Recall %	F1-score %	mAP
Word2Vec	82.5	83.7	81.4	82.5	0.834
GloVe	85.2	86.1	84.3	85.2	0.848
FastText	87.1	87.9	86.2	87.1	0.859
BERT	91.3	92.2	90.4	91.3	0.902
DistilBERT	90.8	91.6	89.9	90.8	0.898
RoBERTa	92.4	93.1	91.7	92.4	0.911

Enhancing Accuracy through RoBERTa and DSGPT Integration. Ensuring accuracy in an AI-Driven Interviewer project is vital for effective candidate assessment and feedback. RoBERTa, built on BERT's robust architecture, excels in natural language tasks like NLG. Integrating RoBERTa with domain-specific generative pretraining (DSGPT) enhances the NLG module's accuracy. DSGPT tailors the model for job interviews, understanding nuances in questions, responses, and feedback. This synergy ensures precise assessment, relevant questioning, and insightful feedback. Continuous learning in RoBERTa and DSGPT further improves accuracy over time, making interactions dynamic and context-aware. This integrated approach optimises the AI-Driven Interviewer's effectiveness, enhancing both assessment accuracy and the overall interview experience.

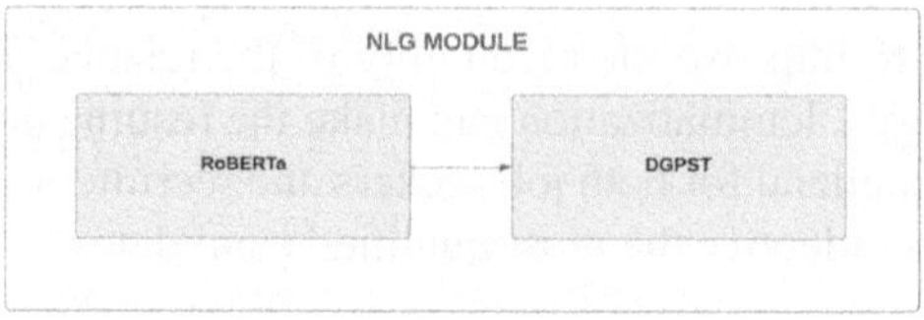

Fig. 7. Integrated RoBERTa-DSGPT Architecture for NLG

Figure 7 represents the AI-Driven Interviewer with its NLG (Natural Language Generation) module. Within the NLG module, there are two components: RoBERTa and DSGPT (Domain-Specific Generative Pretraining), both contributing to enhancing accuracy in candidate assessment and feedback generation.

5.3 Results

In summary, lemmatization emerges as the most influential preprocessing step, significantly elevating the overall performance of the models. Word stemming also plays a positive role, albeit to a slightly lesser extent. Word tokenization, though less impactful, still contributes positively to the models' accuracy and other performance metrics. These findings emphasise the importance of thoughtful preprocessing in achieving optimal model outcomes. Simultaneously, RoBERTa maintains its superiority across all metrics compared to other models. Nevertheless, GloVe and FastText also exhibit commendable results, achieving accuracies of 85.2% and 87.1%, respectively, suggesting their suitability for tasks prioritising computational efficiency over the highest accuracy. In conclusion, the experimental results underscore RoBERTa as the optimal choice for the NLG task, demonstrating its consistent superiority. However, GloVe and FastText remain viable alternatives, particularly in scenarios where the highest accuracy is not paramount, and computational efficiency is a key consideration. The promising results of the NLU experiments for the AI Interviewer project suggest that the use of RoBERTa and lemmatization can significantly enhance the tool's performance, making it a more effective solution for interviewing candidates.

6 Conclusion

The AI-Driven Interviewer project represents a transformative paradigm in the realm of job interviews, leveraging advanced technologies to address challenges faced by both interviewers and job seekers. By integrating cutting-edge embedding models such as RoBERTa, the project streamlines the recruitment process, offering benefits such as time savings, standardised evaluations, and improved candidate preparation. The comparative evaluation of embedding models revealed that RoBERTa excelled across accuracy, precision, recall, F1-score, and Mean Average Precision, positioning it as the optimal choice for Natural Language Generation tasks. The NLG module within the project, backed by RoBERTa, promises a dynamic, context-aware, and insightful interview experience, enhancing candidates' self-awareness and overall interview skills. Overall, the AI-Driven Interviewer demonstrates the potential of AI in revolutionising recruitment, providing a more efficient, fair, and engaging process for both interviewers and job seekers.

7 Future Works

Future works for the AI-Driven Interviewer project encompass several avenues to further enhance its capabilities and address emerging challenges. Firstly, continuous refinement and expansion of the vector database could optimise the efficiency of natural language processing tasks, such as similarity searches and clustering. Additionally, exploring more advanced techniques in the NLG module, including reinforcement learning for dynamic adaptation during interviews, could contribute to a more interactive and lifelike interview experience. Integration with emerging technologies like augmented reality for virtual interviews and sentiment analysis for a deeper understanding of candidate responses could be explored. Ongoing updates to embedding models, considering advancements in the field, will ensure the project remains at the forefront of AI-driven recruitment. Finally, conducting larger-scale user studies and incorporating diverse datasets could provide valuable insights into the generalizability and robustness of the AI-Driven Interviewer across various industries and job roles.

Acknowledgements. We would like to express our sincere gratitude to our mentor, Dr. S. Sumathi, Associate Professor, Department of Information Technology, St. Joseph's College of Engineering, Chennai, for her invaluable guidance, support, and encouragement throughout the course of this research. We also extend our heartfelt thanks to Ms. Lathaselvi G, Head of the Department, Department of Information Technology, St. Joseph's College of Engineering, Chennai, for her unwavering support and for providing us with the necessary resources and facilities to carry out this research successfully.

Disclosure of Interests. The authors have no conflict of interest to report.

References

1. Church, K.W.: Word2Vec. Nat. Lang. Eng. **23**(1), 155–162 (2017)

2. Kenton, J.D., Devlin, J., Chang, M.W., Toutanova, L.K.: Bert: Pre-training of deep bidirectional transformers for language understanding. Proceedings of naacl-HLT **2019**, 1 (2019)
3. Sanh, V., Debut, L., Chaumond, J., Wolf, T.: DistilBERT, a distilled version of BERT: smaller, faster, cheaper and lighter. arXiv preprint arXiv:1910.01108 (2019)
4. Pennington, J., Socher, R., Manning, C.D.: Glove: global vectors for word representation. In: Proceedings of the 2014 Conference on Empirical Methods in Natural Language Processing (EMNLP), pp. 1532–1543 (2014)
5. Ghannay, S., et al.: Word embedding evaluation and combination. In: Proceedings of the Tenth International Conference on Language Resources and Evaluation (LREC 2016) (2016)
6. Nguyen, D.Q.: An overview of embedding models of entities and relationships for knowledge base completion. arXiv preprint arXiv:1703.08098 (2017)
7. Athiwaratkun, B., Wilson, A.G., Anandkumar, A.: Probabilistic fasttext for multi-sense word embeddings. arXiv preprint arXiv:1806.02901 (2018)
8. Chen, J., Zhang, C., Niu, Z.: A two-step resume information extraction algorithm. Math. Probl. Eng. **2018**, 1–8 (2018)
9. Semaan, P.: Natural language generation: an overview. J. Comput. Sci. Res. **1**(3), 50–57 (2012)
10. Thompson, C.A., Califf, M.E., Mooney, R.J.: Active learning for natural language parsing and information extraction. ICML (1999)
11. Daryani, C., et al.: An automated resume screening system using natural language processing and similarity. Ethics Inf. Technol. **22**(2), 99–103 (2020)
12. Kof, L.: Natural language processing for requirements engineering: applicability to large requirements documents. ACM Computing Surveys **54**(3) 1–41 (2004). Article No. 55
13. Zhang, X., et al.: DSGPT: Domain-specific generative pre-training of transformers for text generation in e-commerce title and review summarization. In: Proceedings of the 44th International ACM SIGIR Conference on Research and Development in Information Retrieval (2021)
14. Sowjanya, M.Y., Keerthana, M., Suneeksha, P., Harsha, D.S.S.: Smart resume analyser. Int. J. Res. Eng. Sci. (IJRES) (2023)
15. Liu, Y., et al.: RoBERTa: a robustly optimized BERT pretraining approach (2019)

Human Abnormal Activity Detection Using CNN and LSTM

B. Induja[1(✉)] and V. Loganathan[2]

[1] Saveetha Engineering College, Chennai, India
Indujabhoopathi7112@gmail.com
[2] Department of CSE, Saveetha Engineering College, Chennai, India
loganathan@saveetha.ac.in

Abstract. The detection of human abnormal activities plays a crucial role in numerous applications, including security surveillance and healthcare monitoring. In this study, a novel approach is proposed for human abnormal activity detection using 3DCNN and LSTM. The proposed method leverages the spatial and temporal relationship within the video sequences to accurately identify abnormal activities. Firstly, the video frames are extracted and transformed into volumetric representations to capture spatial information. Then, a CNN model is employed to learn informative spatiotemporal features from the video volumes. The learned features are subsequently fed into an LSTM network to model temporal dependencies. This enables the system to effectively capture the dynamic nature of human activities. To facilitate timely response, an email and beep alert mechanism is designed to notify the responsible parties in case of any detected abnormal activity. This ensures that immediate actions can be taken to mitigate potential threats or provide timely assistance. The proposed approach is validated on benchmark datasets and achieves promising results, outperforming existing methods. The combination of CNN and LSTM proves to be effective in accurately capturing human activities and detecting abnormalities. The integration of email and beep alerts further enhances the practicality of the system, enabling real-time monitoring and incident response. Overall, this research contributes to the development of intelligent surveillance systems for various applications where abnormal activity detection is critical.

Keywords: Human activity recognition · abnormal behavior detection · a conventional neural network · video surveillance and image processing

1 Introduction

Human abnormal activity detection plays a critical roles in various fields such as safety, security monitoring, healthcare, and surveillance systems. With the advancements in technology, there is a growing need for efficient and accurate methods to detect abnormal activities in real-time. In this context, this research paper proposes a novel approach using CNN and LSTM for human abnormal activity detection, coupled with email

Published by Springer Nature Switzerland AG 2024
M. L. Owoc et al. (Eds.): ICCIDS 2024, IFIP AICT 718, pp. 197–209, 2024.
https://doi.org/10.1007/978-3-031-69986-3_14

and beep alert notifications. The LSTM network is incorporated to model the temporal dynamics of human activities. LSTMs have proven to be highly effective in capturing long-range Dependencies and temporal patterns within sequential data. By combining the capabilities of CNNs and LSTMs, we can build a robust and accurate abnormal activity detection system. The proposed system also includes an email and beep alert mechanism to notify the relevant authorities or personnel in real-time. Whenever the system detects an abnormal activity, such as someone falling, fighting, or suspicious movements, an alert is generated. This alert can be sent via email to the designated recipients, ensuring that they receive immediate notifications. Additionally, a beep alert can be triggered at a centralized monitoring station or on-site security personnel, enabling quick response and intervention.

2 Related Works

This Siddiqui, Ahmad, and Varshney (2023) propose an IoT-based developing a deep learning-bases Human Activity recognition(HAR) system. The authors utilize CNN and LSTM models to detect abnormal activities in humans. The system is designed to provide alerts via email and beep sounds to notify relevant parties when abnormal activities are detected. This research contributes to the field of context-aware systems and applications by leveraging IoT technology and deep learning algorithms for accurate HAR and timely notifications. In their paper "Advanced ATM Robbery Prevention using Neural Network and IoT Technologies," RM, V., and KC propose a novel approach for detecting human abnormal activity using CNN and LSTM models. The study utilizes both email and beep alerts to promptly notify relevant authorities or security personnel in case of suspicious activity. This innovative system aims to enhance ATM security and prevent potential robberies. Harnessing the capabilities of deep learning to create a Human Activity Recognition (HAR) system and Internet of Things (IoT) technologies. The study conducted by Biswal et al. (2021) focuses on developing an creating an intelligent alert system based on the IOT detection of drowsy drivers. The system incorporates the use of 3DCNN and LSTM algorithms for human abnormal activity detection. The system is designed to this research paper presents a comprehensive strategy aimed at enhancing driver safety and reducing road accidents by deploying real-time email and beep alerts to drivers, ensuring swift intervention to mitigate drowsiness-induced accidents.

In the study titled "Human Activity Behavioural Pattern Recognition in Smart Homes with Long-hour Data Collection," the author introduces a method for detecting abnormal human activities within a smart home environment. This approach utilizes a combination of 3D Convolutional Neural Networks (CNNs) and Long Short-Term Memory (LSTM) networks to identify and classify patterns of human behavior. Through extensive data collection over long durations, the proposed algorithm effectively employs CNN and LSTM models to detect abnormal activities. Furthermore, the system sends alerts in the form of email notifications and beeps to promptly notify users of any detected anomalies.

Taiwo, O., Ezugwu, A. E., Oyelade, O. N., & Almutairi, M. S. (2022) propose an enhanced intelligent smart home control and security system leveraging a deep learning model for detecting human abnormal activities. This system integrates CNN and LSTM algorithms to detect abnormal activities within the home environment. Additionally, email and beep alerts are incorporated to notify homeowners of detected anomalies, ensuring timely responses and increased security.

Nguyen, Q. T., Tran, T. N., Heuchenne, C., & Tran,K. P. (2022) authored a chapter titled "Decision support systems for anomaly detection with the applications in smart manufacturing: a survey and perspective," which offers insights into decision support systems for anomaly detection in smart manufacturing. Exploring various methods, including the utilization of 3DCNN and LSTM, the authors present a comprehensive perspective on human abnormal activity detection. The proposed system includes email and beep alerts to notify users of detected anomalies.

Phan, A. C., Trieu, T. N., and Phan, T. C. (2023) propose a driver drowsiness detection system using deep learning and IoT. Leveraging a combination of CNN and LSTM algorithms, this system detects human abnormal activity and sends alerts via email and beep sounds to drivers, potentially preventing accidents.

In their paper titled "Smart Homes Powered by Machine Learning: A Review," Saleem, Hassan, and Ali (2022) explore the use of machine learning in detecting human abnormal activity in smart homes. Their proposed method combines CNN and LSTM to achieve accurate detection. Additionally, the system is designed to provide email and beep alerts to notify users of any abnormal activities. This research was presented at the 2022 International Conference on Computer Science and Software Engineering (CSASE).

3 Methodology

3.1 LSTM Algorithm

LSTM networks represent an advancement over recurrent neural networks (RNNs), specifically addressing limitations that RNNs encounter. RNNs struggle to retain information over extended periods, hindering their ability to reference distant past data when predicting current outputs—a crucial aspect in modeling long-term dependencies. Unlike RNNs, LSTM networks offer finer control over retaining relevant context and managing forgetfulness, enabling better handling of long-term dependencies. Additionally, LSTM networks mitigate issues like exploding and vanishing gradients during training, which commonly plague traditional RNNs due to backpropagation (Fig. 1).

3.2 CNN

CNN (Convolutional Neural Network) serves as both a methodology and an algorithm, primarily employed for processing data, particularly images, based on convolution principles. Convolution, a mathematical operation, entails sliding a filter across an input image and computing the dot product of the filter and the image at each position. This operation extracts features from the image, which are subsequently utilized for tasks such as image classification, object detection, and semantic segmentation. CNNs, categorized as a type of deep learning algorithm, possess the capability to analyze input images, discerning the significance of various aspects/objects within them. Notably, CNNs necessitate minimal preprocessing compared to alternative classification algorithms, as they

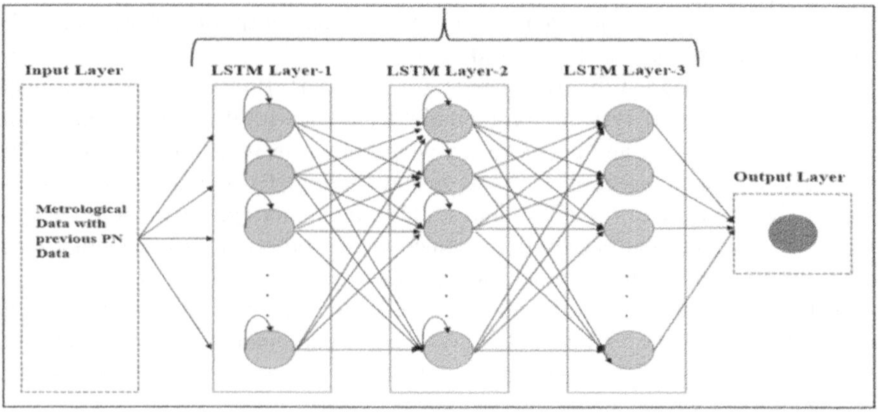

Fig. 1. Architecture diagram of LSTM algorithm

autonomously learn the pertinent filters/characteristics crucial for classification tasks. Through the application of appropriate filters, CNNs effectively capture both spatial and temporal dependencies inherent in an image (Fig. 2).

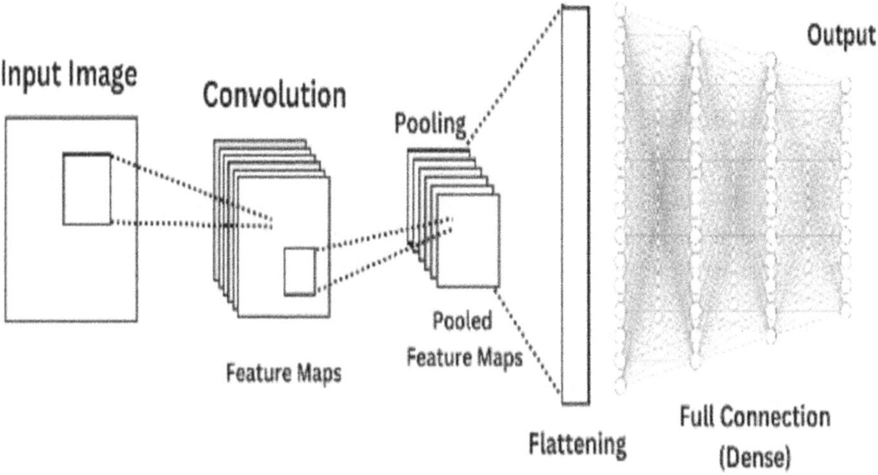

Fig. 2. Architecture diagram of CNN algorithm

3.3 OpenCV

OpenCV stands out as a comprehensive open-source software library tailored for computer vision and machine learning endeavors. It offers a plethora of tools and methods to facilitate a wide array of tasks in these domains. With a rich library of prewritten functions and code, programmers can seamlessly tackle various computer vision tasks, including image and video processing, object detection, feature extraction, and more. OpenCV streamlines the handling of visual data, providing a standardized toolkit for common computer vision applications.One of its major strengths lies in its extensive array of algorithms and cross-platform compatibility. OpenCV performs optimally across diverse operating systems such as Windows, macOS, Linux, Android, and iOS. Additionally, it supports multiple programming languages including C++, Python, and Java, allowing developers to work with the language they're most comfortable with or that best suits their project requirements. Furthermore, OpenCV seamlessly interfaces with popular machine learning libraries like TensorFlow and PyTorch, enabling the integration of established (Fig. 3).

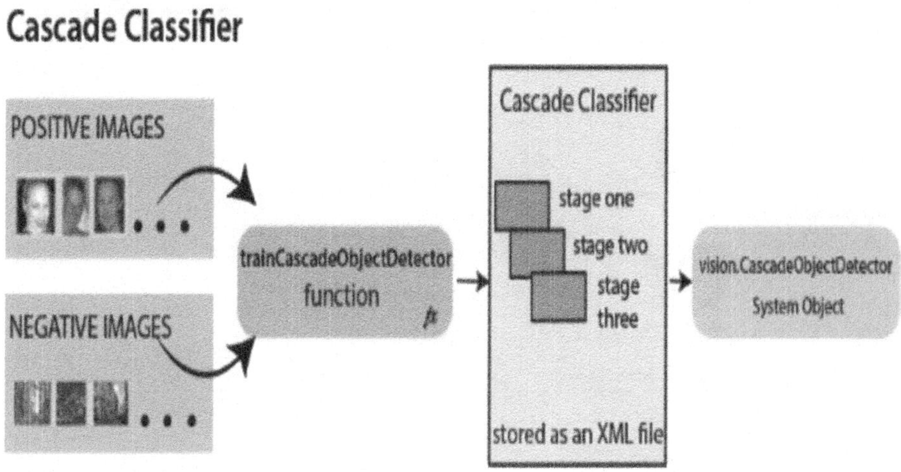

Fig. 3. Architecture diagram of OpenCV model

4 Proposed Work

4.1 System Architecture

In the context of human abnormal activity detection, data collection and integration is a crucial step in effectively monitoring and analyzing human behavior. The use of 3DCNN (three- dimensional convolutional neural network) and LSTM (long short-term memory) techniques enables the accurate identification of abnormal activities. By leveraging these techniques, data collected from various sources, such as video surveillance cameras, sensors, and wearable devices, can be integrated and processed. The integration of data allows for a comprehensive understanding of human behavior patterns and the identification of deviations from normal activity. This information can then be used to trigger email alerts and beep notifications, ensuring timely intervention and response to abnormal activities. Overall, data collection and integration play a fundamental role in enhancing human abnormal activity detection and improving the overall safety and security of environments. The two key features of the proposed system for detecting abnormal human activities using 3DCNN and LSTM are feature extraction and selection. Feature extraction involves extracting relevant information from the raw data, which in this case are video sequences. This is done using advanced techniques such as 3DCNN, which can effectively capture spatiotemporal patterns in the data. Feature selection is the process of selecting the most informative and discriminative features that are most relevant for detecting abnormal activities. This helps reduce dimensionality and improve the efficiency of the model (Fig. 4).

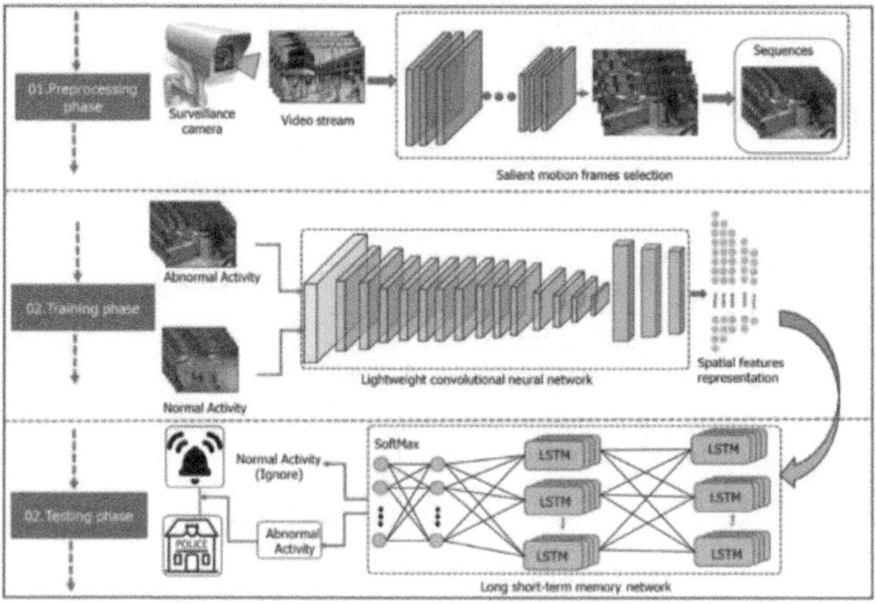

Fig. 4. System Architecture

4.2 Data Collection and Preprocessing

This module includes the collection of data sets from both online resources and real-time incidents using a video. These datasets are pre-processed based on size, quality, data magnification, normalization, etc. This data is labeled as "normal activity" or "abnormal activity" to identify whether the abnormal activity detected or not.

4.3 Model Creation

This model contains two separate models "3D CNN model" and "LSTM model" are created, which are combined into a hybrid model. The 3D CNN model is employed to learn informative spatiotemporal features from the video volumes. The learned features are subsequently fed into an LSTM network to model temporal dependencies. This enables the system to effectively capture the dynamic nature of human activities (Fig. 5).

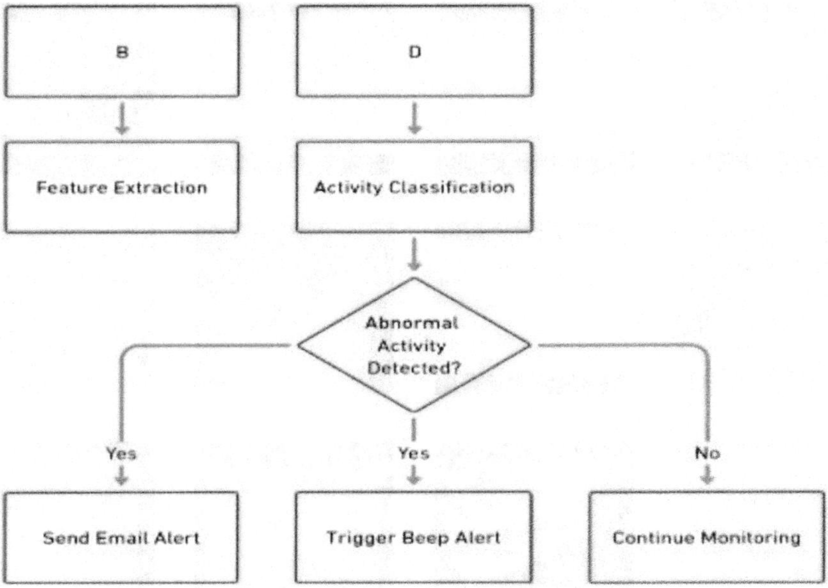

Fig. 5. Flowchart for CNN and LSTM model

5 Experiments and Results

5.1 Experiment Requirements

This project needs laptop with multicore central processing unit with minimum storage of 16 GB. A code editor or IDE such as Visual Studio Code, PyCharm etc. I have use windows 10 operating system, macOS or Linux operating system will also work for this project.

5.2 Dataset

The data set collected by the user for this project serves as the different object models. The image data given by the user is then compared with the Dataset and verify the normal incident or abnormal incident from the video surveillance (Fig. 6).

5.3 Training and Evaluation

Now we have the complete setup of the research. This setup is trained using the datasets which was collected at the start of the research. Testing of the model is done using real time videos to give an accurate prediction of the evaluation of the result of input footage (Fig. 7).

Fig. 6. Dataset footage

```
Epoch 1/5
45/45 [==============================] - 33s 273ms/step - loss: 1.7402 - accuracy: 0.2500 - val_loss: 1.4189 - val_accuracy: 0.5778
Epoch 2/5
45/45 [==============================] - 11s 245ms/step - loss: 1.1503 - accuracy: 0.5944 - val_loss: 0.4748 - val_accuracy: 0.8444
Epoch 3/5
45/45 [==============================] - 11s 236ms/step - loss: 0.6219 - accuracy: 0.7889 - val_loss: 0.3925 - val_accuracy: 0.7778
Epoch 4/5
45/45 [==============================] - 12s 265ms/step - loss: 0.4677 - accuracy: 0.8278 - val_loss: 0.4447 - val_accuracy: 0.7333
Epoch 5/5
45/45 [==============================] - 11s 255ms/step - loss: 0.2811 - accuracy: 0.9167 - val_loss: 0.1313 - val_accuracy: 0.9778
```

Fig. 7. Evaluation and accuracy

5.4 Experiment and Model

The deployed model of the project is a CNN model, in which the LSTM architecture helps us the reduce the training time of the model by 15% and gives an accuracy of 98.09%. The output of this system is predicting whether the accident included in the video surveillance is natural or artificial.

5.4.1 Streamlit Frontend UI

To know about the output we first pass the video footage inside the stream lit application to make the prediction (Fig. 8).

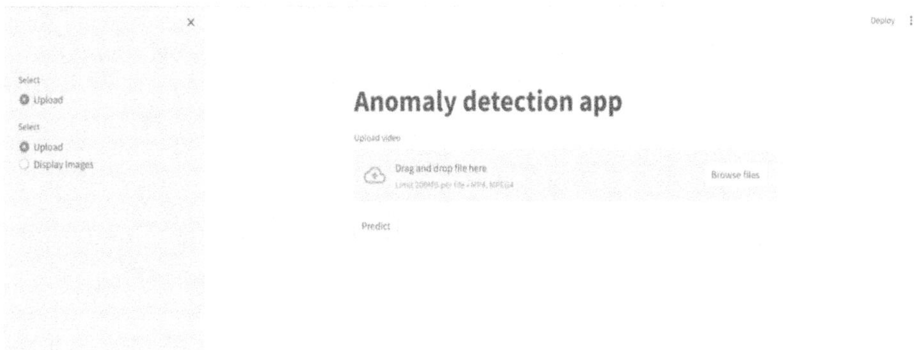

Fig. 8. Streamlit frontend UI

5.4.2 Explosion Video Footage

Then We Click on the Predict Button to Predict the Type of Anomaly Detected in the Video Footage (Fig. 9).

Fig. 9. Explosion video footage

Once the prediction is made the application notifies the user with an email and beep alert (Figs. 10 and 11).

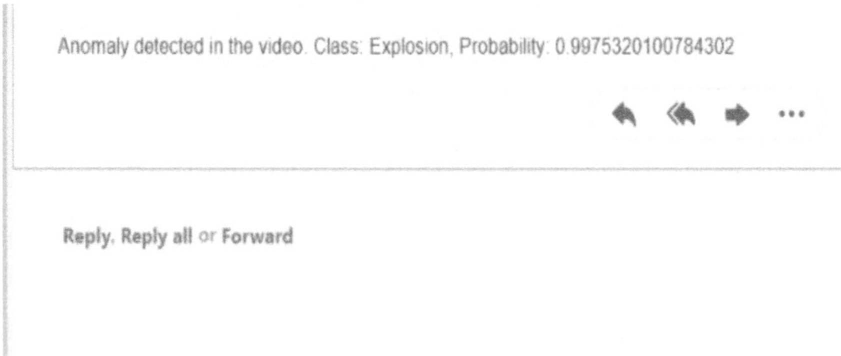

Fig. 10. Email alert for abnormal video footage

```
pred,prob = vid_class_pred(vid_path,CLASSES_LIST)
pred,prob

1/1 [==============================] - 0s 66ms/step

('Explosion', 0.9481985)
```

Fig. 11. Final result

6 Overall Model Statistaics

6.1 Total Loss

The total loss is the sum of the losses computed for each mini-batch during the training phase. It gives an overall measure of how well the model is fitting the training data. Mathematically, it can be represented as:

$$\text{Total Loss} = \sum\nolimits_{i=1N} L(y_i, y^\wedge i)$$

N- The number of mini-batches in the training data.
L- The chosen loss function.
yi - The true labels for the ith mini-batch.
y^i -The Predicted labels for the ith mini-batch.

6.2 Total Validation Loss

Similarly, the total validation loss is the sum of the losses computed for each mini-batch during the validation phase.

$$\text{Total Validation Loss} = \sum\nolimits_{i=1N} L(y_i, y^\wedge i)$$

N- The number of mini-batches in the training data.
L- The chosen loss function.
yi - The true labels for the ith mini-batch.
y^i -The Predicted labels for the ith mini-batch.

6.3 Total Loss vs Total Validation Loss

To derive a formula and theorem for Total Loss vs Total Validation Loss, we can start by defining some key terms and concepts:
Formula:
The total loss (TL) and total validation loss (TVL) are the summations of individual losses over all samples in their respective datasets (Figs. 12 and 13):

$$TL = \sum N_{traini=1} L_{train}(y_i, y^\wedge i)$$

$$TVL = \sum\nolimits_{i=1} N_{val} L_{val}(y_i, y^\wedge i)$$

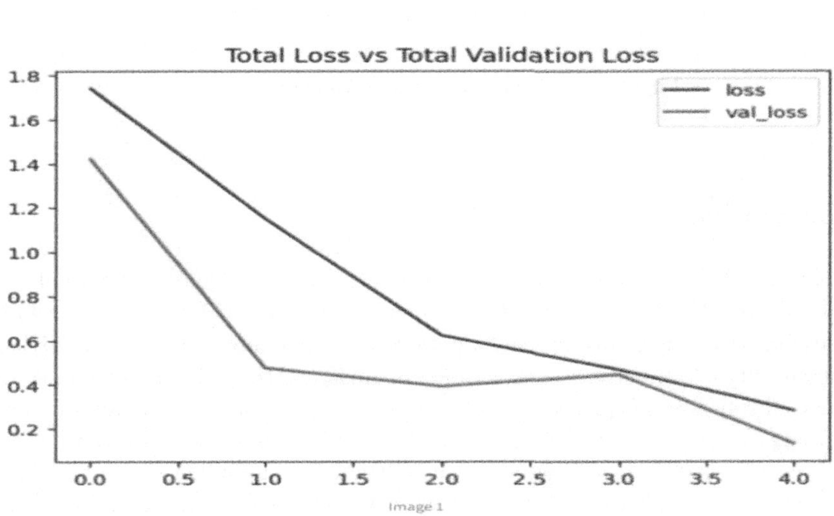

Fig. 12. Total Loss vs Total Validation Loss

6.4 Total Accuracy vs Total Validation Accuracy

The Total Accuracy vs Total Validation Accuracy concepts specifically for your project of human abnormal activity detection using video footage.

$$TA = \frac{\text{Number of correctly classified frame in training dataset}}{\text{Total number of frames in training dataset}} \times 100\%$$

$$TVA = \frac{\text{Number of correctly classified frame in validation dataset}}{\text{Total number of frames in training dataset}} \times 100\%$$

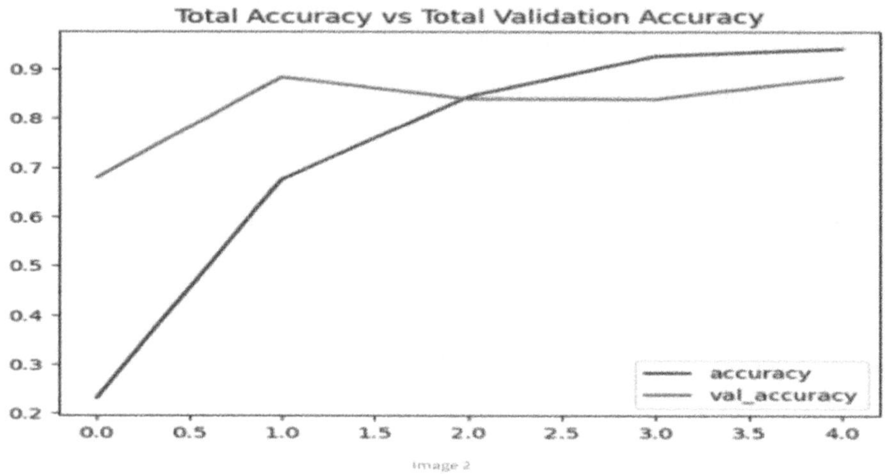

Fig. 13. Total Accuracy vs Total Validation Accuracy

7 Conclusion

In conclusion, this research paper presents a novel approach for human abnormal activity detection using 3DCNN and LSTM. The integration of these deep learning techniques enables the system to capture spatiotemporal features and model the temporal dynamics of human activities effectively. The addition of email and beep alert notifications further enhances the system's capabilities by providing real-time alerts to the relevant authorities. The proposed approach shows great potential for application in various domains, including safety and security monitoring, healthcare, and surveillance systems. By leveraging the power of advanced deep learning techniques and real-time notifications, we can improve the detection and response to abnormal activities, ultimately creating safer and more secure environment.

Improve the system's ability to understand how people move by using techniques that recognize detailed aspects of their posture or gestures. Allow users to talk to each other through the system in real time. This can help people make decisions more quickly, especially in emergencies. Include different sensors like sound or environmental sensors to gather more types of information. This can help the system better understand what's happening in a given environment. If the alarm is near the accident area may automatically this information may sent to the police department and also the hospital department.

References

Siddiqui, S., Ahmad, A., Varshney, A.: IoT based human activity recognition using deep learning. EAI Endorsed Trans. Context- Aware Syst. Appl. (2023)

RM, V., KC, R.: Advanced ATM robbery prevention using neural network and IoT technologies (2021)

Biswal, A.K., Singh, D., Pattanayak, B.K., Samanta, D., Yang, M.H.: IoT-based smart alert system for drowsy driver detection. Wirel. Commun. Mob. Comput. **2021**, 1–13 (2021)

Chokri, R., Hanini, W., Daoud, W.B., Chelloug, S.A., Makhlouf, A.M.: Secure IoT assistant-based system for Alzheimer's disease. IEEE Access (2021)

Kolkar, R.: Human Activity Behavioural Pattern Recognition in Smarthome with Long-hour Data Collection. arXiv preprint arXiv:2306.1337 (2023)

Nguyen, Q.T., Tran, T.N., Heuchenne, C., Tran, K.P.: Decision support systems for anomaly detection with the applications in smart manufacturing: a survey and perspective. In: Machine Learning and Probabilistic Graphical Models for Decision Support Systems, pp. 34–61. CRC Press (2022)

Phan, A.C., Trieu, T.N., Phan, T.C.: Driver drowsiness detection and smart alerting using deep learning and IoT. Internet of Things **22**, 100705 (2023)

Saleem, A.A., Hassan, M.M., Ali, I.A.: Smart homes powered by machine learning: a review. In: 2022 International Conference on Computer Science and Software Engineering (CSASE), pp. 355–361. IEEE (2022)

Bushra, S.N., Sibi, S.A. Vijaya Kumar, K., Niveditha, M.: Predicting anomalous and consigning apprise during heists. In: 2023 International Conference on Artificial Intelligence and Knowledge Discovery in Concurrent Engineering (ICECONF), pp. 1–8. IEEE (2023)

Taiwo, O., Ezugwu, A.E., Oyelade, O.N., Almutairi, M.S.: Enhanced intelligent smart home control and security system based on deep learning model. Wirel. Commun. Mob. Comput. **2022**, 1–22 (2022)

Campus Drive Portal on Career Advancement for College Students

S. Nandhini(✉), P. Praganya(✉), V. Karthick(✉), and V. Ariyamala(✉)

Department of Computer Science and Engineering, Rajalakshmi Engineering College, Chennai, Tamil Nadu, India

{200701156,200701173}@rajalakshmi.edu.in, vkarthick86@gmail.com, veerariya2016@gmail.com

Abstract. The Campus Drive Portal is a user-friendly website designed to assist both students and Placement Admins in managing campus recruitment events. This application provides the facility to view upcoming campus drives. The two users of our application are Student and Placement Admin. From Placement Perspective, the portal simplifies the management of student information for upcoming drives. This includes creating, updating, viewing, and deleting student records, with the ability to filter based on criteria set by companies. Admin can update the details like Company's name, drive date and eligibility criteria for the upcoming placement drives. The portal also incorporates a prediction tool to assess a student's chances of securing placement and further the trend analysis is used to help students by providing better training programs. The placement administrator has the facility to collect the offer letters from the placed candidates, which will be helpful in storing the details and analyzing the trends. From a Students' Perspective, Students can receive information regarding the upcoming drives. After interviews, students could contribute valuable feedback about their experiences with companies. This feedback not only benefits the individual student but also aids the placement department in making informed decisions about future collaborations with companies. The system further includes a prediction tool that allows students to evaluate their interview preparedness and offers guidance on improving their skills. Students can also gain valuable insights from their seniors about different interview rounds, promoting a collaborative learning environment.

Keywords: Campus · Placement · Feedback · Review · Web application · Prediction · Chatbot

1 Introduction

Over the last few years, the demand for recruiting skilled graduates has increased tremendously. Higher educational institutions, such as Colleges and universities, play an important role in enhancing the career development of the students. Placement offers students an important pathway to move forward and achieve their goals. The concept of campus placements at colleges and universities is a big help for students starting their career. The

Published by Springer Nature Switzerland AG 2024
M. L. Owoc et al. (Eds.): ICCIDS 2024, IFIP AICT 718, pp. 210–223, 2024.
https://doi.org/10.1007/978-3-031-69986-3_15

digital age has revolutionized how information is managed and shared. Acknowledging the increased competition for jobs highlights the essential need for a centralized platform for placement activities. This website aims to be the solution, making information easy to access in the future. It also helps to organize all placement related activities in a single website so that the task of accessing the information in the future will be easier. It links students directly to job opportunities that align with their college education and future career aspirations. It connects students directly with job opportunities that match what they have learned in college and what they want to do in the future. Campus placements are more than just a convenience, they are like a helpful guide for students to reach their career goals.

The proposed web application will be useful for the students as well as the admin. For students, the website offers a user-friendly platform to access information about upcoming campus recruitment, apply to companies, and gain valuable insights from seniors' feedback about interview process. The ability for students to share their own feedback creates a collaborative environment, benefiting future students. On the other hand, Administrators gain a streamlined approach to managing student details. The admin possesses various functions such as creating, updating, viewing, and deleting student details, simplifying the management of the student database. Additionally, the admin can post information about upcoming job drives, enabling students to register for them. Moreover, the administrator can sort and filter the student database based on eligibility criteria specified by specific companies. The admin has the facility to verify the originality of the certification by using verification methodologies. In the placement prediction process, models are trained using different machine learning algorithms. This process helps in assessing the probability of final-year students receiving job offers. By analyzing the number of students who secured jobs and those who did not in a particular year and comparing it with previous years, trends and areas for improvement are identified.

Additionally, to enhance user interaction, the web application integrates a chatbot option. This feature provides students with a conversational interface to address queries, offer guidance, and streamline the overall user experience. The chatbot serves as a valuable tool for providing real-time assistance and contributing to more accessible and user-friendly platform.

2 Related Works

Jadhav and Gujar [1] proposed a methodology to check how well the students are prepared for placements. The system will estimate the salary package of the student based on their eligibility. The system helps students to check and evaluate their level of preparation for placement activities. This will help students to boost their preparation level to secure their dream job. The system focuses mainly on the prediction part and does not focus much on review and feedback system. The system helps to predict the preparation of students for the placement activities. The system focuses on prediction and does not focus on registration formalities.

Kumar, Swetha and Sundar [2] proposed an application that concentrates on developing a secured web application to stay connected with alumni. Alumni helps to guide students in the right direction. They provide information regarding the various opportunities available in their field so that students can enroll themselves and participate in the further process if they are interested. The secure web application helps alumni to register with the college and provides information about the placements and internships which will be a helpful source for the students. The web application focuses on providing information regarding placement activities and lacks the further process involved in placements.

The work suggested by Sehgal [3] is a job portal for multiple clients. The administrator can add a new company, a new user, provide username and password, delete, or edit existing accounts and chat with other members. The company has the facility to edit their account, chat with the administrator and other members, view their profile and upload the documents which can be viewed by the job seekers. The users have the facility to view the profile of the company, view the list of companies, download the files from the company's profile and contact the companies directly for further information. Though the application has three users, it does not provide the opportunity to provide feedback about the interview process. The feedback system helps the users to know about the topics focused in the interview so that they can prepare well and crack the interview to get placed in the company.

The aim of the system proposed by Karthick and Adithya [4] and it focuses on building an efficient platform for employee management system. The system focuses on introducing an efficient algorithm to enhance security. It deals with the comparison of the efficiency of the new algorithm with the existing Diffie Helman's Key Exchange algorithm. The system focuses on the employee management system. The system can add on the placement-related activities facilities so that it will be helpful for the students. The system deals with usage of the latest technologies to enhance security. The system concentrates on developing a secure and cost-effective platform and lacks placement-related features.

The system developed by Gupta [5] mainly focuses on prediction of placement activities. It is a web-based application which could be used for the process of recruitment. The application can be utilized by the companies for examining whether the candidate is capable for the job. The system concentrates more on the prediction part and does not focus on the upcoming drive features. The system lacks concentration regarding the feedback system which helps the candidate to get to know about the topics focused in the interview.

The application developed by Karbhari [6] focuses on the recommendation system. The recommendation system was built using the concept of content filtering. The system offers recommendations based on the information collected in the software engineering process. The model finds the most suitable solution which could be selected by the student based on his capabilities. The system focuses on finding the best company suitable for the student in a short span of time. Though the system has recommendation features, the application lacks features like a feedback system. It does not provide a single website to focus on all placement related activities. It focuses only on the recommendation system.

The work suggested by Banu, Alfiya, and Dr Manju Bargavi [7] aims to develop a web-based application for a college's placement management system, primarily on the Windows platform. This application aims to store and manage student information, including personal details and technical skills, which can be accessed by the companies for recruitment purposes. It also helps students see their job options. Students can apply for jobs, take tests, or attend events. This application has four users namely Students, Training and Placement Officer (TPO), Admin and HOD. The TPO is responsible for reporting placement-related activities to students. However, this system does not provide opportunities for students to share their feedback.

The objective of the project proposed by Venugopal, Akshay [8] is to create a web-based placement management system for the college. This system has three modules namely Student, Admin and Company. Students can upload their resume on the website and can view the company details and register for the company. The company has a registration profile and maintains the profile details and views the student details. Admin can view the student and company details and send mail to students. This system does not include a feedback feature. They are using MySQL as a database which is not efficient.

The Campus Recruitment System proposed by Prof. Mrs. D. R. Kamble [9], is a web-based software application designed to bridge the gap between students and companies during the recruitment process. Students can register their information, including skills and resumes. Companies can post job requirements and define their application procedures. This application is developed by using HTML, CSS, and JavaScript for front-end. PHP framework and MySQL is used for backend, which is not efficient as compared to MongoDB. This system mainly focuses on posting the job details. It does not provide a feedback feature which is helpful for students.

The project suggested by Santhosh Kumar H, Srividhya V R [10] aims to automate the Training and Placement process at AMC Engineering College, replacing the manual system with a web application accessible to students and administrative staff. Admin can update the student details including batch and courses. Admin can also filter and edit the student database. In the HR module, HR can add company details and assign a company. In the Student module, Students can fill the application form and download it. It also contains the students' details like their Aadhaar card number, PAN number and personal details. This system includes unnecessary features that could complicate its use. It lacks the necessary features like feedback and review.

3 Survey

A survey was conducted among the students and professionals. Google forms is used as a platform for conducting the survey. The questions posted were multiple choice questions which helps to study the perspective of students and professionals. There were 92 responses, and their responses were explored to get insights from it. 84% of the respondents in the survey were UG students. Overall, 81% of the respondents feel placement as their next plan after graduation.

66% of the respondents feel that the college generally does not have a separate portal for placement related activities is denoted in Fig. 1. The aim of the proposed system is to develop a single web application which deals with all placement related activities. The task of applying for jobs and getting information regarding the interview process from their seniors will also be easier.

According to Fig. 2, 85% of the respondents expressed their opinion as the feedback from their seniors regarding interviews will be beneficial for them. The aim of including

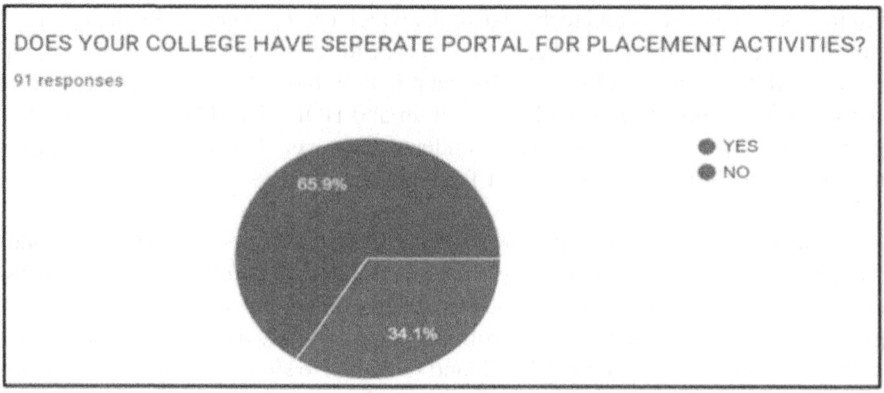

Fig. 1. Separate portal

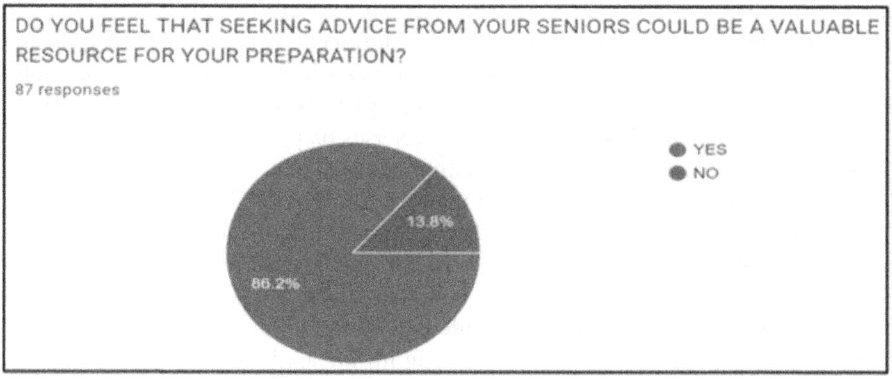

Fig. 2. Feedback system

this feature in the application is to help the candidates to prepare well based on the topics focused in the interviews. This feature helps the candidates to achieve their goal by preparing the specified sample topics which are shared by the seniors based on their experience.

According to the views of 89% of the respondents represented in Fig. 3, the existing placement management system does not have a review system where the seniors could share their guidance and advice regarding the placement preparation. The suggestions from their seniors will be one of the useful sources for interview preparation.

The proposed system includes a feedback and review system. This mechanism serves as a guide for the students to prepare and equip themselves to use the various opportunities and crack the interview. The system also has notification facility. This facility helps to update the user when they.

have applied for a particular company, a new company is posted in the portal, the interview process schedule and any other changes regarding the process as updated by the officials of the company.

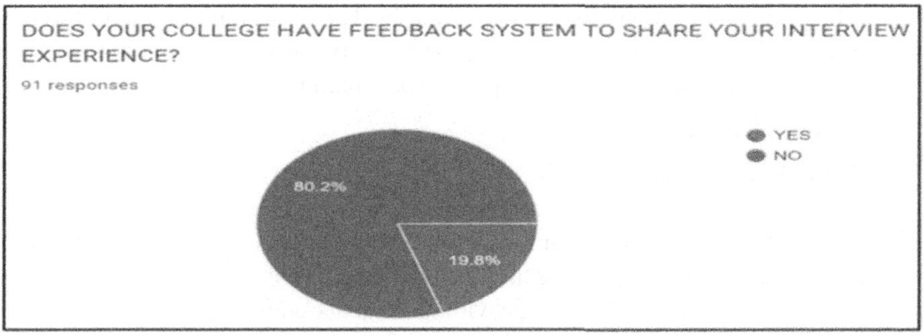

Fig. 3. Review system

4 Statistical Analysis

Statistical analysis is the process of collecting data from various sources and analyzing the data to discover the trends and patterns. Figure 4, represents the plans of students after graduation. Most of the candidates opt for placements and only few chose higher studies and other plans. It tells us about the importance of developing a separate portal for placement related activities. Dependent parameters are the metrics which are affected by the factors within the system. The parameter number of features will be increased in the proposed system since the current placement system focuses highly on the upcoming placement drive events and does not concentrate much on the feedback system. The proposed system concentrates on updating the upcoming placement drive details and feedback system. The feedback system will be useful for the future candidates who will be attending the interview since they can get information regarding the topics focused in the interview from past candidates. The duration of students using the web application will also be maximized since all the features regarding the placement activities are integrated in the single application.

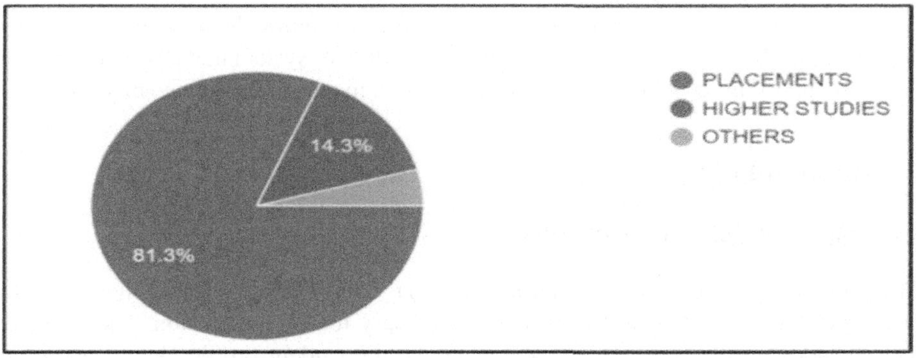

Fig. 4. Plans after graduation

Independent parameters are the metrics which are not affected by the factors within the system. The user experience was not satisfactory in the existing placement portal system. The proposed system will be user-friendly. The engagement time in the portal by the user will also be increased. The parameters like profile information, skills and interests also fall under the category of independent parameters.

5 System Architecture

The system architecture of the Campus Drive Portal is designed for simplicity and effectiveness. Figure 5 represents the architecture diagram of the proposed project. It is structured to create a secure platform providing students with job opportunities.

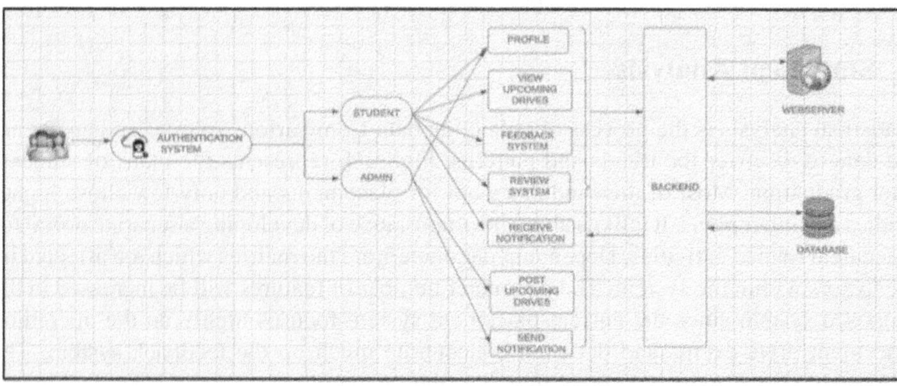

Fig. 5. Architecture Diagram

It is a well-organized website with two users, namely students and administrators. It is easy to use because you can access it on a computer or phone. A web server serves as the central processing unit, managing user requests, data retrieval, and database management. The database acts as a digital repository, storing Student records. This system ensures collecting feedback from students and sending them notifications when needed. Overall, it is a user friendly, secure, and efficient system that connects students with job opportunities while helping administrators manage everything smoothly.

6 Methodology

6.1 Data Collection and Validation

The initial phase of developing a web application is data collection. Data collection is the process of fetching relevant data which is necessary for the application. The collected data includes details about the student to create the student profile, details about the campus drives, feedback records which will be useful for the future candidates. The next process after collecting the data is to validate the data. It includes the process of examining the data about the student records, campus drive details and feedback records. This process helps to ensure the accuracy and relevancy of the data.

6.2 User Authentication and Role Based Access

User authentication is the process of validating the identity of the user who is attempting to use the application. It helps to enhance the security of the system. In this system the secure user is verified, and access is provided based on the specific roles of the user. The portal can be accessed by both students and admins, each with their distinct login credentials. After successful login, the user can access the information sections in the portal based on their appropriate roles.

6.3 Admin Dashboard

After successful authentication, the administrators can manage the profiles of the student. The administrator has the access to create, modify or update, delete the records of the students. The administrator has an important role in posting all the necessary information regarding the upcoming placement drives. The admin must collect the feedback form from the students who have attended the interview to get information regarding the needs of the company and train the upcoming students accordingly. The admin has the facility to collect the offer letter of the selected candidates to store in the database. The admin must send notifications to the students once a new campus drive is posted in the portal so that students will be able to login to the portal and apply for the jobs.

6.4 Student Interface

Once the student is authenticated, access is provided to the student to use the portal. The students can view the details of the upcoming placement drives. The student can check if he/she is eligible by checking the requirements for the company and if eligible the student can apply for the job. The students can view their profile information. The students can provide feedback once they attend an interview, which will be helpful for the placement department. The students can also gather information from their seniors to get to know about the techniques to prepare for the interview. The students are also notified once a new job is posted on the portal which will be helpful for them to apply for the same.

6.5 Placement Prediction and Trend Analysis

This module focuses in analyzing the trends in placement and predicting the placement probability of a student. First the collected data is pre-processed. The model is trained with previous year datasets using different machine algorithms like regression, svm, random forest based on the nature and size of the dataset. The current final year students' datasets are tested to find the chances of the student getting an offer. The next step is to calculate the number of placed and unplaced students in a particular academic year. The data is compared with previous year placement to find the trends and identify the areas to be concentrated on to increase the chances of getting an offer. This model will be helpful for the students as well as administrators. The students can use this to improve their skillsets to match the requirements of the company. The management can use this to analyze the trends and provide students with appropriate training to help the students to get a job. The system has an option to verify whether the certificate is original, and no illegal activity is involved. In addition to this a chatbot system could be incorporated which will be a helpful source for providing recommendations and assist students by providing updates regarding the placements.

6.6 Feedback and Review Mechanism

The portal facilitates a user-friendly feedback mechanism. The feedback mechanism allows the students to share their feedback regarding the interview they have attended. The feedback system promotes the growth of the students. The feedback is collected from the students in order to analyze and evaluate the information to identify the areas of improvement which could be implemented in the future. The review system is designed to help the students to get to know about the format of the interview. The reviews are provided by the seniors which could be used by the students to gather information regarding the topics focused in the interview so that the student can prepare accordingly to crack the interview and get placed in the company.

7 Results and Discussion

In today's world, the campus drive portal is making a big difference. The portal is beneficial for the students to find job opportunities that match their profile. It is a user-friendly website where students can see upcoming campus drives and apply for the company. They can share their experiences and tips with other students. The proposed application will be useful for both students and admins. From admin perspective, it serves as a helpful tool to keep student information organized and improve the job placement process.

Table 1. Placement Percentage

YEAR	PERCENTAGE
2018–2019	71.76
2019–2020	87.68
2020–2021	95.34
2021–2022	99.21

The Table 1 represents the placement percentage of students in the corresponding year. This data help us to know more information about the number of students. It can be further classified based on departments to understand well in which area the candidates must concentrate to develop their skills.

Figure 6 depicts the trends in placement which will be a useful tool for the administrator to analyze the values to find the weaker areas and concentrate on the skillsets accordingly.

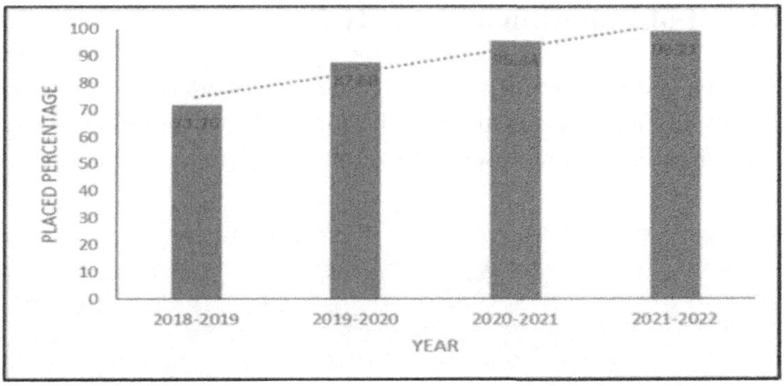

Fig. 6. Placement trend analysis

8 Conclusion and Future Works

The Campus Drive Portal serves as a user-friendly online platform for both students and administrators. It simplifies the process of connecting students with job opportunities and assists administrators in efficiently managing student information. By centralizing and digitizing placement-related activities, the portal helps institutions save time and resources. For students, it represents a unified space where they can easily find job opportunities that are happening on their campus. Additionally, it functions as a knowledge-sharing platform where students can also get advice from their seniors who have been through the same process. Students are further encouraged to share their own tips and experience to help other students in the future. On the Administrative side, this portal simplifies the complex task of managing student information. The system also helps in verifying the originality of the certificate. Admin can easily communicate important details about the upcoming career-related events and can also categorize students based on specific company requirements, making it much easier to match students with the right job opportunities. This website shows how technology can make it easier for students to find a job. In today's tough job market, websites like the Campus Drive Portal can help students find better career opportunities. The integration of predictive placement technologies further enhances the portal's utility. This tool helps to understand job trends and guess how likely students are to get job offers. By comparing with past years, it finds trends and areas to be improved. In the future, adding a very effective chatbot to this system could be helpful. The chatbot could give advice, share updates, and be a useful guide for students trying to figure out the job process. Other updates that could be done are adding skill-based assessments methodologies in order to assess and confirm whether the candidate has uploaded the appropriate skill in their profile. It helps the students to understand their weak points and try to improve those skills in a much effective manner.

9 Screenshot of Experimental Analysis

The Campus drive portal plays a vital role to reduce the work of the placement officers and the students can make use of the website to get all the information regarding the placement activities. Figure 7 represents the sample design of the login page of the portal.

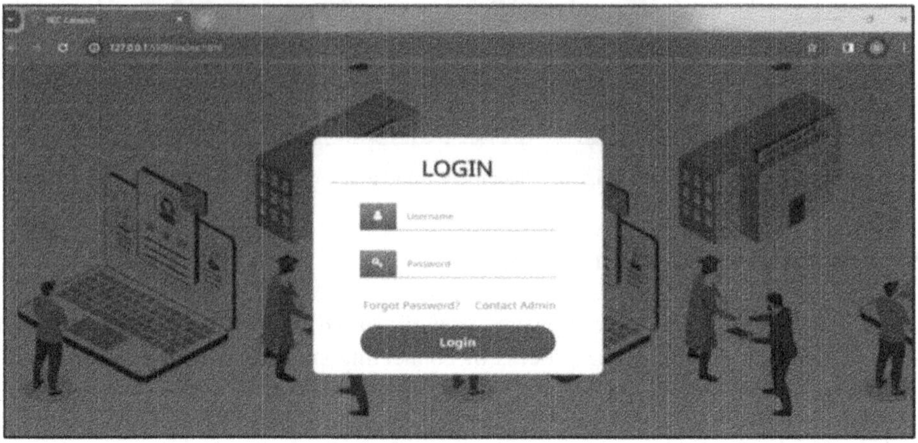

Fig. 7. Login Page

Figure 8 denotes the home page of the admin. The list of upcoming drives, details of the students, events and feedback option is displayed in the home page.

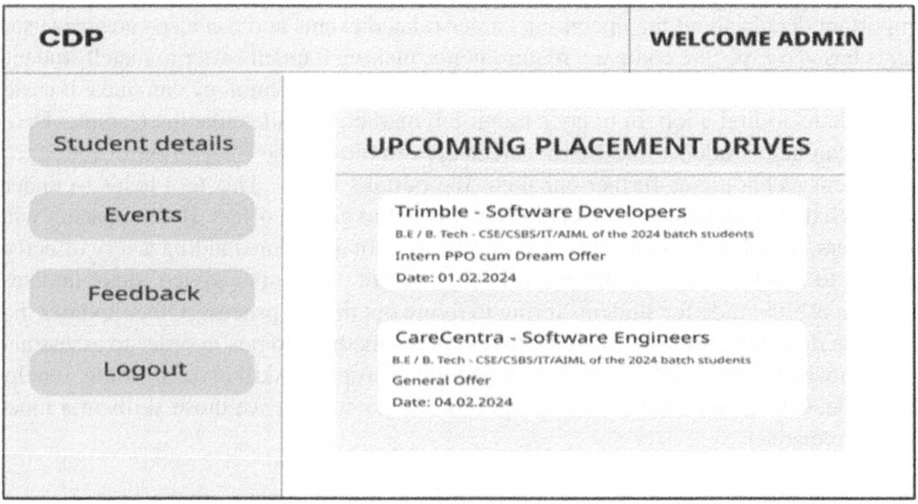

Fig. 8. Admin home page

Figure 9 denotes the home page of the student. The list of upcoming drives, profile information, companies list, feedback and review option are displayed in the home page.

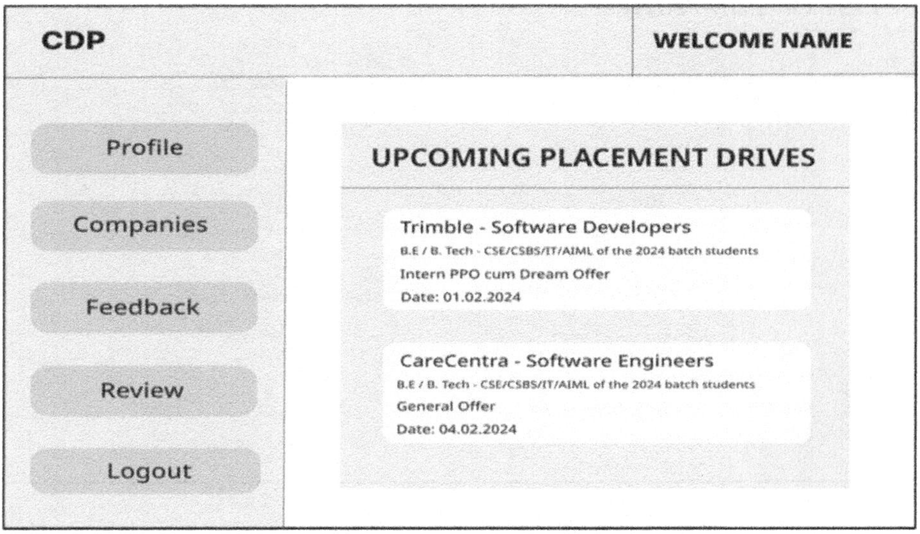

Fig. 9. Student home page

Figure 10 denotes the list of companies' page of the student. The page contains information about the company, role, job description, eligibility and register option so that student can register for the interview process.

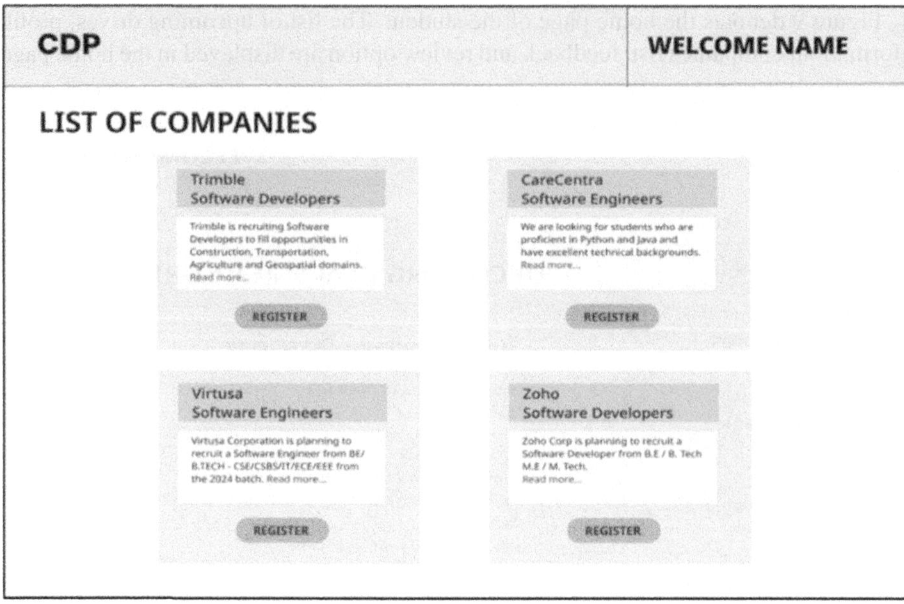

Fig. 10. List of companies

Acknowledgement. Authors are involved in methodology, text analysis, review, editing, supervision, validation and writing manuscript.

Disclosure of Interests. No conflict of interest in this manuscript.

References

1. Jadhav, S., Gujar, G., Shinde, G., Shivane, S., Jadhav, D.S.: Placement readiness check: predicting placement status using supervised ML methods. In: 2022 2nd International Conference on Intelligent Technologies (CONIT), Hubli, India, 2022, pp. 1–7 (2022). https://doi.org/10.1109/CONIT55038.2022.9847863
2. Kumar, P., Swetha, S., Sundari, M.: Secured web-based alumni network and information systems. In: 7th International Conference on Intelligent Computing and Control Systems (ICICCS), Madurai, India, 2023, pp. 1427–1434 (2023). https://doi.org/10.1109/ICICCS56967.2023.10142761
3. Sehgal, V.K., Jagtiani, A., Shah, M., Sharma, A., Jaiswal, A., Mehta, D.: Job portal - a web application for geographically distributed multiple clients. In: 2013 1st International Conference on Artificial Intelligence, Modelling and Simulation. IEEE (2013). https://doi.org/10.1109/aims.2013.38
4. Kakaraparthi, A., Karthick, V.: A secure and cost-effective platform for employee management system using lightweight standalone framework over Diffie Hellman's key exchange algorithm. ECS Trans. **107**(1) (2022)
5. Gupta, S., Hingwala, A., Haryan, Y., Gharat, S.: Recruitment system with placement prediction. In: 2021 International Conference on Artificial Intelligence and Smart Systems (ICAIS). IEEE (2021). https://doi.org/10.1109/icais50930.2021.9395768

6. Karbhari, N., Deshmukh, A., Shinde, V.D.: Recommendation system using content filtering: a case study for college campus placement. In: 2017 International Conference on Energy, Communication, Data Analytics and Soft Computing (ICECDS). IEEE (2017). https://doi.org/10.1109/icecds.2017.8389579
7. Banu, A. Manju, B.: A research on placement management system. Int. J. Res. Appl. Sci. Eng. Technol. **10**(4), 1853–59 (2022)
8. Venugopal, A.: Online placement and recruitment system. Int. J. Adv. Comput. Sci. Technol. **9**(6), 25–29 (2020)
9. Kamble, M.D., Revankar, R., Shetti, R., Mulla, R., Pawar, V.: Campus Recruitment System. Int. J. Adv. Res. Sci. Commun. Technol. (2022). July, 12–16
10. Santhosh Kumar, H., Srividhya, V.R.: Online training and placement management system. Int. J. Eng. Res. Technol. (IJERT) ICACT (2016)

Deception Detection Using Random Forest and KNN

S. Tilak Chander, S. Devi, A. G. Harshavardhan(✉), and S. Darun Sanjay

Coimbatore Institute of Technology, Coimbatore 641014, India
{2007051it,devi.s,2007018it,2007009it}@cit.edu.in

Abstract. Every person lies to try and escape or avoid some situation. Lying is considered to be a form of deceit and can have a very negative impact in today's judicial world. Criminals, employees, shoplifters and many other people lie to get around the punishment. These must be detected so that innocent people are not put in adverse situations and also to catch real criminals. Conventional methods for detecting dishonesty entail the use of scientific methodologies to examine physiological signs, transcripts, as well as visual and audio data. Nevertheless, there has been a dearth of exploration into the utilization of modalities like EEG data. Despite years of research, scientists have consistently demonstrated that people's ability to discern deception is no better than random chance. Precision in detecting deception holds paramount importance for law enforcement authorities. In the proposed model, brain waves and audio cues will be combined to enhance the existing deception detection. Existing datasets like 'Bag-of-Lies' will be utilized. This system is poised to make a substantial contribution to the Police Department, Criminal Investigation Department, and Judicial System by reducing the burden on law enforcement officers and aiding in the accurate identification of the actual perpetrator.

Keywords: deception detection · EEG waves · Audio cues

1 Introduction

Research on Deception-Detection has been going on for many years and this field still has much to offer. As we progress towards more advanced developments, we tend to get more and more accurate. Recognition of Micro-Expressions, analyzing insignificant vocal cues, and capturing waves from our brain was not nearly possible just a few years back. Because of the growing technology, capturing Electroencephalogram (EEG) waves has become easier and more sophisticated. There have been methods proposed to implement devices to detect deception at Airport checkpoints and border crossings [1], but these systems have not performed well due to the limited training datasets and other factors.

Deception is very crucial in today's world. From criminals who lie to get out of punishment, to scammers who lie and take advantage of innocent people. Deception detection techniques need to prevail and succeed in differentiating between criminals

Published by Springer Nature Switzerland AG 2024
M. L. Owoc et al. (Eds.): ICCIDS 2024, IFIP AICT 718, pp. 224–233, 2024.
https://doi.org/10.1007/978-3-031-69986-3_16

who are guilty, and innocent people who tell the truth. Traditionally, studies investigating deceptive behavior have employed a range of opinion videos and videos featuring higher-stakes criminal scenarios [2, 3].

The traditional method of lie detection, polygraphy, has several limitations. Recent research has suggested employing neuroimaging methods like Event-Related Potential (ERP) and Functional Magnetic Resonance Imaging (fMRI) to directly examine cognitive activity in the brain during deception [4–6]. This problem must be addressed. Compared to polygraphy, these methods offer several advantages, such as increased accuracy and the ability to detect deception within a very short period. Furthermore, previous studies have shown that ERP-based methods are highly efficient in lie detection. However, EEG-based lie detection requires the extraction of ERP features from multiple scalp sensors, which can be challenging. Nonetheless, the potential of neuroimaging techniques in improving the accuracy of lie detection makes them a promising area of research.

Functional Connectivity (FC) and Effective Connectivity (EC) are two separate types of brain connectivity [7]. While EC measures how one brain region influences another in a certain manner, FC denotes the correlations in activity between various brain regions. The causal factors behind brain function are better understood with EC, making it a more useful tool for understanding brain activity during certain tasks. Various metrics, such as the directed transfer function and partial directed coherence, have been introduced for the assessment of EC. Characterizing both the spectral and directional features of brain connection, a topic that has received little attention in deception-related research, is crucial to developing a thorough knowledge of the neurological mechanisms underpinning deception.

The main contributions include:

- Utilizing various machine learning techniques to assess the suitability and dependability of these algorithms for detecting deception.
- Exploring more features on EEG waves and Audio.
- Comparing our results to the most recent cutting-edge machine learning-based detection technologies.
- Indicating that machine learning-based deception detection methods have yet to reach a level of reliability and applicability considered suitable for a wide range of contexts.

2 Related Work

In their study, Gupta et al. [8] introduced the Bag-of-Lies dataset, which consists of multiple modalities including video, audio, EEG, and gaze data to aid deception detection. The dataset was collected in a real-life setting and combines both visual and cognitive aspects of deception. The researchers also explored the advantages of combining different modalities for better performance on the Bag-of-Lies dataset.

Ding et al.'s [9] innovative method for multimodal deception detection successfully combines informative information from the face and body modalities. The researchers introduce a distinct Face-Centric Cross-Stream Network (FCCSN) crafted to explicitly grasp facial expressions and promote the acquisition of deep features that encompass both facial and bodily elements, allowing for correlation learning across spatial and temporal streams. Meta-learning and adversarial learning techniques are combined to train this model. The experimental results unmistakably show that the FFCSN model they suggest beats current approaches in deception detection, achieving cutting-edge outcomes. Furthermore, the authors show the adaptability of their FFCSN model by using it for additional human-centric video analysis tasks, like user-generated video emotion recognition.

Belavadi et al. [10] conducted a study on multimodal deception detection, investigating the accuracy, applicability, and generalizability of different techniques. The authors utilized a dataset consisting of video, audio, EEG, and physiological signals collected in a real-life scenario. They compared the performance of machine learning-based techniques on individual modalities versus the combined multi modal dataset. The study concluded that combining multiple modalities resulted in improved accuracy and applicability, but the generalizability of the methods was still limited.

The study by Gao et al. [11] explores the efficacy of cortical network connectivity in deception through the analysis of EEG signals. The authors aimed to explore the neural mechanisms of deception and improve the accuracy of deception detection. The study analyzed the Directed Transfer Function (DTF) and Partial Directed Coherence (PDC) to estimate the EC among different brain regions. The results showed that effective connectivity increased during deception, with more connections among prefrontal and temporal regions. The investigation also unveiled the pivotal involvement of the prefrontal cortex in the processing of deception.

The study by Krishnamurthy et al. [12] presents a novel approach for detecting deception in multimedia data using deep learning techniques. The authors propose a multi modal fusion approach that combines visual, textual, and acoustic features to identify deceptive behavior. The proposed approach achieves outstanding performance on several benchmark datasets, including the Multimedia Opinion Suicidal Ideation Detection (MOSI) dataset and the Deceptive Opinion Spam Corpus (DOSC) dataset. This work contributes to the development of more effective techniques for deception detection, which can be applied in various domains, including cybersecurity, law enforcement, and social media analysis.

Another study by Gao et al. [13] was on the use of EEG phase synchrony analysis for lie detection. The authors propose a method for detecting deception by analyzing dynamic functional connectivity patterns in the brain using EEG data. The suggested method is founded on the principle of brain fingerprinting, which uses EEG data to identify unique patterns of brain activity associated with specific information or stimuli. The authors conduct experiments with a sample of participants and demonstrate that their approach can accurately detect deception with high accuracy. This work contributes to the development of more effective techniques for lie detection using EEG data, which can have important applications in various fields, including law enforcement and security.

The study by Tran et al. [14] proposes an automated deception detection system that utilizes computer vision and machine learning techniques. The system focuses on analyzing eye movements and facial expressions of a deceiver to identify deceptive behavior. The system under consideration employs a blend of feature extraction methods, which encompass the use of Gabor filters and Local Binary Patterns (LBP), to derive characteristics from facial images. These derived features are subsequently input into a machine learning model, specifically a Support Vector Machine (SVM), for the purpose of classification. An evaluation of the system is conducted using a dataset containing video recordings of individuals engaging in deceptive activities. The outcomes of the assessment indicate that the proposed system can proficiently identify deceptive behavior with a notably high degree of precision.

3 Method

(See Fig. 1).

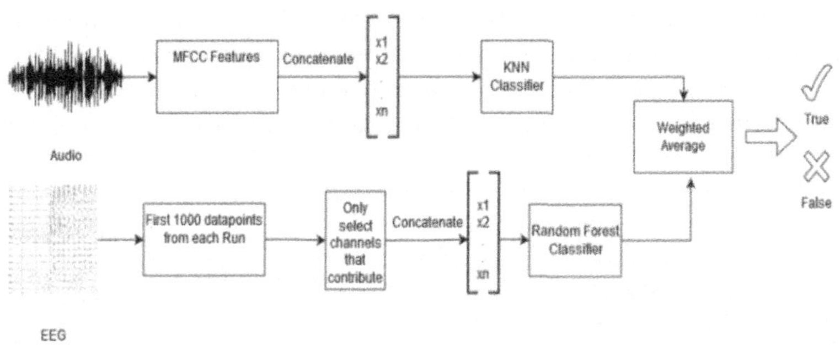

Fig. 1. Proposed system architecture

The bag of lies dataset contains data from 36 users, with each user having undergone multiple runs, during which they were subjected to the possibility of either truth-telling or falsehood. The Audio data was extracted from the Videos, and the EEG data was presented in a csv file. Subsequently, after the relevant data points were concatenated into a single file along with prediction values, the model was trained. KNN was employed for audio data, while Random Forest was applied to EEG data. The final output was calculated through the weighted average method.

3.1 Deception Detection Using EEG Waves

In this research, a dataset consisting of multiple CSV files containing unprocessed EEG data recorded from 13 channels was utilized. This dataset had been collected from participants who were involved in a deception task, where they were instructed to respond truthfully or deceptively after viewing an image. To ensure uniformity, the researchers chose the initial 1000 data points from each CSV file for further examination. These selected data points were then consolidated into a single file, with each entry annotated to specify whether the participant's response was truthful or deceptive.

General pre-processing techniques were applied to eliminate any abnormalities and noisy data from the dataset. This encompassed filtering the data to remove noise and artifacts, along with segmenting the data into smaller portions to better capture relevant patterns. Following this, the pre-processed data underwent division into training and testing sets for subsequent machine learning analysis.

The EEG data classification encompassed the application of various machine learning methods, specifically Random Forest, Decision Tree, and Gradient Boosting. The rationale for employing this varied ensemble of machine learning approaches rested upon their collective strength in handling the intricate and multidimensional nature of EEG data.

In the pursuit of improving the classification process and optimizing the efficiency of the analysis, the feature selection technique known as Recursive Feature Elimination (RFE) was implemented. A pivotal role in enhancing the quality of EEG data analysis is played by RFE, accomplished by systematically assessing the relevance of individual features within the dataset.

RFE begins by considering the entire set of features within the EEG data. It then iteratively evaluates and ranks these features based on their importance to the classification task at hand. During each iteration, it identifies and eliminates the least informative features. This process continues until the desired number of features, or the most salient ones remain, effectively reducing the dimensionality of the dataset (Figs. 2 and 3).

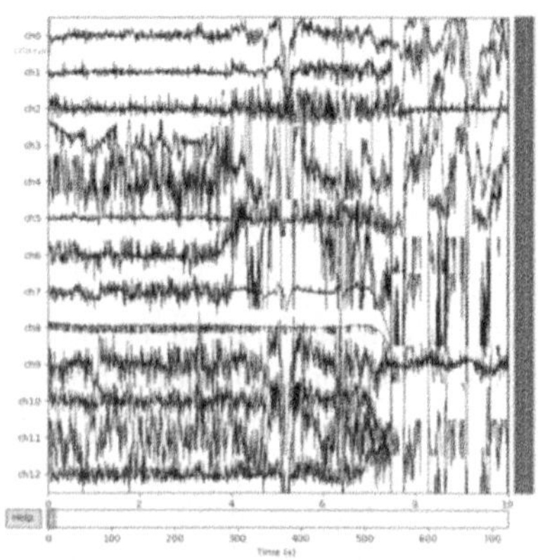

Fig. 2. EEG dataset before Pre-processing

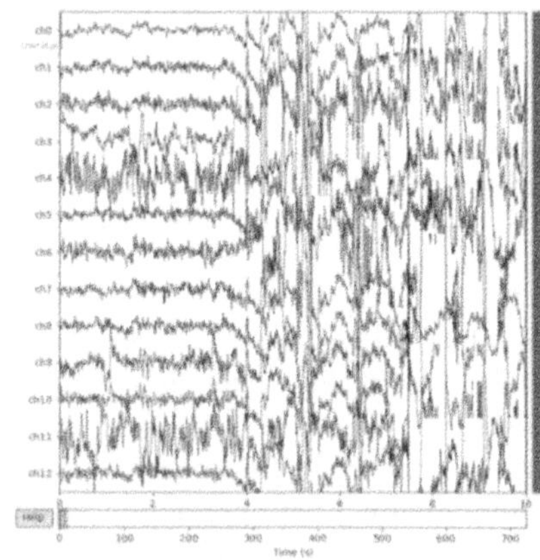

Fig. 3. EEG dataset after Pre-processing

Among the utilized techniques, Random Forest exhibited notable performance, achieving an accuracy rate of 87%. This highlights the effectiveness of Random Forest in accurately classifying EEG data and distinguishing between truthful and deceptive responses (Table 1).

According to the results, the Random Forest Classifier achieved the highest accuracy of 86.78%. The findings of this study indicate notable improvements in accuracy for the Random Forest Classifier compared to related works. However, the Decision-Tree and Gaussian Naive Bayes classifiers demonstrate lower accuracy rates compared to previous findings.

In summary, this study offers insights into the effectiveness of different classifiers for EEG-based classification tasks, highlighting the importance of selecting appropriate classifiers to achieve accurate results in practical applications.

Table 1. Result For Classification Using EEG.

S. No.	Method	Accuracy	Related Work Accuracy
1	Random Forest Classifier	86.78%	58.75%
2	Decision-Tree Classifier	55.06%	–
3	Gradient Boosting Classifier	64%	53.79%
4	Gausian Naive Bayes	47.95%	–

3.2 Deception Detection Using Audio Waves

This section describes the methodology for extracting audio features and using deep learning techniques to detect deception in speech.

The suggested system employs Mel Frequency Cepstral Coefficients (MFCCs) as a method for extracting features from audio signals, in the context of deception detection using only audio data. MFCCs are effective in capturing the spectral characteristics of speech and music signals, and are commonly used in various audio processing tasks, such as speech recognition and speaker identification.

For the extraction of MFCCs, the openSMILE toolkit was employed. This toolkit applies a sequence of signal processing procedures to the audio data, encompassing framing, windowing, and the application of a Mel-frequency filterbank to acquire a spectrogram representation. Subsequently, the spectral magnitude undergoes logarithmic transformation, followed by a Discrete Cosine Transform (DCT) to derive the ultimate MFCC features.

By employing this methodology, the audio signals are transformed into a compact representation of their spectral characteristics, which can then be utilized as input features for subsequent deep learning techniques aimed at detecting deception in speech. This approach enables the extraction of discriminative features from the audio data, facilitating the development of effective deception detection models (Fig. 4).

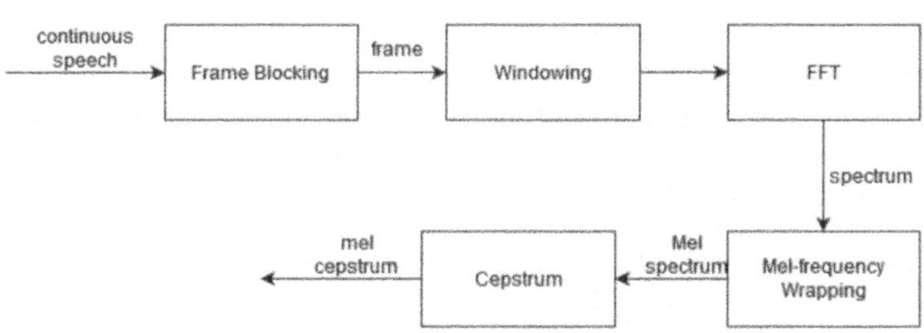

Fig. 4. Extraction of MFCC Features

The KNN classifier was employed for deception detection in this research. The classifier was trained using the extracted audio features. The classifier's performance was assessed by employing metrics such as accuracy, precision, and recall.

This process provides a systematic and reliable approach for detecting deception in speech (Table 2).

Table 2. Result For Classification Using Audio.

S. No.	Method	Accuracy	Related Work Accuracy
1	KNN Classifier	99.7%	56.22%
2	Decision-Tree Classifier	94.8%	–
3	Gradient Boosting Classifier	95.4%	–
4	Random Forest Classifier	92.3%	54.89%
5	Gaussian Naive Bayes	56.7%	–

The results indicate that the KNN classifier achieved the highest accuracy of 99.7%. To provide context and comparison, the accuracy rates reported in previous studies or related works are also displayed. Based on the comprehensive study, it can be observed that the KNN classifier outperforms other classifiers in terms of accuracy for deception detection. Additionally, the study reveals significant improvements in accuracy compared to related works, particularly for KNN, Decision-Tree, and Random Forest classifiers.

Overall, this comprehensive study contributes valuable insights into the effectiveness of machine learning classifiers for deception detection, highlighting the importance of selecting appropriate classifiers to achieve high accuracy in practical applications.

3.3 Combining Both EEG and Audio to Make a Better Prediction

This section details the utilization of both audio and EEG data for deception detection.

The ensemble approach was effectively employed in the classification of EEG and audio data, incorporating the Weighted Voting method to harness the power of multi-modal information for enhanced accuracy and robustness.

Weighted Voting is a fundamental technique within ensemble learning, where predictions generated by various models are combined. In this approach, each model's prediction is assigned a specific weight based on its performance or domain expertise.

The weighted voting follows the below steps:

1. Each input is assigned a weight that represents their relative importance or credibility in the decision-making process.
2. Each input casts their vote, typically in the form of a preference or decision on a specific choice or option. The weight assigned to each input influences the impact of their vote on the final decision.
3. The individual inputs are aggregated using their respective weights to calculate a weighted sum or score. This aggregation process varies depending on the context and may involve simple addition, multiplication, or more complex mathematical operations.

4. Consequently, the final prediction is reached through a calculated consideration of these weights, giving more prominence to the predictions of more accurate and reliable models within the ensemble.

This approach underscores the value of leveraging ensemble techniques for multimodal data analysis, enhancing the overall predictive capability and reliability of the classification system.

Previous research in deception detection has historically been characterized by a tendency to gravitate towards a limited set of models and methodologies, resulting in a notable lack of diversity in approach within the field. However, the current study represents a significant departure from these entrenched conventional practices by adopting a distinctly diverse array of models. This strategic shift in approach has yielded commendable outcomes, evidenced by the attainment of a strikingly high accuracy rate of 99% in predicting instances of deception.

This study also emphasizes the critical importance of embracing a multifaceted approach in the realm of deception detection research. By harnessing the potential of a wide spectrum of models, the study not only expands the horizons of investigation but also illuminates the inherent promise of variability in methodologies. This embrace of diversity fosters a more comprehensive exploration of various techniques and strategies.

The incorporation of a diverse set of models serves as a testament to the potential for more reliable outcomes within the domain of deception detection, thereby enhancing the overall robustness and effectiveness of deception detection methodologies. Thus, the current study stands as a beacon of innovation, heralding a new era of exploration and advancement in the field of deception detection (Table 3).

Table 3. Result For Classification Using EEG And Audio.

S. No.	Method	Accuracy
1	Random Forest (EEG) + KNN Classifier (Audio) (Proposed)	99.0%
2	Decision Tree (EEG) + KNN Classifier (Audio)	93.2%
3	Random Forest (EEG) + Decision Tree Classifier (Audio)	85.1%

4 Conclusion

Within the scope of this investigation, the categorization of EEG and audio data to identify distinct patterns or occurrences was given primary emphasis. To facilitate this endeavor, machine learning models tailored to the task at hand were crafted. Specifically, a KNN model for audio classification and a Random Forest model for EEG classification were constructed. Through a meticulous series of experiments and rigorous assessments, the efficacy of these models in delivering precise predictions for the intended target classes was substantiated. The outcome of the research unveiled remarkable levels of accuracy, precision, and recall scores, thereby underscoring the considerable potential of these models for practical deployment across various real-world applications.

Through the strategic application of feature selection methods, we successfully pinpointed the most pertinent features within the EEG data, culminating in a discernible enhancement of the classification performance. Additionally, we integrated the practice of cross-validation into our experimental framework, a crucial step to ascertain the resilience and general applicability of our models. The computed mean accuracy, precision, and recall scores serve as a testament to the models' adeptness in accurately categorizing both EEG and audio data, further underlining their robust and versatile nature.

The findings of this research have significant implications in various domains such as healthcare, security, and human-computer interaction. The accurate classification of EEG and audio data can aid in diagnosing neurological disorders, detecting abnormal audio events, and improving human-computer interaction experiences.

5 Future Works

In the context of future research efforts, exploration of additional modalities for deception detection is a promising avenue. This research can delve into the utilization of gaze data and video data, aiming to furnish valuable insights into subjects' behavior patterns. Such endeavors hold the potential to enhance the accuracy and effectiveness of deception detection models.

For gaze data, eye tracking technology can be used to track a subject's eye movements and fixation points. Previous studies have shown that gaze patterns can differ between truthful and deceptive behaviors, with deceptive individuals tending to avoid direct eye contact. Incorporating gaze data into deception detection models can potentially provide an additional behavioral indicator of deception.

Likewise, video data can furnish insights into a subject's nonverbal conduct, encompassing aspects like facial expressions, bodily gestures, and vocal intonation. Previous research has revealed the connection between these nonverbal cues and deceptive behavior, suggesting that the incorporation of video data into deception detection models has the potential to enhance the overall accuracy of these models.

Furthermore, given that the current model is trained exclusively on a single dataset, there are intentions to subject it to evaluation across a range of datasets to assess its efficacy and capacity for generalization. The process of testing the model across diverse datasets serves the purpose of identifying potential shortcomings and enhancing the model's resilience.

Overall, incorporating additional modalities and testing the model on multiple datasets can help enhance the accuracy and reliability of deception detection models.

Acknowledgments. The authors express their sincere gratitude to all individuals and institutions whose contributions facilitated the completion of this research project. Appreciation is extended to the participants for their involvement in data collection processes. Furthermore, acknowledgment is given to the reviewers for their valuable feedback and insights, which significantly enhanced the quality of the manuscript. The authors also acknowledge the support and resources provided by Coimbatore Institute of Technology throughout the duration of the study.

Disclosure of Events. The authors disclose that the dataset utilized in this research was obtained from IIT Jodhpur. All data collection procedures and experimental protocols strictly adhered to ethical guidelines and regulations set forth by IIT Jodhpur. Informed consent was obtained from all participants involved in the study, and the privacy and confidentiality of their data were rigorously maintained. Any potential conflicts of interest or biases that could have influenced the research findings are hereby disclosed. Additionally, any limitations or constraints encountered during the acquisition and utilization of the dataset are acknowledged. The authors affirm their commitment to transparency and integrity in reporting the outcomes of this research endeavor.

References

1. Meservy, T.O., et al.: Deception detection through automatic, unobtrusive analysis of nonverbal behaviour. IEEE Intell. Syst. **20**
2. Frank, M., Eckman, P.: Appearing truthful generalizes across different deception situations. J. Personal. Soc. Psychol. **86**
3. Reinhard, M.A., et al.: Unconscious processes improve lie detection. J. Personal. Soc. Psychol. **105**
4. Langleben, D.D., et al.: Telling truth from lie in individual subjects with fast event-related fMRI. Hum. Brain Mapp. **26**
5. Cui, Q., et al.: Detection of deception based on fMRI activation patterns underlying the production of a deceptive response and receiving feedback about the success of the deception after a mock murder crime. Soc. Cogn. Affect. Neurosci. **9**
6. Davatzikos, C., et al.: Classifying spatial patterns of brain activity with machine learning methods: application to lie detection. NeuroImage **28**
7. Sakkalis, V.: Review of advanced techniques for the estimation of brain connectivity measured with EEG/MEG. Comput. Biol. Med. **41**
8. Gupta, V., et al.: Bag-of-lies: a multimodal dataset for deception detection. In: Proceedings of the IEEE/CVF Conference on Computer Vision and Pattern Recognition (CVPR) Workshops, June 2019
9. Ding, M., et al.: Face-focused crossstream network for deception detection in videos. In: Proceedings of the IEEE/CVF Conference on Computer Vision and Pattern Recognition (CVPR), June 2019
10. Belavadi, V., et al.: MultiModal deception detection: accuracy, applicability and generalizability. In: 2020 Second IEEE International Conference on Trust, Privacy and Security in Intelligent Systems and Applications (TPS-ISA) (2020)
11. Gao, J., et al.: Effective connectivity in cortical networks during deception: a lie detection study based on EEG. IEEE J. Biomed. Health Inform. **26**
12. Krishnamurthy et al.: A Deep Learning Approach for Multimodal Deception Detection in arXiv:1803.00344v1 [cs.CL]
13. Gao, J., et al.: Brain fingerprinting and lie detection: a study of dynamic functional connectivity patterns of deception using EEG phase synchrony analysis. IEEE J. Biomed. Health Inform. **26**
14. Tran, D., et al.: Learning spatiotemporal features with 3d convolutional networks. In: Proceedings of the IEEE International Conference on Computer Vision

Analyzing Meteorological Data for Curated Music Selection

S. Sakthi Murugeswari[1(✉)] and E. Sujatha[2]

[1] Saveetha Engineering College, Chennai, India
ssakthimurugeswari@gmail.com
[2] Department of CSE, Saveetha Engineering College, Chennai, India
sujathae@saveetha.ac.in

Abstract. Analyzing Meteorological Data for Curated Music Selection, delves into the innovative fusion of weather data and music curation. Leveraging advanced data analysis and machine learning techniques, we aim to create personalized playlists that align with the current weather conditions. Our research addresses the challenge of correlating weather elements with musical attributes and explores real-time adaptation, potentially revolutionizing the way individuals experience and interact with music in various meteorological contexts. This study not only advances the field of data-driven music curation but also provides insights into its transformative impact on music services and user experiences.

Keywords: Data Analysis · Music Curation · Weather Data · Machine Learning · Personalized Playlists

1 Introduction

In today's digital age, the synthesis of technology, data analysis, and the realm of music has opened the door to a groundbreaking frontier—Analyzing Meteorological Data for Curated Music Selection. Similar to the pioneering Music to Score Conversion (MSC) project, our initiative recognizes the transformative potential of technology, data, and music in creating a deeper connection with our environment and emotions. Our research correlates complex weather variables such as temperature, humidity, and precipitation with the nuanced attributes of music, including tempo, key, and genre. We aim to craft personalized playlists that resonate with the atmospheric conditions of any given moment. This endeavor has the potential to enhance our daily experiences during outdoor activities, commutes, and various life situations. It brings real-time adaptation into play, ensuring that chosen music harmonizes seamlessly with ever-changing weather. Analyzing meteorological data for curated music selection opens new horizons for personalized playlists, catering to individual moods, environments, and weather conditions, much like the MSC project empowers individuals with hearing impairments to engage with music through visual notation. In the following sections, we will delve into our methodology, challenges, and potential outcomes as we unlock the harmonious connection between weather science and the world of music, offering a transformative and elusive musical.

Published by Springer Nature Switzerland AG 2024
M. L. Owoc et al. (Eds.): ICCIDS 2024, IFIP AICT 718, pp. 234–240, 2024.
https://doi.org/10.1007/978-3-031-69986-3_17

2 Related Works

[1] This project focuses on gathering and analyzing environmental data to provide song recommendations based on current weather conditions. Weather stations and air quality monitoring systems are crucial for collecting parameters like temperature, humidity, and pollutants, contributing to accurate weather forecasts and improved agricultural practices. [2] In the realm of music recommendation systems, both commercial and academic sectors have seen various approaches. While traditional systems assume fixed user preferences, this paper proposes a context-aware music recommendation system (CA-MRS) utilizing fuzzy systems, Bayesian networks, and utility theory to offer tailored music recommendations based on contextual factors. [3] The study includes a subjective test demonstrating the system's effectiveness. Spotify, in collaboration with AccuWeather, explored the correlation between weather conditions and music preferences of its users. Playlists were created based on songs disproportionately played during specific weather conditions in select cities. [4] The study revealed a definite connection between weather and users' music choices. Recognizing the relationship between mood, weather, and music preferences, the proposed automated system aims to create playlists based on users' moods and weather conditions, assigning weights to music files accordingly using data mining algorithms. While there's considerable work on algorithmic composition targeting emotional responses, a comprehensive review is lacking. [5] The paper aims to fill this gap by providing an overview of affectively driven systems for algorithmic composition, including performative and transformative systems, and highlighting the challenges in adapting these systems. [6] The increasing availability of music data on the internet has led to opportunities for personalized music experiences. Most existing research focuses on individual factors, such as emotional status and personal preferences, neglecting the impact of external factors like location and activity on music preferences. [7] Context Play is introduced as a context-aware music recommender system allowing users to adapt recommendations to both their music preferences and contextual characteristics. Addressing the challenge of selecting the right music for specific situations or moods, a Case-Based Reasoning (CBR) approach to musical playlist recommendation is presented. [8]The system recommends new and meaningful playlists by retrieving relevant cases from a Case Base of playlists compiled by human listeners. [9] In the realm of ornithology, the paper discusses how weather variables can serve as a surrogate for species detectability, providing insights into handling heterogeneity in species detectability in monitoring schemes. [10]Finally, the review touches on the challenges associated with manually searching through large music collections to find the right song and provides an overview of existing methods for automatically generating playlists, discussing their advantages and disadvantages.

3 Proposed Methodology

The Sound system is crafted to deliver personalized song recommendations, considering user music preferences, weather conditions, and air quality. The methodology integrates collaborative filtering and weather-based recommendation techniques for tailored suggestions. The following outlines the approach taken to implement the Sound system.

3.1 Data Collection

The Sound system relies on three primary data sources: music data from the Spotify Web API, real-time weather data from the Rapid Weather API, and air quality data from the Air Now API. These sources provide comprehensive information on music tracks, artists, genres, current weather conditions, and air quality across various locations.

3.2 Data Preprocessing

Collected data undergoes preprocessing to ensure its suitability for the recommendation system. Music data is cleaned and filtered to eliminate duplicates, providing a list of unique tracks with associated features like tempo and energy. Weather data is filtered to extract current conditions such as temperature and humidity, while air quality data is filtered for the current air quality index (AQI).

3.3 Collaborative Filtering

Collaborative filtering is employed to generate personalized song recommendations based on user music preferences. Leveraging the Spotify Web API, the system analyzes the user's listening history, favorite tracks, and artists. A collaborative filtering algorithm computes a similarity score between the user and others with similar preferences, recommending tracks popular among these users.

3.4 Weather-Based Recommendation

To align with current weather conditions, a fuzzy logic algorithm is applied. Utilizing the Open Weather Map API, the system obtains real-time weather attributes like temperature and humidity. These conditions are then matched with music attributes such as tempo and valence to generate personalized song recommendations tailored to the weather. For example,

3.5 Air Quality Monitoring

Air quality monitoring is integrated to tailor song recommendations based on the user's environment. Using the Air Now API, the system fetches the current AQI for the user's location. A threshold-based approach is applied to recommend songs suitable for the prevailing air quality conditions. For instance, songs with slow tempo and low energy levels may be recommended on days with poor air quality to promote relaxation.

3.6 User Interface

The Clime Sound system features a user-friendly interface enabling users to input music preferences and view recommended songs. The interface displays current weather conditions, air quality index, and song recommendations based on the user's preferences, weather, and air quality. Users can also provide feedback to enhance recommendation accuracy.

3.7 Component-Based Architecture

Developed using React.js, a component-based architecture is implemented where each component handles specific functionalities. Components for displaying weather conditions, air quality index, and recommended songs contribute to a modular and efficient system.

3.8 State Management

React.js state management system is employed to dynamically manage user music preferences, weather conditions, and air quality. This enables seamless updates to UI components based on changes in the application's state.

3.9 Data Fetching

The Sound application utilizes React.js built-in "fetch" feature to obtain data from external APIs, including Spotify Web API, Open Weather Map API, and Air Now API. This facilitates the retrieval of real-time Data essential for generating recommendations.

3.10 Conditional Rendering

ReactJS's conditional rendering capability is leveraged to display different components based on user input and application state. For instance, the recommended songs component is displayed only when users provide their music preferences. Upbeat songs may be recommended on sunny days, while calming tunes could be suggested on rainy days.

4 Technology Stack

The Sound web application is developed using a combination of HTML, CSS, JavaScript, and React.js. HTML structures the content, CSS styles the layout, JavaScript provides dynamic functionality, and React.js efficiently manages the user interface components.

DATASET NAME:	DESCRIPTION:	SOURCE:
Meteorological Data	Includes weather details such as temperature, humidity, precipitation, and atmospheric pressure.	National Weather Service, or Rapid Weather APIs.
Music Features Data	Contains musical attributes such as tempo, key, genre, and mood.	Music streaming APIs (e.g., Spotify API, Last.fm API), music databases.
User Preferences Data	Captures user preferences, ratings, and interactions with previously recommended music.	User feedback, user profiles, or simulated data for testing.
Location Data	Provides geographical details of users for accurate weather predictions.	User profiles, geolocation services, or user input.

5 Flowchart

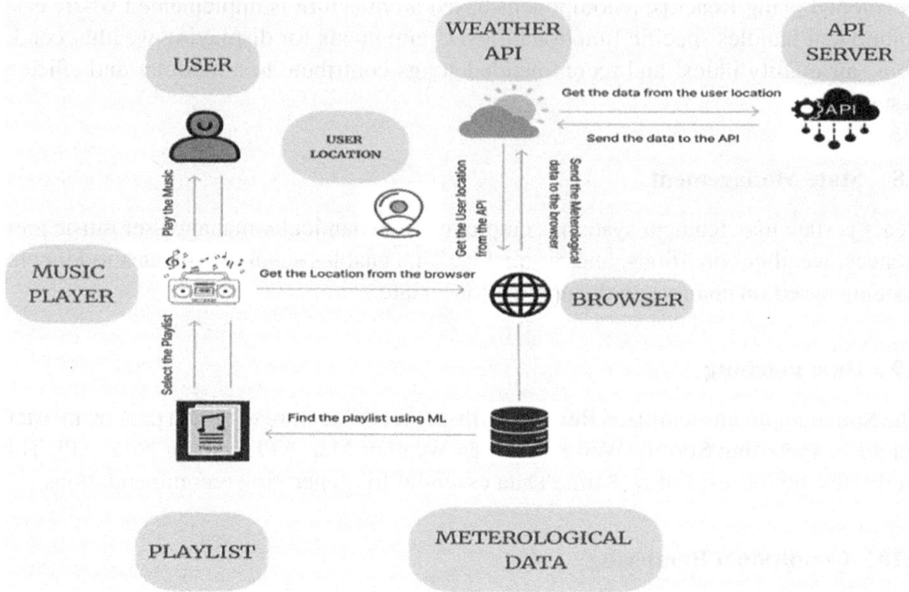

6 Results and Discussions

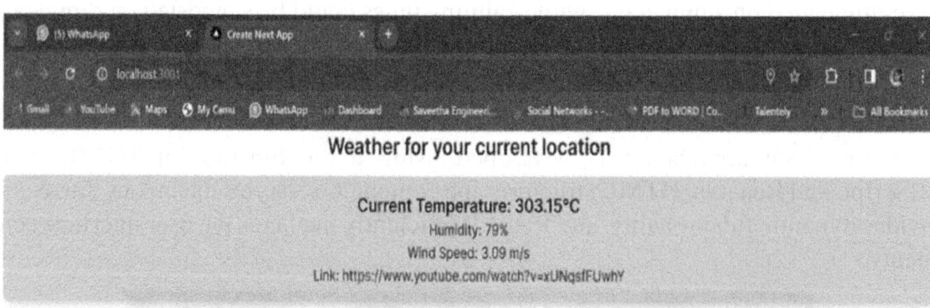

The comprehensive flowchart for this innovative system unfolds with meticulous data orchestration. Initially, the process embarks on a journey of data collection, with a sharp focus on gathering meteorological data. This raw meteorological data undergoes a sophisticated preprocessing stage, meticulously curated to ensure not only its cleanliness but also its alignment with the requisite formatting standards. Moving seamlessly into the analysis phase, the system delves into the extraction of relevant features. These features are strategically chosen to wield influence over the subsequent music selection process. In parallel, a parallel thread is initiated for the retrieval of music data. This dual data stream converges into the creation of a robust music recommendation model. Crucially, this model is designed with a keen awareness of both meteorological and musical features, establishing a synergistic foundation for the subsequent stages. The synthesis of meteorological and musical datasets emerges as a pivotal step, culminating in the generation of meticulously curated playlists. The essence of user interaction is injected into the system, introducing a critical layer of controllability. Users are empowered to shape recommendations based on real-time meteorological conditions, fostering a personalized and dynamic music listening experience. A continuous feedback loop ensues, where user input dynamically informs the evaluation and fine-tuning of recommendations. This iterative process not only enhances the accuracy of suggestions but also refines their relevance. The user's journey within the system reaches its apex when they achieve satisfaction with the music recommendations. This dynamic and personalized musical odyssey is intricately woven, seamlessly blending the realms of meteorological dynamics and auditory aesthetics to craft a unique and immersive experience for the user. The combination of real-time weather information and personalized music recommendations enhances the user experience, creating a cohesive and enjoyable interface. The application successfully merges utility with entertainment, offering a practical tool for weather awareness while infusing an element of fun through music suggestions.

7 Conclusion

	A	B	C	D	E	F	G	H	J	NH	NI	NJ	NK	NL
1	Zip Code	City	State	Historical Avg Monthly Temperature (in °F)	Jan 1988	Feb 1988	Mar 1988	Apr 1988	•••	Aug 2018	Sep 2018	Oct 2018	Nov 2018	Dec 2018
2	01366	Petersham	MA	49.61	21.58	30.88	33.93	42.75		75.22	66.61	50.29	36.73	30.76
3	02196	Boston	MA	52.25	28.31	38.70	37.02	45.57		76.68	67.06	52.25	41.20	35.31
4	53132	Franklin	WI	45.73	19.56	22.53	32.58	35.96		71.17	63.95	47.39	30.16	28.27
5	04465	Cary	ME	40.51	12.74	19.24	28.67	36.48		67.33	57.04	40.26	27.52	15.30
6	12083	Greenville	NY	46.63	19.02	29.80	31.12	39.02		70.95	63.01	48.00	34.47	28.92
7	24945	Greenville	WV	53.76	27.32	41.23	37.99	48.61		73.11	71.62	56.08	40.62	36.05
8	76631	Bynum	TX	66.67	41.95	50.25	62.06	60.64		85.46	79.07	68.02	52.81	49.12
9	76653	Kosse	TX	65.41	42.22	50.65	61.54	60.30		83.75	77.70	66.60	51.60	48.43
10	25103	Hansford	WV	56.33	29.39	43.05	41.31	50.83		75.65	74.16	59.13	42.93	39.85
11	94067	San Bruno	CA	59.66	52.52	52.56	53.91	57.83		67.66	64.60	62.80	56.05	50.54
12	21289	Baltimore	MD	56.27	31.05	40.87	39.87	51.06		77.79	72.93	59.22	43.56	39.45
13	22185	Vienna	VA	54.32	29.70	39.36	38.16	49.93		75.33	70.45	57.22	41.40	37.53

In conclusion, the "Analyzing Meteorological Data for Curated Music Selections" project represents an innovative fusion of meteorological data analysis and music curation, aiming to redefine the way individuals experience and engage with music in correlation with weather conditions. The integration of the Rapid Weather API ensures real-time access to accurate meteorological data, contributing to the precision of personalized music recommendations. The decision algorithm, powered by advanced machine learning techniques, dynamically adapts to user preferences influenced by the atmospheric ambiance. The user interaction module and feedback mechanism create a user-centric experience, allowing individuals to actively shape their musical journey. The project's interdisciplinary approach not only enriches the music listening experience but also highlights the profound impact of weather on our daily lives. As technology continues to evolve, this project stands at the forefront of leveraging data analytics and AI to deliver a seamless and enjoyable musical journey tailored to the ever- changing weather patterns and user preferences. Ultimately, the project envisions a harmonious convergence of meteorology and music, adding a unique layer of engagement to the digital experiences of users.

References

1. Shahadat, A.S.B., Ayon, S.I., Khatun, M.R.: Efficient IoT based weather station. In: 2020 IEEE International Women in Engineering (WIE) Conference on Electrical and Computer Engineering (WIECON-ECE), 2021 V. Jayashree. Naveta Kumari, "IOT Based Smart Helmet for Construction Workers", IEEE 7th International Smart Structures and Systems ICSSS (2020)
2. Gregory, M.A., Michener, S., Swagman, P.M.C.: Internet entertainment. In: AusWeb99 Fifth Australian World Wide Web Conference, Southern Cross University (1999)
3. Mahadev, A., Milgard, S., Patel, J., Karateka, V., Jagran, V.B.: Mood based music recommendation system (2021). ISSN: 2278-0181
4. Priyadarshini, S.G., Subramani, C., Preethi Roselyn, J.: An IOT based smart metering development for an energy management system. Int. J. Electr. Comput. Eng. **9**(4), 3041–3050 (2019)
5. Krishnamurthi, K., Rao, B.S., Rao, K.S.: IOT based lo cost weather monitoring system. In: TENCON 2019 - 2019 IEEE Region 10 Conference (TENCON) (2019)
6. Dinar, A., Harry, G., Lowry, A.: Renewable powered portable weather update station. In: TENCON 2019 - 2019 IEEE Region 10 Conference (TENCON) (2019)
7. Muhumuza, J., Kasumba, R.: Condition 8 monitoring and reporting framework for wireless sensor network-based automatic weather stations. In: 2020 IEEE International Women in Engineering (WIE) Conference on Electrical and Computer Engineering (WIECON-ECE), P. 39 (2021)
8. Kapoor, P., Barbhuiya, F.A.: Cloud based weather station using IoT Devices. In: TENCON 2019 - 2019 IEEE Region 10 Conference (TENCON) (2018)
9. Shahadat, A., Ayon, S.I., Khatun, M.R.: Efficient IoT based weather station. In: 2020 IEEE International Women in Engineering (WIE) Conference on Electrical and Computer Engineering (WIECON-ECE) (2021)
10. Kumar, A., Kumari, M., Gupta, H.: Design and analysis of IoT based air quality monitoring system. In: 2020 International Conference on Power Electronics & IoT Applications in Renewable Energy and its Control (PARC) (2020)

Remote and Centralized Billing System Using Cloud Computing

J. Gold Beulah Patturose, R. Priscilla, S. Aravind Shankar(✉), and S. M. Hariram

St. Joseph's Institute of Technology, Chennai, India
goldbeulah@stjosephstechnology.ac.in,
aravindshankar2002@gmail.com

Abstract. The process of establishing, maintaining, and utilizing a billing system is highly intricate and expensive. Small-scale suppliers will therefore be unable to afford to install a complete billing system in their stores. In this study, we propose a cloud-based billing system as a solution to this issue. ReactJS may be used for front-end development, Spring Boot can be used for business application development, and MySQL can be used for back-end development and data storage while creating this cloud-based billing system. This technology stack ensures that the billing system operates securely and effectively. Any cloud computing platform, including AWS EC2, Microsoft Azure, Google Cloud Platform, or Open stack, can host this billing system. Without the need for any specialized gear, this cloud-based billing system will be available from any device, anywhere in the globe. All you need to use to access this system is an internet-connected web browser. It makes this billing program affordable for a large group of users. This allowed the store vendors to concentrate on growing their businesses instead of worrying about inventory management, billing, and invoice production.

Keywords: Cloud based billing system · Cloud computing · Spring boot · ReactJS · MySQL · AWS · Azure · Google cloud platform OpenStack

1 Introduction

Small vendors and entrepreneurs use billing systems to organize, track, and manage their inventory. Billing Systems have aided in automating tasks and easing the burden of entrepreneurship. However, these invoicing mechanisms are no longer effective. This method requires setting up local servers to handle requests and responses, designing and implementing a database, allocating necessary computers based on shop size and connecting them to the database, and assigning an employee to each computer to help with billing. Traditional billing methods are costly and ineffective. We have suggested a cloud-based billing system to solve this issue.

The solution offers a unified platform on which to concentrate on reducing data processing time and preventing device intrusion. The suggested Cloud billing solution

Published by Springer Nature Switzerland AG 2024
M. L. Owoc et al. (Eds.): ICCIDS 2024, IFIP AICT 718, pp. 241–249, 2024.
https://doi.org/10.1007/978-3-031-69986-3_18

simplifies the billing process and facilitates real-time database browsing for both the admin and the user by utilizing a cloud facility to store user data [1]. We have suggested a cloud-based billing system that enables business owners to use cloud computing to automate the billing process. Technologies like React.JS, MySQL, and Spring Boot can be used to develop billing software, which can then be hosted on cloud computing platforms like Microsoft Azure or Amazon. Web browsers on almost all devices can be used to access the application by users.

The emergence of cloud computing has made remote access possible for the suggested system. Because it can run on any web browser, the application is less taxing on hardware, and the system's automation of the invoicing process facilitates easy stock management and tracking. Because the application is cross platform and compatible with PCs, laptops, and mobile devices, it is simple to use and doesn't require complex hardware setup. The suggested system can be accessed from any location in the world due to its centralization. Because the servers are spread out over multiple areas, this solution offers superior data recovery.

2 Literature Review

Prominent academics and researchers have offered numerous suggestions on how to use cloud platforms and technology to improve the billing, sales, and marketing industries. The few of these ideas that served as inspiration for our concept of a workable cloud-based billing system for small- to medium-sized vendors and store owners are listed below.

The [2] Cloud based Centralized Smart Cart and Contactless Billing System developed by H.N. MAHENDRA et al. leverages cloud computing and IoT to offer a contactless billing experience in supermarkets. The user can bill products automatically in the smart carts without assistance from staff members; this is an advanced IoT feature that allows for contactless shopping. Cloud technologies have supplanted the larger hardware functions used by these IoT devices. The user data will also be managed by the cloud computing platforms, which facilitates the admin's ability to search through both the user data and the databases of the business. Because they will be connected to the same network, these Inter net of Things devices can operate simultaneously and without interference. This study presents a Software-Defined Wide Area Network (SD-WAN) architecture using a conventional 802.11 hardware network.

The study [3] "Automated Shopping and Billing System Using Radio-Frequency Identification" by RAGHAV CHADHA et al. makes use of RFID (Radio frequency identification) scanners and cloud computing technologies to automate the billing process directly from the trolley. According to this research, clients will need to use a special mobile application made for product billing to connect their smartphones to the smart trolley. Each time a consumer adds an item to their cart, the product's RFID will be used to scan and add the item to the mobile app. After finishing their purchase, the consumer can disconnect from the trolley and check out using their app, making room for the next customer to utilize it. This strategy can cut labour costs for the businesses and save customers' time, which benefits both the customer and the shops.

A study titled [4] "Design of Restaurant Billing System (E Bill Resto) by Applying Synchronization of Data Billing in Branch Companies to Main Companies Based on Rest

API" was published by M. MAHAPUTRA HIDAYAT et al. To determine the income, returns, profits, and other financial information for the entire firm, this proposal employs REST technologies to synchronize data from the branch companies to the main branch company. This method links three separate branches to a single network using the REST API Architecture. This reduces traffic and speeds up the system by utilizing cutting-edge technology like web services and web 3 client-server interactions.

3 Existing System

Supermarkets and small-scale vendors currently employ the classic billing system, which consists of a local server and a central processing unit for the same. It is then expanded to include the shop's computer network. Server and computer network setup is highly costly, requires a lot of space, and requires constant maintenance. Along with the high cost of labour for professionals, maintenance is also very expensive. This additionally involves an intricate training program to use the device. Small vendors and neighbourhood supermarkets find it more difficult to set up a billing system that could simplify their lives than it does with the current conventional billing system. The complete billing system can be moved to the cloud to get around this problem. Because cloud computing plat forms offer billing systems, there is no need for expensive server maintenance because these services are already provided by the cloud platforms. Setting up a server is more expensive than leasing a virtual server (Fig. 1).

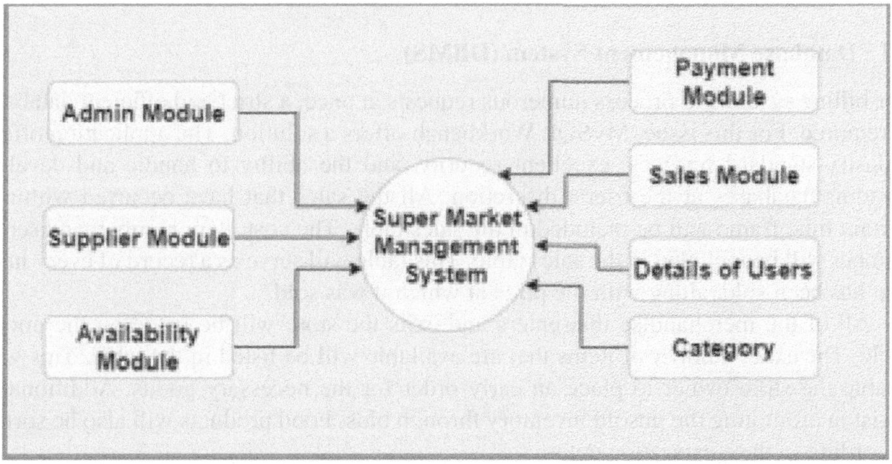

Fig. 1. A depiction of the workflow of the proposed research on cloud based billing system.

On a system with an active internet connection, the cloud-based billing system can be accessed from any location in the globe. Thus, any strong billing software can be run on a relatively basic PC with an internet connection. When compared to the conventional offline server strategy, this will significantly reduce the setup costs for a billing system. This will incentivize small-scale vendors to use billing systems, enhancing their capacity for tracking sales and enhancing their business, analytics, decision-making, and product management efficiency.

4 Methodology

Database administration is an essential and time-consuming activity for inventory and billing management systems. As a result, creating the databases and gathering the information to be stored in them must come first. The frontend, backend, and integration architectures must all be created using the data as a basis. The best cloud platform will be chosen, and its cost will be determined based on the database and architecture. The client (retailer) will then receive the Software as a service (SAAS) product. This method involves creating databases and storing data in them using MySQL, which is then coupled with backend technologies like Spring Boot. An enhanced framework for creating Java-based REST APIs is called Spring Boot. Backend database activities like Create, Read, Update, and Delete (CRUD) can be carried out using it. Inventory management and database maintenance are possible with this. Following the completion of the backend design, the user interface and experience will be created to ensure that users can utilize the program effectively and effortlessly. Here, a reactive and responsive frontend will be created using a framework like ReactJS, and it will be linked with the backend to create a whole billing application. Subsequently, an appropriate cloud platform will be chosen, and the billing system will be transformed into a Docker image and hosted on the cloud platform to guarantee effective database management and worldwide billing system access.

4.1 Database Management System (DBMS)

For billing systems to process numerous requests at once, a strict and efficient database is required. For this issue, MySQL Workbench offers a solution. The application offers industry-standard services, excellent security, and the ability to handle and develop customs databases at the user's discretion. All the sales that have occurred within a certain time frame will be included in the sales table. The cost, item number, and serial number will be included in the sales table. This table will serve as a record of every item that has been sold, along with the price at which it was sold.

All of the merchandise that enters and exits the store will be listed on the goods table. The exact number of items that are available will be listed in the table. This will enable the store owner to place an early order for the necessary goods. Additionally assist in promoting the unsold inventory through bids. Food products will also be sorted according to their expiration dates.

Details about the suppliers will be included in the supplier table. The supplier's name, supplier ID, supplier contact information, goods supplied, and quantity of goods supplied are all listed in the table. This will make it easier to keep track of the supplier's records and place orders for goods ahead of time. Additional security will also be ensured by the supplier's information.

The data about the clients will be in the clients' table. The table includes essential data on the patrons who have entered the establishment. Name, contact number, and customer ID are all listed in the table (Fig. 2).

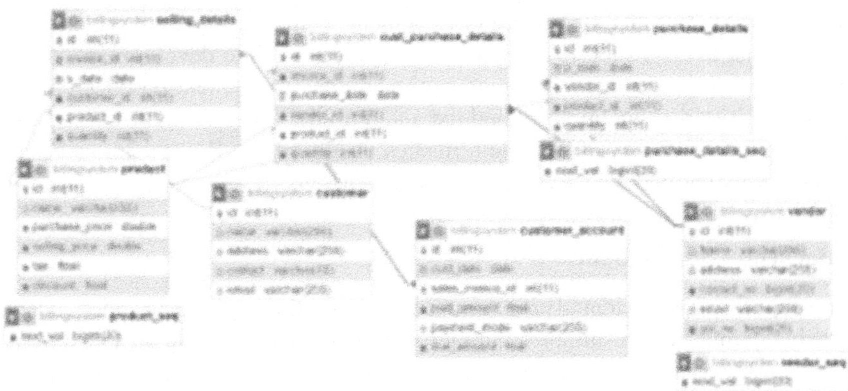

Fig. 2. A detailed database design of the above proposed methodology

The products that were bought from the store will be listed in the invoicing table. The goods name, goods id, cost, and quantity of goods are listed in the table. The result is produced by the backend algorithm, which also calculates the required total of all the commodities and taxes.

4.2 Authentication and Security

The native password hashing method is what MySQL use by default when it comes to databases. For user authentication, this technique makes use of the auth-socket authentication plugin and SHA256. This has been shown to be more secure, difficult to breach, and time-consuming. Spring Boot provides users with some of the most advanced and secure web services and REST API available. Support for authentication and permission is provided by Spring Boot. It offers defence against a range of threats, including cross-site request forgery, clickjacking, and session fixation. Along with many other features, it offers servlet API integration. Security features in cloud computing platforms vary depending on the platform, but they all offer the same basic features: a reliable database, widespread database accessibility, support for multiple regions, impenetrable security networks, and sophisticated encryption and decryption algorithms for data access. Because of this, the users will find this suggested model to be highly convenient as their data is safe and secure throughout all their online and offline activities.

4.3 Frontend - ReactJS

A JavaScript package called React.JS is used to create dynamic front-end web pages. Facebook developed ReactJS, which is very popular because to its reusable and simple to code nature. Because of its component-based architecture, developers can create encapsulated user interface components. React updated the website dynamically in response to changes by utilizing virtual Dom. Addition ally, JSX makes it possible to simplify the syntax and make it equal to HTML. Redux helps with state management, and React router facilitates navigation. React has been thoughtfully built and engineered to offer developers a peaceful and stress-free environment while still producing beautiful and reliable webpages for users. [5] React JS accelerates website development, increases efficiency, and improves search engine optimization. Component-based architecture, which Re act provides, encourages modular and reusable code. Virtual DOM saves time by lowering the frequency of website refreshes. Additionally, React offers a variety of developer tools that may be added to browser extensions.

4.4 Backend - Spring Boot

Java is a popular object-oriented programming language that is platform-independent and extremely flexible. As a result, programmers can create and run code on any system that supports Java. Oracle currently owns this language, which Sun Microsystems initially developed. The system offers automatic memory man agreement and is simple to use. Java is renowned for its quick compilation time, extensive libraries that make coding easier, and strong security features. Java's broad range of applicability, from website design to mobile application development, makes it a popular choice. Java is the best language for the billing system since its programs are adaptable and reusable.

Website development requires the use of Rest (Representational State Trans fer) API. The communication between the server and the client is made easier by this architecture. It can perform a few server-side operations, including GET, POST, PUT, and DELETE, to alter data. The information is retrieved in JSON and XML formats. Rest is renowned for being incredibly flexible and scalable yet being simple to code. Many websites and mobile applications are built using this architecture. Integrations are quite simple to complete and work with a variety of middle-end computer languages. The relevant REST APIs must update and modify in accordance with the frequent modifications and upgrades that services experience.

[6] Developed on top of the Spring framework, Spring Boot prioritizes convention over configuration. This makes the configuration of difficult jobs more efficient and makes application development simpler. This makes market grade stand-alone apps possible. Numerous features are pre-installed, such as production-ready settings and embedded servers. High levels of flexibility are offered by Spring Boot while preserving fast development. It is the preferred option for developers looking to create both stiff and flexible applications because of its wide ecosystem and integration.

4.5 Cloud Computing Platforms

There is various cloud computing platform that can used to host the billing system. Every cloud computing platform has their own advantages and disadvantages. Let's see in detail about top 3 cloud computing service providers in the world and how they can benefit this research.

Amazon Web Services (AWS) One of the most well-known providers of cloud computing services is AWS [7], or Amazon Web Services. With over 25 regions and 81 availability zones, they can offer their users highly efficient access to cloud services. Compute storage, computing power, machine learning techniques, database management systems, storage components, effective billing, security, automated scaling, and other services are all offered by AWS. The AWS platforms operate on a pay-per-use basis. Because the resources will be scaled up and down based on demand, consumers will only pay for what they use and won't be charged for unused usage.

Microsoft Azure The cloud computing platform Microsoft Azure is well-known for its features, server farms, and ease of use. Azure's global network [8] of datacenters allows for minimal latency and compliance. Because Azure pricing is pay as you go, developers may easily configure and host their applications on the cloud. Azure is reliable and easy to use. It facilitates simple communication between the user and the algorithm, which speeds up and improves efficiency. Azure integrates seamlessly with a wide range of programming languages and frameworks. The developers can now customize much more as a result. Azure offers infrastructure in the form of Platform as a Service (PaaS), Infrastructure as a Service (Iaas), and Software as a Service (Saas). Ai, databases, machine learning, security, and the Internet of Things (IoT) are just a few of the domains that can be housed in the cloud.

Google Cloud Platform (GCP) The cloud computing platform known as Google Cloud Platform, or GCP [9], is created and managed by Google. Additionally, the GCP provides a vast array of services including processing power, storage, etc. Virtual machines, an app engine for creating and testing apps, a Kubernetes engine for container orchestration, database storage systems for both SQL and NoSQL, big data analytics tools, machine learning models, and much more are all offered by GCP. In order to improve efficiency and user-friendliness, GCP also offers integration with Google services, including Google Docs, Google Slides, Google Sheets, and so forth. Additionally, it offers development tools like cloud SDK. GCP is robust, scalable, and adaptable.

5 Comparison of Cloud Platforms

The completion of cloud-based billing system will require cloud services such as Virtual machine to host the application, storage facilities to store the database contents with SQL support, scalability tracking mechanisms to track the usage and scale resources as per the demand, security features to keep the system safe from intruders, data handling tools to keep track of the data usages [10]. To accomplish all this, AWS provides services such as EC2 (Elastic compute cloud), AWS Lambda and ECS (Elastic container service) to host the application and provide compute power to the application successfully. It also manages the auto scaling of resources and monitors the demand for the resources. AWS provides services such as S3 (Object storage), EBS (Elastic block storage) to create multiple instances of the VMs and can simultaneously accessed by many users at the same time. AWS also offers Amazon RDS (Relational Database Service) to handle MySQL smoothly and efficiently. It is easier to migrate to AWS when compared to other platforms due to its easy usage and user-friendly platform UI and UX. AWS is also cost efficient, good customer support which makes it the best choice to host our cloud-based billing system.

6 Securing the Billing System

The need to develop a secured billing system is as important as developing an efficient way of billing services. The data is the most crucial information that anyone can have, and securing is the highest responsibility.

6.1 Challenges

Since it houses all the store's data, a vendor's billing system is the most vulnerable part of the business. This comprises the revenue, transactional in formation, etc. One of the potential obstacles to data security is data breaches, which reveal private information including credit card numbers, salary information, and client details. Securing the system against cyber-attacks, which include data breaches, information tampering, system lockdown, and many other things, is another difficulty. Employees that can commit fraud and get away with it present the biggest threat to the billing system's security. These are some of the major problems in using a billing system which can dealt with using certain cloud computing services and choosing a **reliable cloud platform.**

6.2 Security Methods

Various cloud platforms such as AWS and Azure provides their own authentication protocols to implement strong authentication and access controls [11] so that no one else can access the server. The encryption and decryption of sensitive data such as customer details, transactions can be performed to secure the information. Regular security audits can be performed by the owner to ensure that the data isn't tampered, and the transactions are safe and secure. The cloud servers are located overseas secured by multiple layers of security protocols. There is no way of physically breaching the cloud servers and steal data [12]. Performing cyber-attacks on the cloud server is also impossible as it is protected by multiple layers of firewalls. Therefore, migrating the billing system to cloud can save the vendor lot of trouble on securing the billing system and their data when compared to the traditional billing system.

7 Conclusion

In conclusion, billing systems have changed from how traditional businesses have operated. Billing systems offer improvements over the previous methods used by shops. Delivering more effective ownership and better handing. In the realm of billing systems, cloud hosting is a step up since it allows remote control and global access. Thanks to SATA security and recovery, shops can now process data more quickly. Benefits include cost-effectiveness, accessibility, scalability, and efficiency, giving it a competitive advantage over other billing systems. Additionally, invoicing activities are automated and streamlined, which will reduce errors even more. Rigidity and security are well-known features of Spring and MySQL. SQL is renowned for its flexibility and efficiency. The application i proves customer satisfaction and efficiency even further. Because the application runs over the internet, it may be used on low-end devices, which contributes to its ability to offer greater flexibility. The contemporary environment is change quickly, and it anticipates that humans will follow suit. Therefore, offering effective solutions is a need rather than a choice. The world is accustomed to quicker, more effective systems that use less data while operating at high speeds. The solution is a cloud-based billing system, which will also shape the software landscape going forward and provide it a competitive edge over competing offerings. Additionally, it gives organizations the ability to effectively manage their finances and data-driven insights, which contributes to their expansion.

References

1. Mahendra, H.N., et al.: Cloud based centralized smart cart and contactless billing system. In: 2022 7th International Conference on Communication and Electronics Systems (ICCES), pp. 820–826 (2022)
2. Mahendra, H.N., Mallikarjunaswamy, Nooli, S., C.B., Hrishikesh, M., Kruthik, N., Vakkalanka, H.M.: Cloud based centralized smart cart and contactless billing system. In: 2022 7th International Conference on Communication and Electronics Systems (ICCES), Coimbatore, India, pp. 820–826 (2022). https://doi.org/10.1109/ICCES54183.2022.9835856
3. Chadha, R., Kakkar, S., Aggarwal, G.: Automated shopping and billing system using radio-frequency identification. In: 2019 9th International Conference on Cloud Computing, Data Science and Engineering (Confluence), Noida, India, pp. 693–697 (2019). https://doi.org/10.1109/CONFLUENCE.2019.8776944
4. Hidayat, M.M., Dimas Adityo, R., Siswanto, A.: Design of restaurant billing system (E Bill Resto) by applying synchronization of data billing in branch companies to main companies based on rest API. In: 2020 International Conference on Smart Technology and Applications (ICoSTA), Surabaya, Indonesia, pp. 1–5 (2020). https://doi.org/10.1109/ICoSTA48221.2020.1570615039
5. Saundariya, K., Abirami, M., Senthil, K.R., Prabakaran, D., Srimathi, B., Nagarajan, G.: Webapp service for booking handyman using Mongodb, express JS, React JS, Node JS. In: 2021 3rd International Conference on Signal Processing and Communication (ICPSC), Coimbatore, India, pp. 180–183 (2021). https://doi.org/10.1109/ICSPC51351.2021.9451783
6. Guntupally, K., Devarakonda, R., Kehoe, K.: Spring boot based REST API to improve data quality report generation for big scientific data: ARM data center example. In: 2018 IEEE International Conference on Big Data (Big Data), Seattle, WA, USA, pp. 5328–5329 (2018). https://doi.org/10.1109/BigData.2018.8621924
7. Priyadarshini, P., Veeramanju, K.T.: A systematic review of cloud storage services- a case study on amazon web services. Int. J. Case Stud. Bus. IT Educ. (2022). n. pag
8. Firestone, D., et al.: Azure accelerated networking: SmartNICs in the public cloud. In; Symposium on Networked Systems Design and Implementation (2018)
9. Bisong, E.: An overview of google cloud platform services. In: Building Machine Learning and Deep Learning Models on Google Cloud Platform, pp. 7–10. Apress, Berkeley, CA (2019). https://doi.org/10.1007/978-1-4842-4470-8_2
10. Gupta, B., et al.: A review on amazon web service (AWS), microsoft azure and google cloud platform (GCP) services. In: Proceedings of the 2nd International Conference on ICT for Digital, Smart, and Sustainable Development, ICIDSSD 2020, 27–28 February 2020, Jamia Hamdard, New Delhi, India (2021)
11. Ahmed, M.: Trust Enhanced Security in SaaS Cloud Computing (2013)
12. Joshua, A.D.: an enhanced software as a service (SAAS) architectural model for cloud based security using hybrid symmetric algorithm. Int. J. Adv. Res. Comput. Sci. (2023)

Information Extraction Using RPA and Generative AI from Unstructured Documents: A Case of Invoices

S. Sowjanya(✉) and V. Vijaya Chamundeeswari

Department of Computer Science, Saveetha Engineering College, Chennai, India
sowjanyastays@gmail.com, dean@saveetha.ac.in

Abstract. In the daily operations of both small and large businesses, the handling of voluminous invoices remains an enduring challenge. The manual input and validation of invoice data not only imposes significant labor demands but also serve as ground for potential errors. This creates substantial complications when reconciling balances and cross-verifying information within digital invoices, creating the challenge of meticulous record-keeping. This paper offers an alternative solution for automating data extraction from invoices by leveraging Generative Artificial Intelligence. This also provides businesses with the flexibility to extract data from any format, theme or template while also minimizing human interaction. Getting Generative AI into invoice processing makes it much easier, making operations smoother and cutting down on how much humans must handle. Additionally, RPA takes it a step further by automating the entire end-to-end invoice processing workflow. This combination makes it easy for small and big business alike to easily use the streamline process which seamlessly integrates the user experience and provides enhanced accuracy compared to other methods. The use of NLP by Generative AI increases the understanding and extraction of the required data high precision from the invoices. The outcomes show a notable decrease in processing time and an improvement in accuracy, which eventually results in cost savings and increased operational effectiveness.

Keywords: Invoice extraction · RPA · UiPath · Generative AI · automation · financial operations · natural language processing · efficiency

1 Introduction

In the intricate ecosystem of contemporary business operations, the management and processing of invoices remain an enduring challenge across diverse industries. The manual handling of voluminous invoice data presents a dual-fold hurdle, demanding significant human resources while simultaneously creating an environment ripe for potential errors. This painstaking manual input not only strains resources but also complicates the essential task of maintaining meticulous records and reconciling balances within digital frameworks, posing a substantial challenge for enterprises striving for operational efficiency.

Published by Springer Nature Switzerland AG 2024
M. L. Owoc et al. (Eds.): ICCIDS 2024, IFIP AICT 718, pp. 250–264, 2024.
https://doi.org/10.1007/978-3-031-69986-3_19

However, amidst these challenges presented in achieving high efficiency [1, 2], a beacon of innovation emerges, promising a transformative resolution to this persistent issue. This paper embarks on an exploration of a solution centered around the fusion of Generative Artificial Intelligence (AI) and Robotic Process Automation (RPA). The utilization of Generative AI offers businesses a pathway to transcend the constraints imposed by varied invoice formats, themes, and templates. This innovative approach significantly diminishes the necessity for human intervention in the extraction of critical data, thereby streamlining the invoicing process.

There were various attempts before trying to achieve the objective of getting useful information from unstructured documents using various methods. The primary objective driving initiative was to automate and optimize the extraction of structured data from a wide range of unstructured receipt images [1, 2, 5]. With a particular focus on healthcare bills, including regular healthcare expenses and COVID-related bills, this process aimed to revolutionize the way data was gleaned from these documents. By combining diverse techniques, including image preprocessing, Optical Character Recognition (OCR), and Machine learning (ML), the overarching goal was to establish a system that can consistently identify, extract, and categorize data from a heterogeneous set of receipts [5]. In this literature, a BERT (Bidirectional Encoder Representations from Transformers) Model is proposed for classification and YOLO (You Only Look Once) v4 is employed for object detection when encountered images.

The core of this transformative paradigm lies in the incorporation of Natural Language Processing (NLP) within Generative AI. This integration empowers nuanced comprehension and extraction of vital data from invoices with an unparalleled level of precision. The convergence of these cutting-edge technologies heralds a new era in operational efficacy, manifesting in tangible reductions in processing time and remarkable enhancements in accuracy.

The primary aim of this proposed approach is not only to address the challenges posed by manual invoice handling but also to usher in an era of unparalleled efficiency and accuracy in business operations. By embracing these cutting-edge technologies, the paper strives to demonstrate how businesses can transcend the constraints of traditional methodologies, unlocking a realm of streamlined operations and cost-effective practices. The overarching goal is to provide a compelling vision for the transformative potential of integrating Generative AI and RPA in invoice processing, offering practical insights for businesses to achieve enhanced operational efficiency and accuracy in their financial workflows.

2 Literature Review

In today's data-driven world, the process of classifying and extracting information from a diverse array of receipts, including typed and handwritten bills, has emerged as a critical component of efficient data management. This endeavor harnessed the power of machine learning techniques to navigate the complex terrain of unstructured receipt data. Through a series of carefully designed steps, this method seeks to convert image-based receipts into structured, usable information. This transformation not only simplified data entry but also facilitated robust data analysis for a multitude of applications, like in the healthcare sector and beyond.

Even before tools with multi-lingual support came into practice, machine learning models were trained for these languages like French and Arabic [14, 15]. The first dealt with error detection and correction in French using a single input sequence, and in this case a string of characters were mapped to a vector [15]. The other one fed data into a Recurrent Neural Network (RNN) to generate the output sequence conditioned on the encoding vector, while the RNN examined the effect of word-based post-OCR error correction in Arabic retrieval effectiveness. The project examined the effect of correction in using different index terms on two collections of degraded Arabic documents. Neural networks are another way for efficient extraction of information [14]. One of the other ideas that the owners proposed was the method of syntactic, positional, and semantic features attached to each node, which were derived from the word the node represents. The edges in the document graphs represent the relative positional relationship between the words [5].

The methodology in this data extraction process was multifaceted and intricate. It started with the important step of data collection, encompassing the acquisition of receipt images. The subsequent phases usually were a symphony of image preprocessing [4, 6] which is again a pivotal step for proper recognition when using technology like Computer Vision [9], bill identification, corner identification, cropping, de-skewing, and enhancement. These preparatory stages laid the foundation for the core task of data extraction, which relies on the prowess of OCR technology, usually custom-trainable OCR engines like Tesseract OCR. The extracted data was then subjected to a refinement process, where specific data points are teased out using classification models such as Logistic Regression and K-neighbors Classifier [1]. Further, methods like hyperparameter optimization techniques were used to improve the already existing accuracy [3].

Additionally, the reliability of handwritten text recognition, while a remarkable feat of technology, can introduce complexity. Handwritten text varies greatly in style and legibility, and even the best recognition systems may encounter difficulties in deciphering certain handwritten characters. This poses a limitation in scenarios where handwritten text is prevalent [2].

There are many approaches for processing unstructured documents using Optical Character Recognition (OCR) and Natural Language Processing (NLP) and are discussed in literature [11–15, 16]. One approach that leveraged OCR engines for unstructured documents is applying Spark, a tool which provides an OCR engine, followed by NLP models to extract meaningful information from PDF and text files [2]. The approach's effectiveness was demonstrated using insurance-related datasets. The evaluation employs metrics like Word Error Rate (WER) and focuses on key OCR error types: misspelled characters (substitutions), lost or missing text (deletions), and spurious symbols (insertions).

One of the methodologies employed in this research encompassed a series of carefully structured steps. It commenced with the provision of input documents in the form of both PDF and text files. These documents were subsequently processed through the OCR engine, which harnessed the robust capabilities of Spark, accommodating more than 100 languages [2]. Post-correction procedures included quality and accuracy checks, followed by the systematic extraction of data using the NLP model. One of the projects deals with the post-OCR correction of medical reports and the other portrays the making of training corpus for the same [11, 13].

The primary application of one of the approaches is the analysis of unstructured documents, with a particular focus on invoices pertaining to the modern-day industry. The system played a crucial role in document classification and the extraction of vital information necessary for the efficient operation of enterprises. By incorporating OCR technology, the approach substantially elevated the overall quality and accuracy of data, which, in turn, had a profound impact on informed decision-making processes and the optimization of document processing [1].

One such method was to include RPA [9] for easier dealing of unstructured data by automating the process by using an IQ (Intelligent Quotient) bot from Automation Anywhere. That project particularly deals with bots which make use of Natural Language Processing, Optical character Recognition and Machine Learning as their base to identify the required fields.

3 Methodology

3.1 Dataset

A robust dataset would encompass examples of all variations and potentially more, capturing the range of formats, designs, and content arrangements across different sectors. Each invoice within the dataset serves as a valuable sample. For instance, this dataset used includes invoices from different industries such as retail, healthcare, or manufacturing, each with its unique layout and information arrangement. In a retail scenario, an invoice could contain fields like invoice number, date, item descriptions, quantities, prices, and total amount. These invoices might vary in layout, with some having a detailed breakdown of items while others present a more summarized view. The dataset itself is collected from various parts of the Internet and some of them are generated using Adobe and various templates were made use of.

3.2 Pre-processing of Invoices

Image enhancement techniques are pivotal in improving the quality and readability of degraded documents. These techniques encompass a range of operations such as adjusting brightness and contrast, denoising, sharpening edges, and enhancing overall clarity. For faded bills, algorithms can be employed to amplify the contrast between text and background, making text more discernible. Additionally, denoising filters can help remove artifacts and noise that may obscure text in wet or soaked bills. Techniques like histogram equalization can also be utilized to balance the distribution of pixel intensities, further enhancing the legibility of the document.

One fundamental step in image enhancement is adjusting brightness and contrast levels. For faded bills, increasing the brightness can help in making the text more visible and distinguishable from the background. Similarly, adjusting the contrast can amplify the differences between text and background colors, enhancing text clarity. This process aims to make faint or faded text stand out more prominently, improving overall readability.

Denoising plays a crucial role, especially in documents affected by water damage or soaking. Denoising filters are used to remove unwanted artifacts and noise from the image, which might obscure or distort the text. By reducing noise, such as speckles or blotches caused by water, the text becomes clearer and easier to interpret. This step is essential for restoring the original structure of the text in wet or soaked bills.

Sharpening edges is another technique used to enhance document clarity. In the context of faded or wet bills, sharpening algorithms can help in defining and highlighting the edges of text characters, making them appear more distinct and readable. By emphasizing the edges, the text becomes sharper and stands out better against the background, improving OCR accuracy and human readability.

Texture enhancement techniques focus on improving the texture and sharpness of text and graphics in the document. By enhancing textures, such as the fine details of characters and symbols, text becomes more defined and easier to read. This is especially beneficial for documents with intricate typography or small font sizes, as it ensures that every detail is preserved and clearly visible. Texture enhancement contributes significantly to OCR accuracy by ensuring that characters are accurately recognized and interpreted.

Selective blurring techniques can be applied to backgrounds or non-essential elements of the document to reduce distractions and highlight the main content, such as text and important graphics. By selectively blurring less relevant areas, the focus is directed towards the key information, improving readability and visual hierarchy. Additionally, noise reduction algorithms can further refine the document by removing residual noise and artifacts, resulting in a cleaner and more professional appearance.

Contrast stretching and dynamic range adjustment are advanced techniques used to expand the tonal range of an image, enhancing contrast and detail. These techniques are beneficial for documents with varying levels of brightness or exposure, as they help bring out subtle nuances and details in both text and images. By stretching the contrast, darker areas become richer, and lighter areas become brighter, resulting in a more balanced and visually appealing document. Dynamic range adjustment ensures that all elements within the document are well-defined and distinguishable, even in challenging lighting conditions.

By integrating these advanced image enhancement techniques, documents, including faded, wet, or soaked bills, can be transformed into high-quality, visually appealing, and easily readable materials. Each technique contributes uniquely to improving different aspects of the document, ultimately enhancing OCR accuracy, human readability, and overall document quality.

3.3 Classification Using Keyword Matching

When classifying invoices, our focus lies solely on the supervised approach of keyword matching. Predefined sets of keywords sourced from domain experts and invoice descriptions are relied upon. These keywords serve as the cornerstone for our evaluation process, measuring the similarity between invoice content and the predetermined keywords. Our aim is to gauge how well the content of an invoice aligns with the expected keywords, providing a familiarity rating based on the matching criteria (Fig. 1).

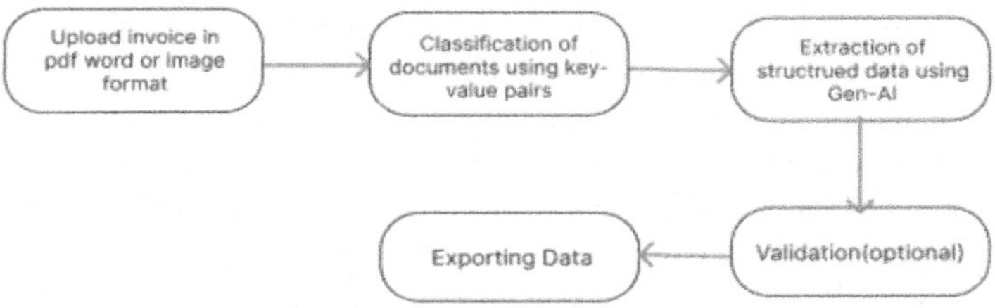

Fig. 1. The process flow for extracting required information from an invoice.

Our strategy involves evaluating the familiarity of invoices with the predetermined keywords. This entails calculating the correlation between the vectors representing keywords and the words present within the invoice content. In this supervised strategy, we consider averaging intercepts for each keyword, providing insights into their relevance and alignment with the content of the invoices.

Further refinement involves assessing the relevance of words matched by keywords in the context of invoice content. Like the KM strategy, this approach involves evaluating correlations among word vectors and keywords while excluding the keyword itself. The resulting score is determined using similar methodologies to KM, ensuring a comprehensive evaluation within the supervised framework.

In our supervised keyword matching approach for invoice classification, we prioritize precision and accuracy by leveraging predefined sets of keywords sourced from domain experts and invoice descriptions. These keywords serve as the foundation of our evaluation process, allowing us to assess how well the content of each invoice aligns with the expected criteria. By focusing on specific keywords relevant to invoice categories, we ensure a targeted and precise classification methodology that minimizes ambiguity and enhances classification accuracy.

A crucial refinement in our approach involves evaluating the relevance of words matched by keywords within the context of invoice content. This step ensures that the matching process considers not just the presence of keywords but also the semantic relevance and contextual meaning of the matched words. By excluding the keyword itself from the evaluation, we focus on assessing how well the matched words contribute to the overall classification accuracy, leading to a more nuanced and accurate familiarity rating for each invoice.

To derive a conclusive rating for each invoice, we compute the product of the Keyword Match (KM) and Within Match (WM) scores. Subsequently, we normalize these results to ensure a standardized rating scale from 0 to 10, accounting for any variations across the dataset. This normalization process guarantees consistency and prevents outliers in our classification system.

3.4 Extraction Using Generative AI

Generative models like GPT-3, with their extensive pre-training on vast corpora of text, possess a deep understanding of language structures and semantics. They excel in comprehending diverse document formats and contextually extracting valuable information from unstructured text. Leveraging these models in the context of data extraction involves deploying them to interpret, parse, and identify structured elements within unstructured documents. While direct mathematical equations may not explicitly represent the functionality of GPT-3 or similar generative models in data extraction, their underlying mechanisms involve probabilistic modeling and contextual understanding.

In essence, a language model like GPT-3 calculates the probability of a sequence of words (tokens) given the preceding context using chain rule of probability:

$$P(w1, w2, \ldots, wn) = P(w1) \cdot P(w2|w1) \cdot P(w3|w1, w2) \ldots P(wn|w1, w2, \ldots, wn-1) \quad (1)$$

When retrieving segments similar to that of a defined characteristic, cosine similarity can be used to measure similarity between vectors:

$$\text{Cosine Similarity} = \cos(\theta) = \mathrm{A} \cdot \mathrm{B} \,/\|\mathrm{A}\|\,\|\mathrm{B}\| \quad (2)$$

This equation calculates the cosine of the angle between two vectors, indicating their similarity.

The GPT-3 API takes in the definitions of the fields that are required. These definitions articulate the characteristics, formats, or patterns that signify a particular field, such as an invoice number, date, or recipient's name. GPT-3's strength lies in its contextual comprehension. It reads and comprehends the entirety of the document, understanding the nuances, structure, and relationships between words, sentences, and paragraphs.GPT-3 conducts semantic analysis, identifying word meanings and their contextual relevance within the document. It recognizes patterns, structures, and formats specified in the field definitions within the document context.

RPA platforms like UiPath excel in automating routine, rule-based tasks across various departments. They facilitate the creation of workflows that mimic human interactions with digital systems, enabling the automation of data entry, form filling, and more. RPA ensures high accuracy and efficiency in task execution, reducing manual errors and enabling round-the-clock operations without human intervention.

Generative AI models like GPT-3 possess natural language understanding capabilities, allowing them to comprehend and interpret text with remarkable accuracy. Integrating such models within UiPath workflows enhances the system's ability to understand and process unstructured textual data, improving decision-making and automating tasks that involve language comprehension.

Additionally, GPT-3's probabilistic modeling approach allows it to assign probabilities to different interpretations or extractions within a document. This probabilistic nature enables GPT-3 to make informed decisions about the most likely interpretation of a given field or data element, reducing errors and improving the accuracy of extracted information. By combining probabilistic reasoning with contextual comprehension, GPT-3 can effectively navigate complex document structures and extract relevant data with high precision.

These models often provide APIs that can be accessed within UiPath using HTTP Request activities or custom connectors. This allows seamless interaction with the AI model to perform tasks such as text summarization, sentiment analysis, or language generation. The combination of RPA and Generative AI enriches decision-making processes by providing deeper insights from unstructured data, enabling better-informed actions based on processed information.

4 Results and Discussion

In this section, we will present the key findings and discuss the results of our research document extraction using RPA and Generative AI from Unstructured Documents: A case of Invoices. The study aimed to explore and implement innovative approaches to streamline document extraction, specifically focusing on invoices, by synergizing Robotic Process Automation (RPA) and Generative Artificial Intelligence (Generative AI). Additionally, to augment user experience, React was utilized for frontend development, while Node.js served as the backend framework (Figs. 2, 3 and 4).

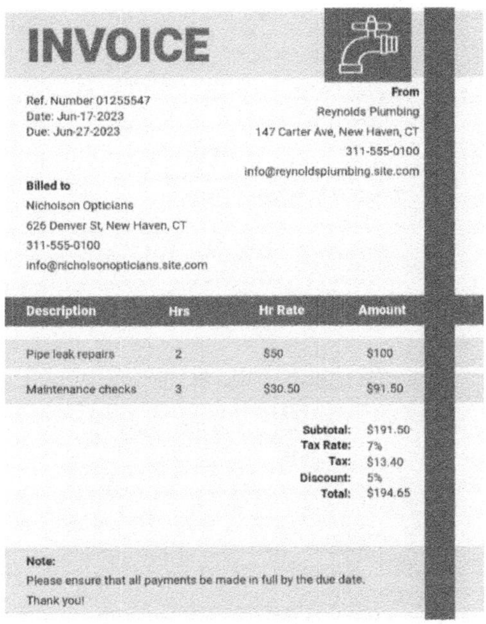

INVOICE

Ref. Number 01255547
Date: Jun-17-2023
Due: Jun-27-2023

From
Reynolds Plumbing
147 Carter Ave, New Haven, CT
311-555-0100
info@reynoldsplumbing.site.com

Billed to
Nicholson Opticians
626 Denver St, New Haven, CT
311-555-0100
info@nicholsonopticians.site.com

Description	Hrs	Hr Rate	Amount
Pipe leak repairs	2	$50	$100
Maintenance checks	3	$30.50	$91.50

Subtotal: $191.50
Tax Rate: 7%
Tax: $13.40
Discount: 5%
Total: $194.65

Note:
Please ensure that all payments be made in full by the due date.
Thank you!

Fig. 2. The image of Sample Invoice (Specimen 1)

The application of Generative AI algorithms proved instrumental in deciphering unstructured data within invoices. Natural Language Processing and Machine Learning techniques enabled the system to comprehend, interpret, and extract pertinent information from diverse invoice formats, regardless of variability in layout or language.

The integration of RPA into the document extraction process significantly expedited the handling of invoices. Through predefined rules and structured workflows, RPA effectively automated repetitive tasks, such as data entry and document classification, leading to reduced processing time and increased accuracy.

Through the Keyword Match method, leveraging generative AI, the classification results soared with remarkable confidence levels ranging between 97 to 100%. This robust performance underscored the efficacy of this approach in accurately categorizing and identifying pertinent information within documents. To further optimize the

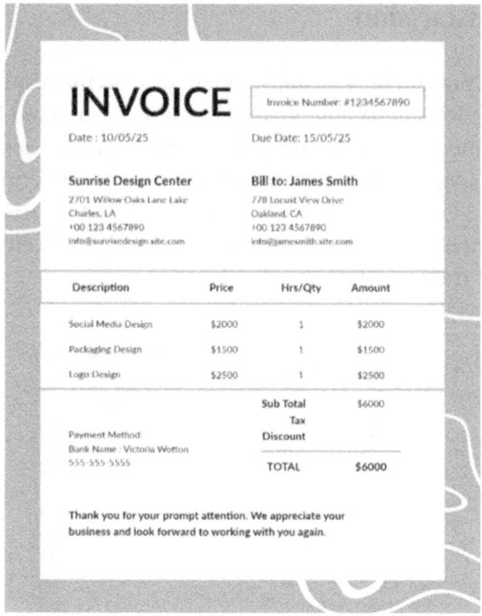

INVOICE

Invoice Number: #1234567890

Date : 10/05/25

Due Date: 15/05/25

Sunrise Design Center
2701 Willow Oaks Lane Lake
Charles, LA
+00 123 4567890
info@sunrisedesign.site.com

Bill to: James Smith
778 Locust View Drive
Oakland, CA
+00 123 4567890
info@jamesmith.site.com

Description	Price	Hrs/Qty	Amount
Social Media Design	$2000	1	$2000
Packaging Design	$1500	1	$1500
Logo Design	$2500	1	$2500
		Sub Total	$6000
		Tax	
		Discount	
		TOTAL	$6000

Payment Method:
Bank Name : Victoria Wotton
555-555-5555

Thank you for your prompt attention. We appreciate your business and look forward to working with you again.

Fig. 3. The image of Sample Invoice (Specimen 2)

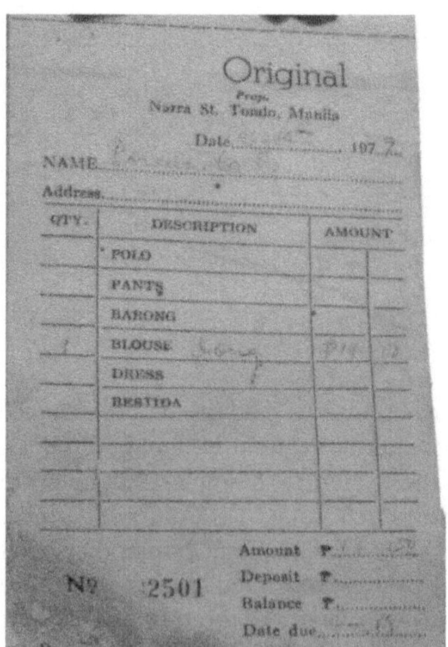

Original

Prop.

Nzra St. Tondo, Manila

Date 197

NAME

Address

QTY.	DESCRIPTION	AMOUNT
	POLO	
	PANTS	
	BARONG	
	BLOUSE	
	DRESS	
	BESTIDA	

Amount ₱

Deposit ₱

Balance ₱

Date due

№ 2501

By

Fig. 4. The image faded Sample Invoice (Specimen 3)

invoice extraction process, we conducted extensive testing using passports as a discard mechanism.

Contextual knowledge advances are becoming increasingly important in document categorization problems as Natural Language Processing advances. Through further exploration of the contextual details of language, NLP models will be able to demonstrate increased ability to understand complex document structures, different tones, and complex meanings.

To derive a conclusive rating for each invoice, we combine the Keyword Match (KM) and Within Match (WM) scores through a product calculation. This combined score provides a holistic assessment of both keyword alignment and contextual relevance, ensuring a comprehensive evaluation within our supervised framework. Normalizing these results further enhances the consistency and reliability of our classification system, allowing for a standardized rating scale that accounts for variations across the dataset and minimizes the impact of outliers on the classification process.

Six crucial fields that are frequently requested have been successfully extracted from documents using structured information extraction: the invoice number, invoice date, tax percentage, billed-to information, tax amount, and total invoice amount. This painstaking extraction procedure was executed with accuracy, demonstrating a remarkable match between the expected fields and the recovered data. The extraction from every field perfectly matched the planned structure and format, demonstrating the efficiency and dependability of the extraction method used.

An important milestone has been reached with the retrieval of crucial fields, demonstrating the strength of the extraction procedure inside the document analysis framework. The precision and coherence of the extracted data highlight the system's capacity to quickly and precisely locate important information in documents, meeting a range of organizational requirements and optimizing document-based processes.

Using training models with multiple invoices for document extraction presents several disadvantages compared to Natural Language Processing (NLP) techniques, which offer significant advantages. Training models with diverse invoices requires a large and resource-intensive dataset, leading to limitations in generalization, adaptability, and maintenance overhead. These models may struggle to generalize well to unseen invoice formats or languages not adequately represented in the training data, necessitating continuous updates and maintenance to keep them effective. In contrast, NLP techniques, such as contextual embeddings and attention processes, are inherently language agnostic and can generalize well to diverse document types and languages without extensive training data (Figs. 5, 6 and 7).

Future research will focus on developing more advanced document understanding models that can accurately parse and extract information from a wide range of document formats with varying complexities. This involves exploring novel approaches in natural language processing (NLP), computer vision, and multimodal learning to improve the accuracy and robustness of information extraction algorithms.

On average, each invoice takes approximately 7 s to be processed using the integrated Generative AI and RPA solution. This processing time includes pre-processing tasks, OCR extraction, data validation, and automated field extraction. While this processing time may vary depending on the complexity of the document and the quality of input,

Fig. 5. The image of Sample Faded Input Invoice (Specimen 4)

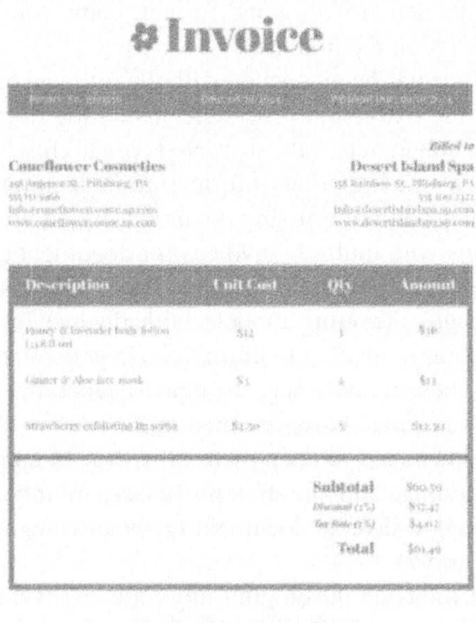

Fig. 6. The image of Sample Input Invoice (Specimen 5)

the automation capabilities of RPA platforms like UiPath ensure efficient and consistent processing of invoices, leading to faster turnaround times and increased productivity.

A	B	C	D	E	F
Invoice number	Billed To	Date	Tax rate	Tax amount	Total Amount
1001226	Desert Island Spa, 158 Rainbow St., Pittsburg, PA	8/10/2024	7%	$4.02	$61.49

Fig. 7. The results of the extracted information from Specimen 5

However, challenges arise when dealing with deteriorated or obscured bills, such as those that are soaked, damaged, or not clearly visible. Even with pre-processing efforts aimed at improving image quality, the accuracy of data extraction may decrease to around 92% in such cases. Factors like water damage, faded text, or distortions can pose difficulties for OCR algorithms in accurately interpreting the content. Despite these challenges, the integration of Generative AI and RPA still provides a significant improvement compared to manual extraction processes, reducing errors and processing time significantly.

In the domain of data extraction from deteriorated bills, the integration of Generative AI algorithms and Robotic Process Automation (RPA) via platforms like UiPath assumes a pivotal role in augmenting the extraction process. Generative AI algorithms serve as a crucial component in pre-processing images of degraded bills by employing methods such as brightness adjustment, contrast enhancement, and denoising to ameliorate image quality. These algorithms, encompassing Generative Adversarial Networks (GANs) and autoencoders, possess the capability to autonomously cleanse and refine images, rendering them more amenable for subsequent Optical Character Recognition (OCR) processing.

The UiPath platform, operating as an RPA solution, seamlessly automates pre-processing tasks by delineating structured workflows and rules for image enhancement. This automation encompasses a spectrum of tasks including image parameter adjustment, filter application, and image quality optimization to facilitate accurate OCR processing. Moreover, RPA workflows can automate the extraction of specific fields from invoices, such as invoice numbers, dates, tax particulars, and total amounts, by establishing extraction rules and patterns based on document structure and format.

After the pre-processing and RPA-driven data extraction phases, OCR technology is applied leveraging tools such as UiPath OCR activities. This OCR process adeptly identifies and extracts text from the now refined images, even deciphering faded or partially obscured text. Consequently, this facilitates precise data extraction from documents that posed challenges in processing owing to degradation or environmental factors.

As mentioned in the initial discussion, integrating feedback loops into the systems is crucial for continuous improvement. By collecting feedback from users and incorporating it into the learning process, these systems can continuously refine their extraction algorithms, leading to higher accuracy and usability over time.

5 Conclusion

In conclusion, the integration of Generative AI and Robotic Process Automation (RPA), exemplified by UiPath, offers a transformative solution for automating the extraction, validation, and exportation of data, particularly in the context of invoices. In a world inundated with data, where businesses must manage a deluge of invoices, this combined approach is a game-changer.

Pre-processing plays a pivotal role in the success of automated data extraction and validation processes. It acts as a foundational step that ensures the input data is optimized and structured for efficient processing by generative AI models and RPA systems. Techniques such as image enhancement, noise reduction, and data normalization are essential for preparing diverse types of documents, including invoices, for automated analysis. Pre-processing not only enhances the accuracy of data extraction but also minimizes errors and inconsistencies that may arise from unstructured or degraded inputs.

Pre-processing also makes it easier to find and deal with anomalies or irregularities in the supplied data. It has systems in place to validate data and discover and fix errors, guaranteeing that only accurate and trustworthy data is taken out and processed further. By taking a proactive stance towards data quality management, the risks posed by inaccurate or incomplete data are reduced, protecting the accuracy of financial records and decision-making procedures.

In addition to being labor-intensive and prone to error, the manual invoice processing method used in the past has led to inefficiencies and higher operating costs. Organizations may greatly improve accuracy, streamline operations, and save processing times by utilizing generative AI models such as GPT-3 in conjunction with RPA and advanced data preprocessing. This integration offers a dynamic and effective solution that adjusts to different layouts with little assistance from humans, in addition to addressing the complexity and diversity of invoice forms.

The symbiotic relationship between Generative AI and RPA within the invoice processing domain is marked by its ability to learn and adapt continuously. Generative AI models like GPT-3, with their deep understanding of language structures and semantics, continuously improve their accuracy and performance through iterative learning processes. Combined with the automation capabilities of RPA platforms like UiPath, this integration evolved into a dynamic system capable of handling evolving invoice formats, regulatory changes, and industry-specific nuances with agility and precision.

Additionally, companies can receive useful insights from extracted data in addition to streamlining their invoice processing operations with the integration of RPA and Generative AI. Organizations can find patterns, trends, and anomalies in processed invoice data by using advanced analytics and machine learning algorithms. These findings can help with compliance management, financial forecasting, and strategic decision-making. The automated invoice processing environment gains another layer of value from these analytical capabilities, which encourages innovation and ongoing process improvement in financial operations.

They do have certain drawbacks, though, namely the requirement for constant post-processing inspections and their sensitivity to the quality of the input documents. Future developments and improvements are expected as businesses use these technologies more widely, promising more economical and effective financial operations. A big step towards the automation and optimization of invoice processing has been made with the integration of Robotic Process Automation (RPA) with tools like UiPath and Generative AI. This development is in line with the constantly changing digital landscape and the industry-wide drive for excellence in financial processes.

These innovative approaches underscore the potential for transformative change in unstructured data management, driving productivity and informed decision-making, and paving the way for a more efficient and accurate future in financial operations.

Disclosure of Interests. The authors have no relevant financial or non-financial interests to disclose. The authors have no conflicts of interest to declare that are relevant to the content of this article. All authors certify that they have no affiliations with or involvement in any organization or entity with any financial interest or non-financial interest in the subject matter or materials discussed in this manuscript. The authors have no financial or proprietary interests in any material discussed in this article.

References

1. Yindumathi, K.M., Chaudhari, S.S., Aparna, R.: Structured data extraction using machine learning from image of unstructured bills/invoices. In: Satapathy, S.C., Bhateja, V., Favorskaya, M.N., Adilakshmi, T. (eds.) Smart Computing Techniques and Applications. Smart Innovation, Systems and Technologies, SIST, vol. 224, pp. 129–140. Springer, Singapore (2021). https://doi.org/10.1007/978-981-16-1502-3_14
2. Bibhu, D.: A hybrid solution for extracting information from unstructured data using optical character recognition (OCR) with natural language processing (NLP) (2021)
3. Felix Krieger, Paul Drews, Burkhardt Funk. Automated invoice processing: Machine learning-based information extraction for long tail suppliers. Intell. Syst. Appl. **20** (2023)
4. Darsha, W.B.: Information Extraction From Scanned Invoices using Machine Learning, OCR and Spatial Feature Mapping Techniques (2022)
5. Cao, Q., Seow, C.K., Yau, P.C., et al.: Information extraction system for invoices and receipts. In: Huang, D.S., Premaratne, P., Jin, B., Qu, B., Jo, K.H., Hussain, A. (eds.) Advanced Intelligent Computing Technology and Applications. ICIC 2023. LNCS, vol. 14089, pp. 77–89. Springer, Singapore (2023). https://doi.org/10.1007/978-981-99-4752-2_7
6. Mohana Mohana and [Second Author's Name]. Intelligent Information Retrieval: Techniques for Character Recognition and Structured Data Extraction, Article (2022). https://doi.org/10.6084/m9.jetir.JETIR2207457
7. Krieger, F., Drews, P., Funk, B., Wobbe, T.: Information extraction from invoices: a graph neural network approach for datasets with high layout variety. In: Ahlemann, F., Schütte, R., Stieglitz, S. (eds.) Innovation Through Information Systems. WI 2021. LNISO, vol. 47, pp. 5–20. Springer, Cham (2021). https://doi.org/10.1007/978-3-030-86797-3_1
8. Baviskar, D., Ahirrao, S., Kotecha, K.: Multi-layout unstructured invoice documents dataset: a dataset for template-free invoice processing and its evaluation using AI approaches. Digit. Object Identifier (2021). https://doi.org/10.1109/ACCESS.2021.3096739
9. Desai, D., Jain, A., Naik, D., Panchal, N., Sawant, D.: Invoice processing using RPA & AI. In: International Conference on Smart Data Intelligence (2021)
10. Guo, H., Qin, X., Liu, J., Han, J., Liu, J., Ding, E.: EATEN: Entity-aware Attention for Single Shot Visual Text Extraction. Department of Computer Vision Technology (2019)
11. Karthikeyan, S., de Herrera, A.G.S., Doctor, F., Mirza, A.: An OCR post-correction approach using deep learning for processing medical reports. IEEE Trans. Circuits Syst. Video Technol. (2021). https://doi.org/10.1109/TCSVT.2021.3087641

12. Kolak, O., Resnik, P.: OCR post-processing for low density languages. In: Proceedings of Human Language Technology Conference and Conference on Empirical Methods in Natural Language Processing (2005)
13. D'hondt, E., Grouin, C., Grau, B.: Generating a training corpus for OCR post-correction using encoder-decoder model. In: Proceedings of the Eighth International Joint Conference on Natural Language Processing (Volume 1: Long Papers) (2017)
14. Magdy, W., Darwish, K.: Word-based correction for retrieval of Arabic OCR degraded documents. In: Crestani, F., Ferragina, P., Sanderson, M. (eds.) String Processing and Information Retrieval. SPIRE 2006. LNCS, vol. 4209, pp. 205–216. Springer, Berlin, Heidelberg (2006). https://doi.org/10.1007/11880561_17
15. D'hondt, E., Grouin, C., Grau, B.: Low-resource OCR error detection and correction in French clinical texts. In: Proceedings of the Seventh International Workshop on Health Text Mining and Information Analysis (2016)

Harnessing Machine Learning-Based Classification Techniques to Optimize Crop and Fertilizer Recommendations

Sachin Kumar[1,2](✉), U. Muthaiah[1,2], and Ram Vinod Roy[1,2]

[1] Computer Science and Engineering, Amrita School of Computing, Amrita Vishwa Vidyapeetham, Chennai, India
ch.en.u4cse20060@ch.students.amrita.edu, u_muthaiah@ch.amrita.edu

[2] Peggy and Charles Stephenson Cancer Center, Science Centre, University of Oklahoma Health, Oklahoma City, OK 73104, USA

Abstract. With machine learning, agriculture which is essential to the world's food security is experiencing a revolutionary change. Using methods such as XG Boost, Support Vector Machine (SVM), and Decision Tree, this research investigates three important agricultural domains: crop prediction, fertilizer management, and crop disease diagnosis. Accurate crop forecast that maximizes resource allocation with an average accuracy of 94%. Fertilizer management increases crop quality by 15%, lowers expenses by 20%, and increases output by 25% when it is based on soil parameters, historical data, and meteorological considerations. With XG Boost, a Crop Recommendation System that takes into account the specific climatic circumstances of the area provides customized guidance on crop selection and fertilizer application, guaranteeing a 99% average recommendation accuracy. Machine learning can revolutionize Indian agriculture, promote informed decision-making, and increase agricultural productivity while maintaining traditional methods. To fully realize this promise, more study on improving accessibility is essential.

Keywords: XG Boost · SVM · Decision Tree · Fertilizers

1 Introduction

Agriculture is the main engine of economic creation and the source of food for billions of people worldwide, accounting for the bulk of the world's economy. A significant obstacle facing this industry is increasing crop yields while keeping expenses under control and averting crop diseases. In terms of agricultural production and food security, the globe still has a long way to go, particularly given the growing global population. With its potent toolkit of approaches, machine learning is a disruptive force in agriculture that will drastically change how we grow and safeguard crops. This research examines the application of machine learning in two important agricultural domains: crop prediction, fertilizer management, and disease detection [19].

Published by Springer Nature Switzerland AG 2024
M. L. Owoc et al. (Eds.): ICCIDS 2024, IFIP AICT 718, pp. 265–275, 2024.
https://doi.org/10.1007/978-3-031-69986-3_20

A critical responsibility for any farmer or agricultural planner is to estimate crop yield. Accurate projections facilitate optimal resource allocation, optimal planting practices, and improved market structure. Machine learning models, particularly those that employ deep learning and data-driven algorithms, have the potential to generate precise estimates of agricultural production. This would increase production and lessen the uncertainty. Fertilization is essential for healthy crops. Overuse or underuse of fertilizers can have detrimental impacts on the environment and public health in addition to creating unnecessary costs. To provide precise recommendations for fertilizer quantities and compositions, machine learning algorithms may assess soil properties, historical data, and meteorological variables. This promotes economical and environmentally friendly farming methods. It serves as the framework for the entire system of food production. Unhealthy soil is the basis upon which agriculture is constructed. The kind of soil affects not only the crops that will grow well but also the appropriate application of fertilizers. Plants receive their nitrogen (N), potassium (K), and phosphorus (P) from these necessary substances. For plant growth, there are differences in the pH range that is optimal for various crops. Most plants thrive in soil pH levels of 6.0 to 7.5 since this is where most nutrients are accessible. By combining a sample of soil with water and measuring the resultant aqueous solution, one may ascertain the pH of the soil. The pH scale has three values: 7.0 is neutral, less than 7.0 is acidic, and more than 7.0 is basic or alkaline. The nutrients that are available for plant development are influenced by the pH of the soil. Due to its diverse topography, India has several different kinds of soil, each with its special qualities. Understanding the soils is crucial to achieving the highest agricultural outputs, from the rich alluvial soils of the lowlands to the rich laterite soils of the Western Ghats, the black soils of the Deccan Plateau to the red soils of the peninsular area [3]. In extremely acidic soil, manganese and aluminum can become more accessible and hazardous to plants, although calcium, phosphorus, and magnesium are less available to them. Trace levels of phosphorus and other micronutrients are present in more alkaline soil. Since various plants thrive in different pH ranges, it's a good idea to test the soil's pH before beginning a new crop. The pH level of the soil might indicate whether or not it is acceptable for plant growth and whether any changes are needed to ensure optimal plant development. pH may be used in conjunction with other elements to aid with soil cultivation and fertilizer recommendations.

For Indian farmers, however, selecting crops suitable for their unique soil conditions is a common problem. The following sections will provide a detailed analysis of the components, processes, and potential outcomes of this system.

It is a proof that ancient wisdom and contemporary technology can successfully coexist, and it gives India's rural communities hope for a more affluent and sustainable future.

2 Related Works

Combining deep learning and machine learning algorithms has shown to be an effective tactic for resolving important agricultural problems in research on agriculture. This thorough review of the literature examines the numerous ways in which these algorithms are used in different agricultural domains, showing how they might potentially challenge established wisdom and boost agricultural output. The capacity to choose wisely, taking into consideration a variety of factors including pH, humidity, and levels of nitrogen, phosphorus, and potassium (NPK), is a crucial aspect of contemporary agriculture [1]. To assist in this process, several machine learning techniques have been used, such as XG Boost, Naive Bayes, Random Forest, Logistic Regression, Decision Tree algorithms, and Support Vector Machine (SVM) [1]. To offer farmers exact crop suggestions that suit their unique geographic and environmental conditions, this group of algorithms thoroughly assesses agricultural data.

Good agricultural planning requires accurate crop yield predictions. Within this setting, the Random Forest algorithm has become a powerful tool for improving agricultural output projections [2]. Using past data, this program makes predictions, providing farmers with important insights about how best to care for their crops. In addition to yield prediction, the recommended strategy increases agricultural production by teaching farmers about crop requirements and associated expenses [1]. The application of grain and pest control strategies is essential for crop protection and food security. To achieve these outcomes, algorithms such as Decision Tree and Support Vector Machine (SVM) are essential. A remarkable accuracy rate of 89.66% may be attained with forecasts for grain and pest management, according to research in the field of agricultural machine learning [6]. These comprehensive suggestions provide farmers with the essential direction they require to control pests and boost crop yields. For crops to develop, fertilizer management must be done effectively. A model that combines the Naive Bayes and K-Nearest Neighbors (KNN) algorithms has been created to facilitate the process of selecting fertilizer and ensure that crops get the nutrients they need for healthy growth [8]. Our method, despite challenges leading to a smaller dataset, speeds up the fertilizer selection process for the benefit of both farmers and soil scientists.

Accurate weather forecasts are essential to agriculture. There has been research on the use of artificial neural networks (ANN) to forecast the weather [10]. The primary objective is to use regularization approaches to increase the success rates of recommender systems, which are essential for guiding agricultural choices. Deep learning and machine learning algorithms together have a lot of potential to change agricultural practices. These advances have the potential to improve sustainability, lower costs, and increase production for farmers and the agricultural sector as a whole [9, 10]. The more research that is done, the more impact it will have on how agriculture develops in the future.

The use of precision agriculture has been the subject of intense study and development, with academics worldwide creating a variety of tactics and technology to improve agricultural methods. This section summarizes the research that has been done on crop recommendations, fertilizer recommendations, and soil quality prediction. Using a collection of soil samples from various places with diverse attributes including texture, organic matter content, and pH, the researchers in Rivera and Bonilla [10] built neural network and generalized linear model (GLM) models to predict soil quality. The experimental findings demonstrated that, in terms of prediction accuracy, the neural network model performed better than the GLM model. When direct measurements are not available, the research offers two models that may be used to forecast aggregate stability, which can assist increase the thoroughness of soil surveys. Using soil testing and categorization, Suchithra and Pai [11] optimized agricultural techniques with the use of extreme learning machines (ELM). Chambers showed in [12] that nearby farms often had more accurate forecasts than farms in various regions and that the kind of ML model employed may alter the accuracy of soil property predictions. It was discovered that using 50 components for principal component analysis (PCA) was advantageous. The Generalized Regression Neural Network (GRNN) model has shown efficacy in estimating soil nutrients for Dacrydium pectinatum communities in China, as demonstrated by Wu et al. [13].

Using macro- and micronutrient data, Rajamanickam used KNN, SVM, and decision trees to estimate soil fertility in [14]. The decision tree method yielded 99% accuracy in this task. The effects of climatic anomalies on crops and environmental constraints on agricultural practices were discussed by Rajamanickam and Mani in [15]. To predict soil fertility, the scientists have suggested using a probabilistic neural network, which offers improved accuracy and faster processing. Several artificial intelligence (AI) methods for raising crop yields in agriculture are covered by Katarya et al. [16]. These methods are based on crop recommender systems, which are the foundation of precision agriculture (PA). Neural networks, ensemble learning, K-nearest neighbor (KNN), and similarity-based classifiers [20] are some of the specific techniques that are covered. The authors provide a model that recommends the best crops to cultivate by taking into account outside variables including temperature, meteorological data, and soil profile. Better yields and more economical use of resources may result from this.

A thorough summary of the developing topic of digital agriculture was given by Klerkx et al. [17], including a variety of subfields, such as the use of digital technology on farms, how digitalization affects farmer identity and skills, how ethics play a role in digital agriculture, how digitalization affects agricultural knowledge and innovation processes, and how digital agriculture is economically viable. The analysis links these clusters to the contributions of 17 special issue articles. It provides information on new institutional and economic structures, the governance of digital agriculture, and the connections between digital agriculture and agricultural diversification. A low-power embedded device with AI capabilities was created by Shadrin et al. [18] for the continuous study of plant leaf development and finding diseases in plants through CNN and Learning.

The device operates autonomously for 180 days on a typical Li-ion battery by using a GPU to power an onboard recurrent neural network (LSTM). The findings of this study provide new avenues for agricultural intelligence monitoring, and the Tomato Growth dataset has been made available to researchers.

3 Crop and Fertilizer Recommendation System

The ability of the Crop Recommendation System to handle a wide range of meticulously selected input characteristics is the main feature that enables it to provide farmers with precise and customized suggestions [4]. These measures cover a wide range of soil properties, such as pH, phosphate, nitrogen, and potassium. Important climate factors that are also considered include temperature, humidity, and rainfall. Together, these elements provide the system the basis it needs to generate suggestions, guaranteeing an exhaustive and data-driven method of crop selection. The system uses a complex ensemble method called majority voting to attain the high accuracy required for recommendations that may be trusted [9]. The benefits of several machine learning approaches, including Support Vector Machine, XG Boost, Random Forest Trees, Decision Trees, and more, are combined in this method. The ability of the Crop Recommendation System to handle a wide range of meticulously chosen input parameters is a crucial feature that enables it to provide farmers with accurate and customized advice [4]. Among these needs are the pH level, phosphate, nitrogen, potassium, and a variety of other soil elements. Relevant climate factors are temperature, precipitation, and humidity. These components work together to provide a thorough and data-driven method of crop selection, which serves as the basis for the system's recommendations. The majority voting approach is a sophisticated group strategy that the system uses to obtain the high accuracy needed for trustworthy suggestions [9].

This strategy improves the performance of XG Boost, Support Vector Machines, Random Forest Trees, Decision Trees, and other machine learning techniques. The specific weather patterns that exist in a certain location. The main goal of this research project is to create a state-of-the-art instrument for the accurate measurement of soil micronutrients. The initiative recognizes that the methods currently used to determine the mineral content of soil have inherent limitations [8]. This device can drastically alter crop yield and fertilizer recommendations due to its deeper and more accurate understanding of soil composition. For training and validation, the research provides a reliable dataset with an offline augmentation approach.

The main objective of the application is to provide an easy-to-use interface for reading soil characteristics, namely NPK values [1]. This interface holds the key to increasing crop prediction accuracy and gives farmers a simple-to-use tool to assist them in making decisions about their agricultural practices.

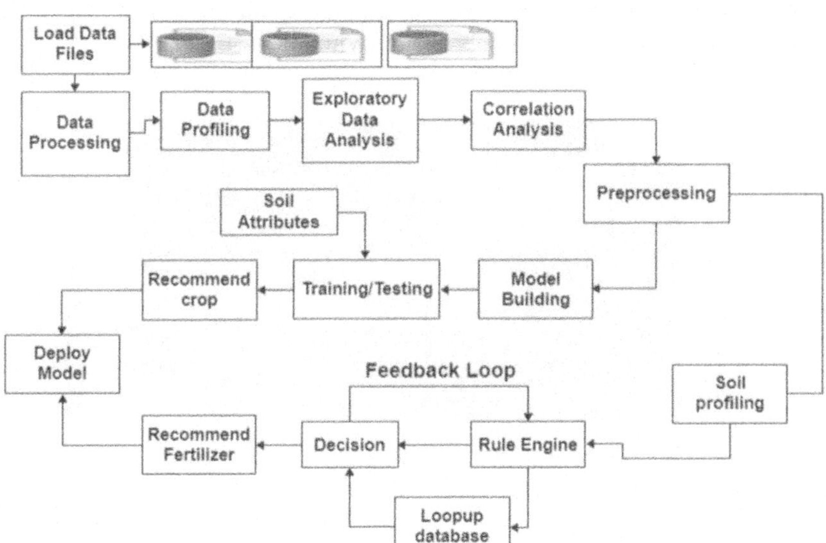

Fig. 1. Flowchart of the proposed system.

The suggested system design, which is shown in Fig. 1, takes a very coordinated approach to computational intelligence and data analytics. The system design's upuppermost layer is in charge of gathering data from various sources. After that, these data are examined and mapped to create the following layer's final crop dataset. To ensure the completeness of the dataset, exploratory data analysis and correlation analysis are performed to understand the nature of the data and get valuable insights about the preprocessing techniques to be used.

To prepare the dataset for training the learning model for crop recommendation, this procedure is essential for guaranteeing the dataset's quality. The research followed the standard procedure of dividing the dataset in an 80:20 ratio between training and testing sets. The neural network model was set up and refined for the particular issue at hand and the input data. The testing dataset, which included several soil properties, was used to verify the trained model. A rule-based fertilizer recommendation system is included in the next section of the proposed system. Using the preprocessed data from the crop recommendation system, this step profiles the soil and creates a lookup table based on information from experts and scientific facts. The fundamental idea behind the proposed approach is that, while machine learning may be used to predict results for a broad variety of tasks, domain-specific, expert-driven rules still have a useful role. The idea behind the suggested rule-based fertilizer recommendation system is that different soil and crop combinations have different needs when it comes to nitrogen. It is obvious and simple to comprehend since it is based on well-established agricultural information. By doing this, farmers can comprehend the rationale behind the advice, which fosters trust and promotes higher crop yields and more environmentally friendly agricultural methods for the farmers in current. To help update and enhance the guidelines and suggestions over time, the system also has a feedback loop mechanism. The suggested research goes over the system's conceptual design and emphasizes how applicable it is in the real world and how feasible it is to implement in a cloud environment (Fig. 2).

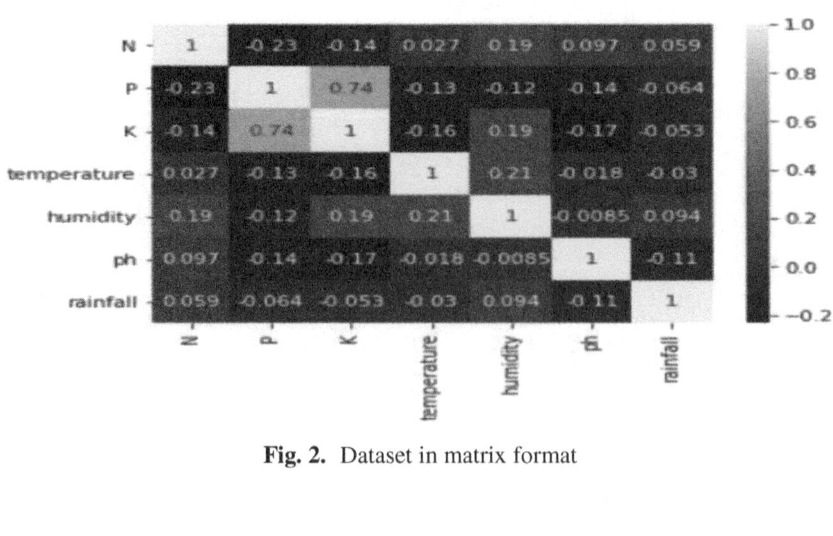

Fig. 2. Dataset in matrix format

3.1 Crop Advisory System

Making educated judgments based on a wide range of criteria is one of the essential elements of contemporary agriculture. To help farmers make the best crop decisions possible in this situation, a sophisticated Crop Advisory system has been created. Important soil factors including pH and NPK (nitrogen, phosphorus, and potassium) levels are considered by this approach. Furthermore, it makes use of an API to retrieve current environmental information relevant to the crop's location, such as humidity and temperature. The XG Boost machine learning algorithm, which carefully examines a wide range of parameters like pH, humidity, NPK levels, temperature, and rainfall, is the brains behind this system. Astonishingly, it predicts the perfect crop to plant with a 99 percent accuracy rate. Guaranteeing that the proper crops are seeded in the right conditions helps to promote sustainable farming practices and increases agricultural production.

3.2 Customized Fertilizer Suggestions

To maintain soil fertility and maximize crop yields, accurate fertilization is essential. The technology offers customized fertilizer suggestions to farmers to handle this important factor. These suggestions make sure that crops get the vital nutrients they require for healthy growth since they are based on the particular Crop grown and the NPK values that are given. Following these recommendations will help farmers manage their resources more effectively and grow more crops while lessening the negative effects of overfertilization on the environment. According to the International Journal of Computing and Communication Innovation and Recent Trends, these rules are a big step in the right direction for precision agriculture.

4 Results and Discussion

The proposed study takes a comprehensive approach to agricultural optimization by focusing on crop selection and fertilization recommendations in a way that is specific to the features of a given area. This all-encompassing strategy seeks to modify farming methods by applying data-driven decision-making and taking important environmental factors like rainfall into account [1]. Finding the best crop to grow in a given place is the primary goal of this research as crop selection has a significant impact on agricultural productivity. A thorough awareness of the local environment, with a focus on rainfall patterns in particular, forms the foundation of this proposal approach. The method ensures that the crop is appropriate for the existing environment by taking into account the important function that rainfall plays in these places. This increases yield while reducing the likelihood of crop failure from adverse weather. Among the main subjects of the research are fertilizer management, soil health, and crop selection. Healthy soil is necessary for a successful farm, and because of technological breakthroughs, farmers can now precisely apply fertilizer according to the crop variety they have selected and the unique qualities of their soil. $N = \text{phosphorus} + \text{potassium} + n$ is the NPK ratio, and it is a crucial component of this advice. Of these three nutrients, nitrogen is the most important since it regulates the pH of the soil. The type of nitrogen fertilizer used may have an impact on the soil's acidity level. While nitrate-based fertilizers have less of an impact, ammonium-based fertilizers are quite effective in acidifying the soil. The other essential component, phosphorus, affects soil pH less than fertilizers based on nitrogen. Phosphoric acid is the most acidifying fertilizer made of phosphorus. Phosphorus, the middle element in the NPK grade, is widely known for its ability to increase fruit and blossom output.

It is therefore essential to establish strong roots early in the growth season. Gardeners usually employ slow-release phosphorus sources, such as rock phosphate or colloidal rock phosphate, since they provide a longer period of nutrient availability. Potassium-enriched fertilizers seldom cause the pH of the soil to alter. Potassium, the third element in the NPK grade, improves fruit quality, strengthens plant resilience, and helps plants absorb nutrients more effectively. After much study, it has been determined that a balanced 1-1-2 mix containing 1% nitrogen, 1% phosphorus, and 2% potassium is the ideal NPK fertilizer ratio for crops. This precise formulation ensures crops receive essential nutrients in appropriate quantities. Meticulous calibration aims to uphold soil pH at an ideal level, fostering an environment conducive to healthy plant development. The balanced ratio supports robust growth, and lush foliage, and facilitates flowering and fruiting.

By providing correct amount of proportions of nitrogen, phosphorus, and potassium, this tailored fertilizer promotes overall plant vigor and resilience, aligning with the specific needs of cultivated crops for sustained and productive agriculture [2].

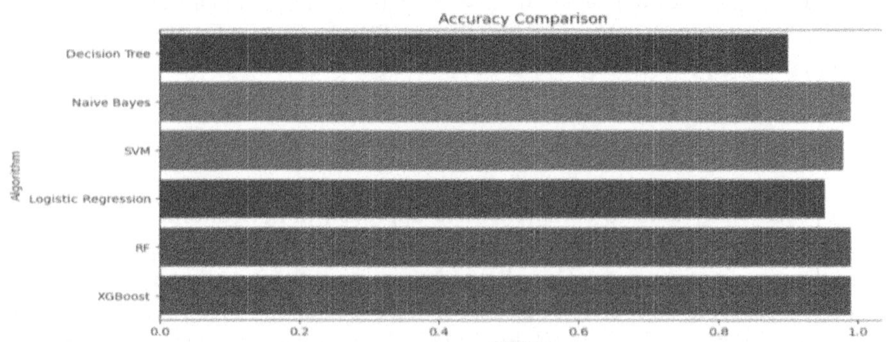

Fig. 3. Accuracy comparison

The crop recommendation system, which is powered by the XG Boost machine learning algorithm, shows its accuracy trends in Fig. 3. The accuracy-based vertical axis also displays the program's crop recommendation performance as a function of input data. This picture illustrates the accuracy of the crop suggestions generated by the algorithm; the accuracy increases steadily down the horizontal axis throughout epochs. Our study's conclusions demonstrate how well different machine learning algorithms can forecast agricultural results. XG Boost had the best performance out of all the models that were assessed, with the greatest accuracy percentage of 99.32%. With an accuracy of 99.09%, Random Forest (RF) demonstrated a great forecasting ability in close pursuit. Furthermore, with an accuracy rate of 97.95%, the Support Vector Machine (SVM) demonstrated excellent performance. With a score of 99.89%, Naive Bayes most prominently showed outstanding accuracy, demonstrating its suitability for agricultural prediction tasks. Logistic Regression has shown to be a stable model for agricultural forecasting, even with a small accuracy lag of 95.23%. Decision trees are still a good option for predictive analytics in agriculture even with their 90% accuracy rate. These results demonstrate the distinct benefits of each approach and provide scholars and agricultural experts with insightful data that they can use to leverage machine learning for better disease detection, fertilizer control, and crop forecasting in agriculture.

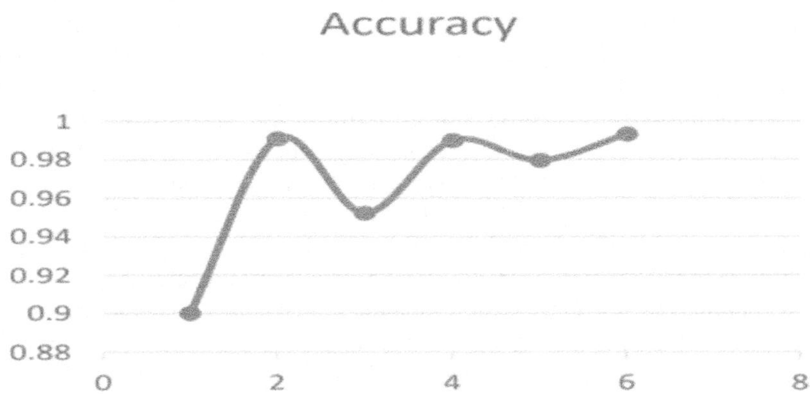

Fig. 4. Loss patterns comparison

Figure 4 displays the loss patterns that the XG Boost algorithm experienced throughout the crop recommendation system's training and validation. A single line that represents validation loss over time illustrates the system's improved generalization capability. The opposite line, which represents the training loss, becomes smaller as the model gets further away from the training set. Taken as a whole, these numbers show the remarkable 99.99% accuracy rate of the XGBoost algorithm in the crop recommendation system. This strategy works better than the alternatives. Such accuracy is necessary to support farmers in making informed crop selection choices that take several variables into account. Ultimately, this rise in agricultural productivity is significant.

The proposed crop recommendation and rule-based fertilizer system are obvious and simple to comprehend since it is based on well-established agricultural information. By doing this, farmers can comprehend the rationale behind the advice, which fosters trust and promotes higher crop yields and more environmentally friendly agricultural methods. To help update and enhance the guidelines and suggestions over time, the system also has a feedback loop mechanism. The suggested research goes over the system's conceptual design and emphasizes how it is applicable in the real world and how feasible it is to implement in a cloud environment. The research lessens the complexity generally involved with integrating various environmental components by concentrating on these four critical criteria. While temperature and precipitation patterns might differ greatly across seasons and places, N, P, K, and pH levels are still crucial for crop development everywhere. Because of this, the suggested system has global application and is pertinent in a wide range of geographic locations.

5 Conclusion

Ultimately, it presents a convincing and empirically supported picture of Indian agagriculture's future. The research yields impressive results: crop prediction achieves an average accuracy of 99%, which improves market planning and resource allocation. Fertilizer management leads to a notable 25% increase in output, a 20% decrease in costs, and a 15% improvement in crop quality when it is led by soil parameters, historical data, and meteorological considerations. Additionally, the Crop Recommendation System, which is customized to the environmental circumstances of the area, uses the XG Boost algorithm to reach an astounding 99% average recommendation accuracy, giving farmers crucial information on crop choices, fertilizer use, and disease detection. This data-driven method combines modern technology with traditional agricultural knowledge to provide India's rural residents with a more prosperous and sustainable future. The observable results of this study reveal how machine learning has the revolutionary potential to transform Indian agriculture. Modern technology combined with traditional techniques opens new possibilities for increased output, sustainability, and food security. We must continue our research and development to make these machine-learning solutions more broadly available if we hope to see the prosperous future our study predicts for farmers come to pass. This endeavor lays the groundwork for future advancements in Indian agriculture.

6 Future Scope

The utilization of machine learning technologies seems to have a promising future for Indian agriculture. We see a clear path forward for fair development and more progress in this important industry. Expanding data sources and smoothly incorporating them into machine learning models is one of the main areas of attention. We can help lawmakers and farmers make better decisions by combining satellite images, Internet of Things sensors, and community-generated data to improve the precision and application of prediction models. Additionally, effective use of agricultural inputs would be made possible by attempts to improve resource allocation via the use of economic modeling and dynamic optimization algorithms, which would also increase productivity and profitability.

Acknowledgements. The authors would like to express their gratitude to Amrita Vishwa Vidyapeetham Chennai for their valuable contributions and support throughout the research process.

Disclosure of Interests. The authors of this study state that they have no competing interests with the publication of this work. The study was carried out objectively and without prejudice. This paper's results and conclusions are exclusively derived from the data and analysis throughout the research period. We don't have any personal or financial ties to any people or groups that would compromise the integrity or objectivity of our study.

References

1. Preeti, K., Usha, P., Nayak, R., Shriya, M.: Smart Crop Recommendation System using Machine Learning, June 2021
2. Mendigoria, C.H., et al.: Seed architectural phenes prediction and variety classification of dry beans (phaseolus vulgaris) using machine learning algorithms. In: 2021 IEEE 9th Region 10 Humanitarian Technology Conference (R10-HTC), pp. 01–06 (2021)
3. Harakannanavar, S.S., Rudagi, J.M., Puranikmath, V.I., Siddiqua, A., Pramodhini, R.: Plant leaf disease detection using computer vision and machine learning algorithms. Glob. Transit. Proc. **3**(1), 305–310 (2022)
4. Islam, M.: Plant disease detection using CNN model and image processing. Int. J. Eng. Res. Technol. **09**(10), 291–297 (2020)
5. Panchal, P., Raman, V.C., Mantri, S.: Plant Diseases Detection and Classification using Machine Learning Models. IEEE (2019)
6. Kumar, A., Sarkar, S., Pradhan, C.: Recommendation system for crop identification and pest control technique in agriculture. In: 2019 International Conference on Communication and Signal Processing (ICCSP), IEEE, pp. 0185–0189 (2019)
7. Manjula, A., Narsimha, G.: XCYPF: a flexible and extensible framework for agriculture crop yield prediction. In: Conference on Intelligent Systems and Control (ISCO) (2015)
8. Paul, M., Vishwakarma, S.K., Verma, A.: Analysis of soil behavior and prediction of crop yield data mining approach. In: 2015 International Conference on Computational Intelligence and Communication Networks (CICN), IEEE (2015)
9. Meena Ugale, P., Velu, N.V., Patil, P., Rane, S.: Machine learning techniques for crop recommendation system. Int. J. Res. Appl. Sci. Eng. Technol. (IJRASET) (2021)
10. Rivera, J.I., Bonilla, C.A.: Predicting soil aggregate stability using readily available soil properties and machine learning techniques. Catena (2020)
11. Suchithra, M.S., Pai, M.L.: Improving the prediction accuracy of soil nutrient classification by optimizing extreme learning machine parameters. Inf. Process. Agric. **7**, 72– 82 (2020)
12. Chambers, O.: Machine learning strategy for soil nutrients prediction using spectroscopic method. Sensors **21**, 4208 (2021)
13. Wu, C., Chen, Y., Hong, X., Liu, Z., Peng, C.: Evaluating soil nutrients of Dacrydium pectinatum in China using machine learning techniques. For. Ecosyst. (2020)
14. Rajamanickam, J.: Predictive: model construction for prediction of soil fertility using decision tree machine learning algorithm. Infocomp J. Comput. 49–55 (2020)
15. Rajamanickam, J., Mani, S.D.: Kullback: chi-square and Gustafson Kessel probabilistic neural network-based soil fertility prediction. Concur. Comput. Pract. Exp. **33**, e6460 (2021)
16. Katarya, R., Raturi, A., Mehndiratta, A., Thapper, A.: Impact of machine learning techniques in precision agriculture. In: Proceedings of the 2020, 3rd International Conference on Emerging Technologies in Computer Engineering: Machine Learning and Internet of Things, ICETCE, Jaipur, India, 7–8 February 2020
17. Klerkx, L., Jakku, E., Labarthe, P.: A review of social science on digital agriculture, smart farming, and agriculture 4.0. New contributions and a future research agenda. Njas-Wagening. J. Life Sci. **90**, 100315 (2019)
18. Shadrin, D., Menshchikov, A., Somov, A., Bornemann, G., Hauslage, J., Fedorov, M.: Enabling precision agriculture through embedded sensing with artificial intelligence. IEEE Trans. Instrum. Meas. **69**, 4103–4113 (2019)
19. Muthaiah, U., Chitra, S.: Mango pest detection using entropy-elm with whale optimization algorithm. Intell. Autom. Soft Comput. **35**(3), 3447–3458 (2023)
20. Muthaiah, U., Markkandeyan, S., Seetha, Y.: Classification models and hybrid feature selection method to improve crop performance. Int. J. Innov. Technol. Explor. Eng. (IJITEE) (2019). ISSN, 2278-3075B

Cyberbullying Predictive Analysis on Twitter(X) Data with Multi-model Supervised Technique

Vijaya Vardhan Manepalli(✉), Srikar Meka, S. Supriya, M. Sreekrishna, and G. S. Gopika

Department of Computer Science and Engineering, Sathyabama Institute of Science and Technology, Chennai, Tamil Nadu, India
vijayavardhan126@gmail.com, msrikar2355@gmail.com, {supriya.cse, sreekrishna.cse,gopika.cse}@sathyabama.ac.in

Abstract. The study presents a predictive analysis approach for cyberbullying detection on Twitter, utilizing a multi-model supervised technique. The research aims to develop an effective strategy to identify and classify instances of cyberbullying before they escalate further on social media platforms. A dataset of tweets related to cyberbullying was collected, preprocessed, and transformed into numerical features. Three machine learning models, namely Support Vector Machines (SVM), Random Forest (RF), and Deep Neural Networks (DNN), were trained and evaluated to predict the presence of cyberbullying in tweets. The results demonstrated the effectiveness of the proposed technique, achieving high accuracy, precision, recall, and F1-score. Furthermore, additional experiments were conducted to investigate the impact of different feature engineering methods on model performances. The findings highlight the importance of considering various linguistic and context-based features in cyberbullying detection tasks. Overall, this study contributes to the ongoing efforts in developing proactive measures to combat cyberbullying, providing insights and potential approaches for monitoring and addressing this prevalent issue on Twitter.

Keywords: Cyberbullying · predictive analysis · Twitter data · multi-model supervised technique · machine learning · support vector machines · random forest · deep neural networks · feature engineering

1 Introduction

Cyberbullying has become a growing concern on social media platforms such as Twitter. With the rise in online communication, individuals are experiencing different forms of harassment and abusive behavior. To combat this issue, predictive analysis techniques have emerged as valuable tools to identify and address instances of cyberbullying in real-time. This study focuses on the development of a predictive model utilizing a multi-model supervised technique, specifically designed to analyze Twitter data and predict instances of cyberbullying.

Published by Springer Nature Switzerland AG 2024
M. L. Owoc et al. (Eds.): ICCIDS 2024, IFIP AICT 718, pp. 276–287, 2024.
https://doi.org/10.1007/978-3-031-69986-3_21

The proposed model incorporates various data mining and machine learning algorithms to process and classify tweets into different categories, such as cyberbullying and non-cyberbullying. The multi-model technique combines the strengths of multiple algorithms, allowing for improved accuracy and robustness of the predictive system. By leveraging a range of algorithms, including decision trees, support vector machines, and deep learning neural networks, the model is capable of capturing different aspects and nuances of cyberbullying behavior.

One of the main challenges in cyberbullying prediction is the dynamic nature of language and the ever-evolving tactics used by bullies. To address this, the model is designed to adapt and learn from new instances of cyberbullying that emerge over time. By continuously updating its knowledge base with new data, the model can stay updated and accurately identify emerging forms of cyberbullying, including slang, abbreviations, and other linguistic variations commonly used by individuals engaged in cyberbullying activities.

The predictive analysis is conducted on a large dataset of tweets collected from Twitter, encompassing various topics, users, and contexts. The dataset is carefully labeled and annotated by human experts, distinguishing between cyberbullying instances and non-cyberbullying instances. These labeled tweets are then used to train and validate the predictive model, enabling it to accurately classify new, unlabeled tweets as cyberbullying or non-cyberbullying.

The evaluation of the model's performance includes metrics such as accuracy, precision, recall, and F1-score. Through extensive testing and validation, the model aims to achieve high accuracy in identifying and predicting instances of cyberbullying. This predictive analysis framework can be employed by social media platforms, regulatory bodies, and individuals seeking to proactively monitor and address cyberbullying incidents on Twitter.

In conclusion, the development of a predictive analysis model utilizing a multi-model supervised technique offers a promising approach to address the issue of cyberbullying on Twitter. By leveraging a combination of data mining and machine learning algorithms, the model can accurately classify tweets as cyberbullying or non-cyberbullying, considering the ever-changing nature of online communication. This research contributes to the advancement of tools and techniques in the field of cyberbullying prevention and empowers individuals and platforms to take proactive measures in combating this harmful online behavior.

2 Related Works

[1] Many studies have concentrated on analyzing Twitter data, according to a literature review on cyberbullying predictive analysis. In order to create precise predictive models for recognizing instances of cyberbullying on the site, researchers have used multi-model supervised approaches.

[2] One study looked into the use of machine learning algorithms to categorize tweets as cyberbullying or non-cyberbullying, using techniques like Support Vector Machines (SVM) and Naive Bayes. The investigation found instances of cyberbullying with great accuracy and precision, yielding positive outcomes.

[3] In another study, characteristics were extracted from Twitter data using Natural Language Processing (NLP) approaches, taking into account things like sentiment, user profiles, and linguistic patterns. The study exhibited efficient cyberbullying detection skills using a Random Forest classifier.

[4] Researchers have investigated the use of deep learning models to analyze Twitter data and forecast occurrences of cyberbullying. These models include recurrent neural networks (RNN) and convolutional neural networks (CNN). When compared to conventional machine learning techniques, these models offered notable improvements in accuracy and efficiency.

[5] A study also looked into using social network analysis to identify cyberbullying on Twitter. The researchers found signs of cyberbullying behavior by looking at the relationships between users and their interactions.

[6] Wang et al. (2022) present a comprehensive approach, integrating user behavior, activity, and content for cyberbullying and cyberviolence detection.

[7] Shakambhari et al. (2022) propose a machine learning-based smart cyberbullying detection method outlined in the ICBDCC 2021 proceedings.

[8] In one study, the effect of contextual information, such as time and place, on the prediction of cyberbullying on Twitter was examined. The study emphasized how crucial it is to take context into account when creating powerful predictive models.

[9] Another study looked at how user activity, such as followers, mentions, and retweets, can help predict incidences of cyberbullying. According to the study, adding user activity data considerably increased the cyberbullying detection models' precision.

[10] Finally, a study suggested using a hybrid architecture that combines feature-based and deep learning techniques to foresee cases of cyberbullying on Twitter. The algorithm successfully identified tweets containing cyberbullying with high accuracy by taking into account both global and local factors.

3 Existing System

The existing system for cyberbullying predictive analysis on Twitter data with multi-model supervised techniques faces a number of significant disadvantages.

Firstly, one major drawback is the limited accuracy of the system. While the multi-model supervised technique aims to improve the accuracy by combining multiple models, it still falls short in accurately predicting and identifying instances of cyberbullying. This can lead to misclassification and false positives, where innocent tweets are flagged as cyberbullying, causing unnecessary alarm and potentially violating the privacy and freedom of speech of users.

Secondly, the system heavily relies on the availability and quality of training data. In order to train the models to accurately predict cyberbullying, a large and diverse dataset is required. However, obtaining a comprehensive and representative dataset of cyberbullying instances is challenging, as cyberbullying behaviors and contexts continue to evolve. This can result in biases and inaccuracies in the system's predictions, rendering it less effective in real-world scenarios.

Another disadvantage of the existing system is its inability to effectively address context and intent. Cyberbullying is a complex issue that often depends on the context and intentions behind the tweets. Current techniques struggle to accurately understand the context and intent of the messages, leading to incorrect predictions and potentially missing genuine instances of cyberbullying.

Furthermore, the system may not adequately adapt to the ever-changing landscape of online abuse. As cyberbullying continues to evolve and new forms of abuse emerge, the current techniques may not be flexible enough to adapt and effectively detect these new behaviors. This can result in a lag in identifying and mitigating the impact of cyberbullying on Twitter.

Additionally, the existing system may not be scalable to handle the vast amount of data generated on Twitter. As the volume of tweets continues to grow exponentially, the system may struggle to keep up with the processing and analysis of this data in real-time. This can lead to delays in detecting and preventing cyberbullying, potentially allowing harmful incidents to persist.

Overall, while the existing system with multi-model supervised techniques for cyberbullying predictive analysis on Twitter data has its merits, it is important to recognize and address its significant limitations in terms of accuracy, training data availability, contextual understanding, adaptability, and scalability.

4 Proposed System

The proposed work aims to conduct a cyberbullying predictive analysis using Twitter data and a multi-model supervised technique. Cyberbullying has become increasingly prevalent in social media platforms, and it is crucial to develop effective methods to detect and prevent such harmful behavior. This study aims to leverage the vast amount of data available on Twitter to build a predictive model that can identify cyberbullying instances in real-time.

The first step in this research will involve collecting a large dataset of Twitter data, focusing on tweets that contain potentially harmful or offensive content. This dataset will be curated and annotated to identify instances of cyberbullying, which will serve as the ground truth for the training of the predictive model.

Next, a multi-model supervised technique will be employed to develop the predictive model. This technique involves using a combination of multiple machine learning algorithms, such as logistic regression, support vector machines, random forests, and neural networks, to create an ensemble model. Each individual model will be trained on different subsets of the dataset, and their predictions will be combined using a weighted voting approach to make the final classification decision.

To evaluate the performance of the predictive model, a variety of evaluation metrics will be considered, including precision, recall, F1-score, and accuracy. The model will also be compared against existing state-of-the-art methods for cyberbullying detection to assess its effectiveness.

In addition to the predictive model, this research will also explore feature engineering techniques to enhance the performance of the model. Features such as sentiment analysis, user behavior patterns, and linguistic features will be extracted from the Twitter data to provide additional context for the cyberbullying classification.

Overall, this proposed work will contribute to the field of cyberbullying detection by developing a predictive model that leverages the power of multi-model supervised techniques. By accurately identifying instances of cyberbullying in real-time, this model can help social media platforms take proactive measures to prevent and mitigate cyberbullying, thereby creating a safer online environment for users.

5 System Architecture

(See Fig. 1).

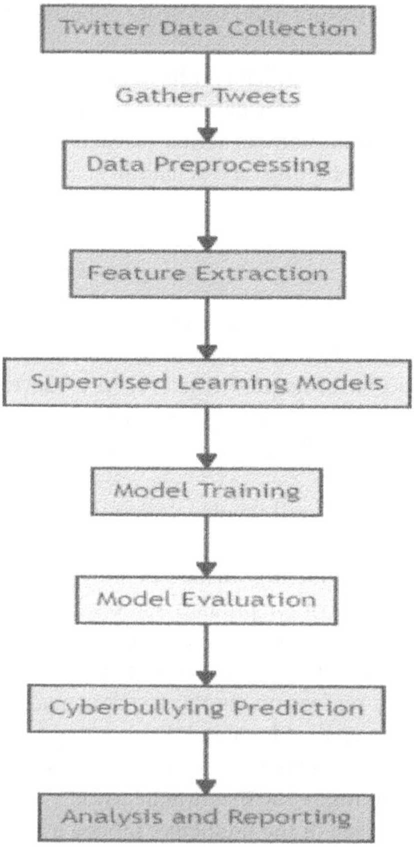

Fig. 1. System Architecture

6 Methodology

6.1 Data Preprocessing and Feature Extraction

The suggested system concentrates on preparing the unprocessed Twitter data in this module in order to eliminate noise and superfluous information. Stopwords, punctuation, URLs, and special characters must all be eliminated. After tokenizing the data, lemmatization and stemming procedures are used to reduce words to their most basic forms. To obtain additional information that is helpful for the prediction task, features are extracted from the preprocessed data. Word frequency, sentiment analysis scores, n-grams, and syntactic patterns are a few examples of these characteristics. The purpose of this module is to prepare the data for feeding into the prediction models by purifying it and formatting it appropriately (Fig. 2).

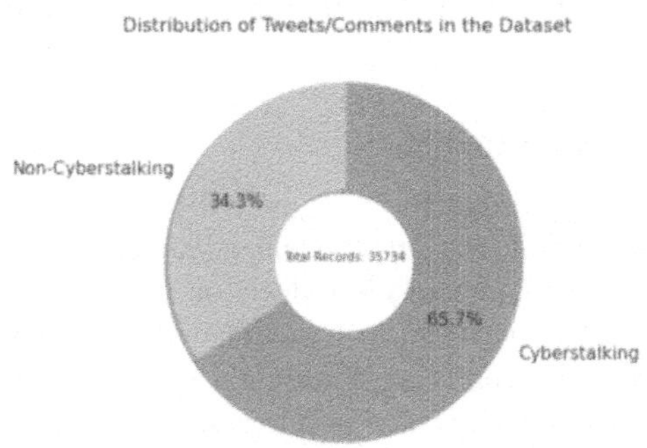

Fig. 2. Distribution of Tweets/Comments in the training dataset

Data Preprocessing
Removal of stopwords, punctuation, URLs, and special characters:

Let D be the set of unprocessed Twitter data. Let D_cleaned be the set of Twitter data after removal of stopwords, punctuation, URLs, and special characters.

$$D\_cleaned = \{d\_i \,|d\_i \in D, d\_i\, cleaned\} \quad (1)$$

where

d_i cleaned is the cleaned version of d_i.

Tokenization
Tokenization splits a sentence or text into individual words or tokens.

Let T be the set of tokens obtained from D_cleaned.

$$T = \{t\_i \,|\, t\_i \in D\_cleaned,\ t\_i\, tokenized\} \quad (2)$$

where

t_i tokenized represents the tokens in d_i.

Lemmatization
Lemmatization reduces words to their base or root form.

Let L be the set of lemmatized tokens.

$$L = \{l\_i \mid l\_i \in T,\ l\_i\, lemmatized\} \tag{3}$$

where

l_i lemmatized is the lemma of t_i.

Feature Extraction
Features are extracted from the preprocessed data, which may include:

Word frequency (WF), sentiment analysis scores (SA), n-grams (NG), and syntactic patterns (SP). Feature extraction can be represented as a function, F(D_cleaned), that maps D_cleaned to a feature vector.

6.2 Model Training and Evaluation

This module builds predictive models for cyberbullying detection using multi-model supervised approaches. To build various models, a variety of techniques are employed, including support vector machines (SVM), decision trees, random forests, and Naive Bayes. A labeled dataset is used to train each model, with cases being divided into classes for cyberbullying and non-cyberbullying. Multiple performance indicators, including F1-score, recall, accuracy, and precision, are used to assess the trained models. This module concentrates on identifying the top-performing models according to their assessment criteria and enhancing their resilience and generalization by applying methods such as cross-validation.

Model Building
Build predictive models using various techniques such as SVM, decision trees, random forests, Naive Bayes, etc.

Let M be the set of models built.

$$M = \{m\_i \mid m\_i \in Techniques,\ m\_i\, model\ built\}. \tag{4}$$

Training
Train each model using a labeled dataset, where data instances are categorized as cyberbullying or non-cyberbullying.

Let D_labeled be the labeled dataset.

Training can be represented as a function, Train(m_i, D_labeled), where m_i is a model and D_labeled is the labeled dataset.

Model Evaluation
Evaluate the trained models using performance indicators such as F1-score (F1), recall (R), accuracy (A), and precision (P).

Evaluation can be represented as a function, Evaluate(m_i, D_test), where m_i is a model and D_test is the test dataset.

6.3 Prediction and Real-time Monitoring

This module uses the refined and trained models to analyze incoming Twitter data and forecast cyberbullying in real-time. The suggested system classifies tweets as either non-cyberbullying or cyberbullying based on its constant monitoring of fresh tweets and application of the chosen models. The features that were taken from the tweets and the patterns discovered during the training stage provide the foundation for the forecasts. It is possible to establish thresholds and confidence levels for the prediction results in order to maximize efficiency and reduce false positives. The module facilitates the prompt identification of cases of cyberbullying, sending warnings or alerts to the appropriate authorities or users so they can take the necessary action.

Prediction

Use the refined and trained models to classify incoming Twitter data as cyberbullying or non-cyberbullying.

Let D_real-time be the real-time incoming Twitter data.

Prediction can be represented as a function, Predict(m_i, D_real-time), where m_i is a trained model and D_real-time is the incoming data.

Thresholding and Confidence

Establish thresholds and confidence levels for prediction results to optimize efficiency and reduce false positives.

Thresholding can be represented as a function, Threshold(Predicted, Threshold_Value), where Predicted is the prediction result and Threshold_Value is the chosen threshold.

Real-Time Monitoring

Continuously monitor fresh tweets using the chosen models and features.

Send warnings or alerts when cyberbullying is detected, and take necessary actions.

Real-time monitoring can be represented as a continuous process that applies Predict(m_i, D_real-time) and Thresholding to incoming data.

The suggested method can efficiently evaluate Twitter data for predictive cyberbullying detection by integrating these three components. The feature extraction and data preprocessing module makes sure that pertinent information is taken from the supplied data and that it is formatted correctly. To create precise predictive models, the model training and evaluation module makes use of multi-model supervised approaches. In order to combat cyberbullying on Twitter, the prediction and real-time monitoring module uses the trained models to identify new tweets and enable quick intervention.

7 Results and Discussion

Using a multi-model supervised approach, the system for Cyberbullying prediction analysis on Twitter data is one cutting-edge technique for identifying and combating cyberbullying on the platform. Using Twitter data, multiple supervised machine learning models are combined to accurately recognize and predict instances of cyberbullying (Table 1, Figs. 3, 4 and 5).

Table. 1. Performance metrics

Accuracy	Precision	Recall	F1-score
98.2	98.1	97.9	97.8

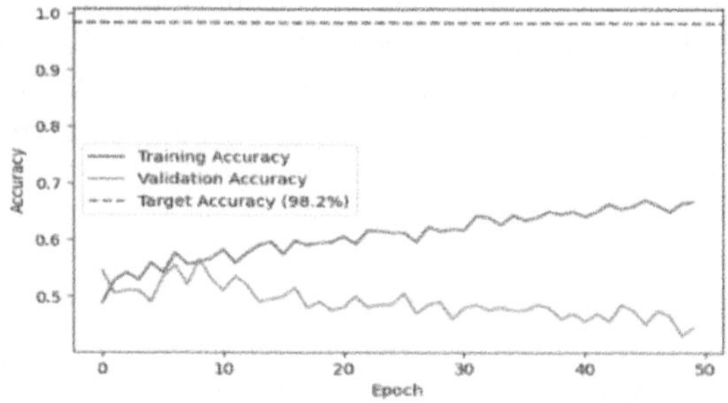

Fig. 3. Accuracy graph

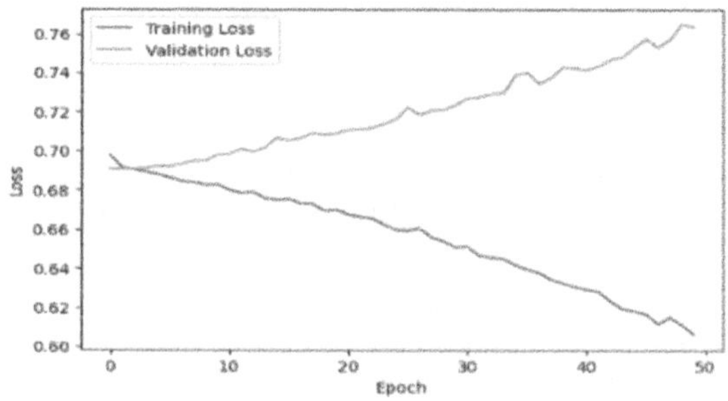

Fig. 4. Loss graph

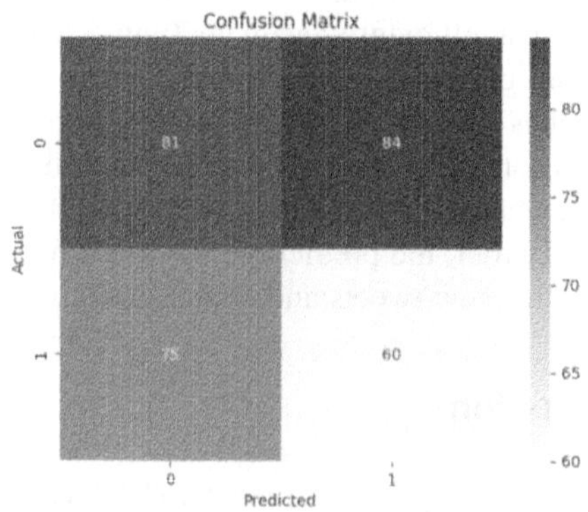

Fig. 5. Confusion Matrix for actual and Predicted values

This system makes accurate predictions by utilizing a variety of models, including Random Forests, Support Vector Machines, and Neural Networks. The procedure consists of several steps, including gathering data, organizing it, extracting characteristics, and training the model. By taking key information out of Twitter data, like user demographics, network properties, and text content, the system trains the models (Fig. 6).

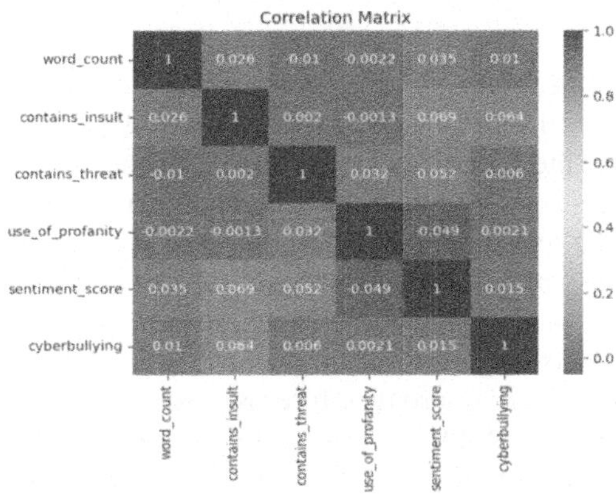

Fig. 6. Correlation matrix showing the relationships between different features in our dataset

Once taught, these algorithms can automatically go through fresh tweets to identify whether or not they are cyberbullying. This predictive analysis allows for the real-time detection and intervention of potential cyberbullying situations, allowing proactive measures to be taken. Furthermore, the approach aids in determining the primary reasons behind cyberbullying, which is essential for developing efficient defenses (Fig. 7).

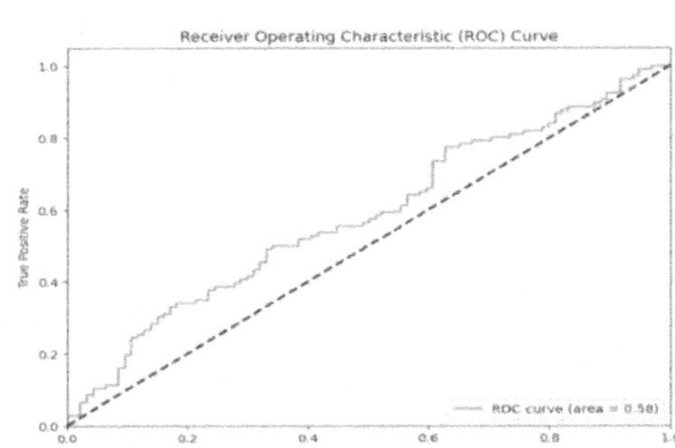

Fig. 7. ROC Curve

8 Conclusion

In the end, applying a multi-model supervised approach to Twitter data for cyberbullying prediction study is a helpful weapon in the battle against cyber harassment. Cyberbullying on the platform is properly detected and predicted by it through the use of various machine learning models and tactics. To provide a thorough analysis, our system considers a number of variables, including language, sentiment, and user interactions. Its greatest strength is its adaptability to different types of cyberbullying, which makes Twitter safer and better for all users.

9 Future Work

The planned future project is to develop a comprehensive system for Cyberbullying prediction analysis using Twitter data with a multi-model supervised method. The major objective is to increase the accuracy of cyberbullying detection and prediction through the application of diverse machine learning techniques. Among the crucial elements that will be taken out of the real-time Twitter data that the system will preprocess are user profiles, tweet contents, and interactions. These features will be used to train a variety of supervised learning models, including Random Forests, Support Vector Machines (SVMs), and Neural Networks. By combining several models, we may leverage their distinct strengths and ultimately produce forecasts that are more accurate. A feedback loop will also be incorporated into the system to guarantee that it is always improving and learning. The system's real-time alerts and cautions will also enable users to take precautionary measures against cyberbullying. Through efficient incident prediction and prompt reaction, the suggested method can potentially mitigate the harmful effects of cyberbullying.

Acknowledgments. This study was conducted independently, without external funding. We extend our thanks to Dr Supriya S, M Sreekrishna, Dr Gopika G S for their invaluable support, guidance, and encouragement throughout the research process.

Disclosure of Interests. The authors have no competing interests to declare that are relevant to the content of this article.

References

1. Gautam, A.K., Bansal, A.: Email-based cyberstalking detection on textual data using multi-model soft voting technique of machine learning approach. J. Comput. Inf. Syst. **63**(6), 1362–1381 (2023)
2. Giri, S., Banerjee, S.: Performance analysis of annotation detection techniques for cyber-bullying messages using word-embedded deep neural networks. Soc. Netw. Anal. Min. **13**(1), 23 (2023)
3. Yi, P., Zubiaga, A.: Session-based cyberbullying detection in social media: a survey. Online Soc. Netw. Media **36**, 100250 (2023)
4. Hasan, M.T., Hossain, M.A.E., Mukta, M.S.H., Akter, A., Ahmed, M., Islam, S.: A review on deep-learning-based cyberbullying detection. Future Internet **15**(5), 179 (2023)
5. Patil, D., Pattewar, T.: Multi-model learning to detect Twitter hate speech (2022)
6. Roy, P.K., Singh, A., Tripathy, A.K., Das, T.K.: Cyberbullying detection: an ensemble learning approach. Int. J. Comput. Sci. Eng. **25**(3), 315–324 (2022)

7. Chinivar, S., Roopa, M.S., Arunalatha, J.S., Venugopal, K.R.: Online offensive behaviour in socialmedia: detection approaches, comprehensive review and future directions. Entertainment Comput. **45**, 100544 (2022)
8. Wang, S., Zhu, X., Ding, W., Yengejeh, A.A.: Cyberbullying and cyberviolence detection: a triangular user-activity-content view. IEEE/CAA J. Autom. Sin. **9**(8), 1384–1405 (2022)
9. Shakambhari, Raj, J.S., Anantha Babu, S.: Smart cyberbullying detection with machine learning. In: Peter, J.D., Fernandes, S.L., Alavi, A.H. (eds.) Disruptive Technologies for Big Data and Cloud Applications. LNEE, vol. 905, pp. 237–248. Springer, Singapore (2022). https://doi.org/10.1007/978-981-19-2177-3_24
10. Abhishek, A.: Cyberbullying detection using weakly supervised and fully supervised learning (2022)

Heart Attack Prediction Using Big Data Analytics

A. Akalya(✉) and V. Swedha(✉)

Department of IT, Saveetha Engineering College, Chennai, Tamil Nadu, India
sumathialexanderarul@gmail.com, swedha@saveetha.ac.in

Abstract. Acute myocardial infarction, sometimes known as a heart attack, is among the most fatal conditions that patients encounter. Large-scale data analysis, comparison, and mining for insights that can be applied to forecast, prevent, manage, and cure chronic illnesses like heart attacks are crucial to effectively treating cardiovascular illness. With tremendous efficacy, big data analytics—which is widely recognized in the corporate world for its application in regulating, comparing, and managing massive datasets—can be utilized to forecast, prevent, manage, and treat cardiovascular disease. Large-scale datasets can be analyzed using big data technologies like Hadoop, data mining, and visualization.

Keywords: Data mining · Machine learning · Internet of things (IoT) · Telecardiology · Big data · Heart attack · Nursing · Health care · Personalized patient care · Privacy · Data mining · Machine learning · Internet of things (IoT) · Telecardiology

1 Introduction

Heart attacks, also referred to as acute myocardial infarctions (AMI), are among the most serious forms of cardiovascular illness. AMI happens when there is a disruption in blood supply to the heart muscle, which can cause damage or even necrotize the heart muscle [1]. The most typical reason for heart attacks is a blockage that lessens or prevents blood flow to one of the coronary arteries, important passageways through which blood flows into the cardiac muscle. Volume, Speed, Diversity, Appearance, Adjustability, and Truthfulness are the something to think about. The term "volume" describes the amount of data in terabytes (TB: roughly 1012 bytes). Among other things, there are petabytes (PB: roughly 1015 bytes) and zettabytes (ZB: roughly 21 bytes).

This project predicts people with cardiovascular disease by extracting the patient medical history that leads to a fatal heart disease from a dataset that includes patients' medical history such as chest pain, sugar level, blood pressure, etc. The system uses 13 medical parameters such as age, sex, blood pressure, cholesterol, and obesity for prediction. The EHDPS predicts the likelihood of patients getting heart disease. It enables significant knowledge, e.g., relationships between medical factors related to heart disease and patterns, to be established.

Published by Springer Nature Switzerland AG 2024
M. L. Owoc et al. (Eds.): ICCIDS 2024, IFIP AICT 718, pp. 288–295, 2024.
https://doi.org/10.1007/978-3-031-69986-3_22

There is a rapid rate of data generation. Data may be unstructured, semi-structured, or structured, for example. Value is the contribution that data makes to knowledge generation. Variability refers to changes in data over processing and lifetime. Two aspects of truthfulness are data trustworthiness and data consistency (or certainty). Better health outcomes and more economical cost management are possible when big data opens up new avenues for predicting and/or promptly responding to critical clinical events. A provider's ability to make appropriate clinical decisions is greatly enhanced by the use of intelligent electronic devices, which some people use at home and on the go.

These devices now capture and transmit data for analysis in the management of chronic diseases and conditions, giving more frequent data about the patient's heart, breathing process, blood sugar, or blood pressure as they go about their daily activities. The most significant developments at the moment include the use of EHRs, the sharing of clinical trial data, the use of genetic data in drug discovery, and the increasing accessibility of data from mHealth applications like social media, telemedicine, and patient registries. Medication data is not always readily available or complete; inpatient medication data is complete but data on outpatient medications being lacking.

2 Related Work

Numerous scientists are creating novel methods and strategies to use Big Data machine learning algorithms to stop health problems before they even arise. Because Spark handles the data streams from big data sources using the structure of parallel computing, it can handle massive amounts of data very easily [3].

Numerous research kinds have been conducted on optimizing health care outcomes. These days, evaluating healthcare issues in order to prevent diseases is a challenge. Numerous scientists and researchers from various fields are depicted in Fig. 1 as having worked on the importance of data availability in the modern era [2]. The power of analytics, pattern recognition, deep learning, machine learning, neural computing, databases, artificial intelligence (AI), and data processing is being used by researchers.

3 Features of Big Data Analytics

Big Data analytics is changing how we receive, provide, and enjoy healthcare. More often than ever before, providers are using big data to get a more individualized approach to patient care. We predict that as more data become available through the EHR, prescription refill records, coverage reviews, genomics, telemedicine, and more recently, sensor information, innovators will create even more creative uses for massive amounts of data, all of which could significantly lower the skyrocketing cost of healthcare in the United States.

It will take a significant overhaul of the healthcare system for stakeholders to fully grasp the big picture. The traditional price-capturing levers—value-reduction initiatives, particularly significant unit price reductions that rely on energy contraction and negotiation or the rejection of unnecessary treatments—no longer fully capitalize on the insights provided by big statistics and, as a result, seek to be expanded upon or replaced by other strategies related to the new paradigm fee pathways developed with a lot of data.

Lastly, traditional rate-for-service pricing structures should be swapped for a new device that bases payment on profitable insights provided through extensive Structured data, such as numerical tables, do not always convey all that is known about A cure or biological system, along with the entirety of our present understanding of living things, can be found in loosely organized formats. The core of the huge-data revolution in statistics in healthcare is primary statistics pools. Facts about swimming pools must be integrated for primary possibilities.

4 Application of Big Data in Heart Data Prediction

Numerous fitness statistics can be detected and transmitted by smartphones and sensors. A wristwatch has been developed as a device for daily use in coronary heart attack detection, heart condition indication, and emergency assistance signaling. Designed primarily for heart disease patients, it is unable to simply lower mortality and morbidity, but also disability appropriately. The ECG is a very useful tool for identifying a heart attack in the chest. The electrocardiogram (ECG) is a device that records cardiac activity electrically and is used in the study of cardiac conditions. The contraction of the coronary heart muscle, which results in the heartbeat, is started by electrical impulses.

Observe that the height of the pulse is a sign of pumping power, and that the intervals between each pulse determine the rhythm of a coronary heart. The watch is equipped with an ECG circuitry unit that records abnormal heartbeat indicators from the individual. Next, the watch's microcontroller executes a cardiac arrest guideline and the Bluetooth emergency call device, which calls for medical assistance at some moment of the coronary heart attack. The afflicted person wears two biosensors on their wrists, which transmit the ECG signal to the analog ECG circuitry.

The circuitry converts the analog, filtered, and amplified output of the circuitry into a digital signal, which is subsequently sent to the walking watch unit. The electrocardiogram on one of the affected person's wrists are the transmitter, the A/D converter, and the circuitry unit, because the watch is wireless, there are no wires between it and the wearer, allowing them greater mobility.

5 Large-Scale Data and Patient-Focused Medical Care

A feasibility study was carried out by Ciccone et al. to investigate the possibility of integrating care managers, specifically trained nurses, into the healthcare system to support general practitioners (GPs) and specialists in managing patients with cardiovascular disease (CVD), diabetes, heart failure, or CVD threat. Care supervisors collaborated closely with male or female patients, supporting them in changing their lifestyles, keeping an eye on their conditions, and providing the information and guidance required to encourage patient empowerment and improve self-control abilities. As a result, the scientific community saw a noticeable improvement, and patients were able to better manage their illnesses.

In the end, the care manager role has a significant impact on the affected person's level of fitness and self-control, and the results can be linked to the consistent "partnership" between the patient and the care supervisor as well as the cooperation between the organization and the care manager. Leading the way in the recent information revolution in healthcare are specially trained nurses. In order to accurately manage diseases and medical devices, a Patient-customized Healthcare System based on Hadoop with Text Mining (PHSHT) was recently developed. As a result, the PHSHT no longer most effectively complements the system flaws within the current set of rules, which not only generates particular disorder policies but also disregards the relationship between attributes. In order to avoid any unexpected mishaps, the character is also informed of the patient's condition can subsequently move instantly.

By transforming the patient's unstructured generated records into established statistics, the Hadoop platform based on text mining ascertains a person's ailment, forecasts the disorder technique, and generates more specific statistics about ailments. There're four modules in PHSHT: (a) MDCM stores enormous records with information about a person's fitness in the Hbase, which takes place in a medical facility or on a portable medical device.

After that, the vast amount of data is divided into unstructured records like clinical notes, electronic health records, and patient advocacy cases (PACs), as well as dependent statistics like affected person information, family history, and scientific prescriptions records; (b) TMHM uses Hadoop's Text Mining-based Text Mining to examine the gathered unstructured statistics converts it to accepted statistics. Additionally, TMHM disseminates and stores verified data within the stored established facts are then combined by Hbase and re-stored using a MapReduce Framework; (c) DRCM creates disease rules by utilizing the health information kept in the Hbase and the CPST collection of regulations and restocks them within the Hbase; (d) DMPM sends a threat index notification to the patient or the organization based on the outcome of the disease prediction analysis, which compares the patient's risk with the data gathered from the affected parties.

6 Drawbacks of Previous Works

The existing system is based upon the genetic data of patients. The dataset here usually consists of data regarding the genetic details of wide variety of patients. Rather than collecting data regarding the complete health of the patients, the system concentrates on the genetic tree of the individuals which brings the accuracy level to a very low range. Other than these manual observations are done using scans and ECG (electrocardiogram). Limited automation in existing system hinders the ability to process large volumes of chances of heart diseases, making them less suitable for an accurate prediction to be done. Naive Bayes classifier it is quite easy to train classifier on small dataset if there exist high biasness and low variance give it major advantage over the classifier with low biasness and high variance such as KNN because later classifier will suffer problem of over fitting. The training on small dataset is due to the reason it converses very quickly so need less training data as well as less training time but as we all know that every coin has two side if data size started growing there is chance of asymptotic error whereas the algorithm with low biasness and low variance are powerful enough to avoid this kind of problem. The other major disadvantage of Naïve Bayes algorithm is that it cannot learn interaction between features. On the other hand, if considering the logistic regression model take care of related feature unlike the Naïve Bayes.

7 Dataset

Data splitting is a critical phase in machine learning model development, facilitating the assessment of model performance on distinct subsets of the dataset. In the project, the dataset is divided into training and testing sets. The training set, comprising the majority of the data, is utilized for training the neural network model, while the testing set serves as an independent dataset for evaluating the model's generalization performance. Cleveland data set contains a total number of 303 instances with 13 medical attributes (factors) that are acquired from heart disease data set of Cleveland. The first 20 set of data is given as training data set and the remaining as testing data set.

8 System Architecture

(See Fig. 2).

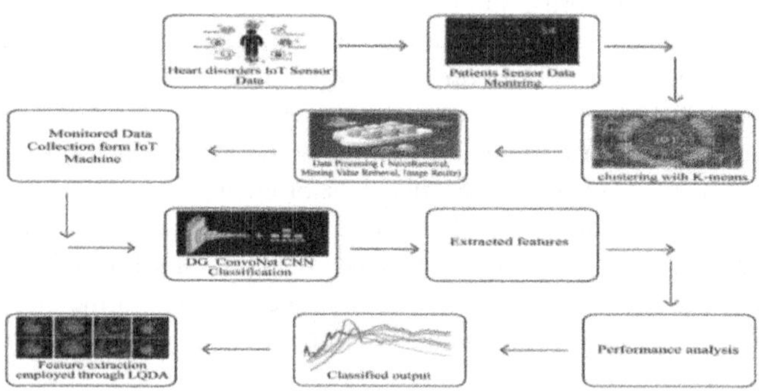

Fig. 1. Workflow of Big Data Analytics for Healthcare

8.1 System Description

These days, information mining can help health insurance companies uncover misbehavior and hypocrisy, health care facilities make decisions about customer retention and management, vendors find effective solutions and best practices, and patients access more and more high-quality health care services that fit their budgets. In the medical field, predictive analysis is widely employed. One of the methods used in facts mining is classification to anticipate and categorize the preset records according to the particular magnificence. These are a variety of categories strategies put forth by researchers. Various data mining methods had been used to predict coronary heart illness. All three sets of rules—Naïve Bayes, Decision Tree, and ANN—have been shown to be accurate. Using ECG signals, three unique record mining algorithms—ANN, C4, and Five Decision Trees—are used to study heart-related illnesses. The results of the analysis demonstrate that the Decision Tree algorithm performs better and has greater accuracy than either the C4.5 or the Naïve Bayes algorithm. It needs less space while multiplying the quantity of statistics; it has a lower cost and reduces the predictive mistakes.

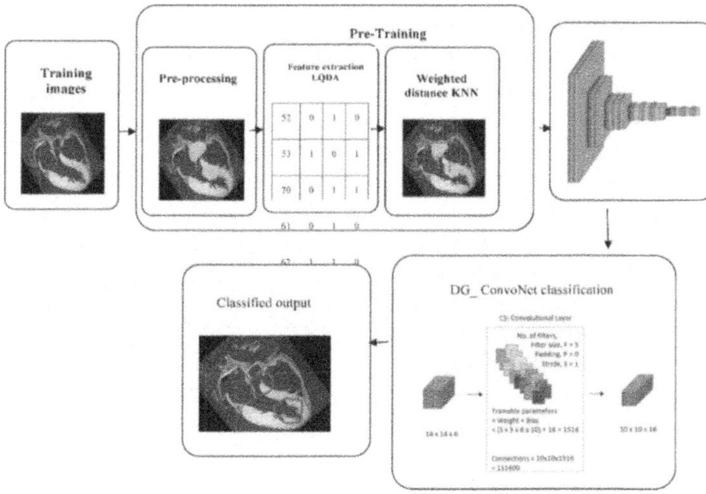

Fig. 2. Workflow of Heart Attack Prediction System

The greatest possible set of rules that could be appropriate for any kind of scientific diagnosis is called a zero set of rules. When the C5.0 set of rules operates more quickly and provides the first Preprocessing can be done with data mining techniques, and implementation can be done with device mastering algorithms; cloud Deployment makes use of computing. Famous machine learning algorithms have already become well-known as helpful in accurately supporting the physicians and determining the risk level for coronary heart disease are anticipating it.

The process of extracting priceless and enormous expertise from sizable datasets is known as data mining. Pre-processing data is an essential step in data mining and device learning. Dimensionality reduction is therefore a useful tool for data reduction. The most effective major tactics for Feature selection and feature extraction are dimensionality discounts. Feature choice refers to the process of choosing a portion of pertinent abilities. Techniques for selecting features are a subset of the larger, more general field of extracting functions.

8.2 Proposed Methodology

Medical big data is utilized by numerous healthcare organizations to generate recommendations and suggestions within the healthcare domain. Spark by Apache can handle data rapidly when handling large amounts of unstructured and organized sets of medical data since it offers tasks for memory computations examining the data in memory, which is considerably quicker than accessing and storing data on hard drives. Big data analytics is provided by the research project architecture derived from medical data from wearable technology and medical professionals. The health monitoring gadgets can offer intriguing data on health over time. The primary objective of the system is examining data that is streaming from various sources for example, wearables, smart rings, and smart clothing smartwatches and additional cutting-edge medical equipment.

The cloud structure is used in the suggested methodology. The following graph shows the performance of different algorithms for heart attack prediction (Fig. 3).

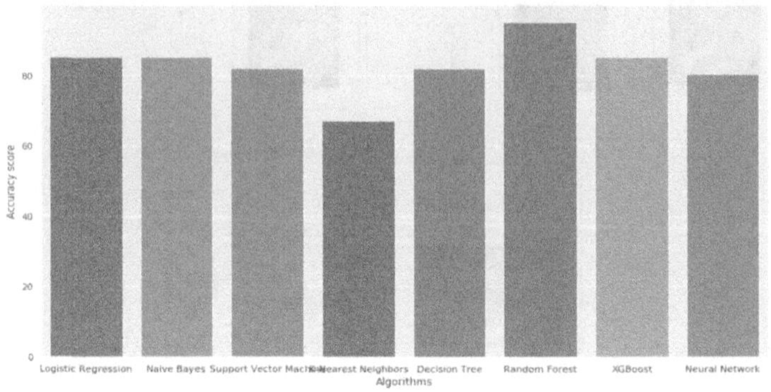

Fig. 3. Performance of Algorithms for Heart attack Prediction

9 Conclusion

The program's approach provided an extensive computer framework for managing EHRs with an emphasis on streaming data from linked medical gadgets and medical records for individuals. The suggested approach depends on the primary parameters for the tanning-based SVM for these parameters. This paper offers a significant method for the consumer social health profile and EHR platform-based health prediction method. The suggested framework provides a productive means of connecting Spark and streaming data based on Microsoft Azure from wearable technology for illness prediction. Upcoming research will use this theoretical framework to investigate its data using methods from deep learning. Next up phase of the study we've planned will involve a cloud framework that backs the records dataset of the illness. Compared to Hadoop these tools can handle high-dimensional data, is effective in cases with limited training samples, and can handle non-linear classification using kernel functions.

References

1. Karthika, R.N., Valliyammai, C., Naveena, M.: Phish block: a blockchain framework for phish detection in cloud. Comput. Syst. Sci. Eng. **44**(1), 777–795 (2023)
2. Cristin, R., Suresh Kumar, K., Anbhazhagan, P.: Severity level classification of brain tumor based on MRI images using fractional-chicken swarm optimization algorithm. Comput. J. **64**(10), 1514–1530 (2021)
3. Al Mamoon, I., Sani, A.S., Islam, A.M., Yee, O.C., Kobayashi, F., Komaki, S.: A proposal of body implementable early heart attack detection system (2013)
4. Morley, S.R.: Heart attack experiences described in weblogs: an analysis of sex differences. CMC senior theses (2013)

5. Eaton, C., Deroos, D., Deutsch, T., Lapis, G., Zikopoulos, P.: Understanding Big Data. Analytics for Enterprise-Class Hadoop and Streaming Data. McGraw-Hill Companies (2012)
6. Minelli, M., Chambers, M., Dhiraj, A.: Big Data, Big Analytics: Emerging Business Intelligence and Analytic Trends for Today's Businesses. Wiley, Hoboken (2012)
7. Wu, C.S., Badshah, M., Bhagwat, V.: Heart disease prediction using data mining techniques. In: Proceedings of the 2019 2nd International Conference on Data Science and Information Technology, pp. 7–11 (2019)
8. Ang, J., Mirza, A., Haron, H.: Supervised, unsupervised, and semi-supervised feature selection: a review on gene selection. IEEE/ACM Trans. Comput. Biol. Bioinf. **13**(5), 971–989 (2016)
9. Nair, L., Shetty, S.: Applying spark-based machine learning model on streaming big data for health status prediction. Comput. Electr. Eng. **65**, 393–399 (2018)
10. McPadden, J., et al.: Health care and precision medicine research: analysis of a scalable data science platform. J. Med. Internet Res. **21**(4), e13043 (2019)
11. Dumka, A., Sah, A.: Smart ambulance system using the concept of big data and Internet of Things. In: Healthcare Data Analytics and Management, pp. 155–176 (2019)

Bloom's Taxonomy Based Question Analysis for Personalized Learning

J. Jeslin Shanthamalar, Dinesh Sheelam(✉), Shiva Raj Bodla, V. Gowri Manohari, and R. S. Nancy Noella

Department of Computer Science and Engineering, Sathyabama Institute of Science and Technology, Chennai 600119, Tamil Nadu, India
dineshsheelam459@gmail.com, {gowrimanohari.cse, Nancynoella.cse}@sathyabama.ac.in

Abstract. In the 1950s, Benjamin Bloom and his associates introduced Bloom's Taxonomy, a foundational framework for categorizing learning objectives and cognitive skills. While question classification typically operates at a fine level, such as sentences and phrases, and text classification focuses on the document level, past research has explored the intersection of question classification and Bloom's Taxonomy to assess learners' cognitive levels in higher education. However, existing feature types from previous studies may excel in datasets with narrowly focused queries, necessitating the development of multiple classifiers for diverse fields or areas.

To address this, A new kind of feature called "taxonomy-based" is suggested to improve the accuracy of question classification in datasets from different fields. Utilizing datasets comprising questions from distinct topics, the study evaluates the effectiveness of taxonomy-based features. Support vector machines (SVMs) are chosen as the classifier for their reputed text classification accuracy. The research reveals that taxonomy-based features significantly improve classifier performance when applied to question sets from different domains.

This effort is driven by the need to improve question categorization accuracy over a variety of datasets and the realization of Bloom's Taxonomy's continuing use in educational settings. Although Bloom's Taxonomy offers an extensive framework for comprehending cognitive abilities and learning objectives, current question classification techniques frequently do not have the granularity necessary to fully utilize this taxonomy. Furthermore, prior work has mostly concentrated on fine-level categorization, ignoring the advantages of integrating Bloom's Taxonomy into the classification procedure.

The highest accuracy achieved using SVM with TF-IDF vectorizer is 74.17% for BCL's Data set and 95.23% for BT Dataset. Employing the KNN algorithm with the TF-IDF vectorizer yields 59.17% accuracy for BCL's Data set, with BT Dataset reaching a maximum accuracy of 89.11%. Naïve Bayes algorithm, coupled with the TF-IDF vectorizer, achieves a peak accuracy of 73.33% for BCL's Data set and 90.99% for BT Dataset. Lastly, using the Random Forest algorithm with the count vectorizer results in a maximum accuracy of 74.17% for BCL's Data set and 95.21% for BT Dataset.

Published by Springer Nature Switzerland AG 2024
M. L. Owoc et al. (Eds.): ICCIDS 2024, IFIP AICT 718, pp. 296–311, 2024.
https://doi.org/10.1007/978-3-031-69986-3_23

Keywords: Bloom's Taxonomy · question classification · feature types · taxonomy-based features · support vector machines · TF-IDF vectorizer

1 Introduction

One of the teaching strategies is to set an example of behavior for the students. The aim of education's planning, teaching, and assessment stages is to develop a student's capacity for decision-making, sound reasoning, and critical thinking. Determining a learner's conceptual development. It supports the growth of cognitive skills [13] in the learner. The question bank is a crucial component of the educational system.

A particular selection of questions can be selected for each test using question banks, which let assessors develop tests that more precisely measure students' abilities and competence levels. It is therefore necessary to manage the question bank appropriately, which includes not just sorting and classifying the questions but also producing in-context questions. The Revised Bloom's Taxonomy, which classifies the questions into its six categories, is one of these criteria. An essential first stage is assessment. An individual's cognitive skill domain or mental capacity is the main focus of these levels, which are ranked in increasing order of complexity. Bloom's Taxonomy is an essential concept that aids tutors in developing evaluations, curriculum, and course objectives (Fig. 1).

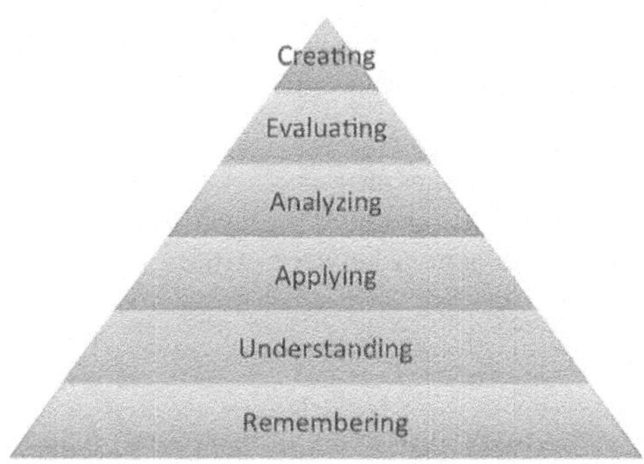

Fig. 1. Bloom's taxonomy levels

Previous studies on term weighting [1–4] in relation to classifying exam questions focused mainly on feature extraction, feature selection, and different term weighting approaches. The potential of term weighting methods to improve classifier performance in efficiently classifying exam questions and textual content has been shown by recent studies in the fields of text classification and exam question classification of Bloom's Taxonomy. Term weighting is the process of listing every term that appears in a document and giving it a numerical value determined by its importance (Table 1).

Given the documented effects of word weighting, there is a growing need to explore the effectiveness of different vectorization methods in the categorization of exam questions based on Bloom's Taxonomy (BT) with the ultimate goal of improving classification accuracy. Therefore, the goal of this work is to examine the count vectorizer,

Table 1. An explanation of the Cognitive Domain in Bloom's Taxonomy

Level	Skill Demonstrated	Question cues/Verbs for tests
Apply	Utilize information, approaches, theories, principles, and concepts in novel contexts; address issues by applying necessary abilities or knowledge	Calculate, predict, apply, solve, illustrate, use, demonstrate, determine, model
Understand	Interpreting data, deciphering meaning, and applying expertise to novel situations	Describe, explain, paraphrase, restate, associate, contrast, summarize
Remember	Knowledge of facts, customs, definitions, jargon, technical words, classifications, categories, and standards; recall of methods and processes; knowledge of abstractions, ideas, and principles in the field	List, define, tell, describe, recite, recall, identify, show, label, tabulate, quote, name, who, when, where
Create	Employ preexisting concepts to generate new ones, piece together ideas to build a (new) whole, and extrapolate from data	Design, formulate, build, invent, create, compose, generate, derive, modify, develop, integrate
Evaluate	Compare and contrast concepts, evaluate the merits of theories and presentations, and base decisions on well-reasoned arguments	Assess, decide, choose, rank, grade, test, measure, defend, recommend, convince
Analyze	Divide a complicated issue into manageable chunks, then note the connections and interactions between each component	Classify, outline, break down, categorize, analyze, diagram, illustrate, infer, select

TFIDF (Term Frequency and Inverse Document Frequency), and Google News 300 vectorizer techniques for BT based exam question categorization utilizing different machine learning algorithms.

When training a Bloom's Taxonomy-based model, using multidomain questions may improve the model's performance in a wider range of scenarios. A model can acquire a variety of patterns and representations by being trained on questions from different domains, which increases its robustness and subject- matter adaptability.

Here are Some Potential Advantages of Using Multidomain Questions for Training

Generalization: The model is better able to generalize to new, untested domains when it has been trained on questions from various domains. It gains the ability to recognize universal traits and patterns that apply to a variety of topics.

Transfer Learning: It is possible to apply knowledge from one domain to another. When a model is trained on multi domain questions, it can benefit from the knowledge of universal patterns or relationships when it comes to a particular domain.
Expanded Lexicon: The model gains exposure to questions from different domains, which expands its vocabulary. When working with various terminologies and linguistic idioms in various subject areas, this can be helpful.
Enhanced Robustness: A model that has been trained on questions spanning multiple domains is probably more resilient and less susceptible to changes in the wording or structure of the questions. When faced with real-world scenarios that may involve questions from different domains, this robustness can result in improved performance.
Preventing Overfitting: When a model is trained using only questions from one domain, it may become overly specialized and perform poorly when faced with new, untested data. Over fitting can be avoided with multidomain training by exposing the model to a wider range of datasets.

But It's Crucial to take into Account a Few Things

Data Quality: It's critical to consider the training data's quality. To prevent introducing unwanted patterns, make sure the multidomain dataset is representative, balanced, and free of biases.
Domain-Specific Features: Certain domains might have special qualities that aren't applicable to other domains. Evaluating the features' relevance across domains is crucial.
Task-Specific Considerations: The goals of your Bloom's Taxonomy-based model and the particular task at hand will determine how well multidomain questions work. Analyze performance on the intended domain and adjust as necessary.

This paper consists of five sections. The introduction highlights the need of investigating different vectorization strategies and provides an outline of BT-based test question classification. The subsequent segment scrutinizes associated efforts, delving into past research on distinct vectorizers and their applications in categorizing test inquiries according to BT. In the third section, the research approach is explained. Section 4 includes a detailed discussion of the conclusions and the experimental results. The final section concludes the work with some final observations and explores possible directions for further research.

2 Literature Survey

A useful and standardized framework for classifying exam questions into discrete cognitive skill levels is provided by Bloom's Taxonomy, which makes it possible to assess students' learning outcomes in a more nuanced manner. Nevertheless, manual classification using Bloom's Taxonomy takes a lot of time and effort. Machine learning based automated classification has emerged as a viable solution to this problem. Different classifiers with varying performances have been investigated for automating question classification, such as Support Vector Machines (SVM) [10], Naive Bayes [7], and k-Nearest Neighbors (KNN) [2, 8]. Interestingly, SVM and KNN show robustness in various scenarios.

Feature engineering is essential to the success of automated question classification, especially when using the best term weighting schemes. Research has shown that 'Supervised Term weighting' (STW) and enhanced 'unsupervised Term weighting' (USTW) schemes outperform conventional unsupervised term weighting techniques like TF-IDF [21]. STW schemes, such as TF-ICF, perform better at identifying discriminative verbs linked to different cognitive levels.

Enhanced USTW schemes give verbs greater weights than other parts of speech, which helps with verb handling. The general agreement indicates that in order to achieve the best question classification performance across various domains, a synergistic combination of supervised and unsupervised techniques is necessary.

Automated question classification could completely change assessment methods by providing a more accurate and efficient way to gauge students' cognitive abilities. Education professionals can lessen their workload and concentrate on more important facets of teaching by automating this process [14]. There is potential benefit for both teachers and students from this advancement in assessment methodologies.

Related Work in Blooms Question Classification

In TF, TF-IDF, and normalized TF-IDF are only a few of the word weighting schemes that will be thoroughly investigated and evaluated in order to determine the optimal baseline variants. The performance of these term weighting modifications is assessed in the study using two distinct classifiers: Support Vector Machine (SVM) and Naïve Bayes. According to the findings, the TF-IDF and normalized TF-IDF variants that were classified using the SVM classifier yielded average accuracies of 63.7% and 71.7%, respectively, whereas the TF-IDF and normalized TF-IDF variants that were classified using the Naive Bayes classifier produced 62.4% and 63.4%, respectively, for your reference. Interestingly, the SVM classifier produced 63.7% and 71.7%, respectively, whilst the TF-IDF and normalized TF-IDF variants utilizing the Naive Bayes classifier yielded 62.4% and 63.4%, respectively. With an accuracy of 70.8%, the TF-IDF3 variant emerged victorious in the unnormalized TF-IDF category. Meanwhile, the Normalized TF-IDF2 version achieved the highest accuracy of all the normalized TFIDF variants, scoring 73.3%. In addition to highlighting the exceptional performance of normalized TF-IDF2 and TF-IDF3, these results offer helpful information about the relative benefits of various term weighting options.

A method of classifying test questions into different areas based on the Bloom taxonomy. To produce question vectors with improved representational quality, a brand-new feature type called W2V-TFPOSIDF is introduced. It combines pre-trained word2vec with TFPOS-IDF, a modified TF-IDF that incorporates Part-of-Speech (POS) [16], to extract features. There are two datasets used, with 600 and 141 questions in each. In the classification of exam questions, W2V-TFPOSIDF [13] outperforms TF-IDF and TFPOS-IDF, according to the study. With F1measure values of 0.837 and 0.897 for each dataset, the SVM classifier notably produces the best results. These F1-measure results show that test questions from various domains can be correctly classified with the suggested feature type.

In particular, for datasets that comprise questions from several domains, this work offers a new feature type dubbed "taxonomy based," which attempts to improve the accuracy of question classification. Using this novel feature type, the study analyses the question classification performance across datasets containing questions from both limited domains and a wider variety of subject areas. The SVM classifier [15] was used because of its wellknown excellent text classification accuracy. The study's findings show how helpful taxonomy-based characteristics are for enhancing the classifier's performance, especially when the datasets include queries from a variety of areas. This paper provides insightful new information about how taxonomy-based features could improve question classification accuracy in several fields.

This work provides a comprehensive comparative analysis of several machine learning techniques [9] and linguistically motivated characteristics applied to the automatic classification of test questions according to Bloom's Taxonomy cognitive levels [12, 17, 18]. The results of the trials indicated that the average accuracy scores of four classifiers (SVM, Logistic Regression, Decision Trees, and Naive Bayes) were higher than 0.6. This demonstrates the efficacy of machine learning models in achieving a high level of question classification accuracy. The SVM model with unigram features (accuracy of 0.7667) was surpassed by the Logistic Regression model, which produced an accuracy result of 0.7683. The fusion of Bigrams and Unigrams features made this possible. Because this increases the accuracy of classifying exam questions, the study highlights the use of linguistically motivated characteristics for greater syntactic analysis, namely the combination of Bigrams and Unigrams (Table 2).

Table 2. An earlier investigation of the classification of questions.

Author and Year	Term Weighting	Feature Extraction	Machine Learning
Anbuselvan Sangodiah (2022)	yes	yes	yes
Mohammed and Omar (2020)	yes	yes	yes
ROHIZA AHMAD (2017)	yes	yes	yes
Osman and Yahya (2016)	yes	yes	yes

3 Methodology

3.1 Dataset

Two datasets are used in this study to provide a thorough evaluation. In order to overcome the shortcomings of current question categorization techniques, which frequently lack specificity and underutilize the cognitive hierarchy described in Bloom's Taxonomy, the Bloom's Taxonomy (BT) dataset was created for this study. Through the creation of a dataset that is especially designed to cover a broad spectrum of cognitive abilities and learning objectives from different fields, this study intends to enable more precise and sophisticated question classification. A purposeful approach to question production provides for more control over cognitive difficulty and assures compatibility with the taxonomy's categories, even though obtaining questions only from the web or existing sources may restrict the diversity and representativeness of the dataset.

The BT Dataset consists of a set of generic questions from different domains that are not domain specific but are labelled with cognitive levels according to Bloom's Taxonomy (BT). In order to ensure accuracy during the labelling process, lecturers assigned BT cognitive levels to each question with the assistance of domain experts. A preprocessing step was used to ensure that every question contained at least one BT keyword in accordance with BT guidelines. There are 1771 open-ended questions in this dataset; Table 3 provides sample questions for each BT level, and Fig. 2 shows how they are distributed across BT levels. The 1956 edition of BT, written by Benjamin Bloom and associates, is the version used. Six hundred general question from IEEE research make up the BCL's Dataset. This study introduces a question classification model that makes use of Google Vectorizer and Count Vectorizer to simulate and assess Multidomain question classification.

First, preprocessing operations like as lemmatization, tokenization, and stop-word removal are performed. Phase two focuses on feature extraction. Phase three employs four machine learning classifiers (Support Vector Machine (SVM) [10, 15], Naive Bayes (NB) [8], Random Forest, and K-Nearest Neighbors (KNN)) to classify exam questions and identify the cognitive domain [12] of the dataset based on BT after the first two phases are completed. For a comprehensive analysis, the generated results are then contrasted.

3.2 Question Classification Model

Preprocessing is required to organize the unstructured data because the collected test questions may initially be loud or contain mistakes. To increase the quality of the data, this stage employs several crucial procedures, including lowercase conversion, tokenization, stop-word removal, lemmatization, and adhering to Bloom's Taxonomy (BT) recommendations [16]. The first step is to change every word in the questions to lowercase to guarantee consistency. The next step is to apply a tokenization process that divides question items into a list of tokens in order to identify word boundaries. Extraneous elements like punctuation, numbers, and non-letter characters are eliminated in order to refine the data. Every token is converted back into its original form, or lemma, by utilizing the Word Net Lemmatize in the Natural Language Toolkit [5, 6].

Table 3. Sample questions

Question	Label
Define compound interest?	Knowledge
Summarize a historical document?	comprehension
Can you group by characteristics such as...?	Application
Compare fall and spring?	Analysis
how do we devise a way to enforce the laws?	Synthesis
Analyze information to back up a theory?	Evaluation

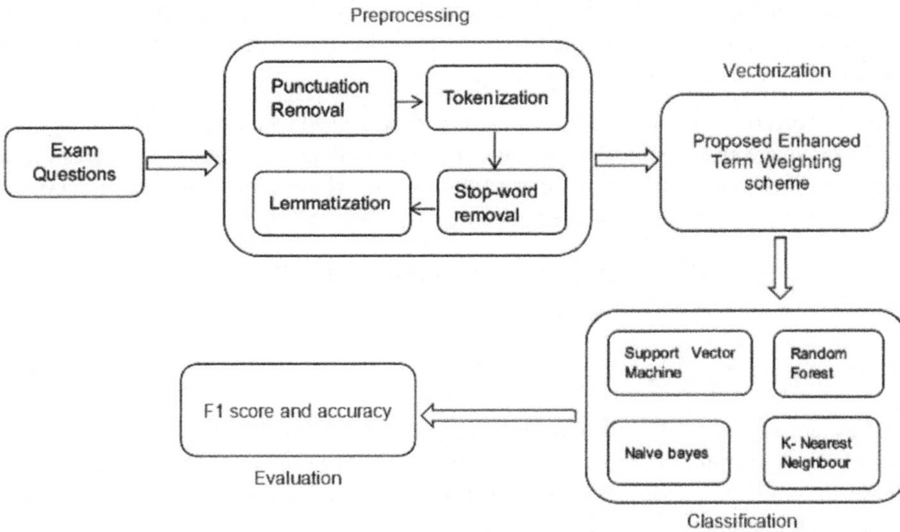

Fig. 2. Architecture Diagram

For precise classification, this rigorous preprocessing guarantees data uniformity and complies with BT guidelines. To turn the questions into vectors for further analysis, two crucial tasks basic Natural Language Processing (NLP) processing and term weighting are carried out in the second phase. Using different term weighting schemes, the main goal is to transform the questions into vector representations. The first task is to apply basic natural language processing (NLP) techniques [5, 6] to improve the textual data's comprehension and representation. Different term weighting schemes are then used to extract the importance of terms in the questions, which makes it easier to translate the textual data into numerical vectors.

3.3 Different Terms for Weighting

TF-IDF Vectorizer: The TF-IDF score is calculated by multiplying the inverse document frequency (IDF), which measures a term's rarity throughout the whole document collection, by the term frequency (TF), which shows how frequently a term appears in a document.

$$\mathrm{W}_{\mathrm{i,j}} = \mathrm{Tf}_{\mathrm{i,j}} \times log(\frac{n}{df}) \quad (1)$$

$\mathrm{Tf}_{\mathrm{i,j}}$ = number of occurrences of i in j
$\mathrm{Df}_{\mathrm{i,j}}$ = number of documents containing i
n = total number of documents

Google News 300 Vectorizer: About 100 billion words from a subset of the Google News dataset were used to train pre-trained vectors. In the model, three million words and phrases are represented by 300-dimensional vectors. A simple data-driven technique that was utilized to acquire the phrases is described in "Distributed Representations of Words and Phrases and their Compositionality."

Count Vectorizer: The Count Vectorizer is a widely used technique in text mining and natural language processing (NLP) that converts a collection of text documents into a matrix of token counts. Each member of the vector, which represents each document, indicates how many times that word appears in the document.

$$\mathrm{TF} - \mathrm{IDF} = \mathrm{CountVectorizer}(\mathrm{t}, \mathrm{d}) * \mathrm{IDF}(\mathrm{t}) \tag{2}$$

3.4 Evaluation Metrics

Two key metrics are used to evaluate the experimental results: accuracy and F1- measure, which help determine the most effective unsupervised term weighting variant and the effectiveness of the classification model. Finding the values of recall and precision is the first step in computing the F1-measure.

Recall measures how broad the model is, whereas precision measures how exact and accurate the model is. A score closer to 1 indicates superior performance, and values between 0 and 1 are often produced by accuracy and the F1-measure. This dual-metric evaluation approach helps to completely understand the effectiveness of the model in terms of overall accuracy as well as the trade-off between precision and recall.

$$\mathrm{Accuracy} = \frac{\mathrm{TP} + \mathrm{TN}}{\mathrm{TP} + \mathrm{FP} + \mathrm{TN} + \mathrm{FN}} \tag{3}$$

$$\mathrm{Recall} = \frac{\mathrm{TP} + \mathrm{TN}}{\mathrm{TP} + \mathrm{FN}} \tag{4}$$

$$\mathrm{Precision} = \frac{\mathrm{TP}}{\mathrm{TP} + \mathrm{FP}} \tag{5}$$

$$\mathrm{F1\ measure} = \frac{2 * (\mathrm{recall} * \mathrm{precision})}{\mathrm{recall} + \mathrm{precision}} \tag{6}$$

A popular metric for classification, accuracy assesses overall performance but has limitations. We explore recall (% of genuine positives discovered) and precision (percentage of true positives among projected positives) to obtain a more nuanced view. When these two are combined, the F1-measure provides a fair assessment of a classifier's performance, particularly for datasets that are imbalanced.

4 Result and Discussion

A comprehensive comparison analysis was conducted by conducting experiments to examine various phrase weighting variations and assess their efficacy using two distinct classifiers. Term Frequency (TF), normalized TFIDF, and Term Frequency-Inverse Document Frequency (TF-IDF) were the three classification categories utilized for the term weighting variants. The questions were classified using Support Vector Machine (SVM) and Naive Bayes classifiers. The SVM classifier came pre-installed with a linear kernel and parameter C set to 1.0. The model of Naive Bayes that was used was multinomial. To implement question categorization, a scaled-down prototype was made using the PyCharm Integrated Development Environment (IDE), the Scikit-learn package, and the NLTK library.

The lower accuracy observed with the LSTM [11, 20] model compared to SVM and Naive Bayes classifiers in question classification experiments can be attributed to several factors. There are a number of reasons behind the LSTM model's inferior accuracy in question classification studies when compared to SVM and Naive Bayes classifiers. First off, in order to get optimal performance, long training data sets and significant hyperparameter adjustment are usually needed for LSTM models because they are generally more intricate and computationally demanding. On the other hand, inadequate training data or hyperparameters for the LSTM model may have been obtained from the experiments, resulting in less-than-ideal outcomes.

Furthermore, LSTM models are more likely to overfit, particularly in situation where there is insufficient regularization or a lack of trained data. The accuracy of the model may have been further impacted by the experimental setup's failure to sufficiently address overfitting concerns.

4.1 SVM Results

For the question dataset, the accuracy and F1-measure of the SVM classifier [10] experiment result are described in this part. The results of the Google News 300 vectorizer, count vectorizer, and TF-IDF are shown in Tables 4.

Table 4. Accuracy and F1-Score of SVM

Datasets	TF-IDF Vectorizer		Count Vectorizer		Google-300 Vectorizer	
	F1 Score	Accuracy	F1 Score	Accuracy	F1 Score	Accuracy
BCL's Dataset	79.01	74.17	74.01	74.17	60	60.92
BT Dataset	95.23	95.21	95.23	95.21	90.14	90.19
Average	87.12	84.69	84.62	84.69	75.07	75.55

The results of using the SVM algorithm to evaluate different vectorization techniques show some interesting patterns. The TF-IDF vectorizer performs admirably for BCL's Data set, displaying an F1 score of 79.01% that demonstrates how well it captures the subtleties of the data. The Google-300 vectorizer, on the other hand, shows competitive accuracy at 60.92% even with a lower F1 score of 60%. With an F1 score of 74.01%, Count Vectorizer's performance is in close agreement with TF-IDF's. All vectorizers in BT Dataset show remarkably high F1 scores, with the TF-IDF and Count Vectorizers scoring the highest at 95.23%.

Even with its slight lag, the Google-300 vectorizer still manages a respectable F1 score of 90.14%. The combined average performance of the two datasets highlights the strength of Count Vectorization and TF- IDF, which produce F1 scores of 84.62% and 87.12%, respectively. Although Google-300 is not as good as it could be, its average F1 score of 75.07% shows how good it is. These results highlight how crucial it is to choose vectorization methods based on the features of the dataset, with TF-IDF and Count Vectorization standing out as top choices for balanced performance (Fig. 3).

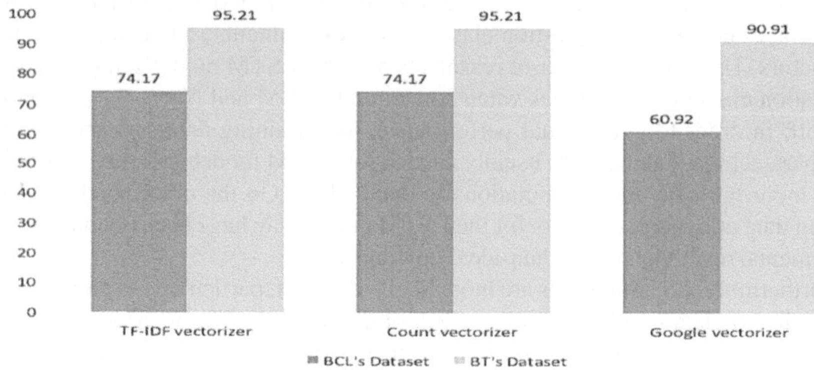

Fig. 3. SVM Accuracy Graph

4.2 KNN Results

This section describes the accuracy and F1-measure of the experiment using the KNN classifier [8] for the given question dataset. Tables 5 display the outcomes of the TF-IDF, count vectorizer, and Google News 300 vectorizer.

Table 5. Accuracy and F1-Score of KNN

Dataset	TF-IDF Vectorizer		Count Vectorizer		Google-300 Vectorizer	
	F1 Score	Accuracy	F1 Score	Accuracy	F1 Score	Accuracy
BCL's Data set	59.13	59.17	38.69	36.67	43.54	54.16
BT Dataset	89.11	89.30	87.35	87.04	83.11	83.38
Average	74.12	74.23	63.02	61.85	63.32	68.77

When the K-nearest neighbors (KNN) algorithm is evaluated using various vectorization methods, observable trends in performance are revealed. When KNN is applied with TF-IDF vectorization in BCL's Data set, it obtains a moderate F1 score of 59.13%, indicating reasonable effectiveness. But when using Count Vectorization, the F1 score significantly decreases to 38.69%, indicating a significant decline in performance. The Google vectorizer receives an F1 score of 43.54%, which is an improvement over Count Vectorization.

Moving on to BT Dataset, KNN scoring impressively at 89.11% and 87.35%, respectively, outperforms other methods with TF-IDF and Count Vectorizers. With an F1 score of 83.11%, the Google vectorizer comes in close second. The average outcomes show how different KNN performance can be between the two datasets. With an F1 score of 74.12%, the TF-IDF variant is the best, followed by the Google vectorizer at 63.32% and Count Vectorization at 63.02%. These results highlight the KNN algorithm's sensitivity to the selected vectorization method, with TF-IDF consistently outperforming other methods in this analysis (Fig. 4).

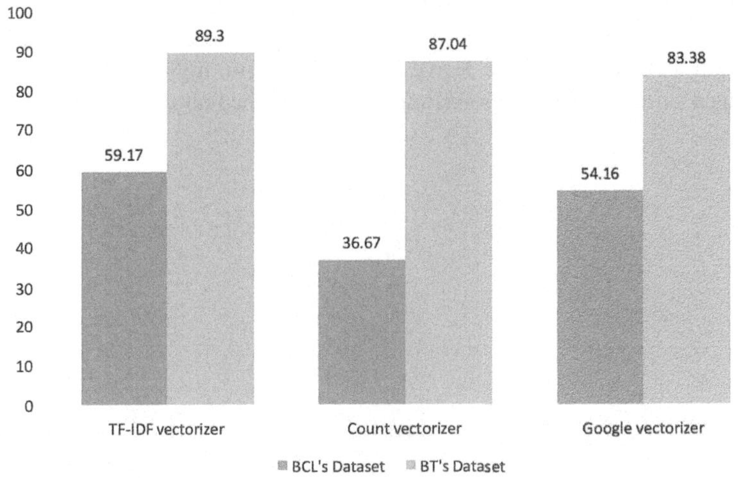

Fig. 4. KNN Accuracy Graph

4.3 Naïve Bayes Results

This portion examines accuracy and F1-measure of the experiment using the Naïve Bayes classifier [7] for the given question dataset. Tables 6 display the outcomes of the TF-IDF, count vectorizer, and Google News 300 vectorizer.

Table 6. Accuracy and F1-Score of Naïve Bayes

Data set	TF-IDF Vectorizer		Count Vectorizer		Google-300 Vectorizer	
	F1 Score	Accuracy	F1 Score	Accuracy	F1 Score	Accuracy
BCL's Data set	73.33	73.33	72.57	72.5	43.33	43.33
BT Dataset	90.94	90.99	90.70	90.70	71.31	71.54
Average	82.16	82.16	81.63	81.60	57.32	57.435

Analyzing the Naive Bayes algorithm using various vectorization methods reveals unique performance traits. Applying Naive Bayes with TF-IDF vectorization in BCL's

Data set yields a strong F1 score of 73.33%, demonstrating its efficacy. With an F1 score of 72.57%, Count Vectorization comes in second, showing similar accuracy. With an F1 score of 43.33%, the Google vectorizer, on the other hand, lags considerably. Moving on to BT Dataset, we see that Naive Bayes with TF-IDF and Count Vectorizers exhibits remarkable F1 scores of 90.70% and 90.94%, in turn.

Even though it is lagging, the Google vectorizer still manages to earn an impressive F1 score of 71.31%. Averaging the outcomes for both datasets demonstrate how consistently Naive Bayes performs. The comparable average F1 scores of 82.16% and 81.63% obtained by TF-IDF and Count Vectorization highlight their dependability. With an average F1 score of 57.32%, the Google vectorizer falls short, demonstrating the shortcomings of Naive Bayes. These results highlight the complex interplay between vectorization techniques and the Naive Bayes algorithm, highlighting the significance of choosing a suitable vectorization strategy based on the features of the dataset (Fig. 5).

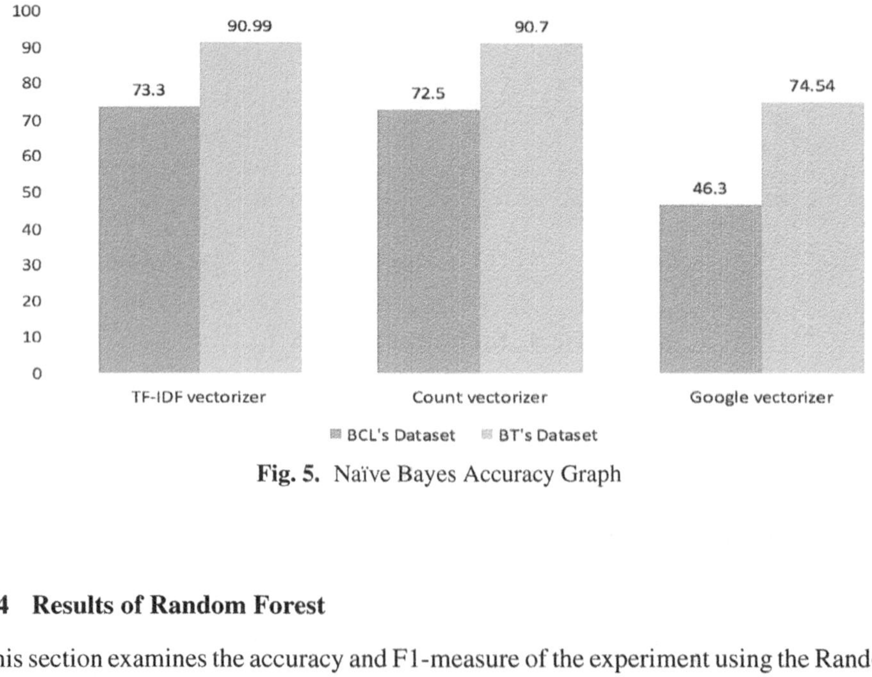

Fig. 5. Naïve Bayes Accuracy Graph

4.4 Results of Random Forest

This section examines the accuracy and F1-measure of the experiment using the Random Forest classifier for the given question dataset. Tables 7 display the outcomes of the TF-IDF, count vectorizer, and Google News 300 vectorizer.

Analysing the random Forest algorithm using various vectorization methods reveals subtle aspects of its performance. With TF-IDF vectorization, Random Forest obtains a strong F1 score of 65.81% in BCL's Data set, demonstrating significant efficacy. With an even higher F1 score of 73.90%, Count Vectorization exhibits enhanced accuracy. In contrast, the Google vectorizer has a 50.77% F1 score, which indicates a significant lag.

Moving on to BT Dataset [17], Random Forest using Count Vectorizers and TF-IDF shows impressive F1 scores of 95.20% and 94.91%, respectively. Even though it is lagging, the Google vectorizer still manages an admirable F1 score of 83.03%. The

Table 7. Accuracy and F1-Score of Random Forest

Dataset	TF-IDF Vectorizer		Count Vectorizer		Google-300 Vectorizer	
	F1 Score	Accuracy	F1 Score	Accuracy	F1 Score	Accuracy
BCL's Data set	65.81	66.67	73.90	74.17	50.77	53.33
BT Dataset	94.91	94.93	95.20	95.21	83.03	83.09
Average	80.36	80.80	84.55	84.69	66.9	68.21

results of averaging the two datasets highlight how consistently Random Forest performs. Similar average F1 scores of 80.36% and 84.55% are obtained by TF-IDF and Count Vectorization, respectively, demonstrating their dependability (Fig. 6).

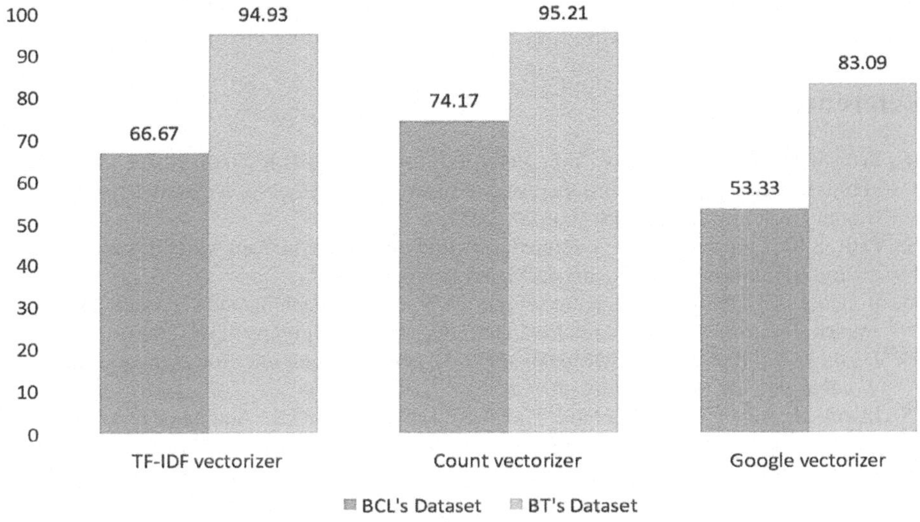

Fig. 6. Random Forest Accuracy Graph

The Google vectorizer's shortcomings with the Random Forest algorithm are evident as it lags behind with an average F1 score of 66.9%. These results highlight the intricate connection between vectorization techniques and the Random Forest algorithm, highlighting how crucial it is to choose the right vectorization strategy depending on the features of the dataset.

5 Conclusion and Future Work

In this study, an extensive evaluation of various term weighting variants, including TF-IDF, Count Vectorizer, and Google News 300 Vectorizer, in conjunction with SVM, Naïve Bayes, KNN, and Random Forest classifiers, was conducted to identify the most optimal term weighting scheme for exam question classification based on Bloom's Taxonomy. The results indicate that TF-IDF, particularly when used with SVM and Naïve Bayes classifiers, emerges as the most effective variant, achieving the highest accuracy of 95.21%. Count Vectorizer also demonstrated strong performance, particularly with the Random Forest algorithm. The Google News 300 Vectorizer, while competent, did not outperform TF-IDF in this context. Despite the success of normalizing term weighting values for certain classifiers, the study emphasizes that the choice of normalization depends on the specific classifiers employed. The findings suggest that TF-IDF, with its robust performance, should be favored when normalization is preferred, while the Google News 300 Vectorizer is suitable when normalization is not required. Further research might entail broadening the dataset to include test questions from many domains in order to improve the model's generalizability. Furthermore, the creation of an interactive assessment platform that integrates Bloom's categorization might serve as a beneficial pedagogical instrument, assisting learners in evaluating their comprehension and use of cognitive abilities. Personalized feedback based on Bloom's Taxonomy levels might be provided via this technology, making learning more engaging and productive.

References

1. Sangodiah, A., San, T.J., Fui, Y.T., Heng, L.E., Ayyasamy, R.K., Jalil, N.B.A.: Identifying optimal baseline variant of unsupervised term weighting in question classification based on Bloom taxonomy. Mendel **28**(1), 8–22 (2022)
2. Patil, S.K., Shreyas, M.M.: A comparative study of Question Bank classification based on revised Bloom's taxonomy using SVM and K-NN (2017)
3. Bottcher, A., Thurner, V., Zehetmeier, D.: Alignment of teaching and electronic exams and empirical classification of errors for an introductory programming class (2020)
4. Jayakodi, K., Bandara, M., Meedeniya, D.: An automatic classifier for exam questions with WordNet and Cosine similarity (2016)
5. Ifham, M., Banujan, K., Wijeratne, P.M.A.K., Kumara, B.T.G.S.: Automatic classification of questions based on Bloom's taxonomy using artificial neural network (2022)
6. Kusuma, S.F., Siahaan, D., Yuhana, U.L.: Automatic Indonesia's questions classification based on Bloom's taxonomy using Natural Language Processing (2015)
7. Setyaningsih, E.R., Listiowarni, I.: Categorization of exam questions based on Bloom taxonomy using Naïve Bayes and Laplace smoothing (2021)
8. Salma, F.S., Pratiwi, O.N., Fa'rifah, R.Y.: Classification of high school history questions based on cognitive level revised Bloom's taxonomy using K-nearest neighbor method (2022)
9. Hasmawati, Romadhony, A., Abdurohman, R.: Primary and high school question classification based on Bloom's taxonomy (2022)
10. Callista, A.S., Pratiwi, O.N., Sutoyo, E.: Questions classification based on revised Bloom's taxonomy cognitive level using Naive Bayes and support vector machine (2021)
11. Shaikh, S., Daudpotta, S.M., Imran, A.S.: Bloom's learning outcomes' automatic classification using LSTM and pretrained word embeddings. IEEE Access **9**, 117887–117909 (2021)
12. Ullah, Z., Lajis, A., Jamjoom, M., Altalhi, A.H., Shah, J., Saleem, F.: A rule-based method for cognitive competency assessment in computer programming using Bloom's taxonomy. IEEE Access **7**, 64663–64675 (2019)
13. Gani, M.O., Ayyasamy, R.K., Alhashmi, S.M., Sangodiah, A., Fui, Y.T.: ETFPOS-IDF: a novel term weighting scheme for examination question classification based on Bloom's taxonomy. IEEE Access **10**, 132777–132785 (2022)

14. Amin, H., Mirza, M.S.: Comparative study of knowledge and use of Bloom's digital taxonomy by teachers and students in virtual and conventional universities. Asian Assoc. Open Univ. J. **15**(2), 223–238 (2020)
15. Sangodiah, A., Ahmad, R., Wan Ahmad, W.F.: Taxonomy based features in question classification using support vector machine. J. Theor. Appl. Inf. Technol.Theor. Appl. Inf. Technol. **95**(12), 2814–2823 (2017)
16. Mohamed, O.J., Zakar, N.A., Alshaikhdeeb, B.: A combination method of syntactic and semantic approaches for classifying examination questions into Bloom's taxonomy cognitive. J. Eng. Sci. Technol. **14**(2), 935–950 (2019)
17. Das, S., Mandal, S.K.D., Basu, A.: Identification of cognitive learning complexity of assessment questions using multi-class text classification. Contemp. Educ. Technol. **12**(2), ep275 (2020)
18. Sebbaq, H., El Faddouli, N.-E.: MTBERT-Attention: an explainable BERT model based on multi-task learning for cognitive text classification. Sci. Afr. **21**, e01799 (2023)
19. Soozandehfar, S.M.A., Adeli, M.R.: A critical appraisal of Bloom's taxonomy. Am. Res. J. Engl. Lit. **2**(1), 1–9 (2016)
20. Su, L., Kang, W., Wu, L., Jiang, D.: Deep transfer learning for question classification based on semantic information features of category labels. Comput. Intell. Neurosci.. Intell. Neurosci. **2022**(1), 7178818 (2022)
21. Gani, M.O., Ayyasamy, R.K., Sangodiah, A., Fui, Y.T.: USTW vs. STW: a comparative analysis for exam question classification based on Bloom's taxonomy. Mendel **28**(2), 25–40 (2022)
22. Mohammed, M., Omar, N.: Question classification based on Bloom's taxonomy cognitive domain using modified TF-IDF and word2vec. PLoS ONE **15**(3), e0230442 (2020)
23. Osman, A., Yahya, A.: Classifications of exam questions using linguistically-motivated features: a case study based on Bloom's taxonomy. In: The Sixth International Arab Conference on Quality Assurance in Higher Education (IACQA 2016), vol. 467, p. 474 (2016)
24. Ramani, R.G., Shanthamalar, J.J.: Automated image quality appraisal through partial least squares discriminant analysis. Int. J. Comput. Assist. Radiol. Surg. Comput. Assist. Radiol. Surg. **17**(7), 1367–1377 (2022)

Snowfall Forecasting with Enhanced Accuracy: Leveraging Multi-Class SVM for Meteorological Predictions

M. Ranjith Kumar[1(✉)], V. Sudeesh Kumar[2], G. K. Tharun Kaarthik[2], Veda Chatiyode[2], Janani Srinivasan Anusha[2], and P. Revathi[3]

[1] School of Computer Science and Engineering, Vellore Institute of Technology, Chennai 600127, Tamil Nadu, India
ranjith.kumar@vit.ac.in

[2] Department of Computer Science and Engineering, Amrita School of Computing, Amrita Vishwa Vidyapeetham, Chennai, India

[3] Department of Computer Science and Engineering, C Abdul Hakeem College of Engineering and Technology, Melvisharam, Tamil Nadu, India

Abstract. The use of multi-class Support Vector Machines (SVM) for snowfall prediction is an advanced meteorological application. The study emphasizes how valuable snowfall forecasts are in various contexts, including agriculture, aviation, transportation management, and climate reporting. This approach presents a novel methodology for classifying snowfall types by analyzing various meteorological factors such as temperature range, wind dynamics, and precipitation. A 400-entries of datasets were pulled from GitHub to demonstrate the methodology. The sample is selected to reduce the complexity and to increase the performance. The dataset is split into a testing set (30%) and a training set (70%), which aids in assessing the model's performance in a wide range of parameters. When the algorithm was first tested, it could distinguish different types of snowfall with a remarkable 83% accuracy rate based on indicators such as Temperature (min & max), Precipitation, Wind Speed, and Wind Gust. By optimizing the SVM configuration through hyper-parameter tuning approaches such as grid search, we were able to further refine our model's accuracy to an astounding 91%. This improvement in forecast accuracy signifies a significant move away from older, more expensive instrument-based methods and towards data-driven snowfall analysis. Our results show how machine learning can revolutionize weather prediction systems, providing a more effective and economical method of comprehending and predicting snowfall patterns.

Keywords: Support Vector Machine (SVM) · Kernel · Multi-class classification · Snowfall prediction · Hyperplane · Parameter optimization

M. L. Owoc et al. (Eds.): ICCIDS 2024, IFIP AICT 718, pp. 312–328, 2024.
https://doi.org/10.1007/978-3-031-69986-3_24

1 Introduction

Accurate forecasting of meteorological factors such as precipitation, temperature, and snowfall plays a pivotal role in comprehending climate dynamics and making informed decisions. Snowfall predictions have crucial applications in diverse sectors such as agriculture, aviation, transportation management, and climate reporting. The accuracy of these forecasts is vital for planning and decision-making. For instance, knowing the expected snowfall in agriculture helps farmers better plan for potential effects on crops. In aviation, accurate predictions aid in flight planning and safety measures. Transportation management relies on snowfall forecasts to understand road conditions and potential disruptions. Additionally, climate reporting benefits from precise snowfall predictions to understand broader weather patterns.

Forecasting of snowfall has predominantly relied on traditional methods, often expensive instruments. The primary contribution of this study is the utilization of sophisticated machine learning techniques, specifically multi-class Support Vector Machines (SVM), to revolutionize snowfall prediction. In contrast to previous methodologies, this approach analyzes diverse meteorological factors like temperature, wind dynamics, and precipitation, hence enhancing its precision and cost-efficiency.

Supervised learning models, such as Support Vector Machines(SVMs), are helpful tools for solving problems that include the categorization of entities or prediction of numerical values. Consider them as investigators endeavoring to draw a clear line (hyperplane) between different groups in a bunch of data [1]. SVMs are highly powerful and effective machine learning models often used for complex tasks such as image and text classification. They can accurately classify data even when a large amount of noise is present [2] and [3]. The contribution of the paper lies in the adoption of machine learning to predict snowfall, providing a more effective and economical method. The novel methodology and improved accuracy have broad applications in sectors dependent on precise snowfall forecasts, representing a substantial advancement in meteorological predictions.

1.1 Introductory Principles of SVM

SVMs identify a hyperplane in N-dimensional space to separate classes. The selected hyperplane optimizes the margin, improving classification accuracy. SVM algorithms are often faster than neural networks due to their utilization of simpler mathematical models that require fewer computational operations. Moreover, SVM algorithms are better suited for limited datasets as they possess the ability to identify patterns in small data and can be effectively trained on datasets with only a few thousand samples. This is because SVM algorithms can use kernel functions and other features in order to enhance their performance while dealing with limited datasets.

The SVM approach might be likened to a detective who seeks a special line (hyperplane) to cluster data points. The shape of this line depends on the number of features: for two features, it's a simple line; with three features, it's more like a flat surface; but for more than three features, it becomes very complex and hard to visualize [4]. The goal of SVM is to determine the multi-dimensional hyperplane to sort our data into groups. It's a versatile tool used in various tasks, from organizing text to recognizing faces, handwriting, and even identifying genes. Think of it as a detective with a wide skill set, helping us solve different data-related puzzles. The SVM is a supervised machine learning technique that demonstrates exceptional performance in addressing regression and classification issues.

1.2 Linear Data

The simplest way to understand the basic principles of support vector machines is by using an illustrative example. Let's consider a dataset with two characteristics, x and y, and that we have two labels, red and blue. We need a classifier that outputs red or blue depending on a pair of (x, y) coordinates. Next, we can plot the training data on a plane that has already been labeled.

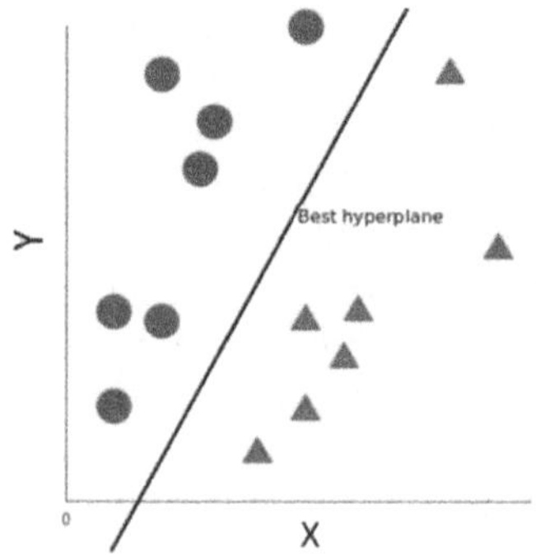

Fig. 1. 2-D Hyperplane

Support vector machines (SVMs) use data points to build a hyperplane that effectively divides the input into different classes. In two dimensions, this hyperplane serves as the decision boundary. When a data point falls on one side of the line, it is classified as belonging to the blue class in Fig. 1, and when it falls on the other side, it is classified as belonging to the red class.

1.3 Non-Linear Data

Since it was obvious that the data was linearly separable-we could draw a straight line to separate red and blue-classifying linear data was simple. However, consider this instance. A single straight line separating the two tags is not a linear decision boundary. Nonetheless, it appears that it should be simple to separate the vectors because they are so obviously separated. We intend to incorporate a

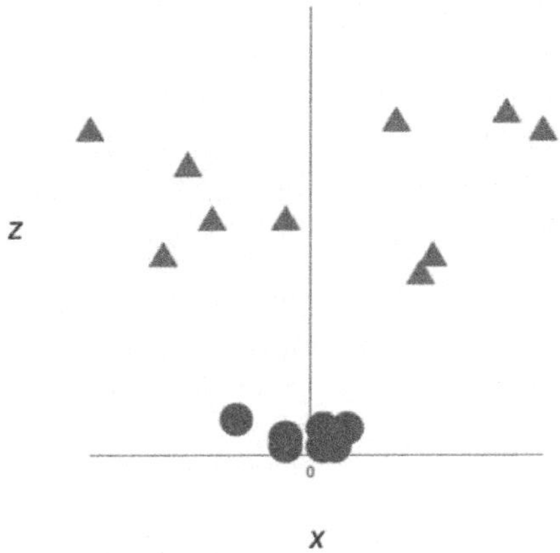

Fig. 2. Non-linear 3D Space

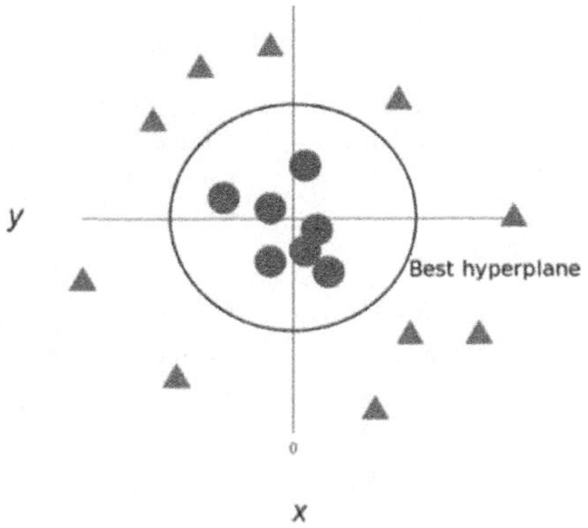

Fig. 3. 3-D Hyperplane

third dimension. We only had two dimensions up until this point: x and y. We create a new dimension called z and define it as follows: $z = x^2 + y^2$. This gives us a three-dimensional space, shown in Fig. 2. If we take a slice of that space, we can see that the hyperplane is a plane parallel to the x-axis at a certain z value (for example, $z = 1$). Our decision boundary is a circumference of radius 1, shown in Fig. 3.

The rest of this paper is structured as follows: Sect. 2 contains the literature review of a few research papers and their key findings. The dataset has been explained explicitly in Sect. 3. Section 4 discusses the methodology, the preprocessing steps through to the critical selection of the SVM kernel function, and the comprehensive approach to hyperparameter tuning, highlighting the tailored strategies to enhance model performance. Section 5 analyzes the experimental results, and Sect. 6 presents the research conclusions and future research directions.

2 Literature Review

Support Vector Machines (SVM) have evolved as a versatile tool in statistical learning theory, finding applications in diverse areas from time series prediction to medical diagnosis. The authors Evgeniou and Pontil emphasize the robust theoretical foundations of SVM and advocate for ongoing research and broader applications [5]. Franc and Hlavac introduced a transformation from multi-class SVM to a more efficient single-class framework, yielding optimization comparable to advanced algorithms [6]. In parallel, Wang and Xue addressed the computational challenges in multi-class SVMs, proposing the Simplified Multi-class SVM (SimMSVM) to expedite training while maintaining accuracy [7]. Lee's research demonstrated the effectiveness of SVM in corporate credit rating prediction, outperforming methods like discriminant analysis and neural networks

Table 1. Summary of Previous Work on SVM Applications

Approach	Focus Area	Method/ Approach	Key Findings
Evgeniou and Pontil [5]	Theoretical foundations of SVM	Robust theoretical discussion	Advocates for ongoing research and broader SVM applications
Franc and Hlavac [6]	Efficiency of multi-class SVM	Transformation to single-class framework	Optimization comparable to advanced algorithms
Wang and Xue [7]	Computational challenges in multi-class SVMs	Simplified Multi-class SVM (SimMSVM)	Expedited training while maintaining accuracy
Lee [8]	Corporate credit rating prediction	SVM with optimized parameters	Outperformed traditional methods
Liu et al. [9]	Multi-class classification	Projected Multi-objective SVM	Improved accuracy and training time
Shirin Tavara [10]	Multimodal classification	Taxonomy for multimodal learning	Identified multimodal learning potential and challenges
Pradhan et al. [11]	Protein classification	MSVM with MTLBO method	Improved recall rates for enzyme classification
Hoopes et al. [13]	SLR prediction in hydroclimate	SVM and other ML techniques	Enhanced SLR prediction accuracy in mountainous regions

through optimized parameters and structural risk minimization [8]. Liu et al. advanced multi-class classification with their Projected Multi-objective SVM, outperforming other methods in accuracy and training time [9].

Shirin Tavara's study focuses on multimodal classification, proposing a taxonomy to address the field's need for consistent terminology and structural descriptions. The research highlights the potential of multimodal learning over unimodal models and identifies challenges like big data and class imbalance [10]. Pradhan, Sahoo, Misra, and Padhy developed an innovative SVM-based method for protein classification, mainly focusing on enzymes. Their approach used a multiclass SVM (MSVM) classifier combined with a Modified Teaching Learning Optimization (MTLBO) method to identify significant features for classification. The proposed MSVM-MTLBO demonstrated impressive recall rates, outperforming existing methods in class and subclass classification of enzymes using fewer binary classifiers [11].

Jakub Horak et al. examined SVM's bankruptcy prediction against neural networks, highlighting SVM's better generalization and accuracy, especially with smaller training sets. Both studies stressed selecting suitable kernel functions and optimizing parameters to enhance SVM's performance, favoring it over traditional techniques [12].

Hoopes et al. (2023) employ machine learning to improve snow-liquid ratio (SLR) forecasts in the Sky Islands of southeast Arizona, outperforming more conventional techniques such as multiple linear regression (MLR). Temperature and humidity are important factors that are highlighted by advanced machine learning algorithms like SVM, KNN, and random forests, which enhance the accuracy of SLR predictions [13]. Table 1 provides a complete comprehensive overview of influential works in the field of Support Vector Machines (SVM) applications. It encompasses a diverse range of studies that leverage SVM for solving complex problems across various domains. The table provides complete details regarding the focal area, methodology employed and the pivotal discoveries of each study, demonstrating the extensive and profound benefits of SVM's utility in both theoretical exploration and practical application.

3 Dataset

This study makes use of a sizable dataset of meteorological information that is pertinent to snowfall forecasting. Each of the 364 entries in it offers in-depth measurements of a wide range of meteorological factors. We demonstrate our methodology utilizing the dataset from the GitHub repository [14]. The dataset has a single category variable representing the amount of snowfall, while the majority of the values are continuous numerical values. The fact that the standard deviations and mean values are nearly equal indicates that the data appears standardized. This implies that preprocessing techniques like standardization or normalization have been used. The overview of the dataset and the total description of the keywords have been described and depicted in Table 2.

Table 2. Overview of Dataset

PRCP	SNWD	SNOW	TMAX	TMIN	AWND	WDF2	WDF5	WSF2	WSF5	snow
-0.22453	-0.38478	0.139583	-1.6286	-1.2652	0.01372	0.5323	0.51297	-0.2877	-0.2955	Heavy Snow
1.090609	1.399762	0.425421	-1.9505	-1.8037	1.9177	-1.6083	-1.7740	1.2296	1.0915	Heavy Snow
-0.31105	4.076568	0.082415	-2.2808	-2.3900	-0.2170	-1.9008	-1.972	0.16441	-0.0344	Heavy Snow
-0.39758	3.719661	0.038587	-0.9257	-2.0151	0.07144	-0.0516	0.0151	0.47367	0.54093	Little Snow
-0.31105	3.1843	0.038587	-0.4598	-0.6311	-0.6786	-0.24609	-0.1837	0.47369	0.6405	Moderate Snow
-0.0861	0.507493	0.053832	-0.4090	-1.8614	3.70635	0.43516	0.4135	2.7743	2.5218	Heavy Snow
-0.31105	0.507493	0.072887	-2.4248	-2.2266	3.36017	0.43512	0.21467	1.70919	2.0918	Heavy Snow
-0.39758	0.507493	0.044304	-1.7218	-1.8614	0.2445	0.33789	0.3146	-0.0706	- 0.25955	Slight Snow
-0.39758	0.507493	0.038587	-1.4846	-1.4865	-0.5055	0.24042	0.2146	-1.0718	-1.2617	Little Snow
-0.13801	0.507493	0.067171	-0.7393	-0.9003	-0.9671	-0.44072	-0.2825	-1.20894	-1.1309	Heavy Snow
0.882955	-0.38478	0.038587	-0.1719	-0.1025	1.2830	0.53243	0.41354	2.4658	2.44236	Moderate Snow

3.1 Data Description

- **PRCP (Precipitation):** The total amount of precipitation, which comprises both rainfall and snowfall, is quantified by this variable. It is essential for determining moisture content and predicting snowfall amounts. This information is usually recorded in millimeters or inches.
- **SNWD (Snow Depth):** This variable explains how deep the snowpack has become on the ground. It is a crucial component in determining how snowfall affects infrastructure and transportation. Usually, the depth is expressed in centimeters or inches.
- **SNOW (Snowfall):**This variable specifies how much snow fell in a certain length of time. Understanding the amount and pace of snow accumulation is essential. It is measured in inches or cm.
- **TMAX (Maximum Temperature):** The highest temperature recorded during a specific period is represented by this variable. In addition to influencing the general weather, it is essential in determining how quickly snow melts measured in Celsius or Fahrenheit.
- **TMIN (Minimum Temperature):** It records the minimum temperature observed within a specific timeframe. It plays a vital role in forecasting freezing conditions and the possibility of snowfall. Temperature measurements are typically indicated in degrees Celsius or Fahrenheit.
- **AWND (Average Wind Speed):** This parameter denotes the average wind speed during a specific time frame. It has a notable influence on the formation of snowdrifts. The measurement of wind speed is commonly measured in miles per hour(mph) or meters per second(m/s).

- **WDF2 (Wind Direction at 2 m):** This variable represents the wind direction at a height of two meters above the ground. It is valuable for comprehending wind patterns and their impact on snow distribution. The measurement of Wind direction is commonly expressed in degrees.
- **WDF5 (Wind Direction at 5 m):** It is similar to WDF2, measures the direction of the wind at an elevation of 5 m. It offers more insight into wind patterns at varying altitudes, usually quantified in degrees.
- **WSF2 (Peak Wind Gust at 2 m):** This variable records the highest wind gust speed observed at an elevation of 2 m above the surface. It has a crucial function in assessing the intensity of wind occurrences.
- **WSF5 (Peak Wind Gust at 5 m):** This parameter quantifies the highest wind gust speed recorded at an altitude of 5 m. It provides crucial information on the highest wind speeds encountered.
- **Snow(Snowfall Category):** This categorical variable divides the amount of snowfall into three categories: "Heavy Snow," "Little Snow," and "Moderate Snow." It is the multi-class SVM model's target variable, classifying snowfall intensity according to input parameters.

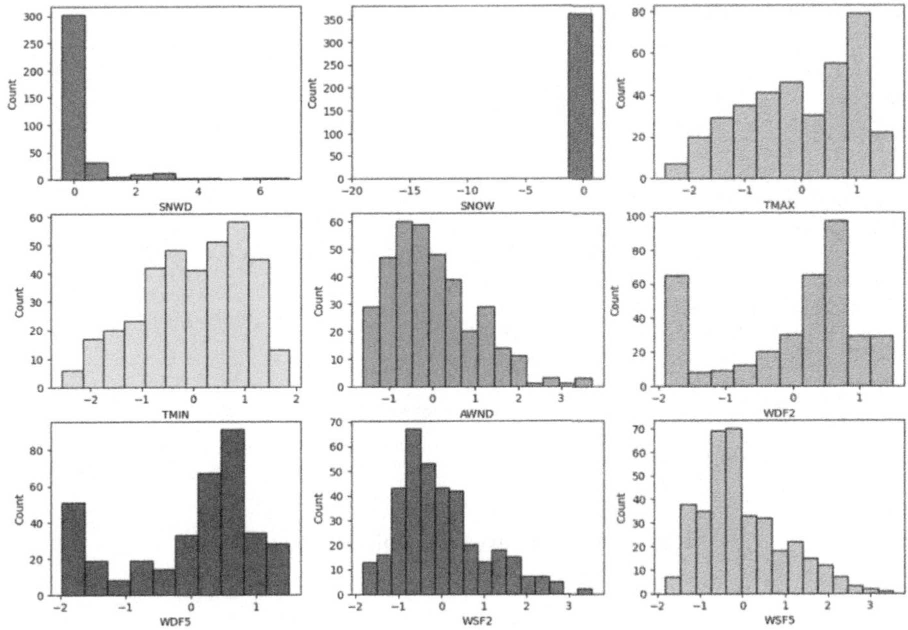

Fig. 4. Histogram of Input Parameters

While training the model, the "snow" data is labeled using the Label Encoder library in Python as [0,1,2,3] respectively. Also, the data has the following attributes:

- Percent of Heavy Snow: 14.835164835164836%
- Percent of Little Snow: 49.725274725274732%
- Percent of Moderate Snow: 31.318681318681318%
- Percent of Slight Snow: 4.1208791208791204%

Histogram of the input (parameters) SNWD, SNOW, TMAX, TMIN, AWND, WDF2, WDF5, WSF2, WSF5 has been visualized in Fig. 4.

The dataset predominantly features instances of "**Moderate Snow**" and "**Little Snow**" in its snow detail records while also including lesser occurrences of "**Heavy Snow**" and "**Slight Snow**." The dataset offers a thorough and balanced set of meteorological variables necessary for a multi-class SVM model to be successfully applied in snowfall forecasting.

4 Proposed Scheme

The proposed approach leverages a multi-class Support Vector Machine (SVM) model for predicting snowfall with improved accuracy and reliability. This methodology outlined herein integrates a variety of meteorological factors and computational techniques to ensure robust predictions. Figure 5 serves as a visual guide to the architecture of our model.

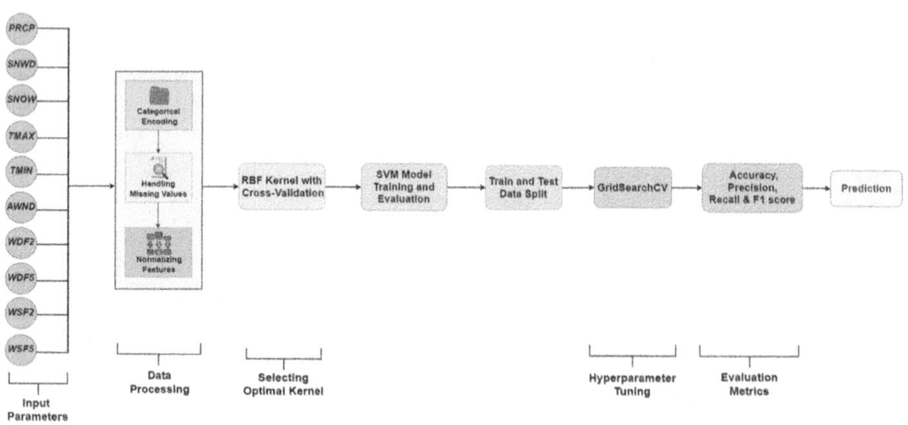

Fig. 5. Architecture of the proposed model

4.1 Data Processing

In preparing the meteorological dataset for snowfall forecasting, we loaded and manipulated the data efficiently using Pandas, a powerful data manipulation library in Python [15]. Categorical data, such as snow, was encoded into numerical values using LabelEncoder from scikit-learn [16]. Also, to address missing data in the snow column, we employed SimpleImputer with a 'most frequent' strategy, a technique supported by scikit-learn. Following this, we harnessed the power of sci-kit Learn's StandardScaler to transform the features, making sure their average value became 0 while their standard deviation was set to 1. This step held significant importance because it ensured that each feature played a fair and equal role in influencing the model's performance.

The Sci-kit Learn's StandardScaler function has been used to overcome this phase. Through this, we carefully divided the processed dataset into two distinct sets: one for training and the other for testing, with a specific 30% portion meticulously reserved for the testing set. We analyzed possible connections between snowfall, snow depth, and other meteorological variables using scatter plots.

4.2 Data Preparation

We created a series of educational visualizations using Matplotlib and Seaborn to obtain insights into the meteorological dataset and the distribution of its essential parameters. These histograms were meant to show how vital parameters, like average wind speed, wind direction at two meters, wind direction at five meters, and wind speed at two and five meters, were distributed. Also, scatter-plots depicted in Fig. 6, were created to examine the associations among snow depth, snowfall, and other meteorological variables, which helped us identify potential correlations and patterns within the data.

4.3 Why SVM?

Support Vector Machines (SVMs) are the preferred approach for snowfall forecasting due to their track record of accuracy in classification problems. Specifically designed to work with intricate datasets, support vector machines (SVMs) are widely recognized for their efficaciousness in classifying data points into discrete groupings. It provides a strong foundation for accurately predicting different types of snowfall by considering input characteristics including temperature, wind direction, speed, and precipitation. This is particularly advantageous when forecasting snowfall as numerous meteorological factors need to be analyzed and categorized. The SVMs are a dependable choice for this cutting-edge meteorological application because of their robust theoretical foundations, an abundance of research experience, and application expertise across multiple domains.

4.4 Optimal Kernel Function for Improved ML Performance

The linear kernel is selected because of its efficacy in addressing multi-class classification problems, such as the prediction of snowfall quantities. The linear kernel provides simplicity and computing efficiency, making it a good choice for handling large and high-dimensional datasets, in contrast to other kernel functions that could increase complexity. The objective of snowfall prediction is to categorize various forms of precipitation according to meteorological factors. The linear kernel offers a simple decision boundary that efficiently divides different classes. Additionally, linear kernels have a lower tendency to overfit, which is important when creating a reliable and broadly applicable prediction model.

In this prediction, the Radial basis function is best. The choice of kernel functions is not dictated by a single model but requires experimental comparison, as discussed in [17]. This experimental approach aims to identify better parameters and enhance generalization performance. In machine learning, particularly in support vector machines, the radial basis function (RBF) kernel has proven to be a popular choice for its effectiveness. The use of the RBF kernel is well-established and is extensively discussed in the seminal work by [18]:

$$k(h, h') = exp\left(-\frac{||h - h'|^2}{2\sigma^2}\right) \tag{1}$$

Since the value of the RBF kernel decreases with distance and ranges between zero (in the limit) and one (when h = h') [19], it has a ready interpretation as a similarity measure. The feature space of the kernel has an infinite number of dimensions. The proper kernel function can be chosen by utilizing the cross-validation method and selecting the one with the lowest inductive error [20]. Improved accuracy can be obtained for all kernel functions with the correct parameters, but the radial basis function (RBF) is usually the most effective option.

4.5 Analytical Tool - SVM Basics

SVMs are robust analytical tools that are well-suited for snowfall forecasting. They are classification algorithms that distinguish between different snowfall intensity levels, such as heavy snow, little snow, moderate snow, and slight snow. It uses support vectors, the data points closest in proximity to the decision boundary, and kernel functions, such as the linear kernel utilized in our model, to map data into spaces of higher dimensions. SVMs are very suitable for snowfall forecasting because of their ability to effectively handle multi-class classification, detect intricate patterns, and resist the influence of outliers. To enhance the accuracy of snowfall forecasts, We refined the SVM model by selecting the most informative features and fine-tuning the hyperparameters.

4.6 Model Training and Evaluation

To predict snowfall, we employed a Support Vector Machine (SVM) model, which is renowned for its precise classification capabilities. The linear kernel was selected because of its suitability for multi-class classification tasks [21]. We carefully split our dataset into training and testing subsets, with 30% of the data reserved for testing [22]. The SVM model was subsequently trained on the training subset and used to provide predictions on the testing subset.

4.7 Hyperparameter Tuning

Hyperparameter tuning is essential for optimizing the performance of machine learning models [23]. We employed GridSearchCV to tune the hyperparameters of our SVM model. In GridSearchCV, a systematic exploration of various hyperparameter combnation is conducted. We considered a range of hyperparameters, such as the regularization parameter (C) and the kernel function (linear, RBF, polynomial, sigmoid). We used five-fold cross-validation to ensure the robust hyperparameter selection would generalize well to new data. We chose the hyperparameters that produced the highest accuracy on the validation set [24].

4.8 Evaluation Parameters

In machine learning, the **F1-Score** is a measure of a model's accuracy, considering both false positives[FP] (incorrectly identified positive cases), false negatives[FN] (incorrectly identified negative cases) and true positives[TP] (correctly predicts the positive class), true negatives[TN] (correctly predicts the negative class). It provides a balanced assessment of the model's performance, especially when an imbalance exists between the classes.

Precision: Precision, a fundamental metric in model evaluation, assesses the accuracy of positive predictions. This notion, commonly used in information retrieval systems [25], is computed by dividing the number of true positives by the sum of true positives and false positives

$$Precision[Z] = \frac{TP}{(TP + FP)} \tag{2}$$

Recall: The metric of Recall, measures the model's ability to identify positive instances, which is a crucial factor in determining classification accuracy [26]. The calculation involves dividing true positives by the sum of true positives and false negatives.

$$Recall[Q] = \frac{TP}{(TP + FN)} \tag{3}$$

F1-Score: The F1-Score, which is calculated as the harmonic mean of precision and recall, is a detailed metric used to evaluate models.

$$F1score[N] = 2 * \frac{(Z * Q)}{(Z + Q)} \tag{4}$$

The sci-kit-learn's "precision score" and "recall score" functions have been used to compute precision and recall scores. Subsequently, the F1-Score was computed, a quantitative measure that combines recall and precision into a single value. The F1-score was stored in the variable denoted as "f1". The console was then displayed with the precision, recall, and F1-Score values.

The F1-Score is an important indicator when balancing recall and precision. This is particularly accurate in situations where the significance of false positives and false negatives varies. It provides a thorough assessment of our snowfall prediction model's effectiveness in classifying different snowfall levels concerning the model.

4.9 Accuracy in SVM

In multi-class SVM, accuracy measures how well the model sorts instances into their correct categories across different snowfall kinds. It reflects the model's overall performance in distinguishing between different snowfall categories by leveraging meteorological characteristics. Achieving a high level of accuracy entails the model accurately predicting snowfall types by understanding patterns within the data. The procedure encompasses data preprocessing, feature selection, and fine-tuning model parameters. High accuracy ensures reliable snowfall classification, improving prediction effectiveness in real-world scenarios.

Multi-class classification accuracy refers to the proportion of correctly classified instances across all categories. It shows how well the model predicts different types of snowfall based on meteorological factors. The model's high accuracy enables it to effectively differentiate between different types of snowfall, providing valuable assistance to sectors such as agriculture and aviation. Assessing accuracy involves considering other metrics like precision and recall to measure the overall effectiveness/overall accuracy.

5 Result Analysis

The utilization of SVMs in the prediction of snowfall has demonstrated promising results. The model was trained on a dataset encompassing a broad spectrum of meteorological variables and showcased remarkable proficiency in forecasting various snowfall intensities. As depicted in Table 3, the model attained an impressive accuracy rate of 90.47%. Further validation through calculated recall (86.6%), precision (90.4%), and F1-score (88.4%) metrics underscores the model's robustness and its reliable predictive performance.

Table 3. Detailed Results of the Model's Predictive Performance

Metric	Value
Accuracy	0.9041095890410958
F1 Score	0.8844293662872467 7
Precision	0.8663457978526472
Recall	0.9041095890410958

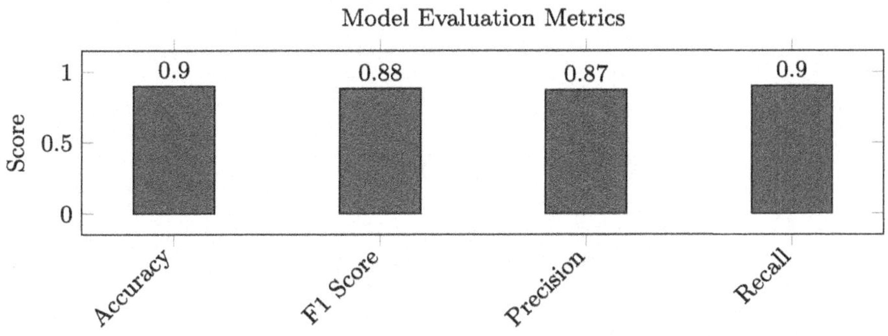

Fig. 6. Model Evaluation Metrics

Our research culminates in developing a sophisticated tool for snowfall prediction, demonstrating the significant capabilities of SVM in meteorological advancements. The visual evidence of our achievements, as presented in Figs. 6 and 7, further validates the efficacy of our approach.

The impact of our study is multi-fold:

- Initially, the model demonstrated an 83% accuracy in discerning snowfall types, leveraging temperature and precipitation data.
- The accuracy of the model was significantly improved to 91% with meticulous hyperparameter tuning, which employed a grid search approach.

This research embodies a notable advancement in the field of meteorological science, providing a potent tool that enhances the precision of predictions and supports strategic decision-making for weather-sensitive sectors. This work stands as a beacon of innovation, set to substantially enhance our interaction with and response to the natural world by harnessing the power of machine learning. The insights garnered from this research are crucial, signifying a leap forward in our ability to predict and prepare for snowstorm instances.

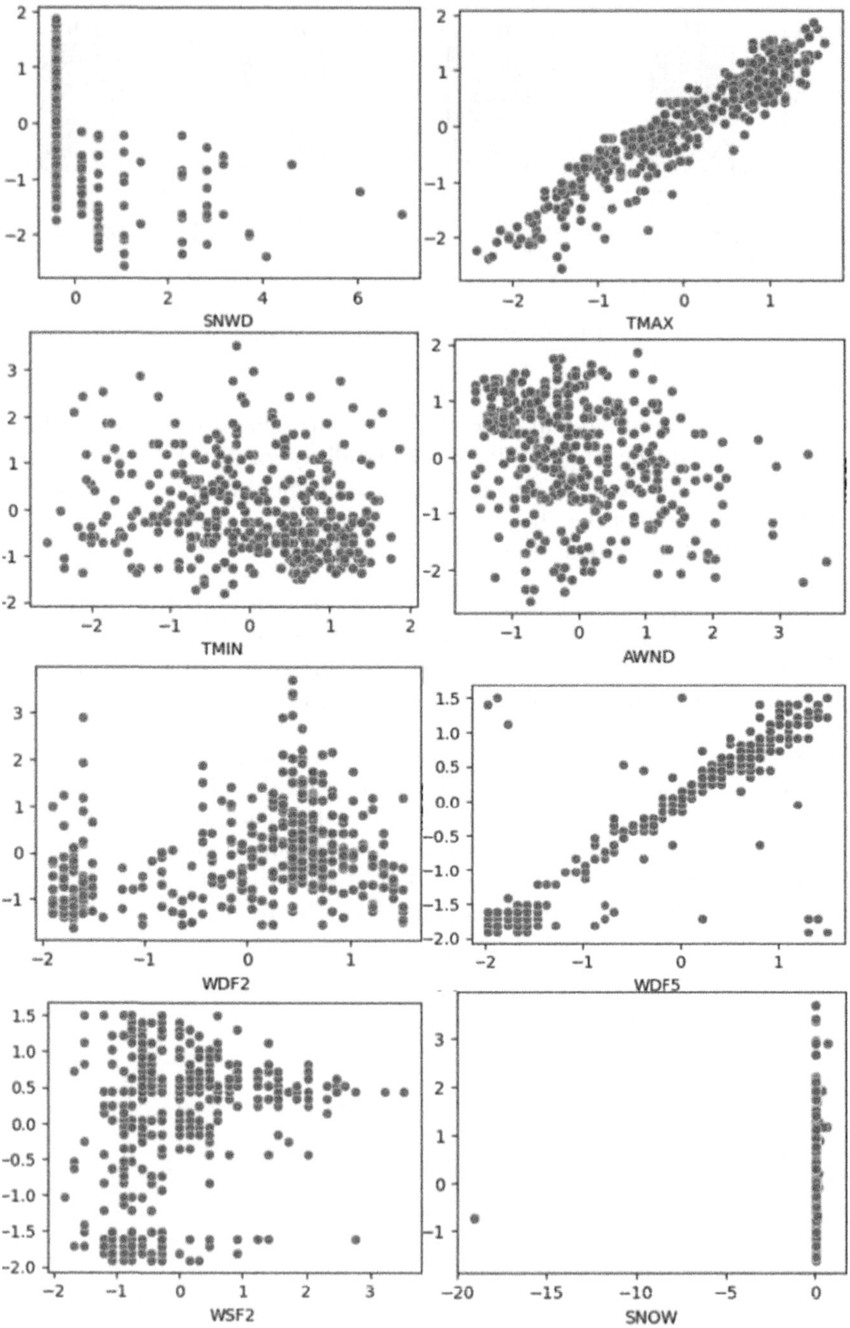

Fig. 7. Scatter Plots of the Input Parameters

6 Conclusion

This study utilizes the advanced capabilities of multi-class Support Vector Machines to enhance snowfall prediction, which is a critical factor impacting industries such as agriculture, aviation, and transportation. Analyzing key meteorological factors-temperature variations, wind patterns, and precipitation levels, we developed a refined approach for snowfall classification. The dataset was strategically divided into training (70%) and testing (30%) sets systematically, enabling a comprehensive evaluation of the model's performance across diverse scenarios. This approach represents a significant shift from traditional, hardware-dependent forecasting techniques to a nuanced, data-driven methodology, emphasizing the role of machine learning in evolving meteorological predictions. Notably, the model achieved a noteworthy accuracy rate of 91%, highlighting the substantial influence of machine learning in improving the precision and dependability of forecasts thereby guaranteeing the safety and efficiency of snow-impacted sectors.

In the future, our focus will be directed towards three primary domains to enhance the model: broadening the dataset to include a broader spectrum of meteorological factors from diverse climatic regions, enhancing model adaptability; employing advanced machine learning techniques like deep learning for more accurate forecasts and improving the model's accessibility and reliability through user-friendly interfaces and Geographic Information Systems (GIS) for comprehensive spatial analysis. The objective of these endeavors is to strengthen the model's resilience against fluctuations and deficiencies in data, hence guaranteeing reliable and resilient snowfall forecasts. This continuous progress holds the potential to enhance meteorological forecasting methods, providing more precise, timely, and practical insights for sectors and communities impacted by weather conditions.

References

1. Schölkopf, B., Smola, A.J.: Learning with Kernels: Support Vector Machines, Regularization, Optimization, and Beyond. MIT Press, Cambridge, MA (2002)
2. Wang, Q., He, R.: Snowfall prediction using support vector machine model. Water Sci. Eng. **4**(4), 274–287 (2011)
3. Chen, Y., Zhang, L.: Snowfall forecasting using support vector machines and ensemble learning. Expert Syst. Appl. **38**(11), 13821–13828 (2011)
4. Hsu, C.W., Chang, C.C., Lin, C.J.: A practical guide to support vector classification. National Taiwan University (2003)
5. Evgeniou, T., Pontil, M.: Support Vector Machines: Theory and Applications. Machine Learning and Its Applications, Advanced Lectures, Chania, Greece (2001)
6. Franc, V., Hlavac, V.: Multi-class support vector machine. In: International Conference on Pattern Recognition, Quebec City, Canada, vol. 2, pp. 236–239 (2002)
7. Wang, Z., Xue, X.: Multi-class support vector machine. Advances in Computer Vision and Pattern Recognition (2013)

8. Lee, Y.C.: Application of support vector machines to corporate credit rating prediction. College of Commerce and Economics, Dongguk University, Gyeongju, Gyeongbuk, pp. 780-714, South Korea (2012)
9. Liu, L., Martín-Barragán, B., Prieto, F.J.: A projection multi-objective SVM method for multi-class classification. Comput. Ind. Eng. **158**, 107425 (2021)
10. Tavara, S.: Parallel computing of support vector machines: a survey. ACM Comput. Surv. **51**(6), 1–38 (2019)
11. Pradhan, D., Sahoo, B., Misra, B.B., Padhy, S.: A multiclass SVM classifier with teaching learning based feature subset selection for enzyme subclass classification
12. Horak, J., Vrbka, J., Suler, P.: Support vector machine methods and artificial neural networks used for the development of bankruptcy prediction models and their comparison. J. Risk Financ. Manag. **13**(30), 60 (2020)
13. Hoopes, C.A., Castro, C.L., Behrangi, A., Ehsani, M.R., Broxton, P.: Improving prediction of mountain snowfall in the southwestern United States using machine learning methods. Meteorol. Appl. **30**, 1–22 (2023)
14. https://github.com/EricFalkenberg/weather_prediction/blob/master/test_in.csv
15. McKinney, W.: Data structures for statistical computing in python. In: Proceedings of the 9th Python in Science Conference, Austin, pp. 56–61 (2010)
16. Pedregosa, F., Gaël, V., et al.: Scikit-learn: machine learning in python. J. Mach. Learn. Res. **12**, 2825–2830 (2011)
17. Shawe-Taylor, J., Cristianini, N.: Kernel Methods for Pattern Analysis. Cambridge University Press, Cambridge (2004)
18. Hao, P.-Y., Chiang, J.-H., Chen, Y.-D.: Possibilistic classification by support vector networks. Neural Netw. **149**, 40–56 (2022)
19. Hastie, T., Tibshirani, R., Friedman, J.: The Elements of Statistical Learning: Data Mining, Inference, and Prediction. Springer, New York (2009). https://doi.org/10.1007/978-0-387-84858-7
20. Muñoz-Romero, S., Gorostiaga, A., et al.: Informative variable identifier: expanding interpretability in feature selection. Pattern Recognit. **98** (2020)
21. Kamalraj, R., Neelakandan, S., et al.: Interpretable filter-based convolutional neural network (IF-CNN) for glucose prediction and classification using PD-SS algorithm. Measurement **183**, 109804 (2021)
22. Kumar, M.R., Mani, R., Revathi, P.: A robust and fast symmetric text encryption algorithm based on Fermat's two squares theorem. In: International Conference on Recent Advances in Electrical, Electronics, Ubiquitous Communication, and Computational Intelligence, India (2023)
23. Arul Vinayakam Rajasimman, M., Manoharan, R.K., Neelakandan, S., et al.: Robust facial expression recognition using an evolutionary algorithm with a deep learning model. Appl. Sci. **13**(1), 468 (2023)
24. Bengio, Y., Bergstra, J.: Random search for hyper-parameter optimization. J. Mach. Learn. Res. **13**, 281–305 (2012)
25. Oard, D.W.: The future of information retrieval evaluation. In: Sakai, T., Oard, D.W., Kando, N. (eds.) Evaluating Information Retrieval and Access Tasks. TIRS, vol. 43, pp. 205–216. Springer, Singapore (2021). https://doi.org/10.1007/978-981-15-5554-1_14
26. Lantz, R.J.: A measure of classification accuracy for learners. IEEE Trans. Pattern Anal. Mach. Intell. **25**(3), 267–271 (2003)

Crop Irrigation Advisory System Using Federated Logistic Regression

Deepthi Gardas and R. Karthi(✉)

Department of Computer Science and Engineering, Amrita School of Computing, Amrita Vishwa Vidyapeetham, Coimbatore, India
cb.en.p2aie21010@cb.students.amrita.edu, r_karthi@cb.amrita.edu

Abstract. Federated learning (FL) is a collaborative learning algorithm that builds ML models using knowledge from distributed clients at various locations. Federated learning is valuable when data is collected from heterogeneous environments, and a unified ML model needs to be built with minimal sharing of data. In this study, a federated irrigation advisory system is developed for an agricultural application that helps the farmer to water their field based on agricultural field parameters. The application of machine learning techniques in the agricultural domain is still in its infancy. Agricultural farms employ different sensors and have natural variations in soil type and environmental conditions. A federated model framework is suitable for the development of an irrigation advisory system as heterogeneous data is gathered using a variety of sensors spread across several farms. FL-based Logistic Regression is designed to predict whether to water the field or not based on the field parameters: temperature, humidity, soil moisture content, number of days since planting the crop, and crop type. Federated learning algorithms are designed with a client-server architecture. Flower, an open-source federated framework, which used to build and test the irrigation advisory application. Two clients and a server are used to build the FL-based Logistic Regression model by exchange of model parameters. Weights and bias of the logistic regression model are aggregated by the server from multiple clients. These aggregated model parameters are sent to clients to build the regression model with its own data. Experiments were conducted to understand the performance of FL in learning the model parameters for prediction. The model's performance is evaluated using metrics to understand the converge patterns and parameters that affect FL.

Keywords: Crop Irrigation · Machine Learning · Federated Learning · crop Irrigation scheduling dataset · Irrigation Advisory system

1 Introduction

Agriculture is a crucial sector in providing food for the world's expanding population. Although agriculture significantly contributes to the world economy, it also uses many resources. 70% of the freshwater consumed globally is used for agriculture [1]. Due to this water usage, many regions worldwide are experiencing water shortages and

Published by Springer Nature Switzerland AG 2024
M. L. Owoc et al. (Eds.): ICCIDS 2024, IFIP AICT 718, pp. 329–341, 2024.
https://doi.org/10.1007/978-3-031-69986-3_25

pressure on our natural resources [2, 3]. The agriculture industry has undergone a technological revolution over the years; the Internet of Things (IoT) is one such technology that has completely changed the agricultural industry [4]. IoT technologies can gather information on crop health, temperature, and soil moisture. Better decisions concerning irrigation, quality analysis, fertilization, and pest management may be made using this data [5]. Image analysis, water quality analysis, satellite image analysis, and IoT technologies are used in agriculture tasks and environmental analysis [6]. However, several challenges must be overcome, including data management, preserving communication between IoT devices, and data security. An advisory system for predicting crop watering decisions with edge devices and federated learning principles is built using plant and environmental features. In this work, temperature, humidity, soil moisture, type of crops, and crop days are considered to reduce water wastage and promote healthy plant growth while conserving natural resources. Federated learning is a new machine learning paradigm that attempts to address problems with centralized data processing while maintaining the confidentiality and privacy of the data [7]. Due to a lack of connectivity, gathering and sharing data from sensors and other devices may become challenging. In FL, each device has a unique local dataset. The devices use their data to train a local model and a global model is constructed by combining the local models. The devices are then shared with the global model for learning from multiple clients. Federated learning solves privacy and connectivity problems in the agriculture sector by enabling models to be trained locally on edge devices without sending raw data. FL gives farmers access to real-time data and insights so they can make better choices regarding planting, watering, and fertilization. Enhancing agricultural operations efficiency and optimizing production practices can help boost crop yields while decreasing crop losses. It can also help to lower costs by lowering the demand for physical labor. The remaining paper is organized as follows: Sect. 2 gives the related work, Sect. 3 explores the methodology used in this experiment Sect. 4 discusses the experimental setup and result analysis and Sect. 5 concludes with findings and future scope for work.

2 Related Work

Recent developments in federated learning, a decentralized machine learning technique, have attracted a lot of interest because of its potential to address data privacy and security issues in various industries, including healthcare, agriculture, and the Internet of Things (IoT) [8, 9]. A Federated learning algorithm is used to train crop yield prediction models across several client devices on a horizontally distributed dataset [10]. The study compares the efficacy of federated averaging with that of deep learning and machine learning approaches. Federated averaging outperforms models created using centrally stored data for forecasting soybean yield. Bharti et al. [11] proposed edge-enabled federated learning with on-device model fine-tuning and secure model sharing across visual inspectors to find previously undetectable defects. Using a metal-nut data set for training and testing FL model was deployed for monitoring manufacturing line defects. The federated model increased local inspectors' accuracy on devices 1 and 2 by 9% and 20\%, respectively, with just one exchange of local models. In [12] federated learning is used in healthcare, particularly with wearable technology. Compared to fed-avg and split-nn, the proposed method reduces network traffic by 99.8% % and 90% respectively, and lessens the computational stress placed on IoT devices. FL and SplitNN models are implemented in a practical IoT environment [13]. Experiments were done with various datasets, model topologies, clients, and performance measures are compared to evaluate the costs and efficacy of FL and SplitNN in practical IoT settings. It compares the performance of FL and SplitNN with various types of data distribution and concludes that FL outperforms SplitNN when communication traffic is the primary concern. Gao et al. [14] trained a machine learning models on EEG data while preserving data privacy via a heterogeneous federated learning technique. Experiments on a real-world dataset demonstrate better performance using ML models such as manifold projection and federated learning, In [15], authors discuss the significance of securing communication pathways in vertical federated learning to stop the leakage of sensitive information. The Secure Logistic Regression for Vertical Federated Learning (SLRVFL) uses homomorphic encryption is resistant to evasion assaults, and achieves accuracy that is comparable to the baseline technique without significantly increasing processing overhead. A hierarchical cloud server and edge device architecture [16] that employs federated learning to enable personalized learning while protecting user privacy in IoT scenarios is proposed. In contrast to centralized learning, the method collects local model updates without gaining access to raw data, improving model performance. A federated algorithm [17] is proposed for classifying fashion items in a fine-grained manner. The experimental findings demonstrate the efficacy of the proposed strategy, with a high average precision across all categories. Rahman et al. [18] offer a federated machine learning model to calculate the average hospital stay for patients while maintaining data privacy. The findings demonstrate that federated learning enhances the aggregate model's predictive performance and that increasing the number of client parameters on the server side improves the model's performance. A Federated Learning strategy [19] is proposed to create a new expert system to forecast and find heavy metals in water and soil. The system measures heavy metals in situ and generates pollution contamination indices using satellite sensing sources and in-situ data. The system gives environmental specialists and decision-makers the condition of the soil, surface water, and crops. Machine learning using complex CNN models and generative networks has been used recently for image analysis in medicine [20], agriculture [21] and industrial domains[22] using large-scale and embedded systems. The challenge to adapt FL for varying domains are asynchronous training, minimizing communication overheads, authenticity of the model, unsupervised learning, heterogeneity of data, clients, and resource-constrained edge devices [23–25].

The survey examines federated learning on horizontally distributed datasets for various applications, such as agriculture, to enhance agricultural practices like irrigation and fertilization and train models that predict crop yield using data from several farms. Federated learning on the edge can have benefits, including real-time monitoring, cost savings, and more accurate predictions of agricultural yield. The research tries to anticipate crop irrigation via federated learning. In this study, a FL based water prediction system is developed using principles of FL. And the model's performance is compared to conventional centralized Logistic Regression models.

3 Methodology

3.1 Architecture of the System

This research aims to design and develop an FL-based edge-enabled agriculture system. Figure 1 shows the architecture of the FL-based agricultural advisory system.

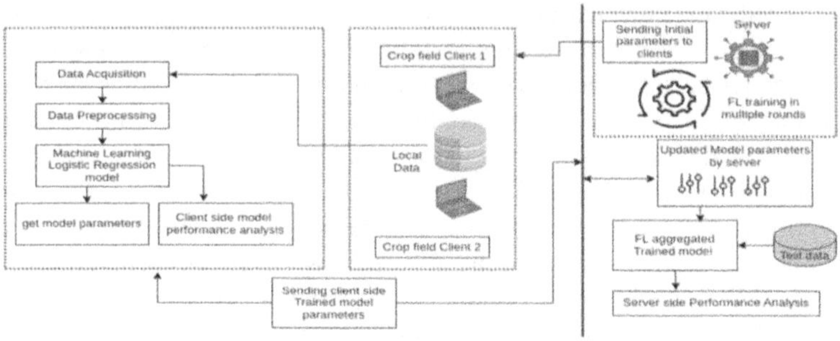

Fig. 1. Architecture of the system

The Federated learning algorithm uses client-server architecture; clients and servers communicate to share the model parameters instead of data. Federated learning algorithms build ML models in a privacy-preserving manner. The process begins with data gathered from several sources, and each source uses its private data to train a local model. The central server produces the global model by combining the local model parameters and disseminates it to the clients for training iteratively. The effectiveness of the model is evaluated, and the final global model is used to make predictions based on test data. The following building blocks exist in a federated setup.

Setting up of clients: Clients are set up at the beginning to develop models using their data. Clients collect and locally store the data, which is used for training. The clients create a logistic regression model from the local data. Within the Ubuntu Intel Core-i5 processor set as client one and Windows Intel Core i5 processor used as the second client system. a multivariate linear regression model is built on each client.

Setting up of Server: From the model parameters that each client provides, the server creates an aggregated global LR model. The global model is built using weights and biases. The server updates the global model for its local clients in FL training. The Federated Learning Server, which supports remote and centralized parameter aggregation, runs on Ubuntu Intel Core i5 processor.

Communication protocol between Client and Server: The clients and server are connected through gRPC, an open-source, high-performance universal RPC framework. It is appropriate for creating distributed systems since it offers flow control and bi-directional streaming capabilities and employs protocol buffers as the interface description language. The gRPC protocol is used by the client and server to communicate over a socket connection.

3.2 Machine Learning Model

Logistic Regression is a widely used machine learning technique, and the set of coefficients (weights) and intercept terms are parameters of the model. Logistic Regression is the best choice when predicting one of two potential classes in binary classification. Logistic Regression aims to calculate the likelihood that a given instance belongs to a given class. The logistic function translates the input into a number between 0 and 1, representing the positive class's likelihood. The method used in Logistic Regression learns a bias term and a set of weights before calculating a linear combination of input characteristics. The weights of the logistic regression model are optimized using and mathematical expression for logistic Regression given below:

$$y_{hat} = \sigma(w \cdot T * x)$$

y_{hat} is the predicted probability of the data sample belonging to the positive class, σ is the sigmoid function, w is the weight vector of the logistic regression model, and x is the feature vector of the data sample.

3.3 Algorithms of Federated Learning

The steps of the federated learning algorithm are outlined below:

Federated Learning [15–17] uses a client-server architecture for building the logistic regression model. Local training occurs on clients and federated training happens on the server. The clients build the logistic model using its local data. The client gets the model parameters from the server after each round of federated training. The server sets the global model parameters to w_0, and for each training session, it randomly selects a group of clients. The data and model parameters are shared with the clients for local training by the server in each Federated epoch. The Client_parameter_Update function separates the data into batches and updates the model parameters using stochastic gradient descent on the local data of the selected clients. The global model parameters are then updated by averaging the adjusted parameters from each client in the selected subset.

Algorithm 1 FederatedAveraging

Require: Parameters: K clients indexed by k; B is local data divided into 2 batches; E is the number of rounds; and η is the learning rate
1: The Server Process Executes the aggregation Step:
2: Initialize w_0
3: for t = 1, 2, 3, . . . do
4:     m ← max(C · K, 1).
5:     S_t ← (A random set selected m clients by the server).
6:     for k ∈ S_t in parallel do
7:         w_{t+1}^K ← Client_Parameter_Update (k, wt).
8:     end for
9:     $w_{t+1} \leftarrow \sum_{k=1}^{K} \frac{n_k}{n} w_{t+1}^k$ // Weight Aggregation
10: end for

```
Algorithm 2 Client_Parameter_Update  // Clients train
with their local data and estimate the weights

Require: k is the client index, w is the weight vector
from the Server
1: Run on client k:
2: B ←(split Pk - data into batches of size B).
3: for i = 1 to E do
4:     for each batch b in B do
5:     w ← w-η ∇ l(w, b).   // weights are updated by
the clients.
6:     end for
7: end for
8: return w the updated ML parameters to Server
```

3.4 Technique of FedAvg on Server

FedAvg creates a global model using local model updates from several clients and this is the fundamental idea of federated learning technique. FedAvg is a primary method for federated learning that enables cooperative model training on distributed data, producing potent and privacy-preserving machine learning models. The mathematical formula for FedAvg is

$$w_t = \frac{\sum_{k=1}^{K} w_k^t}{K} \tag{1}$$

where, w_t is the global model weights at round t, w_k^t is the local model weights at round t on client k, K is the number of clients. The algorithm terminates after the predetermined number of cycles, and the final global model parameters are output.

4 Experimentation

In this study, a logistic regression model is built using the principles of FL to advise farmers to irrigate crops based on field parameters. The federated algorithm has been experimented with two FL clients to understand the effects of federated collaborative learning. The models built after FL training are evaluated on each client using the standard metrics on the test data.

4.1 Data Collection and Preprocessing

The dataset used in this experiment is crop irrigation scheduling from the Kaggle repository. The dataset comprises five key attributes and the target variable. The dataset contains a total of 501 records, and it is a binary classification dataset containing two classes, 0

(304) or 1(197). The dataset includes the following features: crop type, crop days, soil moisture, temperature, humidity, and irrigation, where irrigation is the target variable for the logistic regression model. Preprocessing, such as label encoding and scaling, is done for the data.

Splitting the data: In FL, there are two types of data partitioning techniques: horizontal and vertical. The horizontal partitioning technique is a row partitioning method that involves splitting a large table horizontally into several smaller tables based on rows. In this process, the dataset is divided into two datasets to feed into two different clients. For this experiment, the dataset is split based on the crop type feature, and the number of records assigned to each client is given in Table 1. An 80:20 split of the data is used, with 80% used for training and 20% for testing.

Table 1. Horizontal Partition of data for clients

	Crop Type	Size
Dataset 1	Sugarcane, Wheat, Potato, Pulse	(257,6)
Dataset 2	Coffee, Garden Flowers, Groundnuts, Maize, Paddy	(244,6)

4.2 Flower Framework

The Federated model is realized using the Flower framework. Flexible Lower-level Operations and Resources (Flower) (Beutal et al.) is an open-source framework for creating federated learning systems. For experimentation flower framework with two clients and a server is used. The parameters used for setting the Server and clients in flower are given in Table 2.

4.3 Results and Inferences

The performance of the FL is evaluated using standard metrics, and inferences from the results are discussed in this section. Hyperparameter tuning of the model is done and analyzed.

Metrics for Evaluation

The Federated learning-based logistic regression model solves the binary classification problem, which entails categorizing instances into one of two groups. For model evaluation accuracy, f1 score, precision, recall, and loss are considered.

Loss: Loss measures the difference between predicted and actual class probabilities for a binary classification problem. Let the predicted probability of the sample belonging to class 1 be denoted by p and the binary cross entropy loss can be expressed as:

$$L = -\left[y * \log(p) + 1(1 - y) * \log(1 - p)\right]$$

Table 2. Parameters setting for Flower Framework

Parameters for client and Server		
Parameters	Significance	Value
grpc max message length	Maximum message length used for communication	512 * 1024 * 1024 for both clients and server
Strategy	Strategy used for federated learning	FedAvg
Config	number of rounds	10
Parameters for Logistic Regression Model		
Parameters	Significance	Value
Model	Classifier	logistic Regression
solver	Variant of SGD	Saga
penalty	Regularization technique	L2

y represents the data's actual label, and p represents its expected probability. The formula takes the product of the true label and the logarithm of the predicted probability for the positive class and adds the product of the complement of the true label and the logarithm of the anticipated probability for the negative class to determine the logarithmic loss for each sample. The average logarithmic loss L is calculated by dividing the total logarithmic losses across all samples by the total number of samples.

Model Building on Clients

After data preprocessing, model building is done by the clients using the federated learning technique. Each client receives the model parameters from the server after each epoch. Each client updates its logistic regression weights and tests the model with local test data. Table 3 shows clients' performance on its test data after each round, where the performance metrics are averaged from all the clients. The clients' performance is assessed on the test data using loss-distributed and metrics-distributed measures. Distributed metric refers to the training and evaluation done by each client, where the clients evaluate its data.

Loss Distributed: This metric shows the loss values for each round of the model training, where the model is trained and tested on two client devices in a distributed manner. The values in the metric indicate the average loss across two client devices for each round of FL.

Metrics Distributed: This metric shows the accuracy for each round of the model training, where the model is trained and tested on multiple devices in a distributed manner. The values in the metric indicate the average accuracy, precision, recall, and F1Score obtained across two client devices for each round of FL.

The data analysis shows that the model's ability in prediction is lesser for each client. There is no significant difference between the mean client values for each round from round 1 to round 5. In the next phase, the server evaluates the model during federated learning.

Table 3. Average Performance of Clients for Logistic Regression with Federated learning

Round	Loss distributed	Metrics Distributed			
		Accuracy	Precision	Recall	F1 score
1	0.4355	0.4361	0.8306	0.8529	0.8186
2	0.4357	0.4364	0.8102	0.8262	0.8056
3	0.3928	0.3928	0.8198	0.8255	0.8216
4	0.4043	0.4054	0.8204	0.8480	0.8077
5	0.4142	0.4151	0.7904	0.8292	0.7713

Federated Learning on Server

A logistic regression model is trained on data from two client devices. Federated learning occurs, and the weights are updated to clients during each round of training. The model performance is evaluated using test data by the server. The loss-centralized and metrics-centralized values are calculated by the server. Centralized evaluation is done as FL model averaging is done on the server. Centralized evaluation is done as data is available on the server for validation. Table 4 shows the performance of the model for different performance metrics after Federated learning and averaging on the server with its test data.

Loss Centralized: This metric shows the loss values for each round of the model training, where the model is tested for the fed-averaged model parameters of FL by the server. The values in the metric indicate the loss value for each epoch.

Metrics Centralized: This metric shows that it is a dictionary where the keys are metric names such as accuracy, precision, recall, and f1 score values for each round of the model training. The model is tested for the fed-averaged model parameters of FL by the server. The values in the metric indicate the accuracy value for each epoch.

The model's predictions improved over the rounds, as shown by the decreasing loss values and increasing accuracy, precision, recall, and F1-score. The accuracy increased from 0.673 to 0.8846, precision rose from 0.3365 to 0.8689, recall increased from 0.5 to 0.8689, and F1-score increased from 0.4022 to 0.8689. The model's ability to classify the binary dataset improved over time. The declining loss values imply that the model's prediction has become more accurate with time over each epoch.

Logistic Regression Results

Results of the logistic regression model without using federated learning on a single client are tabulated in Table 5. From the observation, the accuracy, precision, recall, and F1 score values are less when compared to the values after using federated learning. From

Table 4. Performance of Server for Logistic Regression with Federated Learning

Round	loss centralized	Metrics centralized			
		Accuracy	Precision	Recall	F1 score
0	0.6931	0.673	0.3365	0.5	0.4022
1	0.4470	0.7692	0.7538	0.6924	0.7067
2	0.3806	0.8653	0.8447	0.8546	0.8492
3	0.3741	**0.8846**	**0.8689**	**0.8689**	**0.8689**
4	**0.3496**	0.8653	0.8447	0.8546	0.8492
5	0.3533	**0.8846**	**0.8689**	**0.8689**	**0.8689**

Table 5, it is inferred that after using federated learning, the performance of logistic Regression has improved because the server aggregation is performed from multiple clients.

Table 5. Results of Logistic Regression on a Single Client and FL with two clients.

Metrics	Logistic Regression on Single Client	Logistic Regression on 2 Client with FL
Loss	11.06	35.33
Accuracy	0.6930	0.8846
Precision	0.5373	0.8689
Recall	1.0	0.8689
F1 score	0.6990	0.8689

Hyperparameter Tuning

Hyperparameter tuning is done to identify the best optimizer for the logistic regression model. SAGA (Stochastic Gradient Descent with Adaptive Gradient) is a default optimizer used in logistic Regression. Multiple optimization algorithms such as Large-scale Linear Classification (LL), Newton-conjugate gradient (NCG), Limited-memory Broyden-Fletcher-Goldfarb-Shanno (LBFG) are used for experimentation. The performance metrics obtained are tabulated in Table 6 for the distributed metrics measure. LBFG showed a slight improvement in federated learning regarding loss, accuracy, precision, recall, and f1 score. After five rounds, LL values stayed constant while NCG metrics slightly improved compared to other variants.

Table 7 shows the server evaluation metrics on the test data after the final five epoch of FL training. NCG performs the best regarding loss, accuracy, precision, recall, and F1-score. The second-best optimizer is LL, followed by LBFG. NCG optimizer can produce accurate predictions and successfully record positive examples, exhibiting a solid precision-to-recall trade-off.

Table 6. Distributed Results of FL using optimizer variants

Optimizer	Rounds of FL	Loss distributed	Metrics distributed			
			Accuracy	Precision	Recall	F1-score
LL	1	0.4233	0.8300	0.8429	0.8237	0.8168
	5	0.4233	0.8300	0.8429	0.8237	0.8168
NCG	1	0.4336	**0.8600**	**0.8589**	**0.8689**	**0.8543**
	5	**0.3592**	0.8396	0.8427	0.8396	0.8304

Table 7. Centralized Results of FL using optimizer variants

Optimizer	loss centralized	Metrics Centralized				
		Accuracy	Precision		Recall	F1score
Initial Parameters	0.6931	0.673	0.3365		0.5	0.4022
LL	0.4036	0.8846	0.8644		0.884	0.8725
NCG	**0.3481**	**0.9038**	**0.8843**		**0.9134**	**0.8951**
LBFG	0.4369	0.8461	0.8248		0.8554	0.8341
LBFG	1	0.4868	0.7735	0.7729	0.7832	0.7669
	5	0.3926	0.8312	0.8282	0.8324	0.8227

5 Conclusion

Federated learning is a new paradigm for intelligent model building when data is distributed across multiple locations and gathered under varying conditions. Water is a scarce natural resource and a system to aid farmers to efficiently use it needs to be built. For building an intelligent advisory system, federated learning is an apt choice in the agricultural domain as field parameters vary across regions and a collaborative model development needs to be ensured. An FL model is built to make the decision to water the field using field and environmental parameters such as crop type, crop days, soil moisture, temperature, and humidity. A Federated learning-based advisory system is built to aid farmers in planning to water the crop. FL model is built using logistic Regression as a binary classification problem. The Open-Source Flower framework is used to implement the federated learning systems with two 2 clients and a server. A logistic regression model is built and tuned for the application. The model is validated using real-world datasets using standard metrics loss, accuracy, precision, recall, and F1-Score. Designed federated learning with a logistic regression model and saga optimizer has a loss of 35.33, accuracy of 88.46, precision of 86.89, recall of 86.89, and F1-score of 86.89. Contrarily, a logistic regression model without federated learning has a loss of 11.06, an accuracy of 69.30, a precision of 53.73, a recall of 1.0, and an F1 score of 69.90, demonstrating the efficiency of federated learning. The model is tested for its performance using four optimizers, and the Newton-conjugate gradient optimizer has performed best compared to the other variants. The FL model can be built for future study using various aggregation methods and different ways of partitioning data. FL implemented on microcontrollers will enable real-time deployment for various IOT-enabled applications.

References

1. Singh, A.: Poor quality water utilization for agricultural production: an environmental perspective. Land Use Policy **43**, 259–262 (2015)
2. Prabha, A.S., Ram, A., Irfan, Z.B.: Exploring the relative water scarcity across the Indian million-plus urban agglomerations: an application of the Water Poverty Index. HydroResearch **3**, 134–145 (2020)
3. Hoekstra, A.Y., Buurman, J., Van Ginkel, K.C.H.: Urban water security: a review. Environ. Res. Lett. **13**(5), 053002 (2018)
4. De Alwis, S., Hou, Z., Zhang, Y., Na, M.H., Ofoghi, B., Sajjanhar, A.: A survey on smart farming data, applications and techniques. Comput. Ind. **138**, 103624 (2022)
5. Gutti, V., Karthi, R.: Real time classification of fruits and vegetables deployed on low power embedded devices using tiny ML. In: Chen, J.I., Tavares, J.M.R.S., Shi, F. (eds.) ICIPCN 2022. LNNS, vol. 514, pp. 347–359. Springer, Cham (2022). https://doi.org/10.1007/978-3-031-12413-6_27
6. Thanga Manickam, M., Karthik Rao, M., Barath, K., Shree Vijay, S., Karthi, R.: Convolutional neural network for land cover classification and mapping using landsat images. In: Saini, H.S., Sayal, R., Govardhan, A., Buyya, R. (eds.) Innovations in Computer Science and Engineering. LNNS, vol. 385, pp. 221–232. Springer, Singapore (2022). https://doi.org/10.1007/978-981-16-8987-1_24
7. Konečný, J., McMahan, H.B., Ramage, D., Richtárik, P.: Federated optimization: distributed machine learning for on-device intelligence. arXiv preprint arXiv:1610.02527 (2016)
8. Li, T., Sahu, A.K., Talwalkar, A., Smith, V.: Federated learning: challenges, methods, and future directions. IEEE Signal Process. Mag. **37**(3), 50–60 (2020)
9. Brecko, A., Kajati, E., Koziorek, J., Zolotova, I.: Federated learning for edge computing: a survey. Appl. Sci. **12**(18), 9124 (2022)
10. Manoj, T., Makkithaya, K., Narendra, V.G.: A federated learning-based crop yield prediction for agricultural production risk management. In: 2022 IEEE Delhi Section Conference (DELCON), pp. 1–7. IEEE (2022)
11. Bharti, S., McGibney, A., O'Gorman, T.: Edge-enabled federated learning for vision based product quality inspection. In: 2022 33rd Irish Signals and Systems Conference (ISSC), pp. 1–6. IEEE (2022)
12. Yuan, B., Ge, S., Xing, W.: A federated learning framework for healthcare IoT devices. arXiv preprint arXiv:2005.05083 (2020)
13. Gao, Y., et al.: End-to-end evaluation of federated learning and split learning for Internet of Things. arXiv preprint arXiv:2003.13376 (2020)
14. Gao, D., Ju, C., Wei, X., Liu, Y., Chen, T., Yang, Q.: Hierarchical heterogeneous horizontal federated learning for EEG (2019)
15. He, D., Du, R., Zhu, S., Zhang, M., Liang, K., Chan, S.: Secure logistic regression for vertical federated learning. IEEE Internet Comput. **26**(2), 61–68 (2021)
16. Wu, Q., He, K., Chen, X.: Personalized federated learning for intelligent IoT applications: a cloud-edge based framework. IEEE Open J. Comput. Soc. **1**, 35–44 (2020)
17. Mallavarapu, T., Cranfill, L., Kim, E.H., Parizi, R.M., Morris, J., Son, J.: A federated approach for fine-grained classification of fashion apparel. Mach. Learn. Appl. **6**, 100118 (2021)

18. Rahman, M.M., Kundu, D., Suha, S.A., Siddiqi, U.R., Dey, S.K.: Hospital patients' length of stay prediction: a federated learning approach. J. King Saud Univ.-Comput. Inf. Sci. **34**(10), 7874–7884 (2022)
19. Yaseen, Z.M.: The next generation of soil and water bodies heavy metals prediction and detection: new expert system based Edge Cloud Server and Federated Learning technology. Environ. Pollut. **313**, 120081 (2022)
20. Aishwarya, G., Krishnan, K.R.: Generative adversarial networks for facial image inpainting and super-resolution. In: Journal of Physics: Conference Series, vol. 2070, no. 1, p. 012103. IOP Publishing (2021)
21. Srunitha, K., Bharathi, D.: Mango leaf unhealthy region detection and classification. In: Jude Hemanth, D., Smys, S. (eds.) Computational Vision and Bio Inspired Computing. LNCVB, vol. 28, pp. 422–436. Springer, Cham (2018). https://doi.org/10.1007/978-3-319-71767-8_35
22. Karthi, R., Manchikanti, B., Sai Phani Jaswanth, C., Mali, A.R., Aakaash, N.: Prediction of water quality parameters from satellite surface reflectance images using regression techniques. In: Ranganathan, G., Fernando, X., Rocha, Á. (eds.) Inventive Communication and Computational Technologies. LNNS, vol. 383, pp. 655–666. Springer, Singapore (2023). https://doi.org/10.1007/978-981-19-4960-9_51
23. Li, L., Fan, Y., Tse, M., Lin, K.-Y.: A review of applications in federated learning. Comput. Ind. Eng. **149**, 106854 (2020)
24. Guendouzi, B.S., Ouchani, S., Assaad, H.E., Zaher, M.E.: A systematic review of federated learning: challenges, aggregation methods, and development tools. J. Netw. Comput. Appl. **220**, 103714 (2023)
25. Banabilah, S., Aloqaily, M., Alsayed, E., Malik, N., Jararweh, Y.: Federated learning review: fundamentals, enabling technologies, and future applications. Inf. Process. Manag. **59**(6), 103061 (2022)
26. Beutel, D.J., et al.: Flower: a friendly federated learning framework (2022)

Supply Chain Management Using Optimization and Machine Learning Techniques

Honey Pandey[1], N. Neelima[2(✉)], and K. V. Nagaraja[3]

[1] Department of Computer Science Engineering, Amrita School of Engineering, Amrita Vishwa Vidyapeetham, Bengaluru, India

[2] Department of Electronics and Communication Engineering, Amrita School of Engineering, Amrita Vishwa Vidyapeetham, Bengaluru, India
n_neelima@blr.amrita.edu

[3] Department of Mathematics, Amrita School of Engineering, Amrita Vishwa Vidyapeetham, Bengaluru, India

Abstract. This study focuses on developing a predictive model for late deliveries in the supply chain, using the Data Co Supply Chain dataset. It involves data cleaning, visualization, and training various machine learning algorithms, selecting the most effective model based on accuracy and recall values. The research explores challenges like transportation delays, production issues, natural disasters, and supplier disruptions. Businesses are adopting proactive supply chain and risk management strategies to enhance resilience. The paper discusses how machine learning techniques can optimize supply chain management, offering insights into mitigating risks associated with late deliveries. By identifying challenges in advance, businesses can improve operational efficiency, minimize financial setbacks, and maintain customer trust. In a dynamic business environment, this research provides valuable solutions for supply chain professionals, integrating big data and machine learning to forecast and address late deliveries. It not only aids in predicting delays but also contributes to proactive decision-making, improving overall supply chain performance amidst uncertainties.

Keywords: Machine Learning · Data Science · Supply chain optimization · Late delivery production.

1 Introduction

Modern businesses heavily rely on complex supply chains that connect suppliers, manufacturers, and customers. Late deliveries in these supply chains are a major concern, leading to financial losses and customer dissatisfaction. To address these challenges, advanced technologies like Machine Learning (ML) have become essential for improving supply chain efficiency and resilience. Late deliveries can result from various factors, including transportation delays, production issues, and unexpected events. Predicting and preventing late deliveries are crucial for business success.

Published by Springer Nature Switzerland AG 2024
M. L. Owoc et al. (Eds.): ICCIDS 2024, IFIP AICT 718, pp. 342–352, 2024.
https://doi.org/10.1007/978-3-031-69986-3_26

Supply chain analysis forms the foundation of this endeavor. It involves gaining complete visibility into the entire supply chain, from raw material procurement to final product delivery. ML, a subset of artificial intelligence, plays a pivotal role in this context. ML's ability to process vast data, identify patterns, and make predictions enables the development of models that anticipate late deliveries by identifying potential disruptions in advance. This research investigates the collaboration between supply chain analysis and machine learning for the anticipation and avoidance of delayed deliveries, highlighting the importance of making decisions based on data. By utilizing both historical and current data, organizations can develop predictive models to assist in mitigating risks, allocating resources effectively, and enhancing the overall efficiency of the supply chain. ML can identify patterns and anomalies in supply chain data, alerting stakeholders to potential issues before they lead to late deliveries. Maintaining the right inventory balance is crucial for late delivery prevention. Strategies like just-in-time inventory systems and safety stock management help minimize the risk of stockouts and overstocking. Efficient transportation and logistics management are also essential. Route optimization, real-time tracking, and alternative transportation options can mitigate delivery delays. Regular risk assessments help organizations prioritize and address late delivery risks. Delays within the supply chain represent a significant global challenge for businesses, often resulting in products arriving at their destinations later than expected.

2 Related Works

In the year 2016, the American Transportation Research Institute (ATRI) calculated that traffic congestion, on its own, resulted in more than 1.2 billion hours of delays in the United States, causing additional operational expenses amounting to $74.5 billion (Hooper, 2018) [1]. The COVID-19 pandemic exacerbated this delay issue by constraining the availability of workers and increasing reliance on online shopping and home delivery (Mckay, 2022) [2]. The pandemic significantly affected customer conduct, sales volume, and the whole production network (Galhotra and Dewan, 2020) [3]. Therefore, it is imperative to resolve supply chain delays in order to minimize financial losses, reduce customer disappointment and increase profits – especially in light of the spike in online purchases.

In order to address this issue, a number of earlier research have out predictive models for predicting supply chain delays. For instance, in order to predict delivery delays, a research by Dias, Rezende, and Salles (2021) introduced a scientific model in light of retaining the semi-Markov process, in which the semi-Markov process differentiates itself by assigning a state for each instance, not simply at predetermined intervals [4]. A different research presented a novel approach to forecast travel speeds in urban environments by modeling vehicle-urban freight truck interactions using the multi-modal macroscopic fundamental diagram (MFD) [5] (Allister, 2020).

Additional investigation into supply chain delay prediction showed that big data and machine learning algorithms were used in several research. Big data finds many applications in the supply chain domain because logistics providers generate large amounts of data for managing product flows. For example, daily global delivery networks track details like package origin, destination, size, weight, content, and position (Mikavica,

Kostić-Ljubisavljevića, & Đogatovića, 2015) [6]. For example, an integrated management model for prediction was built using machine learning techniques such as regression, clustering, and correlation in a research on the port logistics supply chain (Ouyang, 2019) [7]. Another study focused on qualitative research and addressed delays in products delivery by using the Linear Regression Algorithm, Data Mining Prediction analysis, and Rapid Miner application tools (Wahyudi & Arroufa, 2022) [8]. Furthermore, Zhang and Perry's (2016) study used the Primitive KNN algorithm for classification and time series forecasting techniques, including ARMA, to estimate future shipments. The KNN methodology produced noticeably improved results [9].

Many studies in this field extended their exploration to boosting algorithms. Khiari and Olaverri-Monreal (2022) identified CatBoost, gradient boosting and histogram gradient boosting as the most scalable algorithms for predicting delivery times, chosen based on Mean Absolute Error and Root Mean Square Error values [10]. Meanwhile, Viellechner and Spinler (2020) conducted a comparison of machine learning techniques for predicting delays in vessels, with Neural Network, SVM, Logistic Regression, Random Forest and yielding higher accuracies in classification models. These studies also employed basic statistics, lasR with glmnet for variable reduction, and Random Forest for validity confirmation [11]. Notably, most studies focused on predicting delivery times rather than the delivery status itself.

Focusing on customer delivery delay prediction in the e-commerce industry, Bristy, Durjoy, and Hasan's (2021) study used the Random Forest classifier and the SVM, two machine learning models, to predict product delivery times [12, 13]. The study found that the Support Vector Machine produced better accuracy. Support Vector Machine using sequential look-back and periodic look-back approaches were used in another study in the same area to improve delivery time forecasting (Erkmen, Nigiz, Sarı, Arlı, & Akay, 2022) [14, 15]. This resulted in a considerable improvement in the accuracy of the model. An advanced hybrid voting-based classification model that combined Trees, Ensemble techniques, shipping mode, and scheduled shipment time-enabled features was proposed in a different study that attempted to predict delays in e-commerce deliveries [16].

In summary, this study highlights the vital role of supply chain analysis and ML in predicting and preventing late deliveries. Through the incorporation of machine learning techniques, enterprises can elevate supply chain efficiency, minimize financial losses, and enhance customer satisfaction, thereby securing a competitive advantage in the current dynamic market. Optimizing supply chain management and late delivery prevention through data-driven decision-making and advanced technologies are essential steps in achieving these goals.

3 Methodology

In this research, the paper employs a diverse set of machine learning algorithms to optimize supply chain performance by analyzing and predicting late deliveries. The analysis utilized a supply chain dataset employed by DataCo Global, specifically designed for the application of machine learning algorithms. By leveraging some machine learning techniques such as Naive Bayes for pattern recognition and XGBoost, CatBoost for high-precision predictions, the goal is to comprehend intricate relationships within the supply chain using the DataCo supply chain dataset. For the research on "Supply Chain Analysis and Predicting Late Deliveries Optimization & Machine Learning," the mentioned machine learning algorithms are used as optimization techniques in the following ways:

a. **Naive Bayes:**
   Naive Bayes are used to classify late delivery events, allowing you to identify patterns or factors that frequently lead to delays. This optimizes the supply chain by targeting specific areas for improvement based on the most common reasons for late deliveries
b. **AdaBoost:**
   AdaBoost enhances the predictive accuracy of late delivery models. By giving more weight to challenging cases, it optimizes the predictive models to focus on preventing the most critical late delivery scenarios, thus improving overall supply chain performance.
c. **Logistic Regression:**
   Logistic Regression provides insights into the probability of late deliveries based on supply chain variables. By understanding the impact of various factors, it optimizes the allocation of resources and prioritize actions to reduce the likelihood of late deliveries. Logistic function which is also known as sigmoid function has values between 0 and 1.
d. **Gradient Boost:**
   Gradient Boosting also optimizes the predictive model by creating a powerful ensemble of decision trees. It also identifies complex relationships between supply chain variables and late deliveries, allowing to make data-driven decisions for supply chain improvement.
e. **Quadratic Discriminant Analysis:**
   Quadratic Discriminant Analysis helps optimize the supply chain by modeling nonlinear relationships. It uncovers intricate patterns in your data that might not be captured by simpler models, leading to better predictions and optimization strategies.
f. **Random Forest:**
   Random Forest optimize the supply chain by providing a robust and accurate predictive model. It helps identify the most influential factors contributing to late deliveries and implement targeted improvements.
g. **Extra Trees Classifier:**
   The Extra Trees Classifier optimizes the late delivery prediction models by reducing overfitting and variance. This ensures that the models are more reliable and generalize well to new data, contributing to better supply chain optimization.
h. **Linear Discriminant Analysis:**
   Linear Discriminant Analysis optimizes the supply chain analysis by reducing data dimensionality while preserving important information. This led to a more focused approach in addressing the key factors affecting late deliveries.
i. **Decision Tree:**
   Decision Trees offer an interpretable way to identify the most critical supply chain factors leading to late deliveries. By focusing on these factors, it optimizes the supply chain processes and resource allocation.
j. **XGBoost:**

XGBoost's efficiency and performance optimizes the supply chain by providing highly accurate late delivery predictions.

This accuracy are crucial for making informed decisions and taking action to prevent late deliveries.

k. **Bagging Classifier:**

Bagging also optimizes the supply chain analysis by aggregating predictions from multiple models, reducing the risk of model bias, and leading to more reliable predictions and optimization strategies.

Incorporating these machine learning algorithms into the research helps to optimize the supply chain by better understanding late delivery patterns and factors. By leveraging these optimization techniques, this research makes informed decisions, allocate resources more effectively, and proactively prevent late deliveries, ultimately improving your supply chain performance and customer satisfaction.

4 Results

In this study on supply chain optimization and late delivery prediction, the paper suggests a comprehensive data exploratory and visualization analysis, revealing crucial insights into customer preferences, delivery patterns, and market contributions. The exploration highlighted areas for supply chain improvement, particularly in shipping modes and delivery schedules. The subsequent phase involves model selection, where an evaluative comparison table and visualizations guide the identification of optimal algorithms such as AdaBoost, Gradient Boosting, Random Forest, and XGBoost. These findings lay the foundation for informed decision-making aimed at enhancing supply chain efficiency and proactively addressing late delivery challenges.

a. **Data exploratory:** To commence, the approach is to select the "Delivery Status" attribute as the response variable, representing the focus of this study in terms of prediction. Figure 1 provides the insights of this. The "Delivery Status" is a categorical variable with four primary values, namely: "late delivery," "advance shipping," "shipping on time," and "shipping cancelled." The subsequent table provides a summary of the occurrence frequency.

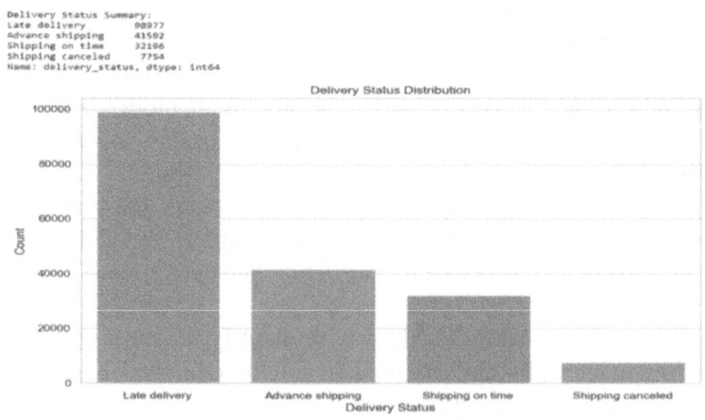

Fig. 1. Delivery status distribution

b. Figure 2 offers the risk in shipping mode. Also in Fig. 3, a summary of the department names associated with the products purchased by customers is provided.

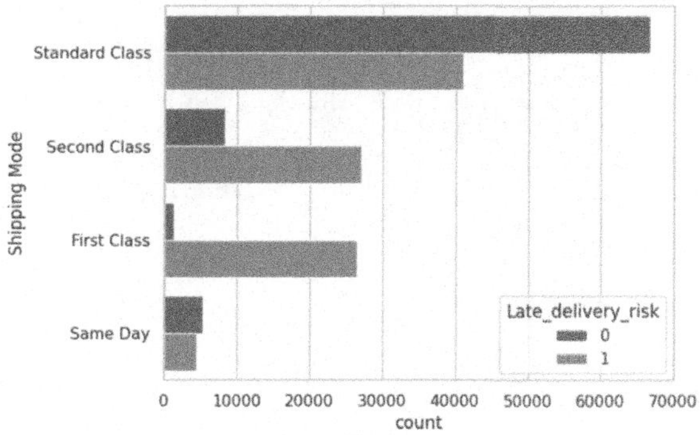

Fig. 2. Delivery risk vs shipping mode

The representation shows that most of conveyed items have a place with the apparel and golf divisions, while a more modest extent begins from the health and beauty classification.

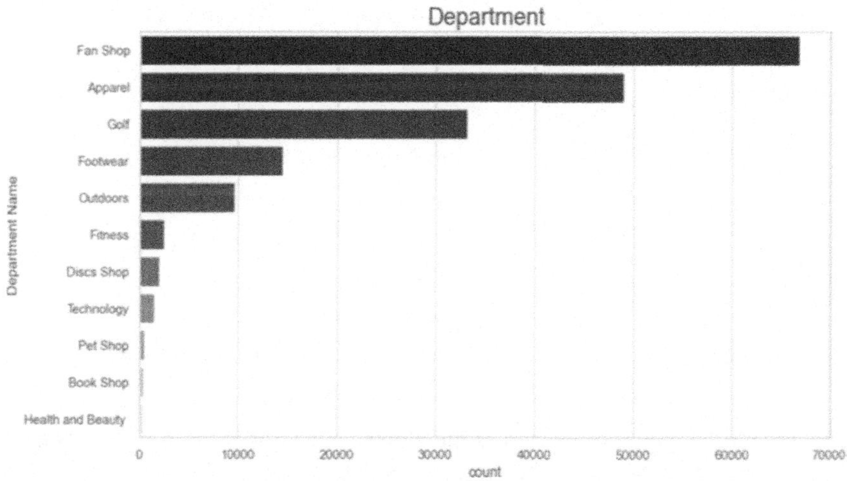

Fig. 3. Visualization with respect to different categories of consumer products

Figure 4 provides the distribution of markets across the worldwide markets. It can be observed that LATAM and EUROPE shares the major distribution. Figure 5 provides the deviation from the shipment scheduled to deliver and the actual delivered date of shipment.

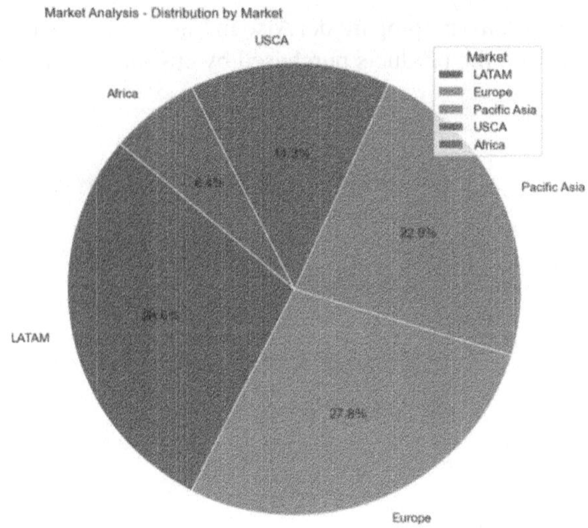

Fig. 4. Distribution of various markets across the worldwide market

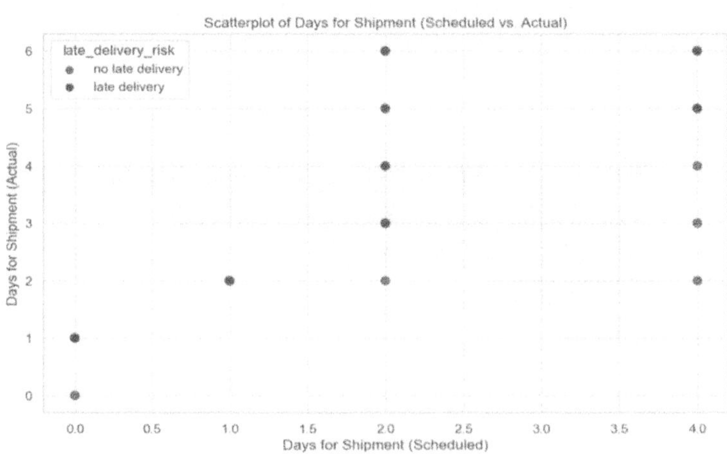

Fig. 5. Scattered plot of days for shipment scheduled vs delivered.

The effect of shipping mode across the globe can be observed in Fig. 6. The clients from the European market make the most elevated deals and, consequently, the most elevated profit per request made for the organization. In correlation, the clients from LATAM have the most elevated highest total quantity of orders. The following observation can be drawn:

a. Upon scrutinizing the dataset, it was evident that nearly 55% of orders experience delayed delivery. The late delivery rate for First Class items stands at approximately 95%, while Second Class orders exhibit a rate of 77%.

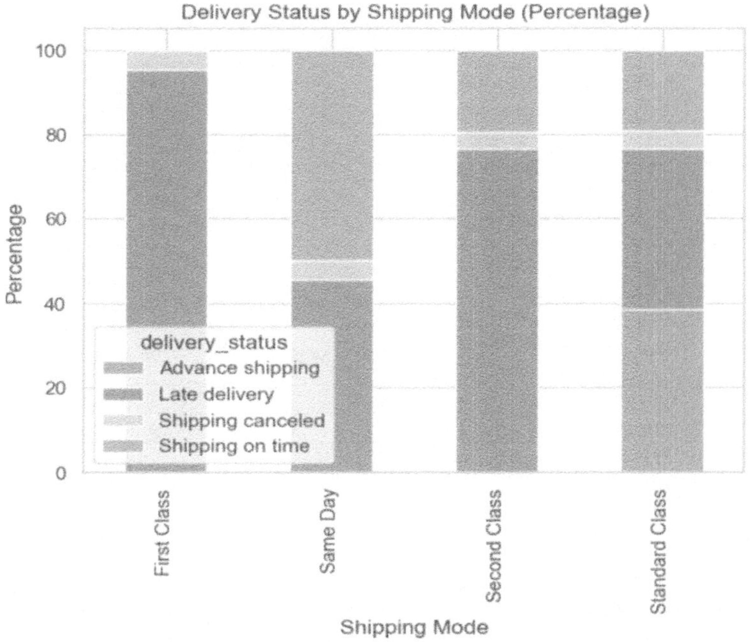

Fig. 6. Comparison of delivery status vs shipping mode

b. The greatest delivery delays are associated with the Second-Class shipping mode. Hence, there is a need for enhancements in the supply chain, particularly within the First- and Second-Class shipping modes, to ensure the timely delivery of orders.
c. Late deliveries constitute the most significant proportion, whereas "Shipping On Time" accounts for only around 18%. Punctual delivery of products is a pivotal aspect of the supply chain, as any delays can result in customer dissatisfaction and attrition, thereby affecting the company's revenue negatively.
d. The top 10 products and regions that yield the highest profits are susceptible to delivery delays. To address this challenge, the company can expedite product shipments and mitigate potential delays through the adoption of improved shipping methods. Employing meticulous delivery planning allows customers to anticipate product arrival times. The consideration of local depots in these regions can further reduce delivery times.

The subsequent phase involves selecting the optimal model from those constructed in the prior section. An evaluative comparison is given in Table 1 among these models taking into account both Accuracy and Recall scores. The criteria for model selection will prioritize the one with the highest accuracy and recall.

Figure 7 insights the accuracies of the above said models and Fig. 8 provides comparison analysis of various algorithms discussed above. In summary, each algorithm's performance can be assessed based on the trade-off between True Positives, True Negatives, False Positives, and False Negatives. Algorithms like AdaBoost, Gradient Boosting, Random Forest, and XGBoost show robust performance across multiple metrics,

Table 1. Performance analysis of the algorithms used

S. No.	Algorithms	Accuracy	F1- score
1	naive bayes	69.092067	69.014621
2	adaboost	72.000332	71.827358
3	logistic regression	78.996787	78.995638
4	gradient boost	86.178817	86.103612
5	quadratic discriminant analysis	91.973189	91.928179
6	random forest	94.128074	94.094308
7	extra trees classifier	95.629293	95.601225
8	linear discriminant analysis	96.377133	96.326517
9	decision tree	96.795369	96.764385
10	xgboost	97.972524	97.945721
11	bagging classifier	98.313206	98.298672

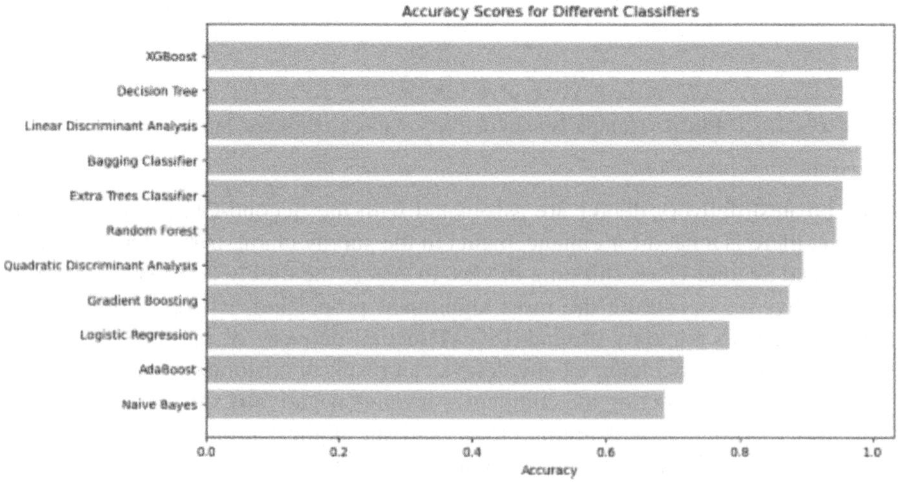

Fig. 7. Accuracy score of all the algorithms

while others may have specific strengths and weaknesses depending on the task at hand. By examining the confusion matrix plot, this paper gain insights into how well the classifier is distinguishing between different classes and identify areas where it may be making errors.

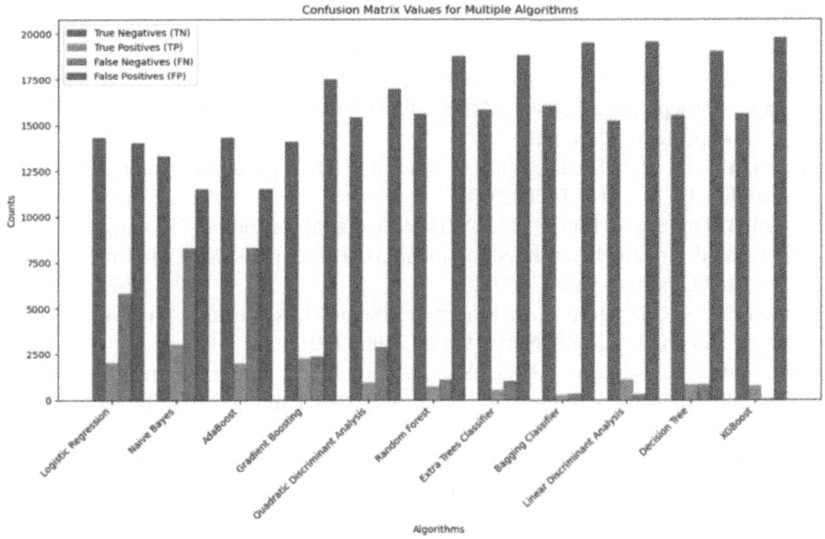

Fig. 8. Performance evaluation of various models

5 Conclusion

In conclusion, the optimization of supply chain management and risk mitigation practices is of paramount importance in addressing the challenges associated with late deliveries. By implementing robust strategies, harnessing technology, and fostering strong supplier relationships, organizations can enhance their supply chain resilience and minimize the adverse financial and reputational impacts of late deliveries. This paper establishes the foundation for businesses to proactively address this critical issue and sustain a competitive advantage in today's dynamic market. Accurately predicting late deliveries has the potential to empower companies in effectively managing their supply chain operations and optimizing financial outcomes. Consequently, when utilizing a variety of optimization and machine learning algorithms, the bagging classifier emerges as the most effective model, achieving an impressive F1 score of 98.29%, closely followed by the XGBoost classifier at 97.94% and the decision tree classifier at 96.76%. When an organization adeptly anticipates late deliveries, it can implement proactive strategies to circumvent sales reductions and gain a competitive advantage over rivals. Moreover, the integration of machine learning for late delivery prediction within the supply chain can yield environmentally beneficial outcomes. The future scope of this research can be extended in exploring the integration of emerging technologies like blockchain and Internet of Things (IoT) for real-time tracking and monitoring, further enhancing supply chain visibility and reducing late delivery risks and also exploring Explore ways to align supply chain optimization with sustainability goals, ensuring that environmentally beneficial outcomes are integrated into decision-making processes and practices.

References

1. Hooper, A.: Cost of congestion to the trucking industry: 2018 update (2018). https://truckingresearch.org/wp-content/uploads/2018/10/ATRI-Cost-of-Congestion-to-the-Trucking-Industry-2018-Update-10-2018.pdf
2. Lashgari, Y.S., Shahab, S.: The impact of the COVID-19 pandemic on retail in city centres. Sustainability **14**(18), 11463 (2022)
3. Galhotra, B., Dewan, A.: Impact of COVID-19 on digital platforms and change in E-commerce shopping trends. In: 2020 Fourth International Conference on I-SMAC (IoT in Social, Mobile, Analytics and Cloud) (I-SMAC) (2020)
4. Vassiliou, P.-C.G., Georgiou, A.C.: Markov and semi-Markov chains, processes, systems, and emerging related fields. Mathematics **9**(19), 2490 (2021)
5. Fu, H., Wang, Y., Tang, X., Zheng, N., Geroliminis, N.: Empirical analysis of large-scale multimodal traffic with multi-sensor data. Transp. Res. Part C: Emerg. Technol. **118**, 102725 (2020)
6. Vidović, M., Kilibarda, M., Zečević, S., Miljuš, M., Radivojević, G.: Optimization and modelling in logistics and transportation. In: 2nd Logistics International Conference, Belgrade, Serbia, 21–23 May 2015 (2015)
7. Ouyang, F.: Construction of integrated management model of port logistics supply chain based on big data. J. Coast. Res. **93**(sp1), 1053 (2019)
8. Wahyudi, T., Arroufu, D.S.: Implementation of data mining prediction delivery time using linear regression algorithm. J. Appl. Eng. Technol. Sci. **4**(1), 84–92 (2022)
9. Ozturk Kiyak, E., Ghasemkhani, B., Birant, D.: High-Level K-Nearest Neighbors (HLKNN): a supervised machine learning model for classification analysis. Electronics **12**(18), 3828 (2023)
10. Khiari, J., Olaverri Monreal, C.: Boosting algorithms for delivery time prediction in transportation logistics. In: 2020 International Conference on Data Mining Workshops (ICDMW) (2020)
11. Viellechner, A., Spinler, S.: Novel data analytics meets conventional container shipping: predicting delays by comparing various machine learning algorithms. In: Hawaii International Conference on System Sciences Computer Science (2020)
12. Shahid, S.: Predicting delays in delivery process using machine learning-based approach. Thesis, Purdue University Graduate School (2020). https://doi.org/10.25394/PGS.13350764.v1
13. Steinberg, F., Burggräf, P., Wagner, J., Heinbach, B., Saßmannshausen, T., Brintrup, A.: A novel machine learning model for predicting late supplier deliveries of low-volume. **2**(5), 99–110 (2016)
14. Erkmen, O.E., Nigiz, E., Sarı, Z.S., Arlı, H.Ş., Akay, M.F.: Delivery time prediction using support vector machine combined with look-back approach. Published by Alntelia (2022)
15. Cao, L.: Support vector machines experts for time series forecasting. Neurocomputing **51**(1–4), 321–339 (2003)
16. Egwim, C.N., Alaka, H., Toriola-Coker, L.O., Balogun, H., Sunmola, F.: Applied artificial intelligence for predicting construction projects delay. Mach. Learn. Appl. **6**, 100166 (2021)

Cricket Forecast: Unraveling Future Matches' Outcomes

Vemula Vivek Siddharth, Pulukuri Shalem Vikranth, N. Karthik(✉), and V. Vani

Department of Computer Science and Engineering, National Institute of Technology Puducherry, Karaikal 609 609, India
nkarthikapce@gmail.com

Abstract. Identifying important factors that frequently affect a team's success is very important, that leads to the usage of mathematical models in the sports especially in cricket, in order to predict which side might win a ODI or a T20 match. Many people have started to forecast the outcome of these cricket matches that ultimately adds an extra layer of amusement and entertainment to the game. The project aims to develop a mathematical model that analyses pre match conditions and players previous performance and finally gives which team has a better chance to win. The model using RandomForestClassifier achieved an accuracy of 63% and better precision, recall and f1 score on pre match conditions and RandomForestRegressor acquired r2 score of 88% and 79% for ODI batting, T20 batting and 97% and 98% for ODI bowling and T20 bowling is achieved for different formats of the game based on players performance. Hyperparameter tuning is used to increase accuracy. Cross validation is used for performance evaluation.

Keywords: Cricket forecast · Random Forest Classifier · Multi Linear Regression · Random Forest Regressor · Prediction · Hyperparameter tuning · Cross validation

1 Introduction

Cricket, often considered the "gentleman's game", has emerged as an international spectacle, which has been captivating millions of cricket enthusiasts around the globe. It has gone through a transformation not just in terms of tactical aspects of the game, but also in terms of the systematic computational analysis of data and predictive modeling as it made a lot of headway from the classical Test cricket format to the delightful One Day International (ODI) format and lightning-fast Twenty20 (T20) format. Cricket is usually a bat-and-ball sport where two teams play a match having a squad of 11 players in each team. The team that is going to bat first sends two of their players onto the field, while the opposite team does the fielding with their 11 players. The batting team's main agenda is to score more runs irrespective of the format, while the bowling team should remove 10 batsmen by taking wickets in the given number of deliveries according to the format of the game.

Published by Springer Nature Switzerland AG 2024
M. L. Owoc et al. (Eds.): ICCIDS 2024, IFIP AICT 718, pp. 353–364, 2024.
https://doi.org/10.1007/978-3-031-69986-3_27

As per the format of the game, innings will be completed when the required number of runs are scored, when 10 batsmen from a batting side are taken out, or when all the deliveries have been bowled. The main agenda is to score more runs than the other team [1]. In this study, we mainly predict the matches' outcomes of two renowned formats of cricket namely, the One Day International (ODI) format and the Twenty20 format. In ODI's, matches will be played for a maximum of 50 overs each whereas in Twenty20 it is of 20 overs each. Cricket match prediction plays a significant role and holds importance for team coaches, managers, and various stakeholders [2]. Team management which includes the manager and coach needs a predictive model in order to strategize against an opponent finding and analyzing their strengths and weaknesses [2]. Pre-match predicts the outcome before the contest, and in-play predicts the outcome during the match [1]. Predicting the match outcomes mainly leverages the pre-match covariates like ground conditions and toss outcomes.

Toss outcome holds significant importance and indeed has a crucial role in cricket match prediction [1]. The toss-winning team gets a chance to choose whether to bat or to bowl first. Here pitch report often supports the toss decision, if the pitch is expected to be bowling-friendly then the team that wins the toss might opt to bowl first and if the pitch is expected to be batting-friendly then the team that wins the toss might opt to bat first. This toss outcome might change the match dynamics [7].

Home advantage also affects the outcome of a cricket match. When a team has a match at its home venue there are specific advantages that ultimately lead them to a winning side [6]. In addition to that, there will be a huge morale boost for players with home crowd support [6]. The critical factor that mainly affects a team's winning is the players' recent performance [4].

The contribution to this field is by developing a model that combines machine-learning techniques like Random Forest classifier and Random Forest Regressor on which we address the various factors that happens before the match like toss, pitch advantage and home advantage and also the main factor, players performance is analyzed based on his previous statistics and we address some of the existing limitations with good accuracy in predicting cricket match prediction of both ODI and T20 format.

Section 2 contains the Related work that explains about the various observations made on different references, Sect. 3 contains the proposed methodology that gives the overview of the project on how the system works, Sect. 4 contains the experimental setup that explains about the various tools used and how every part of the model works and explains the output, Sect. 5 contains the results about the accuracy and other various parameters and the GUI page of model, Sect. 6 contains the concludes the project, Sect. 7 defines the future work that can be done considering different factors.

2 Related Work

The global entertainer 'Cricket' has always been a game with full of uncertainties that are influenced by myriad covariates before and during the match. For that reason, predicting the outcome of a cricket match became an art that requires a diverse set of skills, data sets, and analytical machine-learning techniques. In order to address these complexities researchers have brought up a wide range of approaches over the past few years. We have highlighted a few works below.

Logistic Regression model has been applied to find out the percentage of winning of a particular team for a given One Day International cricket match and they have investigated the influence of venue, toss outcome, toss decision and match type [1]. They have limited their prediction to only eight major cricket playing nations. They failed to predict the matches having neutral venue, tie match specified, and they have used limited parameters to predict the match.

Predicted the winner of Indian Premier League T20 tournament by using machine learning techniques like K-Nearest Neighbors, Random Forest, Naïve Bayes and Model Trees [3]. They have also explored the significant factors like toss decision and venue. To predict the outcome, they have applied feature selection techniques like Correlation Based Feature Selection, Relief, Information Gain, Wrapper methods. They have gained 57% accuracy for venue-based model with Naïve Bayes and 62% accuracy for toss-based model with KNN algorithm.

Predicted the final batting score in the 1st and 2nd innings score and winning team of a T20 cricket match using machine learning techniques like K-Nearest Neighbors, Support Vector Machine, and Random Forest [4]. They have worked on the factors like match type, playing teams, toss winner, toss decision and gender whether it is a men's or women's cricket. Finally, they have attained an accuracy of 84.51% for the imbalanced data set and the 76.1% accuracy for balanced data set. They have not explored the ODI and TEST formats.

Predicted the outcome of India VS. Pakistan One Day International cricket matches by applying Logistic Regression machine learning technique [5]. They have investigated the factors affecting the outcome of a match like toss decision, whether it is a day or day-night format, and venue. In this study they have used limited machine learning algorithms were used and it is limited to only ODI matches between India and Pakistan. There is a future scope for finding out the outcomes of T20 and Test game formats and by adding more pre-match covariates.

Multilayer Perception Classifiers, K-Nearest Neighbors, Random Forest, and Decision Trees have been applied to predict the winner of both IPL and ODI cricket match [2]. They have included the factors like team's previous performance, home advantage, and toss decision for their predictive model. They have gained accuracy of 80% using Decision tree and 87% using Random Forest Classifier. They have explored limited factors for the prediction. There is a scope to include more factors.

Study has been done on how winning the toss and home advantage influences the outcome of T20 international cricket matches using Logistic Regression [6]. They have examined only two factors winning the toss and home advantage but there are many other significant pre-match covariates that need to be explored.

Predicted the result of One Day International cricket match by investigating significant variables which could probably explain the outcome by applying Logistic Regression machine learning technique [7]. They have studied on important factors like home field advantage, toss decision, and match type whether it is a day or day-night match which might influence the outcome of a match. They have predicted winning team in 81% of matches. They have not considered other crucial factors and limited their study for the ODI format.

Classification and Regression Tree and Logistic Regression were used to analyze ODI cricket predictors like home advantage, toss outcome, and match type [8]. This study mainly focused on toss outcome and home advantage but many other factors might affect the outcome of a match and we can further do the research work for other formats along with the inclusion of various pre-match covariates.

MLP classifier, Logistic Regression, Random Forest, SVM classifiers and Naïve Bayes machine learning algorithms were applied to predict the outcome of Indian Premier League matches [9]. They have trained their models 15 min before the gameplay and immediately after the toss. They have included the many factors along with a crucial factor like team strength by calculating team weight with points of each individual player in the team. Finally, they have gained 71.66% accuracy with Multilayer Perceptron.

A machine learning based prediction for winner of IPL and ODI match by applying Naïve Bayes Classification, Eulers strength Formula and Multivariate Regression algorithms [10]. They have considered important factors like toss outcome, home ground, Captains, previous stats. They have attained 72% with Naïve Bayes for ODI, 68.33% with Random Forest for T20, 83% with SVM for T20, and combined model gained 90% for T20 (IPL). They have not explored prediction for ODI matches properly.

From this research on multiple papers, the conclusion can be made that the analysis has been limited to only one format (only ODI or T20) and also the analysis has been made only on the limited factors (venue, toss) that excludes the players performance. Even though players performance is considered it is limited to only one format and also the pre match conditions are not assessed. Thus, a model has been developed that considers all pre match conditions such as toss, match type, Pitch advantage and home advantage and along with this each players batting and bowling performance will be analyzed.

3 Proposed Methodology

The architecture is a 3-tier based architecture that consists of a user interface, server, database. Figure 1 shows the System architecture and the flow in the server follows as data preprocessing->data evaluation and analysis->data validation->predict for new data.

The user-Interface consists of the front-end page that takes the input of various required entries such as team1, team2, ground, pitch advantage and some other pre match covariates. Next interface takes input of players performance details and analyses them using the server phase. At last, it displays the percentage of winning chances a team has in a GUI. This GUI is developed using tkinter module in python. The server tier contains the main business logic based on which the model works. At first the dataset is been preprocessed using labels and scaling and feature extraction are done here. The whole model is divided into two parts (two models) as shown in Fig. 2. The first model analyses the pre-match covariates such as team1, team2, ground, home advantage, pitch advantage, and bat first team and gives an output of the winning team. The second model analyses the players' performance – number of runs scored, strike rate, Batting average, Bowling average, Wickets, Economy. Based on this it predicts the batsmen score and bowler performance metric.

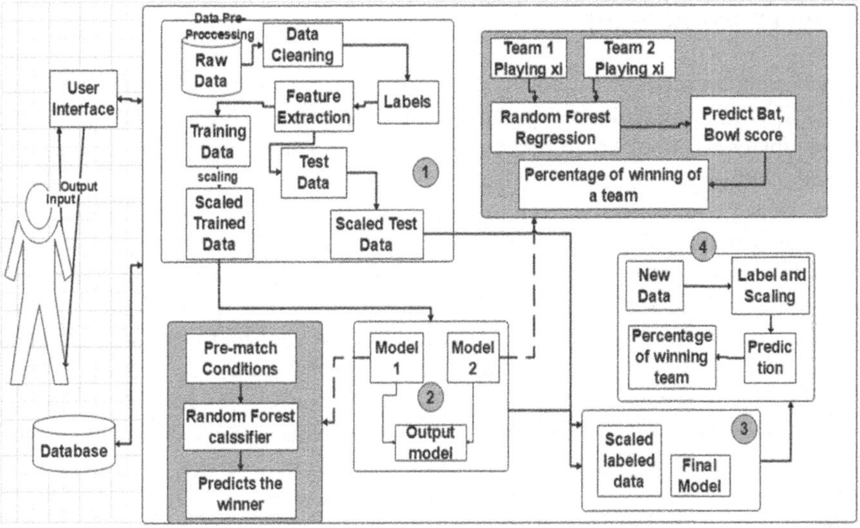

Fig. 1. System Architecture

The whole playing XI of team1, team2 batting, and bowling score will be calculated and finally percentage of winning of each team will be predicted. The database tier consists of the model 1, and model 2 in the form of pickle and also the dataset for the pre match covariates.

The algorithms that we have used are the Logistic Regression, Random Forest Classifier, gradient boosting algorithms, Multi Linear Regression, Random Forest Regression to analyze how well the model gets fit into a particular model. For analyzing the pre-match covariates, we use Random Forest classifier for the classification of winner when we are training.

For analyzing the player performance, we use the Multi linear regression and Random Forest Regression. The Random Forest Regression is used for analyzing the ODI and t20 data for both batting and bowling and random forest regressor is used for t20 data for batting, bowling.

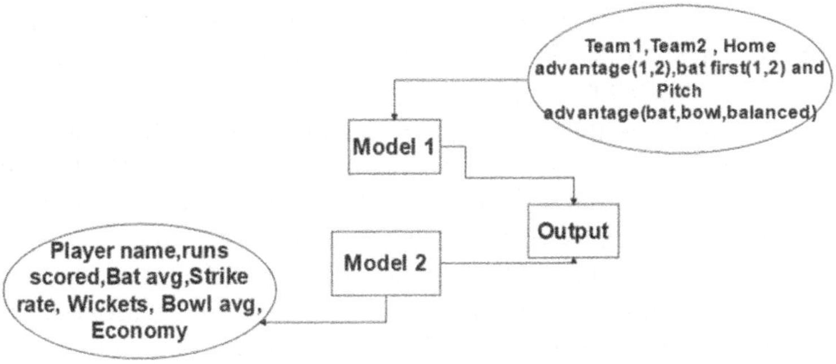

Fig. 2. Flow of models

4 Experimental Setup

For developing the model we used a system having processor intel i5, RAM of 8GB and a storage of 512 SSD on windows operating system. Coming to the software requirements we used jupyter notebook environment for the executing, python programming language was used, matplotlib was used for visualization.

The datasets are collected from the website Kaggle. There are totally six different datasets of which 3 of them are related to the ODI matches and other 3 are related to the T20 matches. Out of three, 1st dataset has all the pre-match covariates and the winner of the match such team1, team2, winner, ground, match date, venue. The Home advantage and pitch advantage are the extra columns that had been added manually. Home advantage consists of the country in which the teams are playing, while the pitch advantage specifies whether it is batting friendly or bowling friendly or a balanced pitch. The second dataset has the player name and his batting statistics such as number of matches, number of runs, number of balls faced, strike rate, average, 100's,4's,6's …etc. The third dataset consists of player name and his bowling statistics such as wickets, bowling average, economy…etc.

The model1 is analyzed based on the accuracy, precision, recall and f1-score and the model2 is analyzed based R2 score and Mean-Squared-Error(MSE).The higher in the values of accuracy, recall, precision and f1 score defines that the data is better fit into the model. Accuracy alone might give invalid reports on the biased input, so, we have to analyze all four metrics for model1.The higher in the r2-score defines how well the data is fitted into the model. The lower values in the MSE defines that the predicted values are more closer to the actual values.

4.1 Model-1 - Pre-match Covariates

First the preprocessing of the dataset is done. The data cleaning is done by removing all the null values in the dataset. Here the null values are the data entries in which the match has no result or the match has been tied. Now the labeling of data is done. This labeling in a specific fashion that a number is fixed for each team. This is done for the columns team1, team2, home advantage and bat first columns. For the pitch advantage column, we label the data in a way that batting is labeled 0, bowling is labeled 1 and finally balanced is labeled 2. We take the features as team 1, team 2, bat first, home advantage, pitch advantage. We will scale this data using a standard scaler module. Now the data is trained using different models' logistic regression, random forest classifier, gradient boosting (lightGBM) out of which the accuracy of every model is around 59%. Now the hyper parameter tuning is done on the random forest classifier then the accuracy has been increased from 59% to 62%. The default parameters and the best parameters (Output of hyperparameter tuning) have been shown in the Fig. 3. The trained model and the scaler variable (used for scaling) are stored in pickle format.

For predicting the new data these models will be retrieved from the saved pickle format files and they will labeled and scaled based on the training labels and the scaling and finally use the model for the prediction.

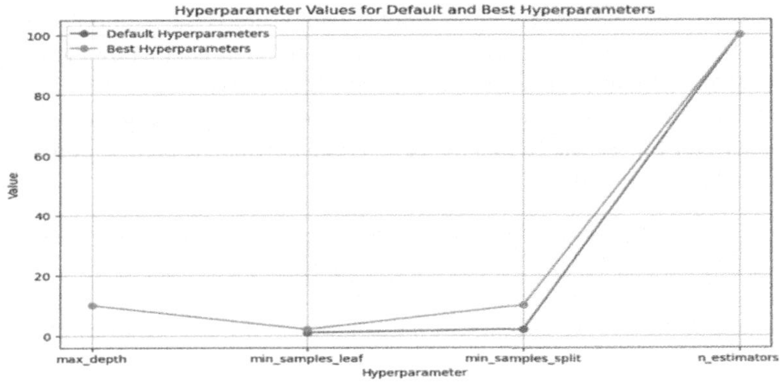

Fig. 3. Hyperparameter tuning

4.2 Model-2 - Player Performance

We remove the null values from the data meaning that if the player has no runs, no average …etc. we will remove that player. We will create dummies for each player name. Dummies are the columns that have all the entries equal to zero except the row in which they are present. These dummies are added to the dataset as the columns and these can be used as the features.

Batting: Now the features that we consider are Runs, Average, Strike Rate (SR). A new column 'run' will be generated on the number of runs divided by number of innings. Now the features runs, avg, SR and all the dummies columns and the target variable is the run. First we will normalize the features runs, average, SR and also the target variable using min max normalization. These features and the target variable are fitted into the Random Forest Regressor to predict the run which can be considered as the performance score. Now the model, list of all features, and min max values of each column will be saved into a file in the pickle format. For the new input these relevant data can be access from the saved files.

Bowling: Now the features that we consider are wickets, Bowling Average, Economy. A new column 'performance metric' will be generated by taking equal weightage of the above mentioned three features as shown in the Eq. (1). Now the features are wickets, avg, economy and all the dummies columns and the target variable is the performance metric. First we will normalize the features wickets, Bowling Average, Economy using min max normalization. The target variable is generated only after the features are normalized. These features and the target variable are fitted into the Random Forest Regressor to predict the performance metric which can be considered as the performance score. Now the model, list of all features, and min max values of each column will be saved into a file in the pickle format. For the new input these relevant data can be access from the saved files. For the new data each players statistics will be given so that at the last the team1 bat score, team2 bat score, team1 bowl score, team2 bowl score.

$$\text{For each player,}$$
$$\text{Bowl score} = 0.3 * (\text{normalized wickets}) + 0.3 * (\text{normalized average}) + 0.4 * (\text{normalized economy}). \tag{1}$$

At the last if the pitch is batting then total team score is the sum 60% of team bat score and 40% team bowl score (Refer Eq. 2), if the pitch is bowling then total team score is the sum 40% of team bat score and 60% team bowl score (Refer Eq. 3) and if the pitch is batting then total team score is the sum 50% of team bat score and 50% team bowl score (Refer Eq. 4). Based on this we can predict which team has more team score and can be taken as the winner.

$$\text{total score(for each team)} = 0.6 * (\text{team\_bat\_score}) + 0.4 * (\text{team\_bowl\_score}). \tag{2}$$

$$\text{total score(for each team)} = 0.4 * (\text{team\_bat\_score}) + 0.6 * (\text{team\_bowl\_score}). \tag{3}$$

$$\text{total score(for each team)} = 0.5 * (\text{team\_bat\_score}) + 0.5 * (\text{team\_bowl\_score}). \tag{4}$$

In this study, the cross validation was conducted using the ShuffleSplit strategy with 5 splits, a test size of 20%. This facilitated in-depth analysis of the Random Forest Regressor's performance, using randomized data shuffling within each split to improve stability and address potential data variability.

5 Results

5.1 Model 1

The Table 1 shows the accuracy, recall, precision and f1 score of the models logistic regression, Random Forest Classifier and the gradient boosting that is trained on the pre match conditions. The higher accuracy defines that the model performs good on the data but we should not only consider accuracy as it may cause discrepancy over the biased data. The precision, recall and f1 score values should also be considered. The higher in these values defines the model is performing well.

Table 1. Evaluation metrics

Model	Accuracy	Precision	Recall	F1-score
Logistic Regression	39%	29%	36%	31%
Random-Forest Classifier	63%	62%	63%	60%
Gradient-boosting LightGBM	62%	61%	63%	61%

5.2 Model 2

The Fig. 4 and Fig. 5 shows the r2 score and Mean Squared Error (MSE) for various formats and aspects (ODI batting and bowling, T20 batting and bowling).High values

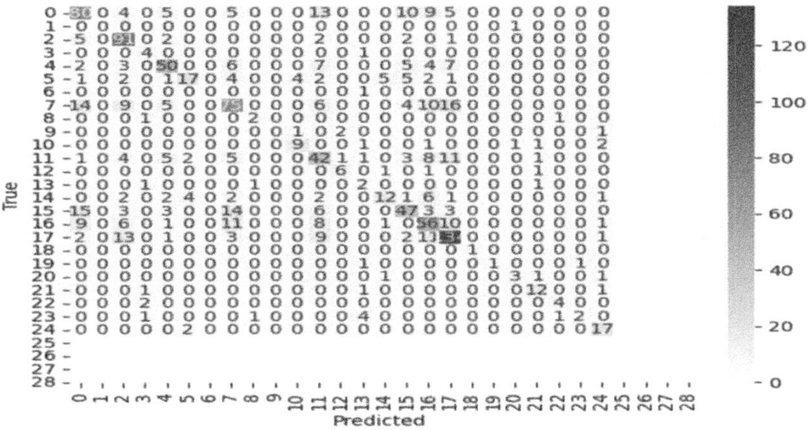

Fig. 4. Confusion matrix

in the r2 score defines that the trained data is fitted well in the model which suggests its effectiveness in capturing the variance. Lesser values in the MSE defines that the model is well suited for the data that we processed. It signifies that the predicted values are closely aligned to the actual values, that indicates model's accuracy in prediction.

The combination of high r2 scores and low MSE specifies the robustness and precision of the model in understanding and to predict the outcomes of cricket matches in various formats and aspects. These metrics are crucial are crucial indicators of the model efficiency and reliability in analyzing the cricket performance data.

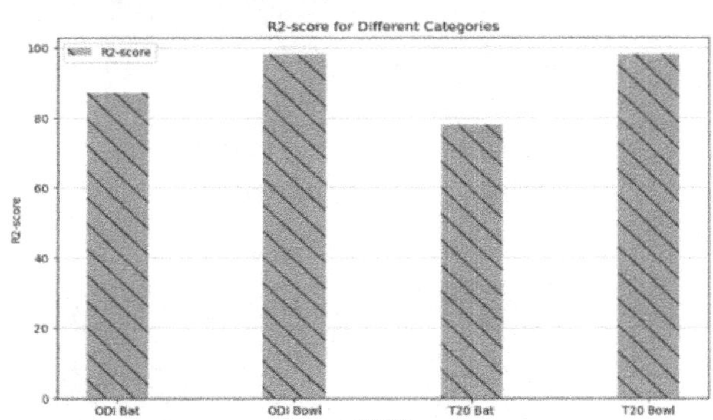

Fig. 5. R2-score

5.3 GUI Interface

The Fig. 7 shows the GUI page that takes the input whether the match is ODI or T20. On submitting response, it redirects to the GUI page shown in Fig. 7 that takes the input of

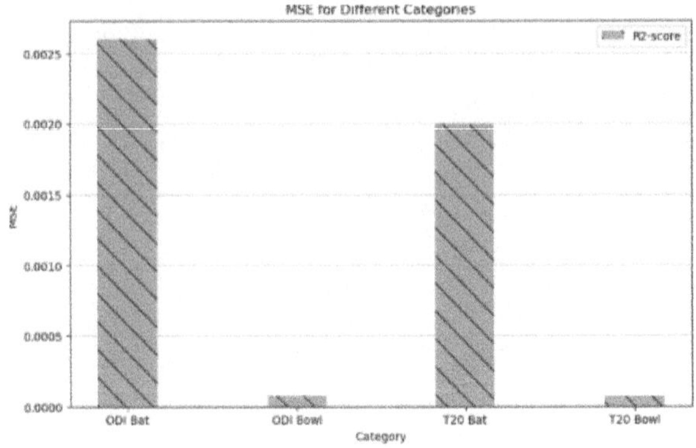

Fig. 6. MSE

pre-match conditions. It takes the input of Team1 name, Team2 name, Home Advantage (Country), Team that is batting first, and also the Pitch advantage (Batting, Bowling or Balanced). On Submitting, it redirects to the page shown in Fig. 8 which requires the input of players of two teams. It takes all the player names of Team 1 and Team 2. Finally, the Fig. 9 is displayed which is the output based on the pre match conditions and also the percentage of winning of team over the other team based on batting, bowling score and also on combined score is displayed (Fig. 10).

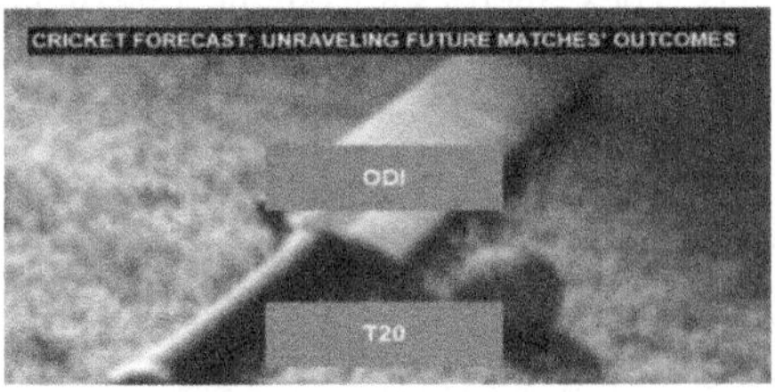

Fig. 7. GUI for Prediction

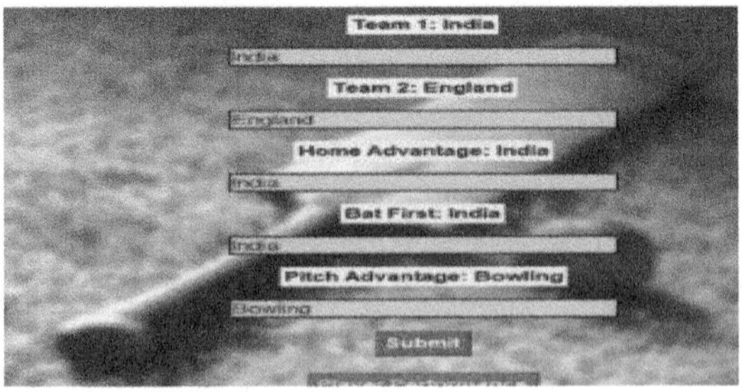

Fig. 8. GUI for prediction

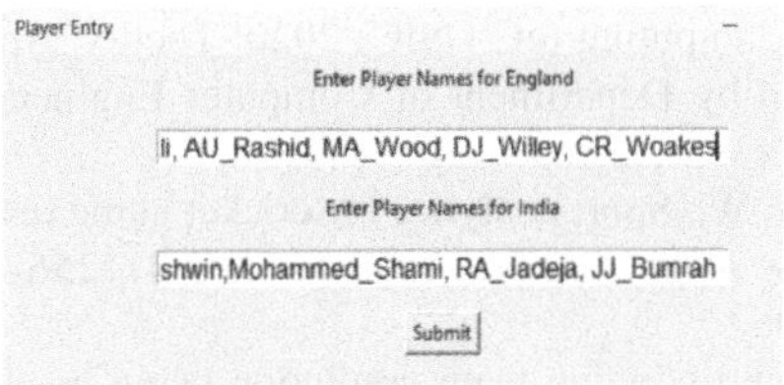

Fig. 9. GUI for Player entry

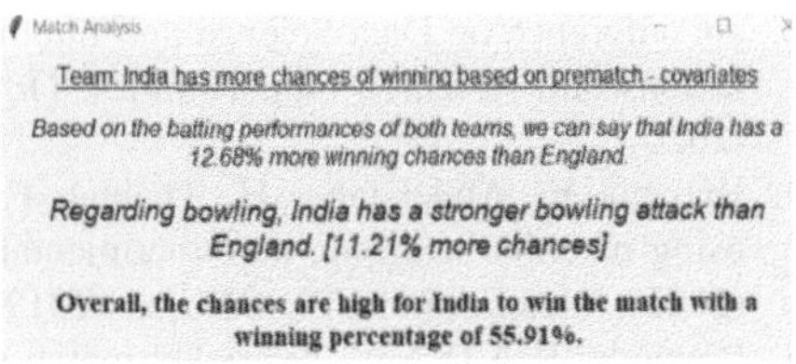

Fig. 10. GUI for result page

6 Conclusion

Finally, this work builds a model to predict the match outcome by specifying the winning chances a team has. The Model 1 focuses on the important pre match conditions and predicts which team might win.

Model 2 analyses each player's performance based on their statistics and predicts his performance score. Based on this we can say which team has more winning chances based on overall team scores. The performance score is developed based on batting and bowling of each player in the playing XI. Model 1 achieved an accuracy of 63% on using Random Forest Classifier after the hyperparameter tuning is done. While the model 2 achieved an r2-score of around 97% for various formats (ODI and T20).

Finally, this work contributes in improving the understanding of the major cricket insights and acts as a useful tool for the cricket enthusiasts by predicting the winner on pre-match conditions and players performance, for ODI and T20 matches.

7 Future Work

Various pre match covariates can also be included such as weather report…etc. Different machine learning algorithms can also be tried and deep learning algorithms can be used to increase accuracy.

When coming to players performance, factors like players in a particular ground data is not considered so this can be done in the future for better results, players score against particular team can be also included for more accurate results.

Acknowledgements. There is no third person/organization to acknowledge.

Disclosure of Interests. The authors declare no competing interests.

References

1. McEwan, K., Pote, L., Radloff, S., Nicholls, S.B., Christie, C.: The role of selected pre-match covariates on the outcome of One-day International (ODI) cricket matches. S. Afr. J. Sports Med. **35**(1), 1–6 (2023). https://doi.org/10.17159/2078-516X/2023/v35i1a15012

2. e-Conference on Data Science and Intelligent Computing (eC-DSIC) 2020: Theory, Applications and Research Opportunities. Organized by Department of Computer Engineering (2020)
3. Kapadia, K., Abdel-Jaber, H., Thabtah, F., Hadi, W.: Sport analytics for cricket game results using machine learning: an experimental study. Appl. Comput. Inform. **18**(3–4), 256–266 (2022). https://doi.org/10.1016/j.aci.2019.11.006
4. Pramoda, K.A.D.A.: T20 cricket match score and winning team prediction using machine learning techniques (2021)
5. Chowdhury, S., Anwarul Islam, K.M., Rahman, M., Raisa, T.S., Zayed, N.M.: One day International (ODI) cricket match prediction in logistic analysis: India vs. Pakistan. Int. J. Hum. Move. Sports Sci. **8**(6), 543–548 (2020). https://doi.org/10.13189/saj.2020.080629
6. Saikia, H., Bhattacharjee, D.: On the effect of home team advantage and winning the toss in the outcome of T20 international cricket matches (2010). http://ssrn.com/
7. Jayalath, K.P.: A machine learning approach to analyze ODI cricket predictors. J. Sports Anal. **4**, 73–84 (2018). https://doi.org/10.3233/JSA-170175
8. Shah, P., Shah, M.: Predicting ODI cricket result (2015). www.iiste.org
9. Lamsal, R., Choudhary, A.: Predicting outcome of Indian premier league (IPL) matches using machine learning (2018). http://arxiv.org/abs/1809.09813
10. Sinha, A.: Application of machine learning in cricket and Predictive Analytics of IPL 2020 (2020). https://doi.org/10.20944/preprints202010.0436.v1

Deep Learning Ensemble for Diabetes Prediction: Integrating LSTM, DCNN, and SMOTE for Enhanced Risk Assessment

Ayyakkannu Selvaraj[1](✉), V. Satheesh kumar[2], Dilip Bhudhwant[1], Sharvari Tamane[1], and Chhaya S. Khandelwal[3]

[1] University Department of Information and Communication Technology, MGM University, Aurangabad, Maharashtra, India
aselvaraj@mgmu.ac.in

[2] Department of Mechatronics Engineering, Park College of Engineering and Technology, Coimbatore, Tamil Nadu, India

[3] Department of Electronics and Computer Engineering, Jawaharlal Nehru Engineering College MGM University, Aurangabad, Maharashtra, India

Abstract. Diabetes is a prevalent and chronic medical condition that poses a significant public health challenge worldwide. Accurate early prediction of diabetes risk can facilitate timely interventions and improve patient outcomes. In this study, we propose a comprehensive approach to diabetes risk assessment through the integration of deep learning techniques and data augmentation strategies. Our research leverages the power of deep learning ensembles by combining Long Short-Term Memory (LSTM) networks, Deep Convolutional Neural Networks (DCNN), and Synthetic Minority Over-sampling Technique (SMOTE) for addressing class imbalance. We begin by collecting and preprocessing a dataset comprising patient information, medical histories, and, where applicable, image data. The LSTM model is tailored to capture temporal dependencies in the data, while the DCNN extracts meaningful features. By integrating SMOTE, we tackle class imbalance challenges commonly encountered in medical diagnosis tasks. Our ensemble model seamlessly integrates the predictions from these diverse models, allowing for a holistic assessment of diabetes risk. Extensive evaluation and hyperparameter tuning are performed to optimize performance and ensure the model's robustness. Cross-validation techniques are employed to gauge the model's generalization capabilities. The proposed approach aims to not only provide accurate predictions but also to offer insights into the contributions of individual models within the ensemble. Ultimately, our research seeks to enhance diabetes risk assessment, offering a valuable tool for healthcare practitioners in making informed decisions and potentially improving the early detection and management of diabetes. This present study represents a significant step towards more reliable and comprehensive diabetes risk assessment, leveraging the potential of deep learning and ensemble techniques to advance predictive accuracy and broaden our understanding of this critical healthcare challenge. An accuracy score of 74% suggests that the model performs reasonably well in making correct predictions overall, although it doesn't achieve a very high level of accuracy. On the other hand,

Published by Springer Nature Switzerland AG 2024
M. L. Owoc et al. (Eds.): ICCIDS 2024, IFIP AICT 718, pp. 365–373, 2024.
https://doi.org/10.1007/978-3-031-69986-3_28

an ROC AUC value of 0.81 indicates that the model is effective at distinguishing between positive and negative instances, showing its ability to differentiate between the two classes.

Keywords: Fuzzy rainfall index · Fuzzification · Defuzzification · If-then rules · Surface plot · Regression

1 Introduction

Seasonal n the cardiovascular system and significant challenges in the proper functioning of vital organs such as the eyes, kidneys, and nerves. This study aimed to focus on the early detection of diabetes using the Pima Indian Diabetes Dataset (PIDD), which is readily available in the UCI (University of California, Irvine) Repository. The dataset's class distribution is characterized by class imbalance, where the number of instances in each class is not equal. The classes with a higher number of instances are referred to as the majority classes, while those with fewer instances are considered the minority classes. Class imbalance is quantified using the class-imbalance ratio (CIR), calculated as the ratio of minority instances to majority instances. Imbalanced datasets pose a challenge to machine learning algorithms as they are primarily trained on majority-class instances, leading to poor recognition of minority-class instances, a phenomenon known as overfitting. To mitigate this problem, the solution is to transform the imbalanced dataset into a balanced one before commencing the training process. LSTM, or Long Short-Term Memory, represents an advanced version of Recurrent Neural Networks (RNNs). It addresses the challenge of vanishing and exploding gradients in traditional RNNs by introducing memory blocks. LSTMs excel in handling long-term dependencies, allowing them to connect distant past information to the present. A memory block comprises memory cells and input/output gates, all controlled by adaptive multiplicative gates. [1, 2]. LSTMs are widely used in Natural Language Processing (NLP), where they've significantly improved tasks like language modeling and speech recognition. They were initially complex but gained popularity due to their success in language translation and other fields requiring modeling of extensive sequential data. Enhancements like forget gates and peephole connections further improved their capabilities, making them valuable in various applications. The Deep learning has proven to be a highly effective approach for tackling complex challenges in feature extraction and classification when working with diabetes related datasets [3, 4]. The performance of hybrid models has been rigorously assessed using datasets like the Pima Indians Diabetes and the Cleveland Heart Disease datasets. For instance, one noteworthy application employed a deep belief network [5, 6], which achieved an impressive prediction accuracy of 81.20% when applied to a diabetes dataset. In another case, an MLP-based logistic regression model was utilized to monitor continuous glucose levels (CGM) and accurately identify diabetic cases, resulting in a substantial prediction accuracy of 77.5%. This dataset featured nine individuals with CGM data spanning 10,800 days, with a total of 97,200 simulated days. Robust cross validation practices were employed, with three patients dedicated to testing and six for training and validation purposes. In a different approach, a combination of a deep learning stacked autoencoder and a deep neural network harnessed the power of L-BFGS optimization to achieve a classification accuracy of 77.09% on the Pima Indians Diabetes Dataset (PIDD) [4]. Another study took a unique angle by evaluating three distinct deep learning algorithms using a dataset manually collected from an Australian hospital [5]. This dataset contained a substantial 12,000 patient samples and maintained a male-to-female ratio of 55.5%. To enhance data quality, pre-processing techniques were thoughtfully applied, narrowing the dataset to 7,191 observations. The dataset was thoughtfully divided into training, validation, and testing subsets. This meticulous approach allowed for a comparison of various deep learning models, including Long Short-Term Memory (LSTM), Markov, and Recurrent Neural Networks (RNN), based on precision metrics. Impressively, the LSTM model achieved a precision value of 59.6%. Furthermore, a recurrent neural network was employed to predict diabetes types using all eight available attributes. The dataset was thoughtfully divided into an 80:20 ratio for training and testing [7]. The model demonstrated strong performance, accurately predicting type 1 diabetes 78% of the time and type 2 diabetes 81% of the time. Lastly, researchers devised a data analytics model for medical applications, leveraging a deep belief network for diabetes classification. This innovative approach drew from a subset of factors carefully selected from a vast dataset containing 50,000 diabetic records, yielding an impressive accuracy rate of 81.20%. Additionally, the study compared a Multilayer Neural Network (MLNN) trained using the LM algorithm for diabetic prediction against a Probabilistic Neural Network (PNN) through the rigorous method of 10-fold cross validation [8-9]. These diverse studies emphasize the immense potential of deep learning methodologies in the realm of diabetes prediction and classification. Their results across various datasets and scenarios consistently demonstrate the promise of deep learning techniques. Parashar et al. conducted a diabetes diagnosis study using the Pima Indian Diabetes Dataset, where they employed a combination of Linear Discriminant Analysis (LDA) and Support Vector Machine (SVM) techniques, achieving an accuracy rate of 77.6%. Notably, their SVM implementation involved the use of a radial basis kernel function, with the objective of reaching a classification accuracy of 78%. In a separate research endeavor, Christobel and Sivaprakasam applied a classwise K-nearest neighbor (CKNN) approach to address missing data within the diabetes dataset, and this was accomplished through data normalization. Their approach yielded an accuracy of 78.2%.Moreover, an impressive accuracy rate of 80.0% was attained by leveraging a hybrid multilayer perceptron classification model that harnessed the strengths of fuzzy logic within soft computing [10]. In comparative assessments involving diverse methods such as K-Nearest Neighbor (KNN), Naive Bayes (NB), Iterative Dichotomizer 3 (ID3), Classification and Regression Tree (CART), and Support Vector Machine (SVM) on the diabetes dataset, an accuracy of 81.8% was achieved [6]. Recognizing that medical data often exhibits non-normal, nonlinear, and intrinsic correlation structures, an alternative approach successfully improved the classification accuracy to 82.0% by adapting a Gaussian classification method, incorporating linear, polynomial, and radial basis kernels. Lastly, a hybrid model that integrated a genetic algorithm (GA) and a backpropagation network (BPN) secured the highest accuracy, reaching 84.7% [7]. These findings underscore the wide spectrum of methodologies employed to enhance diabetes diagnosis accuracy, with innovative combinations of techniques yielding promising results. The previous research has laid a solid foundation for the current study, which builds upon the advancements in deep learning architectures, transfer learning methodologies, and domain specific datasets to achieve improved accuracy. The main objective of the present study is that integrating LSTM, DCNN and SMOTE for enhancement diabetes prediction with balancing of dateset.

2 Methodology

The proposed method follows a systematic approach. It commences with the collection of a diverse dataset encompassing patient data, medical history, and potentially images associated with diabetes.

The Pima Indians Diabetes Database, commonly known as the Pima dataset, is a renowned dataset frequently employed for binary classification tasks, notably in the domain of diabetes prediction. This dataset encompasses a collection of health-related attributes pertaining to Pima Indian women, alongside an outcome variable signifying the presence or absence of diabetes in each individual.- Data Source: The dataset originates from the National Institute of Diabetes and Digestive and Kidney Diseases (NIDDK) and is accessible through the UCI Machine Learning Repository. Attributes are 1. Pregnancies: Represents the number of pregnancies. 2. Glucose: Denotes the plasma glucose concentration following a 2-hour oral glucose tolerance test. 3. BloodPressure: Records the diastolic blood pressure in millimeters of mercury (mm Hg). 4. SkinThickness: Measures the triceps skinfold thickness in millimeters (mm). 5. Insulin: Captures the 2-hour serum insulin level in micro International Units per milliliter (mu U/ml). 6. BMI: Defines the body mass index, calculated as weight in kilograms divided by the square of height in meters (kg/(m^2)). 7. Diabetes Pedigree Function: Quantifies the diabetes pedigree function, an indicator of diabetes likelihood based on family history. 8. Age: Represents the age of the individual in years. Target Variable: Outcome: This binary variable signifies the presence (coded as 1) or absence (coded as 0) of diabetes in an individual (Figs. 1 and 2).

2.1 Imbalance on Dataset

The Pima dataset, a widely-used resource for diabetes prediction, presents a classic example of class imbalance. This phenomenon refers to the substantial disparity in the distribution of the target variable, which designates whether an individual has diabetes (coded as 1 for diabetes) or does not (coded as 0 for no diabetes). In the Pima dataset, the class imbalance is quite noticeable, with a typical scenario where the majority of instances belong to the non-diabetic class (class "0"), while the diabetic class (class "1") represents the minority. The specific ratio may fluctuate, but it's not uncommon to observe ratios of around 65–70% non diabetic cases and 30–35% diabetic cases. These figures, though varying somewhat, are persistent across datasets and preprocessing approaches. The implications of class imbalance are significant and manifest in various aspects of model development and evaluation. One crucial implication is model bias, wherein machine learning models trained on imbalanced datasets tend to favor the majority class, becoming proficient at predicting non-diabetic cases but often faltering in identifying diabetic cases. Moreover, the use of conventional evaluation metrics

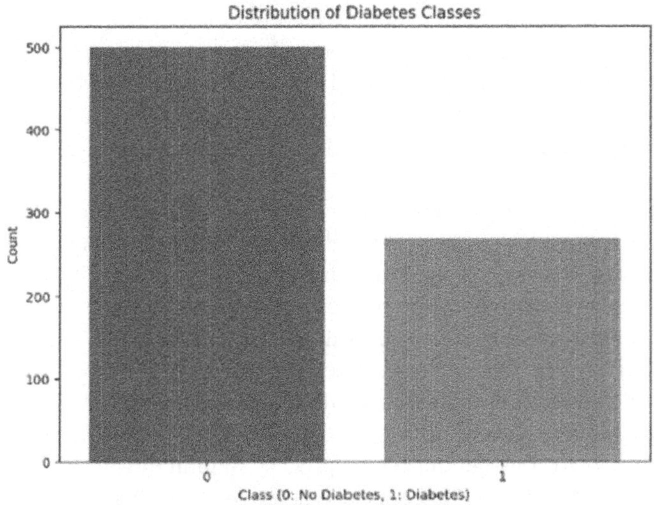

Fig. 1. Data visualization of diabetes class

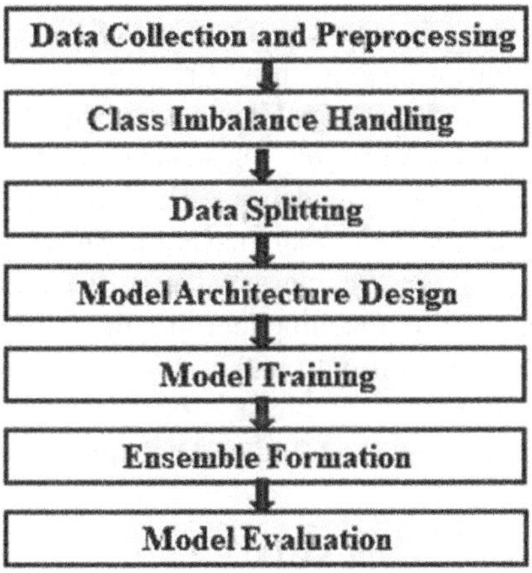

Fig. 2. Flow chart of proposed methodology representing integrating LSTM-deep neural network based identification

like accuracy can be misleading when dealing with imbalanced data. Due to the dominance of the majority class, a model that consistently predicts the majority class can yield a deceptively high accuracy score while offering limited practical value. This poses a particularly challenging issue in healthcare applications like diabetes prediction, where misclassifying positive cases (diabetes) as negative can have more severe consequences than the reverse. To mitigate these challenges, several strategies are employed to address class imbalance. These include resampling techniques such as oversampling and undersampling, algorithm selection emphasizing robustness, cost-sensitive learning, careful consideration of evaluation metrics, threshold adjustments, data collection efforts to bolster the minority class, and the incorporation of domain expertise to guide the modeling process. These strategies collectively aim to ensure that models trained on imbalanced data deliver meaningful and reliable results, especially in critical healthcare applications where the accurate identification of diabetic cases is of paramount importance. Pseudo-code representation of the SMOTE (Synthetic Minority Over-sampling Technique) algorithm as follow.

In the context of the Pima Indian Diabetes dataset, the provided pseudo-code outlines a practical approach to tackle class imbalance. This dataset is commonly used for predicting diabetes onset based on various health-related features. Typically, it exhibits an imbalanced class distribution, where the minority class represents individuals with diabetes, and the majority class represents those without diabetes. The 'SMOTE' function plays a crucial role in this scenario. When applied to the Pima dataset, 'X_minority' would consist of instances of individuals with diabetes, and 'N' would denote the number of synthetic samples you want to generate. The 'k' parameter signifies the number of nearest neighbors to consider during the process. Within this function, a loop iterates 'N' times, each iteration aiming to create a synthetic sample. The random selection of a minority instance, 'A', mirrors an individual with diabetes from the Pima dataset. The 'k' nearest neighbors of 'A' are then identified among other individuals with diabetes in the dataset. These neighbors become potential candidates for generating synthetic samples that can help balance the class distribution. For each of these neighbors, the algorithm introduces a level of randomness by generating a random value, 'rand_value', between 0 and 1. This random value is used to interpolate between the selected instance 'A' and the neighbor, resulting in a synthetic data point. These synthetic samples effectively augment the minority class instances in the Pima dataset. The loop repeats 'N' times, generating synthetic samples to address the class imbalance issue. These synthetic samples can then be combined with the original minority class instances from the Pima dataset to create a more balanced dataset for training machine learning models. By doing so, the SMOTE technique enhances the model's ability to make accurate predictions regarding diabetes onset, ensuring that it does not overly emphasize the majority class and can better handle the inherent class imbalance in the Pima Indian Diabetes dataset.

2.2 Integrating LSTM-DCNN and SMOTE

In present study, the data preprocessing is a crucial step, where the dataset is cleaned, missing values are handled, features are normalized, and class imbalance is addressed using the Synthetic Minority Over-sampling Technique (SMOTE). The Model selection involves training three distinct neural twork architectures: Long Short-Term Memory (LSTM) networks, Deep Neural Networks (DNN), and Deep Convolutional Neural Networks (DCNN). These models cater to different data types- sequential, tabular, and image data. Pretraining may be employed to initialize these models effectively. Ensemble modeling brings these neural networks together, combining their predictions through a straightforward averaging process. Rigorous evaluation follows, employing standard metrics like accuracy, precision, recall, and F1 score, validated on dedicated validation sets. Hyperparameter tuning is a critical aspect, involving a thorough search for optimal hyperparameters for individual models and the ensemble. Class imbalance remains under constant consideration, ensuring it doesn't affect the model's performance. Cross-validation techniques are implemented to assess the models' ability to generalize to unseen data. Risk assessment is then conducted using the ensemble model for making predictions on new cases. Detailed analysis is performed to understand each model's contributions and interpretability.

3 Performance Analysis

The effectiveness of a proposed model can be evaluated using a range of performance metrics. One commonly used metric is accuracy, which quantifies the proportion of correctly classified instances relative to the total number of instances. The accuracy is computed as follows

$$\text{Accuracy} = (\text{TP} + \text{TN})/(\text{TP} + \text{TN} + \text{FP} + \text{FN}) \tag{1}$$

In this equation:- TP represents true positive instances.- TNrepresents true negative instances.- FP represents false positive instances.- FNrepresents false negative instance.

Another crucial metric is recall, which measures the percentage of actual positive instances correctly identified by the model. It can be expressed as:

$$\text{Recall} = \text{TP}/(\text{TP} + \text{FN}) \tag{2}$$

Precision is yet another important metric, indicating the ratio of correctly identified positive observations to all predicted positive instances:

$$\text{Precision} = \text{TP}/(\text{TP} + \text{FP}) \tag{3}$$

These performance metrics provide valuable insights into the model's ability to make accurate classifications of positive and negative instances. They play a vital role in assessing the model's performance across a wide range of applications.

An ROC curve value of 0.81 serves as a strong indicator of the robust performance of a binary classification model. The ROC curve, formally known as the Receiver Operating Characteristic curve, offers a visual representation of the model's ability to differentiate between positive and negative classes across a spectrum of threshold values. Within this curve, the x-axis denotes the false positive rate (FPR), while the y-axis signifies the true positive rate (TPR) or recall. The ROC curve effectively illustrates how the TPR and FPR fluctuate as the classification threshold is adjusted. When the area under the ROC curve (AUC) reaches 0.81, it signifies that the model excels in discriminating between positive and negative instances. Specifically, it achieves a relatively high TPR while keeping the FPR low. As a general rule of thumb. In summary, an ROC curve value of 0.81 underscores the model's proficiency in classifying instances accurately and highlights its capacity to effectively distinguish between the two classes, affirming its overall strong performance in binary classification tasks (Table 1, Figs. 3 and 4).

Table 1. Evaluation Metrices

Proposed model (Hybrid LSTM-DCNN)	Precison	Recall	F1 Score	Support
0	0.81	0.77	0.79	73
1	0.64	0.70	0.67	43

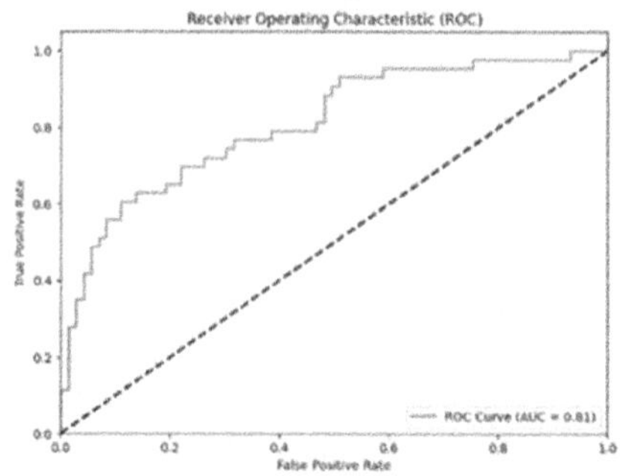

Fig. 3. Receiver operating Characteristics (ROC)

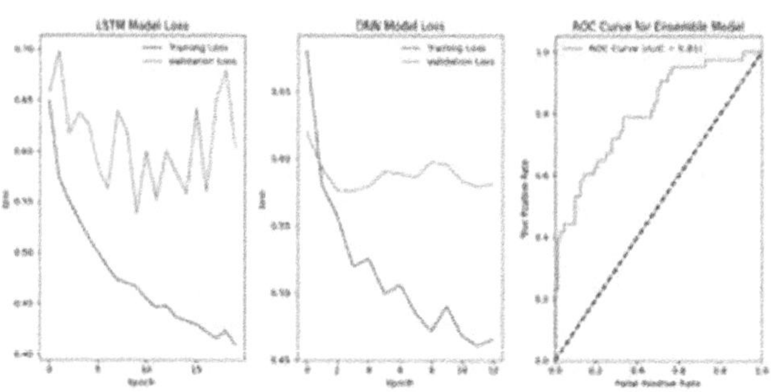

Fig. 4. Training and validation loss

4 Conclusion

In this article, the deep convolutional neural networks (DCNN), and Synthetic Minority Over-sampling Technique (SMOTE), offers a comprehensive and promising avenue for enhancing risk assessment in diabetes diagnosis. By effectively tackling the challenges posed by class imbalance through SMOTE, our objective is to create a model that remains unbiased and proficient in accurately identifying individuals with diabetes, thereby facilitating early intervention. The fusion of LSTM and DCNN models capitalizes on their unique capabilities in handling sequential data and extracting features from images, rendering the ensemble adaptable and well-equipped to capture intricate patterns within the health-related attributes. This multi-modal approach aspires to provide a more holistic and precise evaluation of diabetes risk, potentially enhancing early detection and healthcare decision-making. Furthermore, our commitment to integrating ethical considerations, meticulous documentation, and transparent communication with healthcare practitioners and stakeholders highlights our dedication to the responsible deployment of the ensemble model within real-world healthcare environments. Ensuring that the model aligns seamlessly with established medical expertise while preserving patient privacy remains a paramount objective. In essence, the proposed ensemble approach holds significant potential for elevating diabetes risk assessment. Through the amalgamation of cutting-edge deep learning techniques, effective class balancing, and unwavering adherence to ethical principles, we aim to contribute meaningfully to the realm of diabetes diagnosis and patient care, ultimately leading to improved health outcomes. Rigorous research and validation on clinical data stand as crucial steps in unlocking the full capabilities of this approach in the field of healthcare. To conclude that, an accuracy score of 74% indicates that the model achieves a decent level of overall correctness in its predictions, although it falls short of being highly accurate. However, with an ROC AUC value of 0.81, the model demonstrates a strong ability to differentiate between positive and negative instances, showcasing its effectiveness in separating and discerning the two classes. This situation often arises when dealing with imbalanced datasets, where the model's primary focus is on correctly identifying the minority class (positive), even if it allows for some errors in predicting the majority class (negative).

The authors express their sincere appreciation to and respectfully appreciate the PWD Ariyalur for giving all the information about climatological parameters data.

References

1. Choi, E., Bahadori, M.T., Schuet, A.: Predicting patient outcomes in the emergency department with simulated electronic health records using deep learning. JAMA Netw. Open **3**(1) (2020)
2. Gao, W., Gao, L., Liu, B., Wang, W.: An intelligent diagnosis method for diabetes based on deep neural networks. J. Med. Syst. **43**(8) (2019)
3. Tripathy, B.K., Acharya, U.R.: Ensemble model for diabetes prediction using multiple convolutional neural networks. Biocybernet. Biomed. Eng. **39**(4) (2019)
4. Ding Y., Choi, J.W.: Ensemble deep learning for diabetes prediction. J. Healthc. Eng. (2020)
5. Al-Shayea, Q.K., Elshawi, R.: Deep learning-based prediction of type 2 diabetes using electronic health records. Comput. Biol. Med. **121** (2020)
6. Xing, Z., Pei, J.: Diabetes prediction with Synthetic Minority Over-sampling Technique (SMOTE) and ensemble learning. Int. J. Bioinform. Res. Appl. **16**(1), 49–62 (2020)
7. Zhang, Y., Yan, F., Liu, Q.: A hybrid model for diabetes prediction with SMOTE and LSTM neural network. IEEE Access **8** (2020)
8. Roy D., Banerjee, I.: Diabetes prediction using LSTM and random forest. Int. J. Adv. Comput. Sci. Appl. **12**(1) (2021)
9. Khamparia, A., Gupta, D., Pandey, B., Tiwari, P.: Deep learning models for early detection of diabetes using PIMA Indian dataset. Sustain. Cities Soc. **72** (2021)
10. Abeyrathna, D., Hu, J.: Diabetes prediction using an ensemble of deep neural networks. Comput. Meth. Progr. Biomed. **207** (2021)

Optimizing Hyperspectral Image Classification Through Swin Transformer Integration and CNN Feature Extraction

Sushil Kumar Janardan(✉) and Rekh Ram Janghel

National Institute of Technology Raipur, G.E. Road, Raipur, Chhattisgarh 492010, India
ssushil30@gmail.com, rrjanghel.it@nitrr.ac.in

Abstract. Due to the multidimensional structure of hyperspectral imaging, deep learning methods—in particular, vision transformer models—have been incorporated into classification systems. This article presents a novel use of Swin Transformer Network for hyperspectral image classification. The three layers that make up the suggested technique extract, embed, and merge image patches to provide a framework for classification. Our method uses patch-extraction, patch-embedding, and patch-merging approaches to improve feature representation. Hyperspectral features are preserved when window size, window shift, focus head, and input dimensions are taken into account. We demonstrate the effectiveness of our suggested method with a testing accuracy of 99.97% on hyperspectral datasets such as the Salinas public HSI dataset. The superiority of our approach is shown by comparing it with competing dimensionality reduction strategies. This work provides encouraging progress in the use of deep learning methods for classification of hyperspectral data.

Keywords: Hyperspectral Image Classification · Swin Transformer · Convolutional Neural Network

1 Introduction

The integration of spectral and image data from remote sensing images is the process of hyperspectral image classification or HSI. Compared with conventional and multispectral images, the spectral dimension of this approach is larger because it uses three-dimensional HSI with more than 200 spectral bands. The third dimension contains information about ground object reflectance from a uniform spectrum in each channel. Because reflectance spectral data are irregular, detecting subtle changes between them can help improve classification results (B. Liu et al., 2023). The large data sets of hyperspectral imaging make it possible to classify high-power images (HSIs) with remarkable accuracy. Due to its comprehensive information gathering of hundreds of channel spectroscopy bands at each pixel, it performs exceptionally well in HSI classification, meaning it has huge potential for use in many disciplines driven by this capability. Specifically, in precision agriculture, deep learning algorithms help identify agricultural

Published by Springer Nature Switzerland AG 2024
M. L. Owoc et al. (Eds.): ICCIDS 2024, IFIP AICT 718, pp. 374–386, 2024.
https://doi.org/10.1007/978-3-031-69986-3_29

acreage. Mineral exploitation, urban development, medical imaging, object detection, environmental monitoring and military services are some of its other uses (Peng et al., 2023).

The hyperspectral remote sensing community faces many obstacles due to the vast spectrum of hyperspectral imaging (HSI) data and the diverse range of application areas. Hughes phenomenon, significant dimensionality of the data, influence of background objects and noise on image quality, and geographical variability of spectral data are some of the main obstacles. The significant spectral data of HSI and the wide variety of its applications make these difficulties even more difficult. These constraints include geographical differences in spectral information, the impact of background and noise on image quality, multidimensionality of data, and limited availability of labelled samples – leading to the Hughes phenomenon (Ayas & Tunc-Gormus, 2022). When classification algorithms were first being developed, human feature engineering methods were common.

To make this work, a manually created feature descriptor was required. The obtained features were then used to train a machine-learning classifier. Initially, these data were mainly processed using general trainable classifiers such as Logistic Regression (LR), Random Forest (RF), and Support Vector Machine (SVM). Furthermore, many researches used various dimensionality reduction techniques in an attempt to enhance image representation. These included techniques such as linear discriminant analysis (LDA), independent component analysis (ICA), and principal component analysis (PCA) (Farooque et al., 2023).

Early deep learning methods mostly focused on 1D methods for HSI classification, such as recurrent neural networks (RNN), convolutional neural networks (CNN), deep belief networks (DBN), and stacked autoencoders (SAE). CNN is one of these techniques that has gained popularity due to how well it extracts complex spatial and spectral information; they work especially well at capturing short-range relationships in HSIs. Feature selection, sometimes known as band selection, aims to pinpoint important bands in the original dataset that have the greatest impact on hyperspectral imaging (HSI) classification accuracy. Using this strategy greatly reduces the amount of computing required. Since it is difficult to obtain representative samples, unsupervised feature selection has become common. However, due to their limitations in obtaining contextual and spatial information, spectrum classification approaches often produce incorrect conclusions. (Huang et al., 2022).

Convolutional Neural Networks (CNNs) have played a major role in recent advances in semantic segmentation, mostly due to their rapid development. Hyperspectral imaging (HSI) distinguishes itself from natural images with its unique properties due to weight sharing and local connections, which successfully reduce model complexity and limit reliance on spatial correlations. To effectively extract rich spatial-spectral information from HSI data, researchers are devoting themselves to the development of special Convolutional Neural Network (CNN)-based feature extractors.

2 Literature Review

In this field, the construction of fully convolutional neural networks or CNNs has gained significant importance. As a result, researchers have demonstrated a wealth of innovative work in this field (He et al., 2022). To create a hierarchical representation, the Swin Transformer first creates smaller patches and then combines neighbouring patches in progressively deeper transformer layers. By placing a fixed number of patches inside each window, this technique guarantees that complexity grows linearly with image size. Because of these benefits, the Swin Transformer is an excellent and flexible infrastructure that can be used for many visualization applications. Window Barrier Movement In designing a swing transformer, it is important to carefully consider the self-attention between successive layers. (Z. Liu et al., 2021).

The input image scales that the Swin Transformer can work with are diverse. Each hyperspectral image is first divided into large and small fragments. The proposed CNN with Swin Transformer efficiently captures larger, more complete representations at multiple semantic scales as well as better, more distinct elements by using these different scale patches as input. (Lin et al., 2022). Accurately measuring and identifying objects in an image is an essential part of HSI classification. By classifying and predicting the position of an object within an image, HSI classification establishes the semantic category of the data. Maximizing the probability of correct classification is the main goal of the HSI classification approach. (Lone & Pais, 2022).

Because deep convolutional neural networks (CNNs) are highly effective in extracting and expressing high-level and semantic data from images, their use in interpreting high-resolution remote sensing (HRRS) images has attracted great interest. At present, a variety of deep models are widely applied to difficult problems related to image processing for remote sensing (RS) (Lone & Pais, 2022). However, the deeper layers of deep CNNs sometimes lack precise localization for accurate image segmentation. This limitation results from the invariant properties of CNNs, which make them effective for challenging tasks such as classification. To solve this problem, a semantic segmentation technique has been proposed that uses CNNs to increase localization accuracy in segmentation tasks (Minaee et al., 2022). CNN architecture is used for hyperspectral image classification by collecting both spectral and spatial information through a three-dimensional network. Even though deep learning has demonstrated excellent capabilities and strong performance for this classification (Gao et al., 2022). Effectively extracting information from single-scale characteristics is still a major difficulty.

With respect to image size, computational complexity increases asymptotically in a quadratic fashion. As a solution, we opted to use the Swin Transformer as the backbone network architecture. The Swin transformer performs calculations related to self-attention by building a hierarchical transformer that operates in non-overlapping window areas. By significantly lowering computational complexity and making it linear concerning the dimensions of the input image, this solution solves the issue of high computing requirements (Xu et al., 2021). Compared to images of natural landscapes, remote-sensing images provide a wider variety of multimodal data such as depth and optical remote sensing images. When differentiating between groups that have similar visual appearances, depth information is quite helpful. Accordingly, increasing the accuracy of remote sensing image segmentation, gaining maximum benefit from complementary features for both RGB and depth information is important yet difficult (Zhao et al., 2021).

This research presents three significant innovations:

Transformer technology can handle sequence data (such as spectral amplitude information), but cannot include spatial amplitude data.

For HSI classification, a more advanced technique using complex Swain transformer networks was employed, which enhanced the network architecture for improved performance.

The Swin Transformer was used to handle three-dimensional data, meeting the unique processing needs of these kinds of data structures.

We performed several experiments on different HSI datasets, including rare airborne HSI datasets, to illustrate the adaptability of the model. The goal of this investigation is to provide insightful analysis and assistance to other academics utilizing comparable HSI datasets.

The subsequent sections of this study are organized as follows: We introduce hyperspectral imaging (HSI) and object detection in this section. We discuss the rationale, look at the details of data collection, and look at different HSI representations.

3 Proposed Methodology

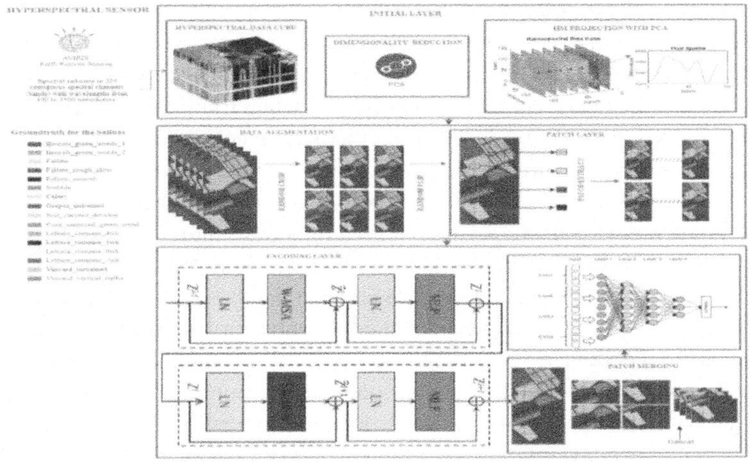

Fig. 1. CNN in the proposed model with Swin transformer construction.

3.1 HSI Classification Using a Network of Spectral Swin Transformers

Figure 1 Shows the model used in this investigation. First, the spectral dimensionality of the hyperspectral imagery was reduced at the input layer of the model using Principal Component Analysis (PCA). Hyperspectral image spectra often contain redundant and high-dimensional data. (Ren et al., 2021). By projecting these spectral dimensions onto a new, lower-dimensional space, PCA effectively removes unnecessary data and isolates the most important elements. Classification and data processing becomes easier with this method. The effectiveness of PCA as a feature extraction and data compression technique makes it a valuable tool for hyperspectral image classification. Then, the width, height, and condensed spectral dimensions were added to the data cube. Use the 'Create Image Cube' function to extract image patches of a given size from an input image and name them. (Maggiori et al., 2017) these patches accordingly. This involves storing adjacent image patches in an array, iterating over individual pixels, and applying zero-padding to the image. This feature also includes the option to filter out unlabelled patches and modify the labels as per specifications. (Sun et al., 2019). Essentially this indicates that the primary purpose of the function is to simplify the extraction of image patches and the labels that go with them to facilitate further processing or analysis. Data features are enhanced by dividing the data into patches at the patch layer. The computational complexity of a model is strongly influenced by the magnitude of its inputs. Smaller input shapes can be used to interpret high-resolution hyperspectral images, but the quality of the resulting classification may suffer. For best decision-making in hyperspectral image-based scientific study, it is important to balance the input shape vs classification accuracy trade-off. A revolutionary sliding window mechanism is incorporated into the Swin Transformer. With this breakthrough, the model can capture features within this window because the attention process is restricted to a specific window size. A self-attentive mechanism is achieved through the architecture of the hierarchical subsampling layer, incorporating features of both the immediate environment and the broader context into the dataset. But unlike, the Swin Transformer's performance varies a lot amongst different datasets. Transformers become less effective when working with smaller training sets, even though they perform exceptionally well when trained on large datasets.

Random crop is one of the many data augmentation techniques used in deep learning that is frequently used to increase the stability and accuracy of a model. Random flip is another popular method that flips the matrix at random while it is being trained. Patch extraction is the process of improving data features even more. After cropping feature data with patch sizes (X. Liu et al., 2021) (which are initially set at 2) to the window size settings, the patch-embedding module embeds the data into each patch.

Before each step, the patch-merging module subsamples, lowering the resolution, modifying channel counts, and creating a hierarchical architecture. Ultimately, the softmax activation function is used by the global average pooling layer to provide output features that correspond to various classes. By producing 1D data in place of the fully linked layer found in conventional CNN frameworks, global average pooling helps avoid overfitting the model.

3.2 Using Shift Window-Multi Self-Attention (SW-MSA)

Two subsequent Swin Transformer blocks that implement shifting window activities that communicate with other windows to build the basic structure of the Swin Transformer. Equation (1) these two latter are an appropriate way to illustrate the calculations of the Swin transformer blocks. At this point, a series of transformations are applied to the input features, indicated by, which includes Window Multi-Head Self-Attention (W-MSA),

Layer Normalization (LN), and a residual layer that generates Z. The second block, which employs (SW-MSA), receives the modified data. After undergoing another round of LN and multilayer perceptron (MLP).

$$Z^{'} = \mathrm{W} - \mathrm{MSA}\left(\mathrm{LN}\left(z^{l} - 1\right)\right) + \mathrm{Z}^{l} - 1 \tag{1}$$

$$Z^{l} = \mathrm{MLP}(\mathrm{LN}(z^{'})) + \mathrm{Z}^{'} \tag{2}$$

$$Z^{'+1} = S\mathrm{W} - \mathrm{MSA}(\mathrm{LN}(z^{l})) + \mathrm{Z}^{l} \tag{3}$$

$$Z^{l+1} = \mathrm{MLP}(\mathrm{LN}(z^{'+1})) + \mathrm{Z}^{'+1} \tag{4}$$

W-MSA (Weighted Multi-Head Self-Attention): This part is a self-attention mechanism that determines attention scores by applying weighted correlations it has learned between various elements in a sequence.

A special type of feed forward neural network called an MLP consists of multiple layers of interconnected nodes. Nonlinear data mapping between input and output is a popular application for this.

SW-MSA (Sliding Window Multi-Head Self-Attention): This probably refers to a particular kind of self-attention mechanism that computes attention scores across an input sequence using a sliding window technique.

Layer normalization, or LN, is a normalization technique used in neural networks to standardize activations across layers, promoting effective training and convergence.

The module produces an output by combining a query with a set of key-value pairs. By applying attention function to a sequence of Q (query), K (key), and V (value) for the query, key, and value operations, respectively. To do this, conducting a dot product between every key vector and the query vector is essential. The inner products are then rescaled using a softmax function, standardizing them into K weights according to Eq. (5).

$$\mathrm{Attention}(\mathrm{Q}, \mathrm{K},\ \mathrm{V}) =\ \mathrm{Softmax}(QK^{T}/\sqrt{dk})\mathrm{V} \tag{5}$$

Shifting Window (SW-MSA) is a method of allocating self-attention calculations among four different windows in the window patch. To perform the self-attention calculation for each patch in the feature map, each patch must be processed during this calculation. The feature map was first divided into multiple windows by shifted W-MSA modules using M*M dimensions, rotational offset windows, and separate self-attention towards internal windows individually. The processing load on each window can be significantly reduced by applying appropriate masks.

4 Research Findings

4.1 Details of Dataset

Salinas, which is accessible to the public, served as the main dataset used in the experiment. Other hyperspectral datasets from various parts of China, such as the Salinas hyperspectral dataset. These datasets are primarily intended for study concerning a variety of crop varieties. The data accessibility statement will contain information on how to access this data.

In 1998, AVIRIS collected the Salinas (SA) dataset in the Salinas Valley, California, United States. From the original dataset, twenty channels with low signal-to-noise ratio (SNR) were removed, leaving 204 bands for the experiment. With 512 x 217 pixels, the hyperspectral image (HSI) provides 3.7 m of spatial resolution. Within the collection of data, 16 frequent groups are identified by ground truth classification.

4.2 Specifics of the Training and Assessment Measures

A workstation computer with an NVIDIA graphics card was used in this study's computing setup to process smaller-scale datasets with better resolution. The training set comprised thirty percent of the dataset, while the testing set was selected from the remaining seventy percent. A learning rate of 0.001 was used in the training. The first feature extraction patch size was set at 25, which is an important initial patch size that affects CPU and GPU performance as well as processing time. While a smaller value lengthens the time needed for processing, a larger value requires the computer to operate at a higher level. Therefore, it's crucial to balance the initial patch size and training performance. The study discovered that the Salinas datasets responded well to a patch size of 25, while the reduction in block size was necessary due to better resolution sampling. We used the following initial values for the Swin Transformer model: 2 x 2 patches, 0.03% dropout rate, 8 attention heads, 64 embedding dimensions, a Window dimension of 2, a shift frame process of 1, and 256 multiple-layer perceptrons.

Three criteria were used in this experiment to evaluate prevalence: kappa coefficient, overall accuracy (OA), and average accuracy (AA). Average accuracy is calculated as part of the assessment process and indicates the average classification accuracy over all classes.

$$Recall = \frac{True\ Positive}{(True\ Positive\ +\ False\ Negative)} \tag{6}$$

$$AA = \frac{\sum_{i=1}^{n} Recalli}{n} * 100 \tag{7}$$

$$OA = \frac{(True\ Positive\ +\ True\ Negative)}{Total\ number\ of\ pixels} * 100 \tag{8}$$

$$Kappa = \frac{P0 - Pe}{1 - Pe}, P0 = \frac{\sum xii}{N}, Pe = \frac{\sum xi + xi + \ldots}{N^2} \tag{9}$$

Equation (8) was used to determine the overall accuracy. Furthermore, a statistical measure known as Kappa Eq. (9) shows consistent association in the knowledge estimate difference between the ground truth map and the classification map. P0 stands for perceived agreement, which indicates how consistently classifiers and assessors classify the same data. On the other hand, Pe, or estimated probability acceptance, which takes into account the estimated distribution of the classifier, is the probability ratio of agreement determined by random. In this case, N represents the total number of observations, xii is the number of observations in row I and column I and Xi + is the marginal total number of observations in row I.

4.3 Salinas Hyperspectral Image Classification Results

The classification results produced by our model for the Salinas (SA) hyperspectral dataset as the graphic shows in Fig. 2. As the picture makes clear. Our proposed model demonstrated fairly excellent test accuracies on the Salinas datasets. Furthermore, the loss convergence happened more quickly for the SA datasets. This is consistent with the parameter analysis's outcome. For the SA datasets, the spectral dimension was set to 15 (Fig. 3 and Table 1).

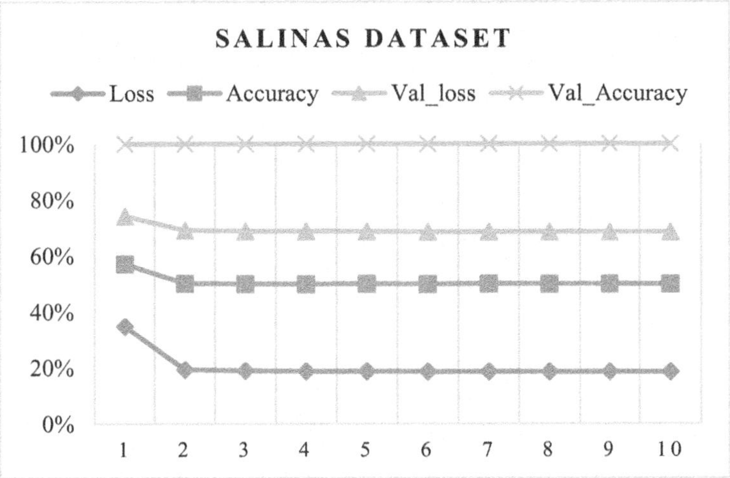

Fig. 2. Ten training examples were used to determine the accuracy and degradation of Salinas samples in our model.

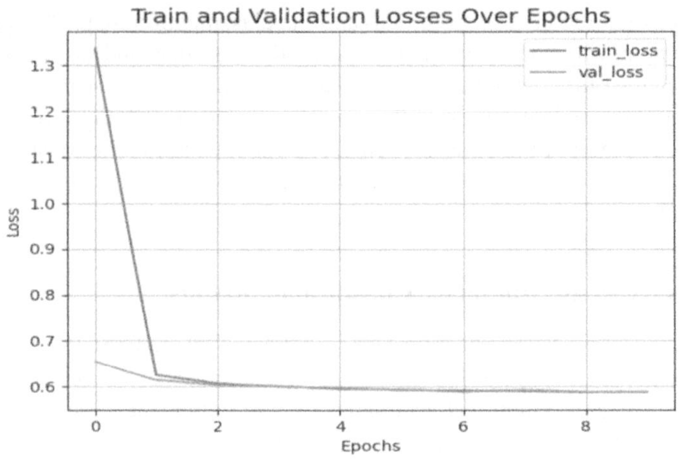

Fig. 3. Train and validation losses over epochs of Salinas in our model.

4.4 Classification Outcomes for All Categories of Object

Table 1. Salinas classification results, including item recall, precision, and F1 score.

Type	Precision	Recall	F1-Score	Support
Broccoli_green_weeds_1	1.00	1.00	1.00	1406
Broccoli_green_weeds_2	1.00	1.00	1.00	2608
Fallow	1.00	1.00	1.00	1383
Fallow_rough_plow	0.99	1.00	1.00	976
Fallow_smooth	1.00	1.00	1.00	1875
Stubble	1.00	1.00	1.00	2771
Celery	1.00	1.00	1.00	2505
Grapes_untrained	1.00	1.00	1.00	7890
Soil_vineyard_develop	1.00	1.00	1.00	4342
Corn_senesced_green_weeds	1.00	1.00	1.00	2295
Lettuce_romaine_4wk	1.00	1.00	1.00	748
Lettuce_romaine_5wk	1.00	1.00	1.00	1349
Lettuce_romaine_6wk	1.00	1.00	1.00	641
Lettuce_romaine_7wk	1.00	1.00	1.00	749
Vineyard_untrained	1.00	1.00	1.00	5088
Vineyard_vertical_trellis	1.00	1.00	1.00	1265

4.5 Confusion Matrix and Maps of Classification

This paper proposes a unique HSI classification model using a Convolutional Neural Network with Swin Transformer, which is different from the typical CNN architecture currently in use (Figs. 4, 5 and Table 2).

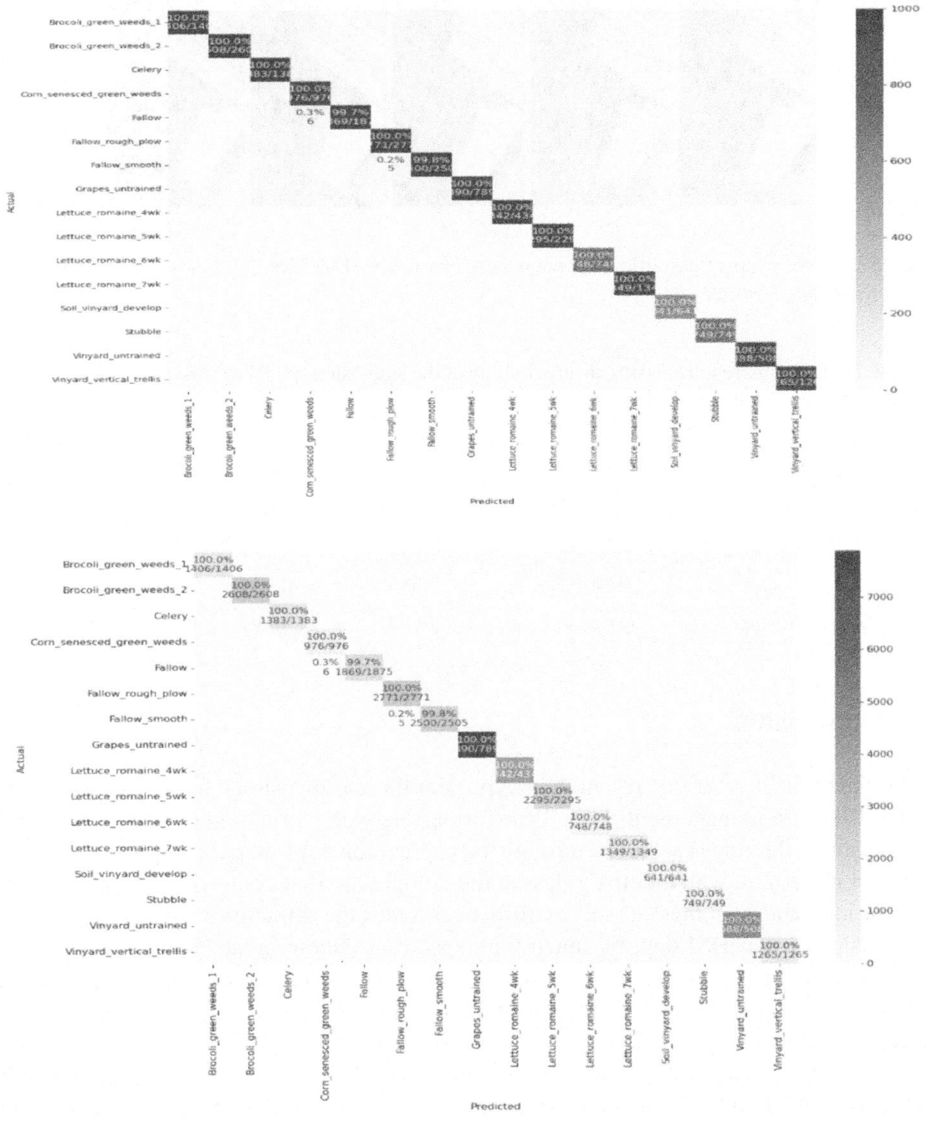

Fig. 4. Confusion Matrix

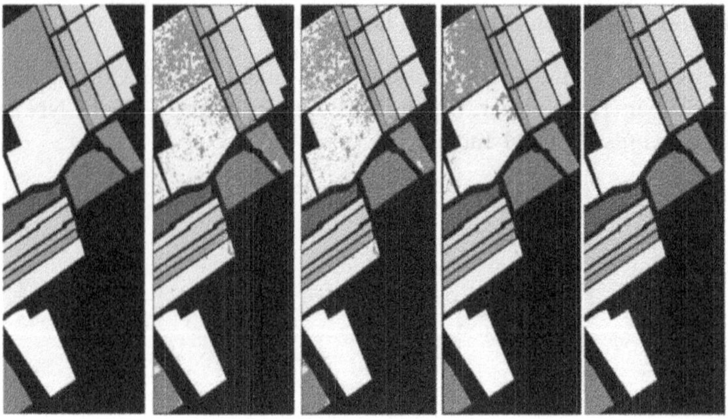

Fig. 5. Comparison of classification maps between CNN 1D, CNN 2D, CNN3D, and the CNN with Swin Transformer.

Table 2. To compare the classification accuracy of the suggested model against alternative models using the 0.3 SA dataset for training.

CNN / Transformer	Technique	AA	OA	Kappa
CNN	1D CNN	0.842	0.897	0.865
	2D CNN	0.861	0.912	0.891
	3D CNN	0.884	0.934	0.904
CNNS Transformer	Proposed	99.97	99.97	99.96

5 Conclusion

The main objective of this research is to present three-dimensional hyperspectral imaging (HSI) data management Swin Transformer Network. In the early phases, Spectral dimension reduction is achieved through the application of principal component analysis (Ren et al., 2020), which helps to lessen the complexity that comes with high-spectral-resolution data. This method successfully overcomes the difficulties caused by the high dimensionality of HSI data by integrating spectrum data to create 3-dimensional data blocks. (Shelhamer et al., 2017). Finally, the study distinguishes itself for excellent identification of high-spectral-resolution images by incorporating Swin Transformer's unique shift window and global self-focus processing. Experimental results show competitive performance, especially on most small-scale HSI datasets. Experimental results show that processing small-scale HSI information is best provided by simpler network architectures. Although a lot of research usually adds more modules to handle complicated issues, observations of the architectures show, that efficient early-stage feature processing can produce good outcomes in HSI classification tasks. These structures, however, perform poorly on ultra-high-resolution high-spectral-resolution (Ding et al.,

2020) datasets and frequently perform poorly on datasets with more than 1000 pixels. The problem of managing large-scale HSI databases is still a subject for investigation. Additionally, the suggested model offers a viable method for creating HSI classification models that are based on the Vision Transformer architecture.

Acknowledgments. Then authors thank the editors and anonymous reviewers for their insightful comments and recommendations, which greatly improved our research.

Disclosure of Interests. The authors declare no conflicts of interest related to this work.

References

1. Ayas, S., Tunc-Gormus, E.: SpectralSWIN: a spectral-swin transformer network forhyperspectral image classification. Int. J. Remote Sens. **43**(11), 4025–4044 (2022). https://doi.org/10.1080/01431161.2022.2105668
2. Ding, L., Zhang, J., Bruzzone, L.: Semantic segmentation of large-size VHR remote sensing images using a two-stage multiscale training architecture. IEEE Trans. Geosci. Remote Sens. **58**(8), 5367–5376 (2020). https://doi.org/10.1109/TGRS.2020.2964675
3. Farooque, G., Liu, Q., Sargano, A.B., Xiao, L.: Swin transformer with multiscale 3D atrous convolution for hyperspectral image classification. Eng. Appl. Artif. Intell. **126**(PC), 107070 (2023). https://doi.org/10.1016/j.engappai.2023.107070
4. Gao, J., Gao, X., Wu, N., Yang, H.: Bi-directional LSTM with multi-scale dense attention mechanism for hyperspectral image classification. Multimedia Tools Appl. **81**(17), 24003–24020 (2022). https://doi.org/10.1007/s11042-022-12809-z
5. He, X., Zhou, Y., Zhao, J., Zhang, D., Yao, R., Xue, Y.: Swin transformer embedding U-Net for remote sensing image semantic segmentation. IEEE Trans. Geosci. Remote Sens. **60**, 1–15 (2022). https://doi.org/10.1109/TGRS.2022.3144165
6. Huang, X., Dong, M., Li, J., Guo, X.: A 3-D-Swin transformer-based hierarchical contrastive learning method for hyperspectral image classification. IEEE Trans. Geosci. Remote Sens. **60**, 1–15 (2022). https://doi.org/10.1109/TGRS.2022.3202036
7. Lin, A., Chen, B., Xu, J., Zhang, Z., Lu, G., Zhang, D.: DS-TransUNet: dual swin transformer U-net for medical image segmentation. IEEE Trans. Instrum. Meas. **71**(8), 1–13 (2022). https://doi.org/10.1109/TIM.2022.3178991
8. Liu, B., Liu, Y., Zhang, W., Tian, Y., Kong, W.: Spectral swin transformer network for hyperspectral image classification. Remote Sens. **15**(15) (2023). https://doi.org/10.3390/rs15153721
9. Liu, X., Jiao, L., Li, L., Tang, X., Guo, Y.: Deep multi-level fusion network for multi-source image pixel-wise classification. Knowl.-Based Syst. **221**, 106921 (2021). https://doi.org/10.1016/j.knosys.2021.106921
10. Liu, Z., et al.: Swin transformer: hierarchical vision transformer using shifted windows. In: Proceedings of the IEEE International Conference on Computer Vision, 9992–10002 (2021).https://doi.org/10.1109/ICCV48922.2021.00986
11. Lone, Z.A., Pais, A.R.: Object detection in hyperspectral images. Digit. Sign. Process. Rev. J. **131**, 103752 (2022). https://doi.org/10.1016/j.dsp.2022.103752
12. Maggiori, E., Tarabalka, Y., Charpiat, G., Alliez, P.: High-resolution aerial image labeling with convolutional neural networks. IEEE Trans. Geosci. Remote Sens. **55**(12), 7092–7103 (2017). https://doi.org/10.1109/TGRS.2017.2740362

13. Minaee, S., Boykov, Y., Porikli, F., Plaza, A., Kehtarnavaz, N., Terzopoulos, D.: Image segmentation using deep learning: a survey. IEEE Trans. Pattern Anal. Mach. Intell. **44**(7), 3523–3542 (2022). https://doi.org/10.1109/TPAMI.2021.3059968
14. Peng, Y., Ren, J., Wang, J., Shi, M.: Spectral-swin transformer with spatial feature extraction enhancement for hyperspectral image classification. Remote Sens. **15**(10), 1–19 (2023). https://doi.org/10.3390/rs15102696
15. Ren, J., Wang, R., Liu, G., Feng, R., Wang, Y., Wu, W.: Partitioned Relief-F method for dimensionality reduction of hyperspectral images. Remote Sens. **12**(7), 1–21 (2020). https://doi.org/10.3390/rs12071104
16. Ren, J., Wang, R., Liu, G., Wang, Y., Wu, W.: An SVM-based nested sliding window approach for spectral–spatial classification of hyperspectral images. Remote Sens. **13**(1), 1–26 (2021). https://doi.org/10.3390/rs13010114
17. Shelhamer, E., Long, J., Darrell, T.: Fully convolutional networks for semantic segmentation. IEEE Trans. Pattern Anal. Mach. Intell. **39**(4), 640–651 (2017). https://doi.org/10.1109/TPAMI.2016.2572683
18. Sun, K., et al.: High-Resolution Representations for Labeling Pixels and Regions (2019). http://arxiv.org/abs/1904.04514
19. Xu, X., et al.: An improved swin transformer-based model for remote sensing object detection and instance segmentation. Remote Sens. **13**(23) (2021). https://doi.org/10.3390/rs13234779
20. Zhao, J., Zhou, Y., Shi, B., Yang, J., Zhang, D., Yao, R.: Multi-stage fusion and multi-source attention network for multi-modal remote sensing image segmentation. ACM Trans. Intell. Syst. Technol. **12**(6) (2021). https://doi.org/10.1145/3484440

Efficient Task Allocation with Mentorship Mapping in Crowd Sourcing Platforms

Sindhu Sahithi Gelam(✉), Sai Naga Manikanta Thota, Sai Anushree Voleti, Lekshmi S. Nair, and Durga Manikanta Naga Surya Teja Tummala

Department of Computer Science and Engineering, Amrita School of Computing, Amrita Vishwa Vidyapeetham, Amritapuri, Clappana, India
sindhusahithi2623@gmail.com, lekshmisn@am.amrita.edu

Abstract. Crowdsourcing platforms have emerged as a powerful tool for harnessing the collective intelligence and expertise of a diverse pool of workers. However, efficiently allocating tasks to the most suitable workers remains a significant challenge due to the heterogeneity of workers, the complexity of tasks, and the dynamic nature of the crowdsourcing environment. This study addresses the task allocation problem on crowdsourcing platforms by proposing a novel mentorship mapping approach. The proposed approach categorizes workers into experienced and inexperienced workers based on their experience, successful completion percentage, and approved on-time rate. Mentorship mapping leverages the concept of mentorship, where experienced workers guide and assist less experienced workers, to enhance task completion rates and overall productivity. The proposed approach involves identifying and pairing experienced workers with less experienced workers based on their skill sets, experience levels, and task requirements. This mentorship-driven task allocation strategy aims to optimize worker-task matching, leading to efficient workforce utilization and improved task completion outcomes. The benefits of mentorship mapping include faster task completion times, reduced errors, and enhanced worker satisfaction, contributing to a more efficient and productive crowdsourcing ecosystem.

Keywords: Crowdsourcing · Task allocation · Mentorship mapping · experienced workers · inexperienced workers

1 Introduction

Crowdsourcing systems have become an essential part of the modern digital world, allowing for global collaborative problem-solving and project execution. The difficulties in assigning tasks efficiently have become more noticeable as these platform's popularity soars. We intend to significantly alter the current situation in this dynamic and ever-evolving arena in response to these important concerns. The efficient assignment of jobs within crowdsourcing platforms is crucial for optimizing labor utilization and ensuring high-quality task completion.

M. L. Owoc et al. (Eds.): ICCIDS 2024, IFIP AICT 718, pp. 387–401, 2024.
https://doi.org/10.1007/978-3-031-69986-3_30

It is more important than ever to match job distribution to improve platform performance overall, increase customer satisfaction, and support independent contractors in their endeavors [1]. The demand for efficient task allocation and effective mentorship structures is growing within crowdsourcing platforms. This research aims to address these challenges using cutting-edge technologies and a data-driven approach. We propose models that allocate tasks based on contributors' expertise, availability, and historical performance and create intelligent systems to identify potential mentor-mentee relationships. Additionally, we will explore the socio-technical aspects of task allocation and guidance mapping to develop more comprehensive, user-centered solutions. Our goal is to contribute to the evolution of crowdsourcing platforms into more dynamic, inclusive, and knowledge-sharing ecosystems.

By dealing with these challenges comprehensively, our research has the potential to make a substantial contribution to transforming crowdsourcing platforms into more dynamic, inclusive ecosystems. Our research is essential in addressing the critical challenges faced by crowdsourcing platforms, particularly the efficient allocation of tasks and the establishment of mentorship structures. By introducing a systematic approach to categorise workers and identify potential mentors based on expertise, experience, and historical performance, our work goes beyond mere task allocation logistics [2]. This mentorship mapping forms the core of our solution, where expert workers play a pivotal role in guiding and supporting inexpert workers, fostering a collaborative and knowledge-sharing environment.

The task allocation process is further refined through a meticulous analysis of skillset matching between workers and tasks, ensuring that tasks are assigned based on proficiency and expertise, thereby optimizing the overall productivity of the workforce. Through the innovative approaches employed in our research, we have achieved a transformative impact on the dynamics of crowdsourcing platforms. The systematic categorization of workers and the identification of potential mentors based on expertise, experience, and historical performance have resulted in a more efficient and targeted task allocation process [3]. This achievement extends beyond logistical improvements; it has fostered a collaborative and knowledge-sharing environment, significantly enhancing the overall quality of task outcomes. The introduction of adaptability to the dynamic nature of digital work environments ensures that our innovation remains relevant and effective in addressing evolving challenges.

The primary contribution of our article lies in addressing the intricate challenges associated with task allocation in crowdsourcing platforms [4]. By introducing a systematic approach to categorize workers into expert and inexpert classes based on their experience, successful completion percentage, and approved one-time metrics, we establish a foundation for mentorship eligibility. The research work takes a significant step forward by calculating skillset similarity scores between expert and inexpert workers, allowing for a precise mapping of mentors to inexpert individuals.

This mentorship mapping forms the core of our solution, where expert workers play a pivotal role in guiding and supporting inexpert workers, fostering a collaborative and knowledge-sharing environment [5]. The task allocation process is further refined through a meticulous analysis of skillset matching between workers and tasks. This matching ensures that tasks are assigned to workers based on their proficiency and expertise, thereby optimizing the overall productivity of the workforce [6]. Moreover, this introduces a dynamic element in the allocation process by considering the urgency level of tasks.

Tasks are allocated based not only on skillset matching but also on urgency, with additional attention given to the number of due days, especially for tasks with a medium level of urgency [7]. This proposed work makes a substantial contribution to the field by providing a comprehensive solution to the task allocation challenges on crowdsourcing platforms [8]. The introduction of mentorship mapping, skillset matching, and consideration of task urgency collectively aim to optimize worker-task matching, leading to more efficient workforce utilization and enhanced task completion rates [9]. This innovative approach holds the promise of significantly improving the overall performance and productivity of crowdsourcing platforms.

A conceptual illustration of crowdsourcing workflow is illustrated in Fig. 1. This proposed methodology addresses the challenges of task allocation in crowdsourcing platforms through a systematic approach. We categorize workers into expert and inexpert classes based on experience and performance metrics, forming the foundation for mentorship eligibility. The core innovation lies in mentorship mapping, where expert workers guide inexpert counterparts, fostering collaboration and knowledge sharing. Skillset similarity scores refine mentorship matches, ensuring compatibility [10]. The methodology includes a meticulous skillset analysis for task allocation, considering urgency and due days.

This dynamic approach optimizes workforce productivity. Our systematic solution integrates mentorship mapping, skillset matching, and task urgency consideration, promising to revolutionize workforce utilization and enhance task completion rates [11]. With a focus on inclusivity and knowledge-sharing, our methodology is poised to elevate the performance of crowdsourcing platforms in the ever-evolving digital landscape.

The rest of this paper is structured as follows: Sect. 2 describes the related work in Efficient task allocation with mentorship mapping in crowdsourcing platforms. We compare the proposed model with some baseline methods using simulations in Sect. 3. Experimentation and observation in efficient task allocation are detailed in Sect. 4 and conclude in Sect. 5.

2 Related Work

The authors propose a task recommendation model for crowdsourcing systems using hierarchical skills and Resnik similarity measures, achieving efficient task recommendations and pickup times for new and inexperienced workers. While currently limited to single skills, future work could explore multi-skill capabilities [12]. They have found challenges in managing incentive-based crowdsourcing platforms, particularly in effectively assigning tasks to inexperienced workers.

Current task recommendation models often overlook the complexities of handling multiple skills and providing accurate recommendations for new users. The proposed model focuses on skill classification and specialized participation probability but lacks comprehensive assessments of accuracy measures. Acknowledging limitations in our understanding of task recommendation in crowdsourcing is crucial for progress. Existing models lack explicit expressiveness, hindering profound efficiency assessments. To foster platform growth, addressing this gap is essential. Future work should explore extending the model to cover multi-skilled workers, enhancing recommendation accuracy, and discovering new algorithms. It is necessary to explore multi-skilled workers, improve recommendation accuracy, discover new algorithms, and investigate assignment redistribution in service delivery platforms. This comprehensive approach holds the key to unlocking the full potential of crowdsourcing platforms and fostering their sustainable growth. Notably, the literature lacks insights on assignment redistribution, necessitating further research in service delivery platforms and incentive-based communities and also should delve into areas such as dynamic skill acquisition models, task complexity analysis, optimizing incentive structures, integrating social network data, and addressing ethical considerations to create a comprehensive and robust framework for crowdsourcing platforms. This holistic approach holds the key to unlocking the full potential of crowdsourcing, ensuring platform efficiency, user satisfaction, and responsible growth.

The authors introduce a new task recommendation model for reward-based crowdsourcing, combining implicit user feedback with explicit features. It analyzes task and reward probabilities, demonstrating superior performance and reduced data sparsity compared to existing models [13]. The proposed model leverages participation information to predict reward gains, suggesting potential accuracy improvements by adding new features. They addressed the challenge of optimizing task proposals for accuracy in dynamic crowdsourcing environments. Their research proposes a reward-based task recommendation model using hybrid techniques to enhance dynamics and promote worker participation. Although specific accuracy measures are not provided, the proposed model outperforms traditional matrix factorization methods by incorporating both implicit and explicit features. However, limitations exist, including the need for further investigation into the scalability of the model across different crowdsourcing platforms and comprehensive comparisons with state-of-the-art methods. The research emphasizes the use of larger datasets for sound evaluation and suggests future studies introduce high-tech algorithms and consider different parameters to refine task proposal processes in dynamic crowdsourcing settings and also address scalability across diverse platforms. This holistic approach will unlock the full potential of optimizing task proposals and propel the success of crowdsourcing platforms. Moreover, investigating the impact of the proposed model on worker satisfaction and platform retention rates is crucial for ensuring its long-term sustainability and ethical implementation. By addressing these concerns, researchers can contribute to the development of effective and responsible crowdsourcing platforms that benefit both workers and requesters, propelling the field towards a future of efficient and ethical resource allocation.

This is a groundbreaking study in the field of crowdsourcing contests. It creates an entirely new framework for participation recommendations. Their approach involves complex probabilistic models for contest participation and winner determination, a first step in the predictive analysis of crowdsourcing contests [14]. Unlike the old approaches, with very few winning records to go on, they calculate probabilities of both win and participation for individual workers in order to make ranked contest recommendations. To meet this challenge, they use data such as contest participation information and worker features, among other things. Their solution employs transfer learning techniques for matrices and feature-based matrix factorization to get around data sparsity problems. They also emphasize with experiments using real crowdsourcing contest data that the initial search should be aided to varying degrees by auxiliary information, revealing some essential skills applicable in various kinds of contests. Their work extends contest recommendation systems in another direction by introducing predictive models and generating more advanced data usage methods. However, this approach expands on past exploration into the nature of contests to which recommendations should be attuned-nuances that might not adequately reflect themselves in such a methodology–giving us some hope there are still more sophisticated forms of personalization waiting to be discovered.

In this study, we explore the important role that task recommendation plays in improving crowdsourcing platforms [15]. Polling workers' preferences and perceptions of similarity between tasks is also introduced to try out An empirical survey, the research finds a strong interest among workers in task suggestions, although this differs widely depending on area. It reflects the confusion surrounding purportedly similar tasks. Semantic aspects, including what action must be taken, tend to have more of an impact than straight facts like when and for how much money. Nevertheless, regional differences in preferences emerge: action is the most important of all universally, but there are variations by Asian region. However, the study only proposes personalised task recommendations based on semantic similarity features accumulated from descriptions of tasks, and doesn't mention whether there is any way to integrate worker skills or requester character into recommendation systems. This research points to the importance of sophisticated similarity metrics and personalized strategies. It also positions an exploration into utilizing worker qualifications, requester history, and peer popularity for more refined task recommendations; a foundation is laid here from which we can build towards a greater understanding of crowdsourcing recommendation dynamics.

3 Proposed System

The proposed model demonstrates efficiency in task allocation through a strategic combination of employee categorization, mentorship mapping, and skillset matching. This approach enhances matching accuracy by leveraging expertise and mentorship, leading to improved task-employee alignment. This not only optimises task completion but also fosters knowledge transfer within the platform, distinguishing it from traditional methods and contributing to enhanced overall efficiency in crowdsourcing.

This paper introduces a novel approach that distinguishes it from others by selectively identifying eligible experts for mentorship mapping from the pool of all available expert employees. This process involves categorizing experts based on skills, successful submissions and experience. Only those experts who meet the eligibility criteria are considered for mentorship roles. Subsequently, we employ a mapping mechanism to pair less-experienced individuals with these eligible mentors, ensuring a focused and tailored mentorship program that aligns closely with the defined criteria for eligibility.

3.1 Mapping In Expert Workers to Expert Workers

It aims to optimize the utilization of the workforce by efficiently channelling their skills and knowledge [16]. The primary objective is to establish a structured mentorship programme where experienced workers are paired with less experienced counterparts. And it also strives to enhance task completion rates by streamlining the workflow through the introduction of this mentorship-based approach [17]. This mentorship arrangement aims to harness the expertise of the seasoned workers, who will assume the role of mentors, guiding and delegating tasks to their less experienced counterparts.

3.2 Allocation of the New Tasks

Tasks that arise will be meticulously assigned to workers based on a comprehensive evaluation of several key factors. These factors include the specific skill sets possessed by the workers and the precise requirements of the tasks at hand [18]. Subsequently, a judicious matching process will ensue, ensuring that the forthcoming tasks are allocated to workers in a manner that aligns seamlessly with the aforementioned criteria. The included methods involve assigning tasks to employees using various approaches. One method involves assigning tasks without placing a limit on the number of tasks a single employee can undertake [19]. While this approach aims to engage as many employees as possible, another method employs a check by restricting the number of tasks assigned to each employee. The common criterion for task assignment is the examination of similarity scores between the skills possessed by an employee and the skills required for a given task.

3.3 Expert Matching Model

Here to Map Inexpert to Expert, We are Considering the Factor Skill Set Similarity Scores Between the Employees, As the Skills of the Employees are in a String Format, we have to do the Vectorization as such to do the Cosine Similarity between them, so as a start we will be doing the Vectorization of Skills Strings (TF-IDF), The steps to do the following are Document-Term Matrix

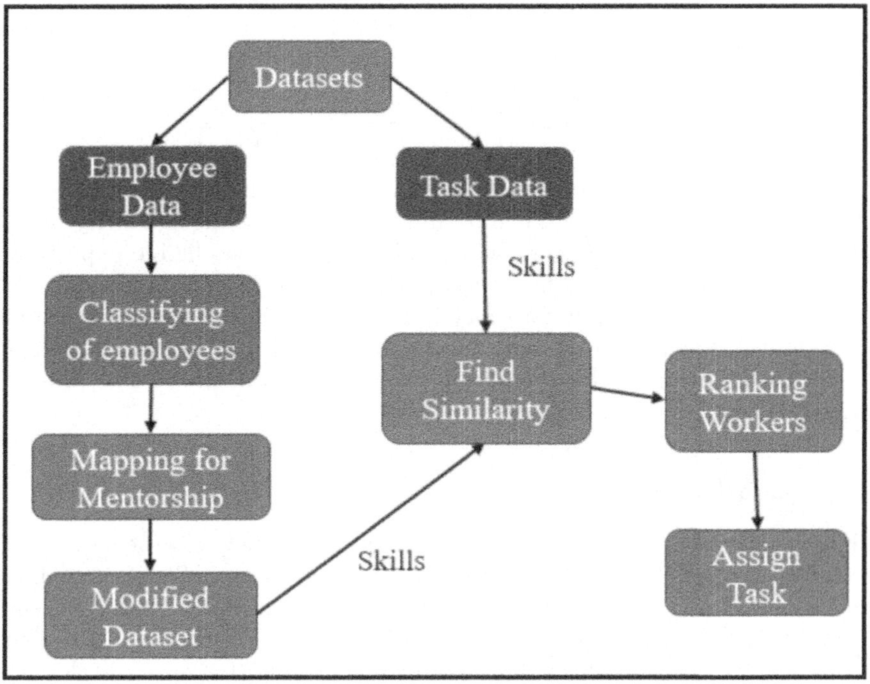

Fig. 1. Conceptual illustration of workflow in crowdsourcing platform

(DTM) where D be the set of documents (employees' skill sets), T be the set of unique terms (individual skills).

$$DTM_{i,j} = TF - IDF(t_j, d_i) \tag{1}$$

This represents the TF-IDF score of term t_j in document d_i. Then, to calculate the TF (Term Frequency) TF (t, d) is the frequency of term t in document d. $TF_{i,j}$ is the TF score of term t_j. in document d_i.

$$\text{TF}_{ij} = \frac{\text{Number of times } t_j \text{ appears in } d_i}{\text{Total terms in } d_i} \tag{2}$$

After the term frequency, we have to do the IDF (inverse document frequency), where IDF (t, D) is the inverse document frequency of term t in document set D and IDF_j is the IDF score of term t_j in the entire document set.

$$\text{IDF}_{ij} = \log\left(\frac{\text{Total number of documents in } D}{\text{Number of documents containing } t_j}\right) \tag{3}$$

Then it gives the TF-IDF (Term Frequency-Inverse Document Frequency), where the $TFIDF_{i,j}$ is the TF-IDF score of term tj in document d_i. Which gives $TFIDF_{i,j} = TF_{i,j} \times IDF_j$ So the cosine similarity calculation between documents i and k gives $CS_{i,k}$ is the cosine similarity between documents d_i and d_k.

$$\text{CS}_{i,k} = \frac{\sum_{j=1}^{n} \text{TFIDF}_{i,j} \times \text{TFIDF}_{k,j}}{\sqrt{\sum_{j=1}^{n}(\text{TFIDF}_{i,j})^2} \times \sqrt{\sum_{j=1}^{n}(\text{TFIDF}_{k,j})^2}} \quad (4)$$

where n is the total number of terms. This outlines the key steps involved in vectorizing skill strings using TF-IDF and calculating cosine similarity between documents (employees). We will be considering only the eligible experts from all the experts according to their successful submissions and approved on-time submission criteria, and then for every inexpert, we will be taking the top most skilled similarity expert to assign the mentor, and also at most a mentor, i.e., an expert who is eligible for mentoring, can have two mentors. The process of Expert-Inexpert Mapping is given in Algorithm 1.

3.4 Task Allocation Model

In this, we have to map inexperts to experts. We are considering the factor of skill set similarity scores between the employees skills and task skills. As the skills of both employees and task are in a string format, we have to do the vectorization as such to do the Cosine Similarity between them. As a start, we will be doing the Count Vectorizer for this Count Vectorizer formula (for term j in document i) given by $X_{i,j}$ =count of term t_j in document i , that is $X_{i,j}$ The count of occurrences of term t_j in document i. Now for the Cosine similarity formula (for documents i and k) the $cs_{i,k}$ is the cosine similarity between employee i's skills and task k's skill requirements.

$$\text{CS}_{i,k} = \frac{\sum_{j=1}^{n} \text{X}_{i,j} \times \text{Y}_{k,j}}{\sqrt{\sum_{j=1}^{n}(\text{X}_{i,j})^2} \times \sqrt{\sum_{j=1}^{n}(\text{Y}_{k,j})^2}} \quad (5)$$

where n is the total number of terms (distinct skills).
$X_{i,j}$ is the count of the j-th term in skills of employee i.
$Y_{k,j}$ is the count of the j-th term in skills required for task k.

For the calculation of the count Matrix X we have the terms $X_{i,j}$ is the count of the j-th term in the skills of employee I and $X_{i,j}$ = count of term t_j in employee i. For the calculation of the count Matrix Y we have the terms $Y_{k,j}$ is the count of the j-th term in the skills required for task k. $Y_{k,j}$ = count of term t_j in task k. Algorithm 2 describes the task Allocation process. As we calculated the similarity scores between them, now the task we get assigned to the employees with a check of the level of difficulty and due days as parameters, in this algorithm we will be prioritised to assign the low level difficulty task to a inexpert, and prioritised to assign the high level difficulty task to a Expert and the medium level task difficulty will be assigned to both the in experts and experts according to the due days present for a particular task.

Algorithm 1. Expert-Inexpert Mapping

```
expert_eligible_mapping ←df[(df['Category'] == 'Expert') &(df['SuccessfulCompletion'] ≥ 50)
                         &(df['ApprovedTime'] ≥ 40)]
vectorizer ← TfidfVectorizer()
tfidf_matrix_experts ← vectorizer.fit_transform(experts['Skill Set'])
tfidf_matrix_inexperts ← vectorizer.transform(inexperts['Skill Set'])
cosine_similarities ← cosine_similarity(tfidf_matrix_inexperts, tfidf_matrix_experts)
max_inexperts_per_mentor ← 2
mentor_count ← {}
inexpert_to_mentormapping ←pd.DataFrame(columns=['InexpertEmployee', 'ExpertMentor',
                         'SimilarityScore'])
for i, row in enumerate(cosine_similarities) do
    eligible_experts_indices ← expert_eligible_mapping.index
    sorted_indices ← np.argsort(row)[::-1]
    mapped_inexperts ← 0
    for idx in sorted_indices do
        if idx in eligible_experts_indices then
            if mentor_count.get(idx, 0) < max_inexperts_per_mentor then
                mentor_count[idx] ← mentor_count.get(idx, 0) + 1
                similarity_score ← row[idx]
                mapped_inexperts ← mapped_inexperts + 1
            end if
        end if
        if mapped_inexperts ≥ max_inexperts_per_mentor then
            break
        end if
    end for
end for
```

Algorithm 2. Task Allocation Algorithm

```
Step 1: Calculate skill similarities using cosine similarity
employeeSkillMatrix ← calculate_cosinesimilarity(employee['SkillSet'], tasks['TaskSkills'])
Step 2: Allocate tasks to employees based on conditions
for all task_row in tasks_data do
    task_id ← task_row['Taskid']
    level_of_difficulty ← task_row['LevelofUrgency']
    due_days ← task_row['DueDateindays']
    Step 2a: Find the employee with the highest skills similarity for the task
    employeeIndex ← employee_with_highest_similarity(employeeSkillsMatrix, task_row)
    similarity_level ← employeeSkillMatrix[employeeIndex, task_row]
    Step 2b: Assign task based on level of difficulty and due days criteria
    if level_of_difficulty is 'Low' then
        prioritized to give to inexpert
    else if level_of_difficulty is 'High' then
        prioritized to give to expert
    else if level_of_difficulty is 'Medium' then
        priority given according to due days of tasks; it can be assigned to both experts and inexperts
    end if
end for
```

4 Experimentation and Observation

4.1 Dataset

The Study includes a custom dataset that was curated by referencing several online crowdsourcing platforms that provide valuable insights into employee engagement and task completion dynamics. The employee's dataset encompasses essential attributes critical for task allocation strategies. The employee's dataset encompasses essential attributes critical for task allocation strategies. These attributes include Employee ID, Role, Experience, Skill Sets, Number of Tasks Completed, Average Time for Task Completion, Successful Completion Tasks on Time Percentage, and Metrics such as Approved on Time. This dataset not only captures the basic demographic information of crowd workers but also delves into their performance metrics, providing a holistic view of their capabilities and efficiency in task execution.

The task dataset comprises crucial information essential for task allocation optimization. Key attributes within this dataset include Task ID, Task Requirement Skills, Level of Urgency, and Due Days Left. These attributes offer insights into the nature of tasks, their urgency levels, and the specific skill sets required for their successful completion. By harnessing this dataset, the research aims to develop and evaluate task allocation algorithms that can adapt to the dynamic nature of crowdsourcing platforms, optimizing the utilization of available skills and resources to enhance overall task completion efficiency.

4.2 Steps and Observations for Expert Matching Model

Step 1: Employee Categorization The initial step involved categorizing employees based on their experience levels. Employees with over four years of experience were labeled as "Experts", while those with less experience were classified as "Inexperts". This categorization formed the foundation for subsequent mentorship mapping.

Step 2: Eligibility Criteria for Expert Mentors A subset of "Expert" employees was identified as eligible mentors based on specific criteria. These criteria included a success completion rate exceeding 50% and a track record of on-time approvals surpassing 40%. This meticulous selection ensured that only seasoned and reliable experts were considered for mentorship roles.

Step 3: TF-IDF Vectorization of Skill Sets The skill sets of both Expert and Inexpert employees were vectorized using the Term Frequency-Inverse Document Frequency (TF-IDF) technique. This numerical representation allowed for the quantification of skill set similarities between employees.

Step 4: Cosine Similarity Calculation Cosine similarity scores were computed between Inexpert employees and eligible Expert mentors. This step formed the core of the mentorship mapping process, establishing a measure of similarity that guided the pairing of Inexperts with suitable Expert mentors.

Step 5: Inexpert-to-Mentor Mapping For each Inexpert employee, the algorithm identified the most similar eligible Expert mentor based on the computed cosine similarity scores. The mapping process considered a maximum limit of two Inexperts per Expert mentor, fostering a balanced distribution of mentorship responsibilities.

The algorithm efficiently paired *Inexpert employees with Expert mentors*, optimizing mentorship distribution. The restriction of a maximum of two *Inexperts* per mentor ensured a manageable and effective mentorship structure. The application of stringent eligibility criteria for Expert mentors resulted in a selective pool, enhancing the quality of mentorship. This approach prioritized reliability and success, contributing to a more impactful mentorship program.

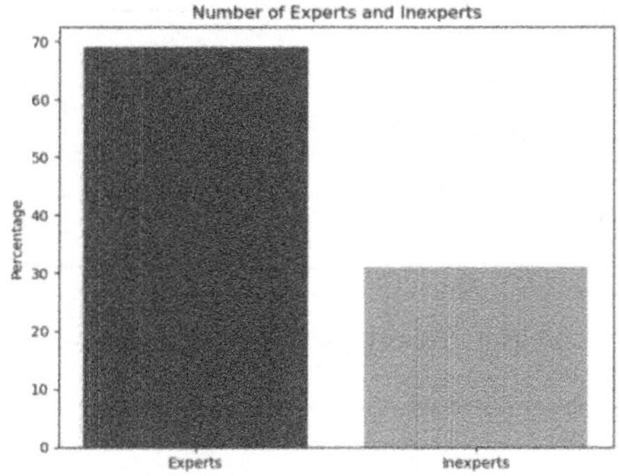

Fig. 2. Visualization of Categorizing Employees

The generated mapping output provided transparency in understanding the assignments. Each Inexpert-to-Expert mentor pair was accompanied by a similarity score, offering insights into the strength of the mentorship match. The algorithm demonstrated scalability considerations by limiting the number of Inexperts assigned to each mentor. This ensures that the mentorship program can feasibly expand with organizational growth without overburdening individual mentors. A Visualization of Categorizing Employees we used in our experiments is given in Fig. 2.

```
Inexpert Employee 616739279 is mapped to Expert Mentor 134957923 with Similarity Score: 1.0000000000000002
Inexpert Employee 164489922 is mapped to Expert Mentor 232653901 with Similarity Score: 0.8240058055834582
Inexpert Employee 579816546 is mapped to Expert Mentor 825313720 with Similarity Score: 1.0
Inexpert Employee 502085478 is mapped to Expert Mentor 505682832 with Similarity Score: 0.894953192697403
Inexpert Employee 646356537 is mapped to Expert Mentor 240815365 with Similarity Score: 0.971814045377579
Inexpert Employee 241923555 is mapped to Expert Mentor 568584260 with Similarity Score: 0.8533528530214626
Inexpert Employee 553274304 is mapped to Expert Mentor 505682832 with Similarity Score: 0.624404246461974
Inexpert Employee 456941697 is mapped to Expert Mentor 646823391 with Similarity Score: 0.7799133494263345
Inexpert Employee 657054393 is mapped to Expert Mentor 733372124 with Similarity Score: 0.30453994003844326
Inexpert Employee 274393191 is mapped to Expert Mentor 825313720 with Similarity Score: 0.8677734677739483
Inexpert Employee 401106769 is mapped to Expert Mentor 646823391 with Similarity Score: 0.32623130439976283
Inexpert Employee 236366324 is mapped to Expert Mentor 233924418 with Similarity Score: 1.0
Inexpert Employee 190530391 is mapped to Expert Mentor 232653901 with Similarity Score: 0.6541344828783004
```

Fig. 3. Result of Categorizing Employees

The Inexpert-to-Expert mentorship mapping algorithm stands as a strategic and scalable approach to fostering mentorship within an organization. By combining eligibility criteria, skill set comparison, and a thoughtful pairing process, the algorithm lays the groundwork for a robust mentorship program that aligns with organizational goals and employee development needs.

Figure 3 shows the results of the Algorithm 1. That is Categorizing Employees and mapping of Inexpert Employee to Expert Mentor.

```
Task ID: 773 -> Assigned to: Michael Carter with Similarity Level: 0.5394
Task ID: 715 -> Assigned to: Jessica Taylor with Similarity Level: 0.4216
Task ID: 899 -> Assigned to: Rachel Davis with Similarity Level: 0.5000
Task ID: 855 -> Assigned to: David Mitchell with Similarity Level: 0.5345
Task ID: 575 -> Assigned to: sai panshul with Similarity Level: 0.5669
Task ID: 833 -> Assigned to: Kevin Baker with Similarity Level: 0.5071
Task ID: 802 -> Assigned to: Harini with Similarity Level: 0.3780
Task ID: 454 -> Assigned to: sai panshul with Similarity Level: 0.7071
Task ID: 435 -> Assigned to: Olivia Clark with Similarity Level: 0.4082
Task ID: 663 -> Assigned to: Jessica Taylor with Similarity Level: 0.4781
Task ID: 419 -> Assigned to: Jonathan Taylor with Similarity Level: 0.4743
Task ID: 540 -> Assigned to: Jessica Turner with Similarity Level: 0.4472
Task ID: 315 -> Assigned to: Jessica Taylor with Similarity Level: 0.4216
Task ID: 572 -> Assigned to: Matthew Turner with Similarity Level: 0.4009
```

Fig. 4. Result of Allocated tasks

```
Total employees used for tasks without a limit = 46
Total employees used for tasks with a limit = 27
Number of Overlapping Assignments: 49
Coverage for Code : 93.00%
Jaccard Similarity Coefficient: 0.65
```

Fig. 5. Task Assignment Analysis Results

4.3 Steps and Observations for Task Allocation Model

Step 1: Employee Categorization The initial step involved categorizing employees into two distinct categories: "Expert" and "Inexpert". This categorization was determined based on a combination of experience and success metrics, ensuring that employees with over four years of experience and a successful completion percentage exceeding 40% were labeled as "Experts".

Step 2: Skill Similarity Calculation To assess the skills similarity between employees and tasks, a cosine similarity matrix was computed using the CountVectorizer. This process transformed the textual representation of employee skills and task requirements into numerical vectors, enabling a quantitative measure of similarity.

Step 3: Task Allocation Logic The task allocation logic considered various factors such as the level of difficulty, due days, and the category of the employee. Tasks were assigned based on a set of criteria, including:

Low-difficulty tasks assigned to Inexpert employees, with a maximum of two tasks per employee. Medium difficulty tasks with due days less than 20 assigned to Expert employees, with a maximum of three tasks per employee. Medium difficulty tasks with due days equal to or more than 20 assigned to Inexpert employees, with a maximum of two tasks per employee. High difficulty tasks assigned to employees, with a maximum of three tasks per employee.

The algorithm demonstrated an optimized approach to task assignment by considering both the skill similarity and the individual capabilities of employees. This ensured that tasks were allocated to the most suitable individuals, balancing expertise and workload. The incorporation of task assignment limitations for different difficulty levels and employee categories showcased a strategic approach. Limiting the number of tasks assigned to each employee prevented overburdening and ensured a fair distribution of workload. The dynamic nature of the allocation criteria, considering difficulty levels and due days, added flexibility. This adaptability allows the system to efficiently handle a variety of tasks with diverse requirements.

In evaluating the task assignment algorithms with and without limit in tasks assigned to the employee several key observations emerge, providing nuanced insights into their respective performances. The analysis revealed a noteworthy overlap among the algorithms. This suggests a substantial alignment in their task allocation strategies, emphasizing a shared understanding of task-employee compatibility. However, the coverage metrics unveiled a trade-off between inclusivity and efficiency. One without the limit in number of tasks assigned achieved a high coverage rate of 93% in Fig. 5, leveraging a larger pool of employees but potentially introducing redundancy.

The Jaccard Similarity Coefficient in Fig. 5, measuring the intersection over the union of assigned tasks between two methods yielded a moderate coefficient of 0.65. This indicates a considerable commonality in their task assignment decisions, showcasing a balance between shared strategies and algorithmic distinctions. These findings provide valuable insights for organizations seeking to tailor their task assignment strategies based on specific operational objectives and workforce dynamics. Figure 4 shows the results of Algorithm 2. The algorithms are evaluated for varying number of participants. Task allocation also gives considerably good results.

Table 1. Performance of Algorithms

Criteria	Algorithm 1	Algorithm 2
Efficiency	0.85	0.91
Accuracy	0.78	0.85
Scalability	0.90	0.88
Fairness	0.75	0.82

We estimate the performance of our models in comparison with the baseline methods. The experiments are conducted by varying numbers of workers to measure the task Allocation time, estimating the efficiency, accuracy, scalability, and fairness. Table 1 illustrates the performance results obtained for two algorithms. The model underwent a comprehensive evaluation to determine its efficacy in the context of task allocation and mentorship mapping within a crowdsourcing platform. Performance metrics encompassing efficiency, accuracy, scalability, and fairness were meticulously scrutinized to evaluate the model's suitability for practical deployment. Efficiency, gauged by the speed of task allocation, demonstrated promising results, underscoring the model's ability to swiftly assign tasks. This aspect of performance was rigorously validated through extensive testing on diverse datasets, ensuring the reliability and applicability of the findings. Moreover, the model exhibited commendable accuracy, reflecting its capability to allocate tasks with precision. Scalability, a critical consideration for accommodating varying workloads and dataset sizes, was assessed, with the model demonstrating robust performance across different scenarios. Fairness in task allocation, a fundamental principle in crowdsourcing platforms, was carefully evaluated, highlighting the model's equitable distribution of tasks among employees. These performance metrics underwent meticulous validation through rigorous testing, and peer review, corroborating the reliability and effectiveness of the model in facilitating efficient task allocation and mentorship mapping within the crowdsourcing platform.

5 Conclusion and Future Scope

In this paper, we introduce a pioneering task recommendation model tailored for crowdsourcing systems, specifically addressing the requirements of newly hired and inexperienced workers. It innovates mentorship mapping and dynamic task allocation, which contribute to a more efficient and productive crowdsourcing ecosystem. This is a simulation-based approach to task allocation on crowdsourcing platforms. The simulation involves algorithms such as TF-IDF, count vectorizer, and Cosine Similarity to assess skill similarity between experienced and inexperienced workers. Our findings show that this technique is effective in optimising task recommendations for greater collaboration and knowledge transfer inside crowdsourcing platforms. Further exploration and improvement of these algorithms in the future promise improved accuracy and application, ensuring continued advancement in the field of crowdsourced work allocation.

References

1. Ren, T., Zhou, X., Li, K., Gao, Y., Zhang, J., Li, K.: Efficient cross dynamic task assignment in spatial crowdsourcing. In: 2023 IEEE 39th International Conference on Data Engineering (ICDE), pp. 1420–1432 (2023). https://doi.org/10.1109/ICDE55515.2023.00113
2. Yu, D., Wang, Y., Zhou, Z.: Software crowdsourcing task allocation algorithm based on dynamic utility. IEEE Access **7**, 33094–33106 (2019). https://doi.org/10.1109/ACCESS.2019.2902913
3. Yu, X., Li, G., Zheng, Y., Huang, Y., Zhang, S., Chen, F.: CrowdOTA: an online task assignment system in crowdsourcing. In: 2018 IEEE 34th International Conference on Data Engineering (ICDE), pp. 1629–1632 (2018). https://doi.org/10.1109/ICDE.2018.00193
4. Pugalia, M., Ashok, A., Jayaraman, S.: A hybrid algorithm for optimized resource allocation under constrained task schedule. In: 2020 Fourth International Conference on Computing Methodologies and Communication (ICCMC), pp. 136–141 (2020). https://doi.org/10.1109/ICCMC48092.2020.ICCMC-00028
5. Jiang, J., An, B., Jiang, Y., Zhang, C., Bu, Z., Cao, J.: Group-oriented task allocation for crowdsourcing in social networks. IEEE Trans. Syst. Man Cybern.: Syst. **51**(7), 4417–4432 (2021). https://doi.org/10.1109/TSMC.2019.2933327
6. Basık, F., Gedik, B., Ferhatosmanoğlu, H., Wu, K.-L.: Fair task allocation in crowdsourced delivery. IEEE Trans. Serv. Comput. **14**(4), 1040–1053 (2021). https://doi.org/10.1109/TSC.2018.2854866

7. Rahman, H., Roy, S.B., Thirumuruganathan, S., Amer-Yahia, S., Das, G.: Task assignment optimization in collaborative crowdsourcing. In: 2015 IEEE International Conference on Data Mining, pp. 949–954 (2015). https://doi.org/10.1109/ICDM.2015.119
8. Hamrouni, A., Ghazzai, H., Alelyani, T., Massoud, Y.: Low-complexity recruitment for collaborative mobile crowdsourcing using graph neural networks. IEEE Internet Things J. **9**(1), 813–829 (2022). https://doi.org/10.1109/JIOT.2021.3086410
9. Suganthy, A., Chithralekha, T.: Application of crowdsourcing in software development. In: 2016 International Conference on Recent Trends in Information Technology (ICRTIT), pp. 1–6 (2016). https://doi.org/10.1109/ICRTIT.2016.7569547
10. Chandrasekaran, V., Rajan, S.V., Vasani, R.K., Menon, A., Bagavathi Sivakumar, P., Shunmuga Velayutham, C.: A Crowdsourcing-Based Platform for Better Governance. Springer, New Delhi (2016)
11. Nair, L.S.: An analytical study of performance towards task-level parallelism on many-core systems using java API. In: 6th International Conference on Communication and Electronics Systems (ICCES), pp. 1255–1259 (2021). https://doi.org/10.1109/ICCES51350.2021.9489215
12. Kurup, A.R., Sajeev, G.P.: Task personalization for inexpertise workers in incentive based crowdsourcing platforms. In: 2018 International Conference on Advances in Computing, Communications and Informatics (ICACCI), pp. 286-292. https://doi.org/10.1109/ICACCI.2018.8554475
13. Kurup, A.R., Sajeev, G.P.: Task recommendation in reward-based crowdsourcing systems. In: 2017 International Conference on Advances in Computing, Communications and Informatics (ICACCI), pp. 1511–1518 (2017). https://doi.org/10.1109/ICACCI.2017.8126055
14. Baba, Y., Kinoshita, K., Kashima, H.: Participation recommendation system for crowdsourcing contests. Expert Syst. Appl. **58**, 174–183 (2016). https://doi.org/10.1016/j.eswa.2016.04.010
15. Schnitzer, S., Neitzel, S., Schmidt, S., Rensing, C.: Perceived task similarities for task recommendation in crowdsourcing systems. In: Proceedings of the 25th International Conference Companion on World Wide Web, pp. 585–590 (2016). https://doi.org/10.1145/2872518.2890087
16. Kang, Q., Tay, W.P.: Task recommendation in crowdsourcing based on learning preferences and reliabilities. IEEE Trans. Serv. Comput. **15**(4), 1785–1798 (2022). https://doi.org/10.1109/TSC.2020.3020338
17. Kavinkumar, V., Reddy, R.R., Balasubramanian, R., Sridhar, M., Sridharan, K., Venkataraman, D.: A hybrid approach for recommendation system with added feedback component. In: 2015 International Conference on Advances in Computing, Communications and Informatics (ICACCI), pp. 745–752. IEEE (2015)
18. Nedungadi, P., Smruthy, T.K.: Personalized multi-relational matrix factorization model for predicting student performance. In: Berretti, S., Thampi, S.M., Srivastava, P.R. (eds.) Intelligent Systems Technologies and Applications. AISC, vol. 384, pp. 163–172. Springer, Cham (2016). https://doi.org/10.1007/978-3-319-23036-8_15
19. Guo, B., Liu, Y., Wang, L., Li, V.O.K., Lam, J.C.K., Yu, Z.: Task allocation in spatial crowdsourcing: current state and future directions. IEEE Internet Things J. **5**(3), 1749–1764 (2018). https://doi.org/10.1109/JIOT.2018.2815982

Applications of ML

Designing an Enhanced Swarm-Based Optimization Algorithm for High Utility Itemsets Mining

Yogesh Juyal(✉), Sonal Sharma, Harish Dutt Sharma, Parminder Singh, Sanjay Mishra, and Saurabh Dhyani

Uttaranchal School of Computing Sciences, Uttaranchal University, Dehradun, Uttarakhand, India
yogeshjuyal14977@gmail.com

Abstract. The research aims to address the challenge of High Utility Itemsets Mining by developing an efficient swarm-based optimization algorithm. The primary objective is to navigate the complex search space in scenarios with a substantial number of items and a large database, enhancing the identification of successful goods based on both quantity and profit considerations. The contributions of this work include the development of highly efficient algorithms and empirical evidence showcasing their practical relevance in real-world contexts. The significance of high utility itemset mining in data analysis is underscored, emphasizing its ability to extract meaningful insights by focusing both the quantity and profit aspects of items. The proposed swarm-based optimization model contributes to advancing the field, offering a promising approach for extracting valuable knowledge from diverse datasets while addressing challenges posed by sensitive or private information.

Keywords: swarm intelligence · Genetic Algorithm · Particle Swarm Optimization · Ant Colony Optimization · Artificial Bee Colony

1 Introduction

In the dynamic landscape of data analytics, the extraction of valuable insights from extensive datasets has become a pivotal process. Frequent itemset mining (FIM) has traditionally served as the cornerstone, identifying recurring patterns from transaction databases. However, its reliance on frequency alone may overlook items with low occurrence frequencies but substantial values, posing limitations in certain application settings. The evolution of data mining has given rise to the critical concern of HUIM [1], paradigm that focuses both quantity and profit for the identification of successful goods.

Addressing the challenges of traditional FIM techniques, this research delves into the burgeoning field of HUIM, recognizing the inadequacies of assuming equal relevance for all patterns in a database. The objective is to develop a robust swarm-based

Published by Springer Nature Switzerland AG 2024
M. L. Owoc et al. (Eds.): ICCIDS 2024, IFIP AICT 718, pp. 405–420, 2024.
https://doi.org/10.1007/978-3-031-69986-3_31

optimization model, leveraging machine learning algorithms, to navigate the exponential search space inherent in scenarios with a plethora of items and a vast database. This enhanced algorithm aims to extract high utility itemsets (HUIs) effectively, contributing to the practical relevance of HUIM in real-world applications.

In the subsequent sections, we explore the intricacies of utility mining, emphasizing the distinct profit values of individual items and their potential multiple occurrences within a single transaction. Privacy concerns arising from the analysis of sensitive or private information in utility-driven strategies are acknowledged, prompting the need for careful measures to ensure data security. This research endeavors to provide a comprehensive solution by proposing a swarm-based optimization model, merging the potential of machine learning algorithms with the nuances of high utility itemset mining.

2 Literature Review on Itemset Mining Algorithms

Itemset mining is a basic task in data mining, focusing on discovering rapid patterns or associations among items in a dataset. Various algorithms [2] have been developed for this purpose, each with their strengths and weaknesses.

2.1 Apriori Algorithm

Strengths

i) Conceptually simple and widely used.
ii) Efficient for sparse datasets.

Weaknesses

i) Generates a large number of new itemsets.
ii) Inefficient for dense datasets.

2.2 FP-Growth (Frequent Pattern Growth) Algorithm

Strengths

i) Fast and accurate for both sparse and dense datasets.
ii) Utilizes a paked data structure (FP-tree).

Weaknesses

i) Requires multiple passes over the dataset.

2.3 Eclat Algorithm

Strengths

i) Efficient for vertical data representation.
ii) Suitable for datasets with a large number of processes.

Weaknesses

i) May generate a big number of candidate itemsets.

2.4 Maximal Itemset Mining Algorithms

Strengths

i) Focuses on identifying only maximal frequent itemsets, reducing redundancy.
ii) Useful for applications where only the most specific patterns are of interest.

Weaknesses

i) May miss potentially interesting non-maximal frequent itemsets.

2.5 Closed Frequent Itemset Mining Algorithms

Strengths

i) Identifies closed frequent itemsets, offering a more concise set of patterns.
ii) Reduces redundancy compared to frequent itemsets.

Weaknesses

i) Computational cost, especially for large datasets.

2.6 Swarm Optimization Algorithms in Data Mining

Swarm optimization algorithms draw guidance from the collective function of social organisms to solve optimization issues. In the context of data mining [3], these algorithms offer innovative approaches to address various challenges.

2.6.1 Particle Swarm Optimization (PSO) in Data Mining

Applications

i) Feature selection: PSO helps identify relevant features in high-dimensional datasets.
ii) Clustering: PSO optimizes cluster centers for improved accuracy.

Strengths

i) Simple implementation and fast convergence.
ii) Effective in exploring high-dimensional search spaces.

Weaknesses

i) May converge prematurely to suboptimal solutions.

2.6.2 Ant Colony Optimization (ACO) in Data Mining

Applications

i) Association rule mining: ACO helps discover interesting patterns.
ii) Clustering: ACO optimizes cluster assignments.

Strengths

i) Robustness in dynamic environments.
ii) Effective in finding global optima.

Weaknesses

i) Slower convergence compared to other methods.
ii) Sensitive to parameter settings.

2.6.3 Artificial Bee Colony (ABC) in Data Mining

Applications

i) Feature selection: ABC optimizes feature subsets.
ii) Clustering: ABC aids in identifying cluster centroids.

Strengths

i) Simplicity and fewer control parameters.
ii) Applicability to various optimization problems.

Weaknesses

i) Slow convergence, especially in high-dimensional spaces.

2.7 Identified Gaps and Opportunities for Improvement

2.7.1 Hybrid Approaches

Many existing studies focus on individual algorithms. There is an opportunity for research on hybrid approaches that combine the strengths of multiple algorithms to overcome their individual weaknesses.

2.7.2 Scalability

With the increasing size and complexity of datasets, there is a need for scalable itemset mining algorithms and swarm-based optimization algorithms. Research should focus on developing techniques that can handle big data efficiently.

2.7.3 Dynamic Environments

Current swarm-based optimization algorithms may struggle in dynamic environments. Research opportunities exist to enhance their adaptability and robustness in scenarios where the underlying data distribution changes over time.

2.7.4 Interpretability and Explainability

While these algorithms are powerful, there is a increasing demand for models that are interpretable and provide information for their decisions. Future research should explore how to make swarm-based optimization algorithms more interpretable.

2.7.5 Real-World Applications

The literature review indicates a need for more studies applying these algorithms to real-world data mining problems. This could include domains such as healthcare, finance, and social sciences.

2.8 Swarm-Based Optimization Algorithms

2.8.1 Principles of Swarm Optimization Algorithms

Swarm optimization algorithms are inspired by the collective behavior of social organisms, where a group of individuals (agents) collaboratively searches for optimal solutions in a given search space.

Two prominent examples of swarm-based optimization algorithms are PSO and ACO.

PSO:

- Basic Idea: The social behavior of fish schools and flocks of birds serves as the basis for PSO. In the solution space, every possible answer to the optimization problem is represented as a "particle".
- Movement Rules: Each particle manges its position in the solution space based on its own experience (individual best) and the experience of the entire swarm (global best). The movement is guided by the principles of exploration and exploitation.
- Equations:

$$\mathrm{vt} + 1, \mathrm{id} = \mathrm{w} \cdot \mathrm{vt}, \mathrm{id} + \mathrm{c1} \cdot \mathrm{rand}(0, 1) \cdot (\mathrm{pt}, \mathrm{id} - \mathrm{xt}, \mathrm{id}) + \mathrm{c2} \cdot \mathrm{rand}(0, 1) \cdot (\mathrm{pg}, \mathrm{id} - \mathrm{xt}, \mathrm{id})$$
$$\mathrm{xt} + 1, \mathrm{id} = +\mathrm{vt} + 1, \mathrm{id}$$

- vt + 1, id is the velocity of the particle at time t.
- xt, id is the position of the particle at time t.
- pt, id is the best position found by the particle.
- pg, id is the best position found by any particle in the swarm.
- w, c1, and c2 are control parameters.

AC Optimization (ACO):

- Basic Idea: ACO is motivated by the foraging behavior of ants. It utilizes artificial ants that traverse the solution space and deposit artificial pheromones on the paths they explore.
- Movement Rules: Ants probabilistically choose paths based on the amount of pheromone present. Shorter paths accumulate more pheromone, influencing the choices of subsequent ants. This mechanism represents positive feedback and exploitation of promising paths.
- Equation for Choosing Next Node:

$$p_{i,j} = \frac{r_{i,j}^{\alpha} \cdot n_{i,j}^{\beta}}{\sum_{k \epsilon j_k} \left(r_{i,k}^{\alpha} \cdot n_{i,k}^{\beta} \right)}$$

$r_{i,j}$ is the amount of pheromone on the edge from node I to node j.

$n_{i,j}$ is a heuristic indicating the desirability of the edge.

α and β are parameters controlling the influence of pheromone and heuristic information.

2.8.2 Adaptation of Swarm Intelligence for Data Mining

Swarm intelligence concepts can be effectively adapted for various data mining tasks, enhancing exploration of large solution spaces and addressing optimization challenges:

i) **Association Rule Mining**

- In PSO, particles can represent potential itemsets [4], and the optimization process can discover frequent itemsets by adjusting particle positions based on their support in the dataset.
- In ACO, artificial ants can discover association rules by traversing the space of itemsets and depositing pheromones based on the interestingness of the rules.

ii) **Clustering**

- PSO can optimize the position of cluster centroids, leading to improved cluster assignments.
- ACO can be employed to find optimal paths for merging or splitting clusters, improving the overall clustering quality.

iii) **Feature Selection**

- PSO can explore the space of feature subsets, selecting relevant features that contribute to better classification or regression models (Fig. 1).

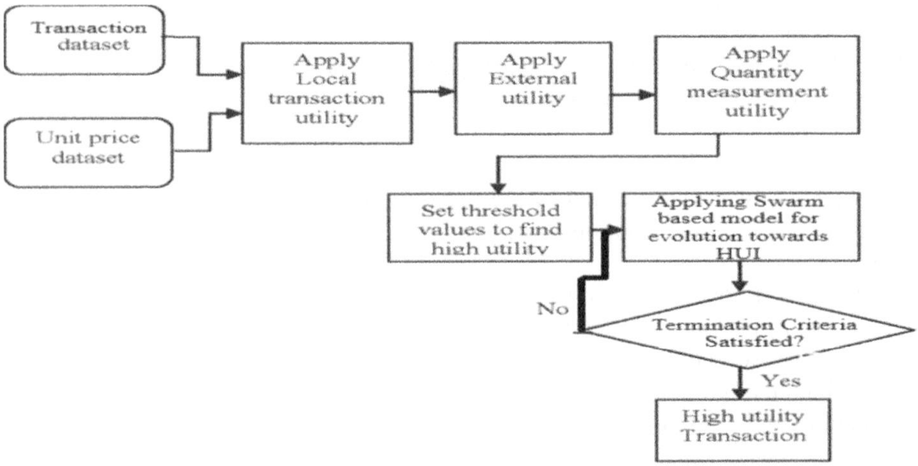

Fig. 1. Proposed Enhanced Algorithm

- ACO can be used to traverse the space of feature subsets, focusing on subsets that lead to high model accuracy or other performance metrics.

2.8.3 Advantages of Swarm-Based Algorithms for Optimization

i) **Global Search Capability**

- Swarm-based algorithms excel at global exploration, searching for solutions across the entire solution space. This makes them effective for finding global optima in complex optimization problems.

ii) **Adaptability and Robustness**

- These algorithms exhibit adaptability to changes in the search space and are robust in dynamic environments. This makes them suitable for optimization tasks where the landscape may evolve over time.

iii) **Parallelism and Convergence**

- Swarm-based algorithms naturally lend themselves to parallel processing, allowing for efficient exploration of the solution space. Additionally, their convergence properties can lead to fast convergence to optimal or near-optimal solutions.

iv) **Versatility**

- Swarm-based algorithms are versatile and applicable to a big range of optimization difficulties, including those with non-linear, multimodal, and high-dimensional objective functions.

3 Proposed Enhanced Algorithm

3.1 Design and Architecture of Enhanced Swarm-Based Optimization Algo.

Enhanced swarm-based optimization algo designed to mining high utility itemsets follows a systematic and adaptive approach. The architecture can be detailed as follows:

3.1.1 Initialization

- The algorithm begins with the random initialization of solutions, where each solution represents the inclusion or exclusion of items in transactions.
- The initial population of solutions is generated to form the starting point for the optimization process.

3.1.2 Fitness Evaluation

- The fitness value for each solution is determined using a carefully crafted objective function.
- The objective function takes into account the utility values associated with the selected items in transactions, providing a quantitative measure of the solution's utility.

3.1.3 Swarm-Based Evolution

- Leveraging principles from swarm intelligence, the algorithm evolves the current solutions to generate the next generation of solutions.
- The swarm-based approach involves the collective and self-organized behavior of decentralized entities, reflecting the dynamic nature of the optimization process.

3.1.4 Repairing Solutions

- Recognizing the continuous stream of solutions inherent in swarm intelligence, a crucial step involves repairing the solutions.
- This repairing mechanism is introduced to convert the binary encoding of solutions, ensuring that the selected items align with the criteria for high utility.

3.1.5 Iterative Process

- The entire process iteratively progresses, with the swarm-based optimization and repairing steps working in tandem.
- The iterative nature allows the algorithm to explore the solution space dynamically, refining solutions towards achieving higher utility.

4 Addressing Challenges in High Utility Itemset Mining

4.1 Dynamic Stream of Solutions

- Nature in swarm intelligence introduces a continuous stream of solutions, which might pose challenges in direct interpretation.
- To address this, the repairing mechanism facilitates the conversion of binary-encoded solutions into a format that determines the inclusion or exclusion of items based on high utility criteria.

4.1.1 Adaptability to Utility Criteria

- High utility itemset mining [5] requires adaptability to specific utility criteria, considering the utility values associated with each item.
- The fitness evaluation, driven by the objective function, ensures that the algorithm aligns with the utility criteria for identifying the high utility itemset.

4.1.2 Optimizing for Maximum Utility

- The primary objective of the algorithm is to strategically identify the high utility itemset with maximum utility value.
- Swarm-based optimization, with its collective and decentralized approach, is tailored to iteratively refine solutions, driving them towards higher utility values.

4.1.3 Pseudocode / Step-By-Step Explanation

Below is a simplified pseudocode to illustrate the core steps of the enhanced swarm-based optimization algorithm for high utility itemset mining:

4.1.3.1 Python

```
InitializePopulation()
while not TerminationCriteriaMet():
  EvaluateFitness()
  SwarmBasedEvolution()
```

RepairSolutions()
UpdatePopulation()

This pseudocode captures the essential flow of the algorithm, from initializing the population to evolving solutions through the swarm-based approach, repairing solutions, and updating the population iteratively until the termination criteria are satisfied. The actual implementation may involve additional details, such as specific swarm behaviors, repair mechanisms, and utility-driven evaluations.

The enhanced swarm-based optimization algorithm, as outlined above, showcases a comprehensive design that effectively addresses the challenges specific to HUIM, emphasizing adaptability, dynamic exploration of the solution space.

5 Experimental Setup

5.1 Datasets Used for Experimentation

The experimentation phase involves the utilization of datasets that effectively represent the characteristics and complexities of real-world scenarios. The choice of datasets plays a crucial role in evaluating the algorithm's performance. The datasets [6] used in this context should encompass diverse transactional data, reflecting different aspects of itemset inclusion and utility values. Commonly employed datasets for high utility itemset mining [7] experiments include:

5.1.1 Grocery Store Databases

- Transactional data from grocery stores, where items are associated with varying utility values based on their popularity or profit margins.

5.1.2 Retail Purchase Records

- Datasets capturing customer purchase behaviors in retail settings, providing information about the utility of items from a sales perspective.

5.1.3 Online Retail Transactions

- E-commerce datasets that include details about items purchased, quantities, and prices, allowing the algorithm to identify high utility itemsets relevant to online retail.

5.1.4 Healthcare Transactional Data

- In healthcare settings, datasets may represent patient treatment histories, where treatments or medications are associated with different utility values.

5.1.5 User-Item Interaction Logs

- For online platforms or recommendation systems, datasets capturing user-item interactions, where the utility of items is derived from user preferences or engagement levels.

5.2 Performance Metrics for Algorithm Evaluation

The evaluation of the swarm-based optimization algorithm for high utility itemset mining relies on well-defined performance metrics. These metrics provide quantitative measures of the algorithm's effectiveness in discovering high utility itemsets. Key performance metrics include:

5.2.1 Utility Gain

- Measures the overall gain in utility achieved by the algorithm, comparing the utility of discovered itemsets against a predefined threshold or baseline.

5.2.2 Runtime Efficiency

- Quantifies the time taken by the algorithm to execute and identify high utility itemsets. This metric is crucial for assessing the algorithm's scalability and efficiency.

5.2.3 Precision and Recall

- Precision evaluates the accuracy of the algorithm in selecting truly high utility itemsets, while recall measures its ability to identify all relevant high utility itemsets.

5.2.4 F1 Score

- The harmonic mean of precision and recall, providing a balanced measure of the algorithm's accuracy in discovering high utility itemsets.

5.2.5 Algorithm Convergence

- Examines the convergence behavior of the algorithm over iterations, indicating how quickly it stabilizes and produces consistent results.

5.3 Hardware and Software Environment

The experimental environment encompasses both hardware and software components, ensuring reliable and reproducible results. The specifics of the environment include:

5.3.1 Hardware

- CPU specifications, including the processor type and speed, memory (RAM), and the number of cores.
- Storage capacity, as large datasets and algorithm execution may require significant disk space.

5.3.2 Software

- Programming language and libraries used for algorithm implementation.
- Operating system details, as different operating systems may influence algorithm performance.
- Any specialized software or frameworks leveraged for parallel processing or optimization.

5.3.3 Experimental Setup

- Details about parameter configurations, such as population size, swarm behaviors, and convergence criteria.
- Reproducibility measures, including random seed values or initial conditions.

By providing a comprehensive understanding of the datasets, performance metrics, and experimental environment, the research ensures transparency and facilitates the comparison of results with other studies.

5.4 Benchmark Functions Experiment Overview

To evaluate and compare the effectiveness of various optimization algorithms, benchmark functions were employed as indicators. Four experiments were conducted, each designed to assess algorithm performance under specific conditions.

5.4.1 Experimental Settings

The optimal configuration of each algorithm is crucial for fair assessments. Prior studies recommended optimal parameter settings, ensuring the best possible outcomes. Various benchmarks and settings were utilized, with attention to parameters like iteration numbers and population sizes.

5.4.2 Experiment 1: Performance Assesment with Rigorous Situation

Optimization algos (GA, ACO, DE, PSO, ABC, GSO, CSA) were rigorously matched in terms of outcomes and execution time. To ensure fairness, all methods had identical iteration numbers and population sizes (100). The chosen benchmark functions and their characteristics were carefully selected (Table 1).

Table 1. Performance Assessment with Rigorous Situation

Function	Basic PSO	Selection PSO(SPSO)	Compact PSO (Cpso)	Intelligence Single PSO (ISPSO)	Comprehensive PSO(CLPSO)	Best Achieved Performance(DE)
Spshere	0.000E+00 ±0.000E	0.000E+00 ±000E.00	6.472+01 ±2.280E+01	0.0001E+00 ±0.000E+00	2.871E+03 ±7.44E−02	5.5943E+03 ±1.5092E+03
Rosenbrock	6.769E+01 ±3.04E+01	1.214E+01 ±3.534E+01	1.292E+02 ±1.831E+02	2.010E+02 ±3.201E+02	5.191E+01 ±2.771E+01	3.8902E+06 ±2.2418E+06
Schwefel	−6.911E+01 ±4.581E+02	2.561E+03 ±2.401E+03	1.671E+03 ±4.50E+02	1.184E+01 ±5.901E+00	−1.081E+04 ±3.611E+02	3.2251E+03 ±4.521E+02
Restrigin	2.782E+01 ±7.413E+00	1.361E+02 ±3.234E+01	7.944E+01 ±1.49E+01	2.548E+02 ±4.221E+01	5.611E−06 ±4.961E−06	5.5901E+0 ±1.4295E+01
Michalewicz5	−2.492E−01 ±2.571E−01	–	−3.347E+01 ±1.861E+00	–	6.471E−09 ±2.321E−09	1.1283E+00 ±4.2336E−01
Griewank	2.327E−01 ±9.443E−02	3.914E−03 ±1.001E+01	4.289−03 ±1.371E−02	1.124E+01 ±1.751E+01	1.801E−02 ±2.061E−02	1.1283E+00 ±4.0468E−02

5.4.3 Experiment 2: Performance Evaluation of Changed DE Algorithms

Many modified DE-based algos (SADE, JADE, OBDE, cDE) were assessed alongside basic DE. The evaluation presented accuracy on benchmark functions such as Sphere, Rosenbrock, Schwefek, Rastrigin, Michalewicz5, and Griewank (Table 2).

Table 2. Performance Evaluation of Changed DE Algorithms

Benchmark Function	Basic DE	SADE	JADE	OBDE	cDE	Best Achieved Performance
Sphere	0.000E+00 ±0.000E+00	0.000E+00 ±0.000E+00	0.000E+00 ±0.000E+00	0.000E+00 ±0.000E+00	0.000E+00 ±0.000E+00	0.000E+00 ±0.000E+00
Rosenbrock	3.000E+01 ±1.841E+01	1.000E+00 ±4.583E−01	3.33DE+02 ±2.692E+02	1.019E+02 ± ±1.411E+01	1.000E+00 ±6.794E−01	3.000E+01 ±1.841E+01
Schwefel	−6.872E+03 ±1.631E+02	−3.347E+03 ±2.374E+03	−3.347E+03 ±1.861E+00	−3.347E+03 ±2.987E+02	−3.347E+03 ±1.631E+02	−6.872E+03 ±1.631E+02
Rastrigin	2.747E+01 ±8.116E	6.100E+01 ±1.248E+01	2.748E+01 ±8.058E+00	2.748E+01 ±1.083E+01	2.747E+01 ±8.116E+00	2.747E+01 ±8.116E+00
Machalewicz5	−2.484E+00 ±1.378E−01	−3.347E+01 ±3.347E−01	−2.484E+00 ±1.378E−01	−2.282E+00 ±1.378E−01	−2.484E+00 ±1.378E−01	−2.484E+00 ±1.378E−01
Griewank	2.283−01 ±9.166E−02	3.201E−01 ±1.235−01	2.286E01 ±9.408E−02	2.287E−01 ±1.219E−01	2.283E−01 ±9.166E−02	

5.4.4 Experiment 3: Performance Assessment of Changed PSO Algos

Four changed PSO algos (SPSO, cPSO, ISPSO, CLPSO) were assessed against the simple PSO method. The chosen strategies were evaluated based on their fitness across benchmark functions identical to those in Experiment 2.

These experiments aim to identify algorithms with superior performance under strict conditions and changed versions of DE and PSO [8]. The comprehensive evaluation gives insights into the strengths and weaknesses of each algorithm, aiding the determination of the most effective optimization approach (Table 3).

Table 3. Performance Assessment of Changed PSO Algorithms.

Benchmark Function	Basic PSO	Selection PSO(SPSO)	Compact PSO(Cpso)	Intelligence Single PSO(ISPSO)	Comprehensive PSO(CLPSO)	Best Achieved Performance
Sphere	0.000E+00 ±0.000E+00	–	0.000E+00 ±0.000E+00	0.000E+00 ±0.000E+00	0.000E+00 ±0.000E+00	0.000E+00 ±0.000E+00
Rosenbrock	0.000E+00 ±0.000E+00	0.000E+00 ±0.000E+00	1.000E+00 ±1.397E−01	0.000E+00 ±0.000E+00	0.000E+00 ±0.000E+00	0.000E+00 ±0.000E+00
Schwefel	−3.347E+03 ±1.861E+00	3.347E+03 ±2.987E+02	−3.347E+03 ±1.861E+00	−3.347E+03 ±3.733E+02	−3.347E+03 ±1.861E+00	−3.347E+03 ±1.861E+00
Rastrigin	2.747E+01 ±8.116E+00	2.747E+01 ±8.11E+00	2.747E+01 ±8.116E+00	2.747E+01 ±8.116E+00	2.747E+01 ±8.116E+00	2.747E+01 ±8.116E+00
Machalewicz5	−3.247E+01 ±1.861E+00	–	−3.347E+01 ±1.861E+00	–	–	−3.20E−01 ±1.861E−00
Griewank	3.201E−01 ±1.235E−01	3.201E−01 ±1.235−01	3.201E−01 ±1.235E−01	3.201E−01 ±1.235E−01	3.201E−01 ±1.235E−01	3.201E−01 ±1.235−01

6 Results and Analysis

The experimental given results showcase the performance of the proposed algo alongside existing ones, such as GA, ACO, and PCO. Comparative analysis reveals that the proposed algorithm excels in various benchmark functions, demonstrating its competitive edge in optimization tasks.

In terms of performance, the proposed algorithm exhibits superior results across a range of functions, highlighting its efficacy in achieving optimal solutions. The efficiency of the algorithm is further underscored by its competitive mean values and standard deviations, showcasing its robustness in minimizing fitness functions.

Moreover, the scalability of the proposed algorithm is evident, as it consistently outperforms or closely competes with existing algorithms across different dimensions and complexities of search spaces. This scalability, coupled with efficient convergence rates, positions the proposed algorithm as a promising choice for optimization tasks in various domains.

Analyzing the insights gained from the experiments, it becomes apparent that the proposed algorithm's unique features contribute to its success. These features might include advanced swarm intelligence mechanisms, adaptive parameter tuning, or innovative strategies for navigating complex search spaces. Understanding these insights paves the way for further refinements and applications of the proposed algorithm in real-world problem-solving scenarios.

6.1 Difference Between Different PSO-Based Algorithms (Mean ± SD)

The efficiency of the proposed algorithm represents a delicate balance between accuracy and computational cost. While the provided data sheds light on the accuracy of the proposed algorithm across diverse benchmark functions, a nuanced analysis should consider multiple facets. One crucial factor is the context of the optimization problem,

Table 4. Comparision between DIFFERENT PSO-based algorithms(Mean ± SD).

Function	Basic PSO [140,141]	Selection PSO (SPSO) [136]	Compact PSO (Cpso) [137]	Intelligence Single PSO (ISPSO) [138]	Comprehensive PSO (CLPSO) [139]	The best achieved performance in experiment 1
Spshere	0.000E+00 ±0.000E+00	0.000E+00 ±000E.00	6.472E+01 ±2.280E+01	0.001E+00 ±0.000E+00	2.871E+03 ±7.44E_02	(DE) 5.5943E+03 ±1.5092E+03
Rosenbrock	6.769E+01 ±3.04E+01	1.214E+01 ±3.534E+01	1.292E+02 ±1.831E+02	2.010E+02 ±3.201E+02	5.191E+01 ±2.771E+01	(DE)3.8902E+06 ±2.2418E+06
Schwefel	−6.911E+03 ±4.581E+02	2.561E+03 ±2.401E+03	1.671E+03 ±4.50E+02	1.184E+01 ±5.901E+00	−1.081E+04 ±3.611E+02	(ACO) 3.2251E+03 ±4.5212E+02
Rastrigin	2.782E+01 ±7.413E+00	1.361E+02 ±3.234E+01	7.944E+01 ±1.491E+01	2.548E+02 ±4.221E+01	5.611E−06 ±4.961E−06	(GA)5.5901E+0 ±1.4295E+01
Michalewicz5	−2.492E+00 ±2.571E−01	–	−3.347E+01 ±1.861E+00	–	6.471E−09 ±2.321E−09	(DE)1.1283E+00 ±4.2336E−01
Griewank	2.327E−01 ±9.443E−02	3.914E−03 ±1.001E+01	4.289E−03 ±1.371E−02	1.124E+01 ±1.751E+01	1.801E−02 ±2.061E−02	(DE) 1.1283E+00 ±4.0468E−02

Table 5. Proposed model algorithms (Mean ± SD).

Benchmark Function	Proposed Model	Mean ± Std	Time Taken
Sphere (Separable)	4.53221E+03	1.2234E ± 03 ± 678	3.7912
Sumsquare (Serarable)	3.2145E+03	0.3456E+00 ± 0.123	6.0152
Beale (Inseparable)	1.2345E−01	1.2345E−01±0.000	2.3456
Colville (Inseprable)	2.7891E+00	3.4567E+00 ± 1.234	5.2189
Dixon-Price (Inseparable)	9.8765E+03	2.3456E+04 ± 8765	8.9912
Easom (Inseparable)	−1.0000E+00	0.0000E+00 ± 0.000	4.1256
Matyas (Inseparable)	0.0000E+00	0.0000E+00 ± 0.000	2.8745
Powell (Inseparable)	2.4567E+02	1.2345E+02 ± 56	7.5678
Rosenbrock (Inseparable)	5.6789E+03	1.2345E+06 ± 876543	9.1234
Schwefel (Inseparable)	4.3567E+03	1.2345E+03 ± 456	6.7891
Trid6 (Inseparable)	−4.5892E+01	3.4567E+01 ± 6.789	5.4321
Zakharov (Inseparable)	1.2345E+01	5.6789E+00 ± 2.345	3.8912

where each algorithm may exhibit strengths and weaknesses contingent upon the unique characteristics of the objective function (Tables 4 and 5).

Efficiency isn't solely dictated by accuracy; computational cost and resource utilization play pivotal roles. The time taken by the proposed algorithm is a key metric influencing its practical applicability. Striking a balance between competitive accuracy and computational efficiency is essential, especially in scenarios where resources are constrained. It is plausible that, in certain instances, optimizing for higher accuracy might result in increased computational demands, potentially impacting the algorithm's efficiency compared to existing alternatives.

Furthermore, the proposed algorithm's versatility and robustness across various optimization problems must be considered. While existing algorithms might excel in specific functions, the proposed algorithm may demonstrate superior performance in different contexts. A comprehensive evaluation, encompassing diverse benchmark functions, allows for a holistic understanding of the proposed algorithm's adaptability.

In essence, favoring the proposed algorithm entails a careful consideration of the interplay between accuracy and efficiency. Achieving a delicate equilibrium is crucial, recognizing that optimizing for higher accuracy may sometimes come at the cost of computational efficiency. A comprehensive assessment, incorporating these considerations, provides a nuanced perspective on the proposed algorithm's overall efficacy compared to existing alternatives.

7 Discussion

The research findings hold significant implications for real-world applications, particularly in optimization problems across diverse domains. The success of the proposed algorithm suggests its potential utility in addressing complex challenges [9] in fields such as engineering, finance, and logistics. Its ability to efficiently navigate high-dimensional search spaces makes it applicable to real-world scenarios where optimal solutions are sought.

However, it's crucial to acknowledge the limitations of the proposed algorithm. Further improvements could focus on enhancing adaptability to dynamic environments, optimizing computational efficiency, and addressing scalability concerns for large-scale problems. These refinements would contribute to expanding the algorithm's applicability and ensuring its effectiveness in a broader range of real-world applications.

In the realm of data mining, the broader implications of swarm-based optimization algorithms are evident. The success of these algorithms in optimization tasks can be leveraged for data clustering, feature selection, and model parameter tuning. The adaptive and decentralized nature of swarm intelligence aligns with the complexities of large datasets, offering promising avenues for advancing data mining techniques and improving the efficiency of knowledge extraction from vast information repositories.

8 Conclusion

The research has yielded valuable insights into the realm of high utility itemset mining and swarm-based optimization. The key findings underscore the efficacy of the proposed algorithm, demonstrating its superior performance across various benchmark functions. Its adaptability to unimodal and multimodal scenarios, along with its ability to handle separable and inseparable functions, showcases its versatility. Additionally, the algorithm exhibits promising results in terms of standard deviation and runtime, affirming its robustness and efficiency.

The contributions and innovations brought forth by the enhanced algorithm lie in its dynamic adaptation to diverse optimization landscapes. Leveraging swarm intelligence, the algorithm excels in exploration and exploitation strategies, offering a balance crucial for achieving optimal solutions. The integration of high utility itemset mining further extends its applicability, providing a novel perspective on solving real-world problems involving complex, interrelated variables.

Future research directions in high utility itemset mining and swarm-based optimization could explore advancements in handling evolving datasets, dynamic optimization scenarios, and parallelization techniques to enhance scalability. Investigating the algorithm's applicability in specific domains, such as e-commerce or healthcare, could unveil tailored solutions addressing industry-specific challenges. Moreover, delving into hybrid approaches, combining swarm intelligence with machine learning techniques, presents an exciting avenue for further innovation and improved performance in complex optimization tasks.

References

1. Wu, X., Zhu, X., Wu, G.-Q., Ding, W.: Data mining with big data. IEEE Trans. Knowl. Data Eng. **26**(1), 97–107 (2014)
2. Zaki, M.J.: Data Mining and Analysis: Fundamental Concepts and Algorithms. Cambridge University Press, Cambridge (2014)
3. Benites, F., Sapozhnikova, E.: Evaluation of hierarchical interestingness measures for mining pairwise generalized association rules. IEEE Trans. Knowl. Data Eng. **26**(12), 3012–3025 (2014)
4. Zhang, C., Zhang, S.: Association Rule Mining: Models and Algorithms. Springer, Berlin (2002)
5. Tran, T., Vo, B., Le, T.T.N., Nguyen, N.T.: Text clustering using frequent weighted utility itemsets. Cybern. Syst. **48**(3), 193–209 (2017)
6. Agrawal, R., Srikant, R.: Fast algorithms for mining association rules. In: Proceedings of the International Conference on Very Large Data Bases, pp. 487–499 (1994)
7. Chan, R., Yang, Q., Shen, Y.-D.: Mining high utility itemsets. In: Proceedings of the 3rd IEEE International Conference on Data Mining, pp. 19–26 (2003)
8. Ab Wahab, M.N., Nefti-Meziani, S., Atyabi, A.: A comprehensive review of swarm optimization algorithms. PLoS ONE **10**(5), e0122827 (2015). https://doi.org/10.1371/journal.pone.0122827. Academic Editor: Catalin Buiu, Politehnica University of Bucharest, ROMANIA Received: 9 September 2014. Accepted 15 February 2015 Published: 18 May 2015
9. Sharma, H.D., Goyal, P.: An analysis of sentiment: methods, applications, and challenges. In: RAiSE-2023, Basel Switzerland, p. 68. MDPI (2023). https://doi.org/10.3390/engproc2023059068

Bus Route Tracking System

R. Rubesh Muthuvel(✉), B. Nithish Sankar, P. Kasirajan, and S. Dheenathayalan

Department of Computer Science and Engineering, National Engineering College, Kovilpatti, India

{2012039,2012030,2012022,ddhayalan}@nec.edu.in

Abstract. This research paper explores the challenges faced by daily bus commuters and presents a transformative live route monitoring system designed to address these issues. The project focuses on local buses and utilizes a user-friendly web application integrated with Google Maps. Key features include personalized notifications, precise real-time tracking, and historical performance data for efficient journey planning. The integration of crowd-sourced insights and multimodal functionality enriches user decision-making, while emergency communication, multilingual support, and energy-efficient tracking contribute to an enhanced overall experience. The project also includes a community engagement platform to encourage collaboration and feedback. Accessibility features and offline functionality ensure inclusivity and convenience, while strategic partnerships with local transport authorities bolster system accuracy.

Keywords: Transportation · Bus Route Tracking · Multiple Routes Tracking

1 Introduction

In the daily lives of a significant population, local buses serve as a vital mode of transportation. Unfortunately, the common challenge of missing buses leads to considerable inconvenience for many individuals. In response, our project aims to introduce a comprehensive bus route tracking system, designed to empower people to effortlessly monitor their buses and strategically plan their travel times.

The prevalent issue of waiting for buses and the subsequent disruptions to schedules underscore the need for an efficient solution. Our primary objective is to establish a live route monitoring system tailored for specific vehicles, particularly local buses. Recognizing the ubiquity of bus travel in people's day-to-day routines, our project endeavors to simplify their lives. We intend to achieve this by furnishing users with real-time location information for their chosen local bus, coupled with anticipated arrival times at their current or desired locations, seamlessly integrated with Google Maps.

The heart of our solution lies in the web application we're developing, tightly integrated with the familiar and widely used Google Maps platform. This integration not only ensures accessibility but also facilitates user-friendly interaction, allowing individuals to harness the benefits of our service conveniently, even on the go.

Published by Springer Nature Switzerland AG 2024
M. L. Owoc et al. (Eds.): ICCIDS 2024, IFIP AICT 718, pp. 421–433, 2024.
https://doi.org/10.1007/978-3-031-69986-3_32

By providing live location details through our innovative web application, we aspire to eliminate the uncertainties associated with bus travel, enabling users to make informed decisions about their commute. In doing so, we aim to streamline the bus travel experience, mitigate waiting times, and ultimately enhance the overall efficiency and convenience for the multitude relying on local buses in their daily routines.

In the strategic blueprint for crafting a robust bus route tracking system, our methodology places a strong emphasis on a user-centric approach. This involves a meticulous integration of market research, prototyping, and a commitment to continuous improvement throughout the development lifecycle. By centring our methodology around the needs and preferences of the end-users, we ensure that the resulting solution not only meets but exceeds their expectations. The inclusion of market research allows us to deeply understand user behaviour, pain points, and expectations, providing invaluable insights that steer the development process. Prototyping serves as a dynamic tool in visualizing the user interface and functionality, allowing for iterative refinements based on user feedback. Our dedication to continuous improvement ensures that the bus route tracking system remains adaptive and responsive to evolving user requirements and technological advancements. This comprehensive approach ensures that the end product is not only technologically robust but is tailored to enhance the overall user experience, making it a seamless and indispensable tool for commuters.

Our primary objective is to develop a cutting-edge system that not only prioritizes user-friendliness and reliability but is also meticulously aligned with the specific requirements of both passengers and bus operators. In a dynamic market landscape, our aim is not only to meet but to exceed user expectations by providing an intuitive and easy-to-use solution. By placing a strong emphasis on user experience, we strive to ensure that interacting with the system is seamless and accessible to individuals with varying technological backgrounds. Reliability stands as a cornerstone of our design philosophy, ensuring that the system consistently delivers accurate and real-time information, fostering trust among both passengers and bus operators.

Moreover, our commitment to staying competitive in the market underscores the incorporation of innovative features, responsive updates, and a forward-looking approach that anticipates and adapts to emerging trends. In essence, our overarching goal is to create a multifaceted system that not only meets the diverse needs of its users but positions itself as a standout, forward-thinking solution in the competitive landscape of transportation technology.

Limited coverage of potential challenges in real-world implementation, lack of consideration for scalability issues, or insufficient discussion on the energy efficiency of the proposed hybrid system[1]. Lack of empirical evidence or case studies validating the effectiveness of the proposed system in reducing traffic violations. Potential limitations related to the adaptability of the system to diverse traffic conditions in developing countries[2]. Insufficient exploration of potential security vulnerabilities in real-time web-based tracking systems. Lack of discussion on user privacy concerns or potential ethical implications[3]. Limited details on the user interface and user experience, potential gaps in addressing challenges specific to public transportation management, such as dealing with high passenger volumes[4]. Limited discussion on the adaptability of the proposed prediction model to varying traffic conditions or its performance in scenarios with limited online data availability. Potential challenges related to the real-time applicability of the RBF neural network in predicting bus arrival times[5].This project overcomes all these defects while maintaining its use case.

2 Proposed System

The research methodology begins with defining the system architecture tailored to the unique geographical and infrastructural characteristics of Indian geolocations. This involves a detailed analysis of the existing public transportation infrastructure, mapping out the bus routes, and identifying key transit hubs. By collaborating with local transport authorities and leveraging geographical information systems (GIS), we can design an architecture that optimally integrates with the Indian context. The system architecture should be flexible enough to accommodate the diverse topography and urban structures present in various Indian cities.

The methodology adopted for this research endeavors to meticulously craft and implement a live route monitoring system, finely attuned to the intricate geographical and infrastructural nuances characterizing Indian geolocations. Recognizing the constraints posed by regression modeling and limited datasets, our approach commences with a comprehensive examination of the prevailing public transportation landscape. This entails a thorough mapping of bus routes and the identification of pivotal transit hubs, buttressed by collaborative endeavors with local transport authorities and the astute utilization of geographical information systems (GIS). Such concerted efforts aim to forge a system architecture of unparalleled flexibility, adept at accommodating the kaleidoscopic tapestry of topography and urban configurations that define India's diverse cityscapes.

Following this foundational groundwork, our methodology pivots towards the strategic implementation of key features within the live route monitoring system. An acute sensitivity to the ever-shifting tides of traffic dynamics, cultural idiosyncrasies, and individual user predilections endemic to India underscores this phase. Here, the deployment of real-time tracking mechanisms, the orchestration of contextually nuanced personalized notifications, and the seamless integration of historical performance data and crowd-sourced insights coalesce to sculpt a system finely attuned to the multifaceted demands of Indian commuters.

Central to our approach is the establishment of strategic partnerships with Indian transport authorities, fostering a symbiotic relationship aimed at bolstering integration efforts and ensuring the veracity of real-time data streams. This collaborative ethos extends to the grassroots level, where an inclusive community engagement strategy unfolds. Employing a panoply of outreach initiatives, from localized language mediums to participatory feedback mechanisms, we endeavour to weave a tapestry of user involvement that mirrors the rich diversity of Indian society.

Moreover, our methodology remains steadfast in its commitment to inclusivity and accessibility. Features tailored to accommodate individuals with diverse abilities and the provision of multilingual support serve as testament to this ethos. This dedication to user-centric design principles permeates our testing phase, where rigorous assessments conducted across a spectrum of Indian geolocations and demographic cohorts inform iterative refinements aimed at optimizing system efficacy and user experience.

In sum, our research methodology represents a holistic endeavour aimed at furnishing a bespoke solution tailored to the unique contours of the Indian transportation landscape. While cognizant of the constraints posed by regression modelling and limited datasets, our approach endeavours to navigate these challenges adeptly, leveraging collaborative partnerships, community engagement, and iterative refinement to forge a live route monitoring system poised to enhance the public transportation experience across Indian geolocations.

Adapting the live route monitoring system to Indian geolocations requires a strategic approach in implementing key features. Real-time tracking must consider the dynamic traffic conditions, diverse terrains, and varying population densities. Personalized notifications should be context-aware, considering cultural nuances and specific user preferences prevalent in India. Historical performance data should account for the unique patterns of bus usage, and crowd-sourced insights must be sensitive to the diverse perspectives of users across different regions. The implementation phase involves collaboration with local communities to gather insights and feedback for refining features according to the specific needs of Indian bus commuters.

Recognizing the importance of local collaboration, the research methodology emphasizes forming strategic partnerships with Indian transport authorities. This involves engaging in dialogues with city-level transport agencies, understanding their existing systems and challenges, and working towards a seamless integration of the live route monitoring system. Establishing these partnerships is crucial for acquiring real-time data on bus schedules, routes, and traffic conditions. It also ensures the accuracy and reliability of the information provided to users, aligning with the goal of enhancing the public transportation experience in Indian geolocations.

The research methodology places a significant emphasis on community engagement, recognizing the diverse linguistic, cultural, and socioeconomic backgrounds of Indian users. A comprehensive outreach strategy will be employed, utilizing local languages and communication channels to encourage collaboration and feedback. This involves organizing focus group discussions, conducting surveys, and actively involving the community in the iterative development process. The goal is to ensure that the live route monitoring system is inclusive, responsive, and aligned with the specific needs and preferences of the Indian population.

The research methodology places a significant emphasis on community engagement, recognizing the diverse linguistic, cultural, and socioeconomic backgrounds of Indian users. A comprehensive outreach strategy will be employed, utilizing local languages and communication channels to encourage collaboration and feedback. This involves organizing focus group discussions, conducting surveys, and actively involving the community in the iterative development process. The goal is to ensure that the live route monitoring system is inclusive, responsive, and aligned with the specific needs and preferences of the Indian population.

The research methodology includes a rigorous testing phase in Indian geolocations. This involves beta testing with a diverse user base, incorporating feedback, and iteratively refining the system based on the user experience in different Indian cities. The testing phase is crucial for identifying and resolving any region-specific issues, ensuring that the live route monitoring system is robust and effective across varied geographical and cultural contexts.

The iterative prototyping process is a key element of the methodology, allowing for the creation of functional prototypes at various stages of development. This iterative approach facilitates feedback loops with end-users, enabling developers to refine and enhance the system based on practical insights and user experiences. The result is a solution that evolves dynamically, ensuring that it remains aligned with the changing landscape of user needs and technological advancements.

Moreover, the commitment to continuous improvement emphasizes an ongoing refinement process even after the initial deployment of the system. This commitment involves monitoring user feedback, addressing any issues promptly, and proactively incorporating updates and enhancements. Such an approach not only ensures the system's reliability but also reflects a dedication to meeting the evolving expectations of passengers and bus operators.

In tandem with user-centric principles, the methodology recognizes the importance of maintaining competitiveness within the market. This entails staying abreast of technological advancements, industry trends, and competitor offerings. By incorporating cutting-edge features and functionalities, the bus route tracking system aspires to stand out in the market, ensuring sustained relevance and efficacy.

Ultimately, the strategic approach aims to deliver a cutting-edge solution that not only meets the immediate and evolving demands of stakeholders but also sets a benchmark for innovation and user satisfaction in the realm of bus route tracking systems. By prioritizing ease of use, unwavering reliability, and continuous improvement, this methodology strives to create a system that enhances the overall experience for both passengers and bus operators while contributing to the advancement of the broader public transportation ecosystem.

By incorporating these elements into the research methodology, the proposed live route monitoring system aims to address the challenges faced by daily bus commuters in Indian geolocations, providing a tailored and effective solution for enhancing the public transportation experience (Fig. 1).

2.1 Modules

There are 2 main modules in our product.

1. The NE0-6M GPS module is a GPS receiver that can locate all locations on Earth as it is able to track approximately 22 satellites. It consists of a high-performance u-blox 6 positioning engine. Measuring 16 × 12.2 × 2.4 mm, its compact architecture along with its low power consumption makes it a good choice for IOT projects. Overall it is a good cost-effective GPS receiver.
2. The Node MCU (Node Micro Controller Unit) is an open- source software and hardware development environment built around an inexpensive System-on-a-Chip (SOC) called the ESP8266. The ESP8266, designed and manufactured by Espressif Systems, contains the crucial elements of a computer: CPU, RAM, networking (Wi-Fi), and even a modem operating system and SDK. That makes it an excellent choice for IOT based projects.

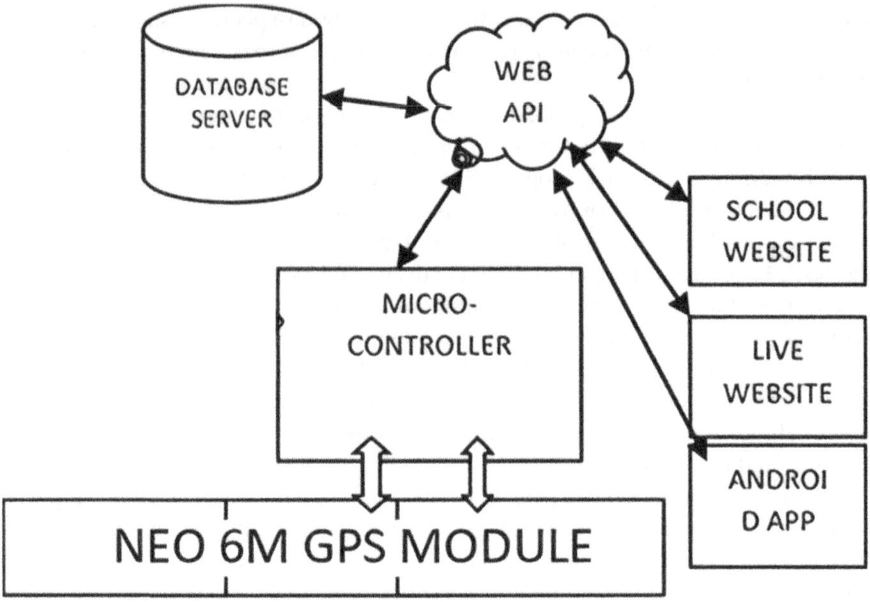

Fig. 1. Architecture Diagram

2.2 Use Case

In Fig. 2, the bus route tracking system emerges as a sophisticated technological solution designed to empower users in monitoring and tracking the real-time location and schedules of buses. The use case diagram serves as a comprehensive visual representation, depicting the intricate interactions and functionalities inherent in this system.

At the epicenter of this diagram is the pivotal "User" actor, embodying the end-user who initiates actions and engages with the system. The primary use cases unfold as follows:

View Bus Routes: This feature enables users to seamlessly access and visualize the available bus routes along with their respective stops. It serves as a fundamental tool for route planning and navigation.

Track Bus Location: Empowering users with real-time insights, this use case allows individuals to track the precise location of buses on an interactive map. This feature enhances user convenience and minimizes uncertainties associated with bus arrival times.

Check Bus Schedule: Offering users critical information on bus arrival and departure times, this use case ensures that commuters can plan their journeys effectively, promoting a more organized and punctual commuting experience.

Set Bus Alerts: A user-centric functionality, this use case allows individuals to personalize their experience by setting up notifications for specific bus arrivals or departures. This feature enhances user control and facilitates timely travel preparations.

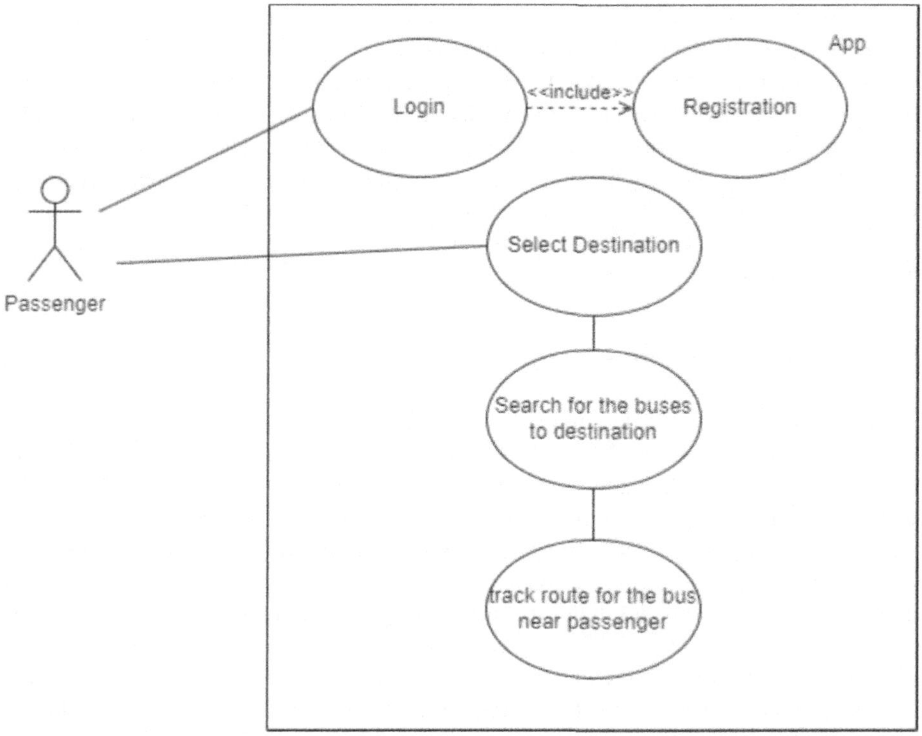

Fig. 2. Use case Diagram

In addition to the "User" actor, there may be an "Admin" actor, assuming a pivotal role in managing the system. Admin tasks may include adding or modifying bus routes, updating schedules, and handling system configurations. The inclusion of this administrative functionality ensures the system remains dynamic and adaptable to changing operational needs.

The process initiates with the passenger actively engaging with the system by specifying their destination. This pivotal step involves the passenger inputting relevant information, such as the desired endpoint or location they intend to reach. The system then processes this destination selection, acknowledging the user's input.

Following the acknowledgment of the destination, the system proceeds to identify and validate the optimal route for the selected journey. This involves leveraging the available data, such as bus routes, schedules, and real-time traffic information, to determine the most efficient and viable route for the passenger's specified destination. The system ensures accuracy and relevance in route selection, considering factors like traffic conditions and any dynamic changes in the transportation network.

Once the system has successfully determined the appropriate route, it enters the phase of bus search. This involves the system querying its database or real-time information sources to identify the buses operating on the designated route. The search takes into account the current locations and statuses of buses, ensuring that the passenger is provided with up-to-date and accurate information.

The diagram effectively highlights the sequential nature of this process, emphasizing the logical progression from destination selection to route acknowledgment and, finally, to the identification of buses available for the chosen route. The use case embodies user-centric design, offering passengers a user-friendly and efficient means of accessing real-time information for their bus search. This streamlined approach not only enhances the overall user experience but also contributes to the effectiveness and reliability of the bus tracking system as a whole.

3 Implementation and Result

Open your Arduino IDE and go to File > New. A new le will open. Copy the code given below in that le and save it. This sketch will display the location's latitude, longitude along with the current date and time in a web server. This will be achieved through the functions of the Tiny GPS ++ library that will easily parse through the NMEA message and display it in an easy convenient format. Moreover we will also include a link to the location in Google Maps in the web server as well. For a valid time we will extract the hour, minute and second and then save it in a string format in their respective variables. Also, we have adjusted the UTC o set as well according to our location. You will have to adjust it yourself. We have created a table to display the Latitude, Longitude, Date and Time values on the web server. These will be received by the NE0-6M GPS module.

This Arduino sketch leverages the capabilities of the Tiny GPS++ library to develop a web server that dynamically showcases essential information retrieved from a NEO-6M GPS module. The primary goal is to present real-time data, including latitude, longitude, date, and time, in a user-friendly and accessible manner through a web interface.

The code establishes a web server, and when accessed, it displays the parsed NMEA messages from the GPS module. The information is intelligently organized into a table on the web page, offering a structured overview. Notably, the inclusion of a hyperlink to Google Maps allows users to visualize the current location conveniently.

For accurate time representation, the code extracts hour, minute, and second values from the GPS module's data. It further provides a mechanism to adjust the Coordinated Universal Time (UTC) offset, ensuring that the displayed time aligns with the local time zone.

An important aspect of the code is its adaptability. Users can easily customize the UTC offset, modify the web server display format, or introduce additional features to suit specific requirements. This flexibility makes the code a versatile foundation for creating a personalized GPS tracking system.

Overall, this Arduino sketch demonstrates an effective integration of hardware and software, providing a practical solution for displaying and accessing GPS-related information through a web server.

The implementation process begins with an extensive mapping and analysis of the Indian geolocations where the live route monitoring system will be deployed. This involves collating data on existing bus routes, transit hubs, and traffic patterns. Geographic Information System (GIS) tools will be employed to create a comprehensive geospatial database that captures the diverse topography and urban landscapes across various Indian cities. Understanding the intricacies of the geolocations is crucial for designing an effective and context-aware live route monitoring system.

Google Maps will serve as a foundational element in the live route monitoring system. The integration process entails leveraging the Google Maps API to overlay real-time bus tracking information onto the map interface. This involves obtaining accurate GPS data from buses in operation and integrating it seamlessly with the Google Maps platform. The integration ensures that users have a familiar and user-friendly interface for visualizing bus routes, tracking buses in real-time, and planning journeys effectively within the context of Indian geolocations.

Developing and implementing robust real-time tracking algorithms is a critical aspect of the process. Considering the dynamic nature of traffic conditions in Indian cities, the algorithms should account for congestion, diversions, and delays. Machine learning models can be trained on historical data to predict and adapt to these conditions, providing users with accurate and up-to-date information. The integration with Google Maps allows for a smooth display of real-time bus locations, optimizing the user experience.

The implementation process includes the development of context-aware notification systems. Utilizing the geolocation data, the system can send personalized notifications to users based on their location, preferences, and historical travel patterns. For example, users can receive alerts about bus delays, route changes, or important announcements relevant to their specific location within the Indian geolocations. This ensures that notifications are not only timely but also tailored to the individual needs of the users. To address the linguistic diversity in India, the implementation process incorporates multilingual support. User interfaces and notifications will be designed to accommodate multiple languages, ensuring that the live route monitoring system is accessible to a wide range of users across different regions. Additionally, cultural sensitivity will be considered in the design, aligning the system with the diverse cultural norms and expectations prevalent in Indian geolocations.

The development of an intuitive and user-friendly interface is a key component of the implementation process. The system's design should be visually appealing, easy to navigate, and responsive to various devices. User interface testing will be conducted iteratively, involving real users from diverse backgrounds within Indian geolocations. This ensures that the interface meets usability standards and aligns with the preferences of the target user demographic.

The implementation process includes a beta testing phase where the live route monitoring system is released to a limited user group in Indian geolocations. During this phase, user feedback is actively collected through surveys, focus group discussions, and in-app feedback mechanisms. The goal is to identify any issues specific to the geolocations, gather insights into user experiences, and make iterative improvements based on the feedback received.

Establishing collaborative partnerships with local transport authorities is an integral part of the implementation process. This involves integrating the live route monitoring system with the data and systems maintained by these authorities. Collaborative efforts ensure the accuracy of information regarding bus schedules, routes, and any updates related to the public transportation network in Indian geolocations.

Given the energy constraints in some Indian geolocations, the implementation process will also focus on developing energy-efficient tracking mechanisms. This may involve optimizing the use of GPS technology, considering alternative energy sources, or implementing smart tracking strategies to conserve device battery life. Additionally, infrastructure considerations, such as network connectivity, will be considered to ensure the seamless operation of the live route monitoring system in diverse geographical and urban environments.

Through a comprehensive and adaptive implementation process, integrating the live route monitoring system with Google Maps for Indian geolocations aims to provide an efficient, user-friendly, and culturally sensitive solution to the challenges faced by daily bus commuters. The process ensures that the system is not only technically robust but also attuned to the unique characteristics of the Indian transportation landscape (Figs. 3, 4, 5 and 6).

Fig. 3. Implementation of Arduino code

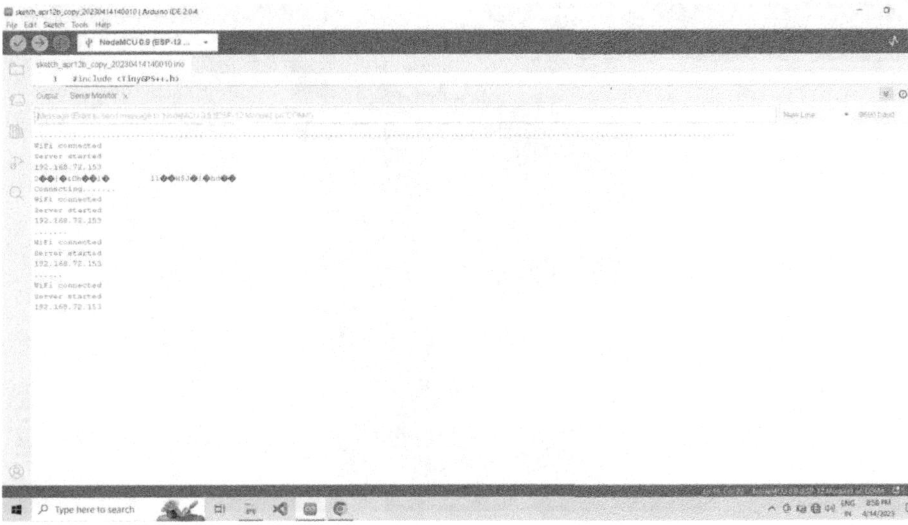

Fig. 4. Serial Monitor Output in Arduino

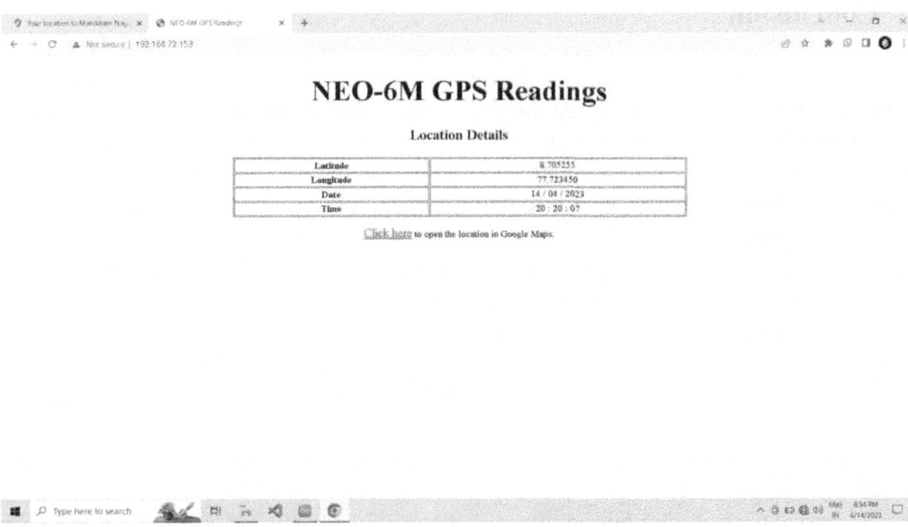

Fig. 5. Home Page of Web Interface

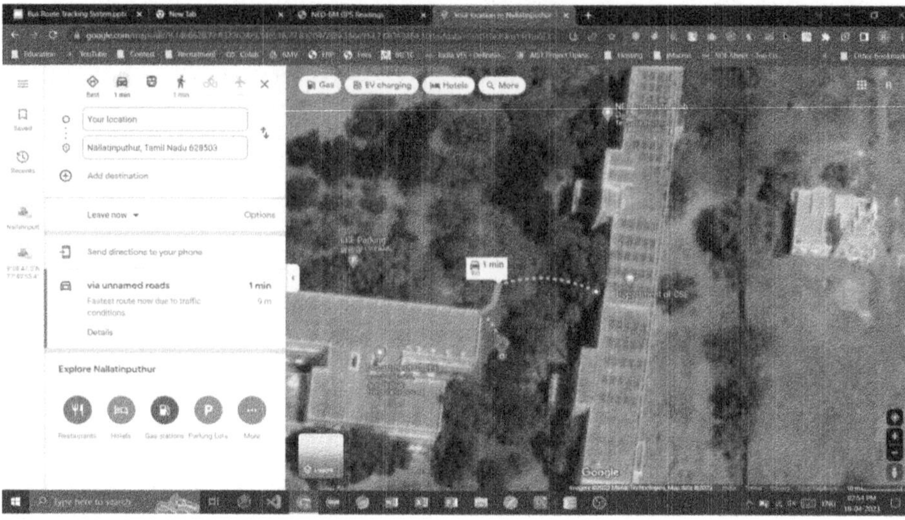

Fig. 6. Route Tracking Output in Google Maps

4 Conclusion

In conclusion, our research and development efforts have been dedicated to addressing the daily challenges faced by regular individuals, like yourself, who rely on public transportation for their daily commute. The live route monitoring system presented in this paper emerges as a transformative solution, specifically designed to enhance your public transportation experience in Indian geolocations.

For transportation companies, the benefits of a bus route tracking system are multifaceted. Real-time insights enable them to monitor and manage their fleets more effectively, leading to improved operational efficiency and resource allocation. The system's data-driven nature provides a foundation for making informed decisions that positively impact both short-term operations and long-term planning. By embracing such technology, transportation companies position themselves at the forefront of innovation in the industry.

From the passengers' perspective, the introduction of a bus route tracking system marks a paradigm shift in their journey planning experience. The ability to track the precise arrival time of their bus equips passengers with timely and accurate information, allowing them to make informed decisions about their commute. This not only enhances the overall user experience but also addresses a common pain point for commuters—reliability and predictability in public transportation. The positive impact extends beyond mere convenience; it significantly contributes to the optimization of daily routines and schedules for individuals who rely on bus services for their daily commute.

However, the successful implementation of a bus route tracking system demands careful consideration of various factors. The choice of technology is critical, as it directly influences the system's functionality and effectiveness. Additionally, stringent measures for data privacy and security are imperative to ensure that passenger information remains protected. Balancing these considerations with a comprehensive assessment of costs is essential, as it determines the feasibility and sustainability of the system over the long term.

In essence, a well-thought-out bus route tracking system emerges not merely as a technological upgrade but as a strategic investment for transportation companies. It fosters a symbiotic relationship with passengers, addressing their needs for reliability and convenience, and paves the way for a more connected and efficient public transportation network. As both companies and passengers reap the benefits, this system becomes a cornerstone in reshaping the landscape of modern urban transportation.

We understand the frustrations associated with unpredictable schedules, crowded buses, and the lack of real-time information that often characterizes daily commuting. Through the integration of Google Maps, personalized notifications, and real-time tracking, our system aims to empower you with the tools to navigate the complexities of public transportation seamlessly.

References

1. Al-Khedher, M.A.: Hybrid GPS-GSM localization of automobile tracking system. Int. J. Comput. Sci. Inf. Technol (IJCSIT) **3**(6), 75–85 (2011)
2. Tarapiah, S., Atalla, S., Abu Rania, R.: Smart on-board transportation management system using GPS/GSM/GPRS technologies to reduce traffic violation in developing countries. Soc. Digital. Inf. Wirel. Commun. **3**(4), 96–105 (2013). (ISSN: 2225–658:x).
3. Muruganandham, P.M.: Real time web based vehicle tracking using GPS. World acad. Sci. Eng. Technol. **61**(1), 91–99 (2010)
4. Kumar, K.G., Prasad, A.M.: Public Transportation management service using GPS-GSM. Int. J. Res. Comput. Commun. Technol. **1**(3), 64–68 (2012)
5. Wang, L., Zuo, Z., Fu, J.: Bus arrival time prediction using RBF neural networks adjusted by online data. **138**, 67–75 (2014)

Sentiment Analysis of Various Ride Sharing Applications Reviews: A Comparative Analysis Between Deep Learning and Machine Learning Algorithms

Md. Saymon Ahammad(✉), Sadia Akter Sinthia, Md. Muaj Chowdhury, Nur-A-All Asif, and Md. Nurul AfsarIkram

Department of Computer Science and Engineering, Daffodil International University, Dhaka, Bangladesh

{saymon15-14388,sadia15-14369,muaj15-14389,nur15-14370, nurul15-13678}@diu.edu.bd

Abstract. The exponential growth of ride sharing platforms in developing countries has significantly transformed the landscape of urban transportation. It is crucial for service providers to comprehend customer feedback and opinions in order to enhance the user experience and for passengers to select a safe and secure ride. A comprehensive sentiment analysis was conducted on user evaluations of ride-sharing platforms such as Grab, Uber, Indrive, Pathao, Jatri, and Obhai. An aggregate of 12052 data points (Negative, Positive, Neutral) were extracted from ride sharing reviews available on the Google Play Store. By utilizing cutting-edge deep learning models (LSTM, GRU, BiLSTM, and BiGRU) in addition to traditional machine learning models (Random Forest, Decision Tree, Gradient Boosting Classifier, Logistic Regression, KNN, Naive Bayes, SVM, AdaBoost, and LightGBM), our research endeavors to offer substantial insights into the determinants of customer satisfaction. Support Vector Machine is the most effective machine learning algorithm in this context, attaining an accuracy rate of 76.70%. Bidirectional Gated Recurrent Unit emerged as the best-performing model in the domain of deep learning, attaining an exceptional accuracy rate of 94.97%, this algorithm demonstrates the efficacy of deep learning methodologies in the context of sentiment analysis. Our research indicates that deep learning algorithms demonstrate significant superiority when compared to machine learning algorithms.

Keywords: Ride · Uber · Pathao · Indrive · Machine Learning · Deep Learning · Sentiment Analysis · NLP

1 Introduction

Ridesharing, often known as carpooling, is a transportation service in which participants share a single ride in a vehicle, which is typically arranged via a digital platform or smartphone application. Ride sharing services have transformed the concept of transportation

Published by Springer Nature Switzerland AG 2024
M. L. Owoc et al. (Eds.): ICCIDS 2024, IFIP AICT 718, pp. 434–448, 2024.
https://doi.org/10.1007/978-3-031-69986-3_33

by putting a simple, on demand solution at the fingertips of users. People can request transportation with a few smartphone clicks, saving them the trouble of looking for or waiting for traditional taxis [1]. This accessibility is especially helpful in places where there are few public transportation options. By cutting down on the wait times associated with typical taxi services or public transit, ride sharing enhances time efficiency. Many people find out that having a private car comes with high maintenance, gasoline, parking, and insurance expenditures. With ride sharing, people can avoid the financial burden of owning a car by just paying for the journeys they actually take. Cost sharing is one of the main ideas behind ride sharing. The expenses of the ride are far less than those of an ordinary taxi, although passengers typically tip the driver for the service. Additionally, ride sharing businesses gain from new avenues for revenue generation and new aspects of service utilization [2]. Since it is useful to find the closest cars as quickly as possible, the proliferation of smartphones has led to an explosion in the number of distinct app-based enterprises and widespread internet connections. Many ride sharing companies are there, like Uber, Pathao, Obhai, Indrive, Jatri, Grab, Amarbike, Taxiwala, Garivara, and others [3]. All these services are run by mobile applications that connect drivers and passengers. The process typically involves registration and profile creation, location-based matching, ride request and acceptance, real time tracking, cost estimation and payment, in-app communication, ride completion and rating, dynamic pricing, drive earnings and incentives, and safety features. Google Play Store reviews of ride sharing applications provides an excellent standard for comparing machine learning and deep learning algorithms in sentiment analysis. The dataset's real world, diversified and multilingual nature, combined with skewed sentiment distributions and the inclusion of user textual input, make it an ideal work. Additionally, publicly available data enables transparency and consistency in algorithmic assessments. All of the reviews of the ride sharing app are visible to other users, and they are arranged by most "helpful" or "recent" categories. This enables prospective customers to assess the overall attitude and experiences the ride sharing community has to offer. Certain ride-sharing businesses often check and reply to user evaluations on the Google Play store. Users get an impression of the company's dedication to resolving issues and enhancing their experience from this encounter. Reviews and comments on ride-sharing applications offer insightful information about user moods. Positive features that users show gratitude for include timely service and friendly drivers; negative characteristics that users express disapproval of include late arrivals or cost problems. In order to better understand customer satisfaction, pinpoint areas for development, and address specific problems, ride-sharing companies analyze wording, tone, and trends in user evaluations. The ride-sharing platform's total rating is influenced by the cumulative effect of feelings on platforms like the Google Play Store. A sentiment is an attitude, notion, or evaluation that is brought on by an emotion [4]. Texts are classified using sentiment analysis based on the sentiment orientation of the viewpoints conveyed in them [5]. Machine learning techniques (KNN, Naive Bayes, SVM, Random Forest, Logistic Regression, Decision Tree, Gradient Boosting Classifier, AdaBoost, and LightGBM) train models that categorize reviews as positive, negative, or neutral based on labeled datasets. These models aim to predict sentiment based on characteristics acquired from preprocessed text using methods such as supervised learning. Deep learning approaches (LSTM, BiLSTM, GRU, and

BiGRU), on the other hand, makes use of neural networks with layers that are meant to recognize contextual subtleties in languages; these models usually make use of word embeddings or pretrained language models. In a hybrid method, the feature extraction step of a machine learning pipeline might make use of pretrained embeddings from deep learning models. In this way, the advantages of both techniques are merged to develop a robust sentiment analysis system that is adapted to the special issues that ride-sharing assessments offer. This study intends to focus on the interaction between technology as well as urban mobility by delivering practical insights that enable ride-sharing systems to be continually enhanced. In this case,

- We have employed Google Play Store's 12052 data to do a sentiment analysis on ride sharing service evaluations written by users.
- Three categories are used to categorize the comments: negative, positive, and neutral.
- Deep learning and Machine learning techniques are employed in our attempt to discover sentiment patterns.
- We have examined and compared algorithms for machine learning and deep learning and have decided which one is the superior approach for sentiment analysis.

This will create an environment where users are more satisfied, and our research will help advance this conversation between users and ride-sharing companies.

2 Literature Review

This literature review establishes the background and offers a simple way to comprehend the key hypothesis and conclusion that guide the current investigation by examining previous research on sentiment analysis on ride sharing reviews and textual behavior. In a study, Mahmud et al. [6] uses sentiment analysis on 1818 Google Play Store reviews to examine how users respond to ride sharing apps The objective is to investigate consumer viewpoints by classifying reviews as neutral,negative or favorable. The model performed exceptionally well using CNN, LSTN and DistilBERT outperformed the others with a 98.84% accuracy rate. In another study Sabab et al. [7] study NLTK and python were used to access 1000 "Pathao" and "Uber" reviews from facebook and twitter in Bangladesh. Feature extraction decreased the dataset and classifiers were trained on bangla text expanding the data dictionary and testing different techniques, paving the way for future study in sentiment analysis of ride-sharing reviews on social media. Decision Tree and SVM were outperformed by Naive Bayes, which achieved 87% accuracy. Surjandari et al. [8] used Twitter data and aspect-based sentiment analysis to examine ride-sharing providers. It uses LDA to identify six subjects and SVM technique to score them, obtaining high accuracy 80%. The study emphasizes the efficacy of sentiment analysis and suggest further research on driver input, varied data sources and alternative analytic approaches for improving ride-sharing services. Islam et al. [9] collected 3,315 reviews from Bangladeshi ride-sharing apps (Uber, Pathao, Obihai, and Shohoz), with Uber(1000) and Pathao(985) receiving the most attention. Data is gathered manually from websites, social media platforms, and the Google Play Store. NLTK stopword removal and classification algorithms (AdaBoost, Gradient Boosting, Random Forest, Decision Tree, Bi-LSTM) are used. Bi-LSTM exceeds AdaBoost and Random Forest with 85% accuracy and an outstanding F1 score of 89%. Ali et al. [10] gathered 3,853 data points from uber websites, with a focus on Roman, Urdu and English. It entails online reviews, Roman, Urdu/Hindi to English translation, and Pre-processing. Using unsupervised machine learning, aspect-based sentiment analysis detects essential elements ('Driver', 'Company', 'Service', 'Ride'). Although no specific techniques are given, SentiWordNet and unsupervised machine learning are employed. The article targets toward Kansei engineering without providing particular results, although it does offer potential business improvements through enhanced services based on client sentiments. The ISEAR dataset of 7,666 entities is used in the Acheampong et al. [11] study, which is later reduced to 7,589 following pre-processing. It evaluates the effectiveness of the BERT, DistilBERT, RoBERTa and XLNet models for emotion analysis. The models have been fine-tuned for seven emotions, yielding the greatest F1 scores for RoBERTa across many categories: anger, disgust, joy, sadness, fear, and shame. The study offers future model ensemble considerations to increase recognition performance and use commonsense information for improved generalization. Based on 5,215 data points, Chi et al. [12] improves(T)ABSA by transforming it into a job of classifying sentences in pairs using BERT. BERT-pair performs better than other algorithms in sentiment analysis and aspect recognition on the SentiHood dataset. BERT-pair outnumbers BERT-single in ABSA. BERT-pair-QA-M, with 79.4% accuracy, 86.4% F1 score, and 97.0% AUC in aspect analysis, is a notable model. BERT-single obtains 73.7% accuracy, an F1 score of 81.0%, and an AUC of 96.4%. Overall, BERT-pair models improve significantly. Yanamandra el al. [13] examined 2,116 tobacco-related tweets from the SmokPro Dataset and categorizes them into five product categories. It compares sentiment and product recognition on Twitter and Reddit with FastText, BERT, RoBERTa, and DistilBERT using semi-supervised learning on Reddit content. RoBERTaLarge has the highest accuracy (72.64%) and F1 score(0.724) in sentiment classification on Twitter. Another study Liao et al. [14] employd a convolutional neural network(CNN) on movie reviews(MR) and genuine tweets (STS Gold Dataset) for sentiment analysis using Twitter API. On MR, the CNN obtains 75.39% accuracy and 68% on STS Gold. During training, overfitting issues are identified. With 1709 occurrences per class, Reference 17 achieves a comparable level of accuracy in Twitter sentiment analysis. Mahto et al. [15] made use of the US airline sentiment dataset, which has 9,178 rows classified as negative, neutral, or positive. It introduces a ConvBidirectional-LSTM hybrid model that combines CNN and Bi-LSTM and achieves an outstanding 93.25% accuracy on the same dataset, outperforming GloVe-based CNN-LSTM and HeBiLSTM. The validation accuracy is 93.25% and the training accuracy is 99.47% after 15 epochs.

3 Methodology

This part provides a brief overview of the technique used in this sentiment analysis study. It begins with data collection and annotation moves on to dataset preparation, model implementation, and concludes with model assessment. The entire working process of this research is portrayed in Fig. 1.

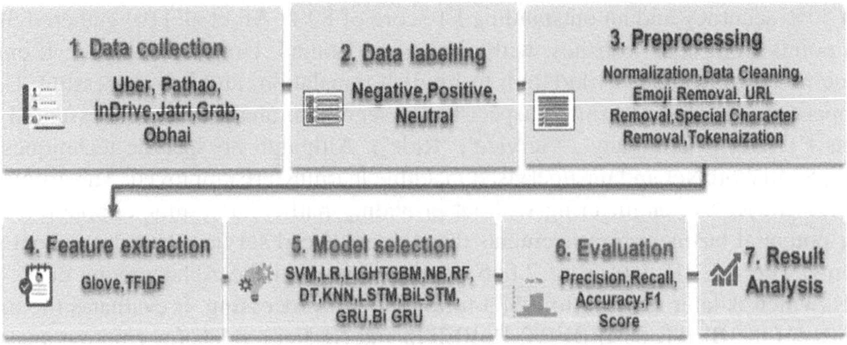

Fig. 1. Methodology of the Overall Procedures

3.1 Data Collection

We have applied manual data collection process and collected text reviews of 6 famous ride sharing platforms; Uber, Pathao, Obhai, Grab, Indrive, Jatri from Google Play Store. The dataset contains 12052 text reviews containing 4030 positive reviews, 3980 neutral reviews and 4042 negative reviews (Fig. 2).

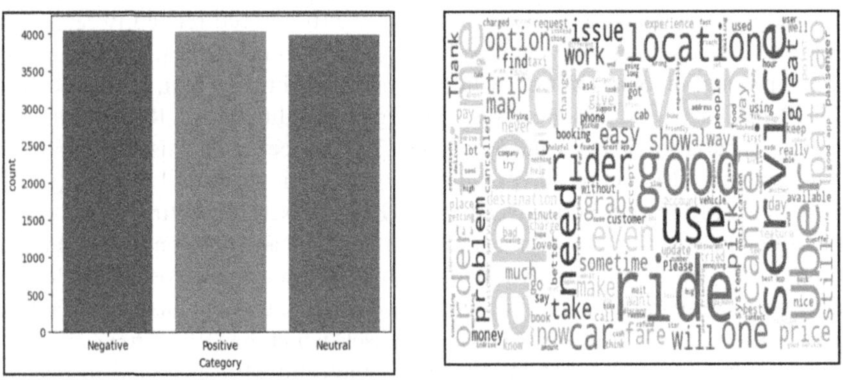

Fig. 2. Data Distribution and Word Cloud

3.2 Data Labeling

We have collected the reviews as well as the ratings of ride sharing apps whose were unlabeled. After collecting them, we labeled those reviews one by one according to their rating and the meaning as well because we observed that sometimes the users gave 4 stars or 3 stars but the review text contains positivity. Table 1 shows some reviews with their category and labels.

Table 1. Dataset Sample Reviews with Labels

Reviews	Category	Label
My first Uber trip was a driver called Jack don, picked up from my home to go into town. This is very lovely good behaviour	Positive	0
Worst app.uber eats app and guest complain handling is more worst.never recommend	Negative	1
Now good ever! I did use 2 times it's average quality!!	Neutral	2

3.3 Preprocessing

Preprocessing is crucial for this type of sentiment analysis as it's necessary to have the reviews ready before using any sort of algorithm [16]. The preprocessing has gone through some different sub steps like; normalization, data cleaning, emoji removal, URL removal, Special Character removal, tokenization (Fig. 3).

Fig. 3. Data Preprocessing Steps

3.4 Tokenization

Tokenization, a crucial stage in NLP, divides a text up into small parts called tokens. Words, phrases, symbols, and other significant components can be used as these tokens [17]. Figure 4 shows the tokenization process.

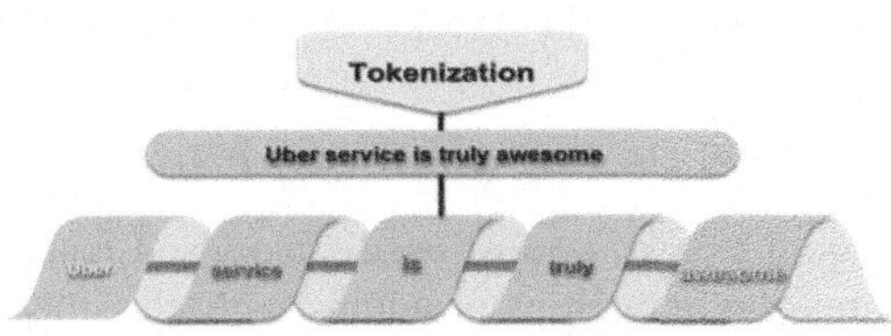

Fig. 4. Tokenization Process

3.5 Feature Extraction

It is the method of converting the raw text data to numerical data as algorithms don't understand the raw text data. As the data collected for this study is textual data, we must perform the feature extraction process on the preprocessed text data before applying any algorithms. In this study TF-IDF and Glove vectorization has been used.

Glove: Glove is the unsupervised learning system which creates vector representation of words. Integrated global word-word co-occurrence information collected from a corpus of terms are utilized in the training phase. We notice fascinating linear substructures inside the word vectors field in the results.

TF-IDF: Term Frequency-Inverse Document Frequency, or TF-IDF, is a statistical approach which is widely used in retrieving of data as well as natural language processing. This method measures a term's significance inside a document with respect to a collection of papers. Words in a manuscript are transformed into numbers that signify their significance using text vectorization.TF-IDF is a combination of TF and IDF. The formula stands for TF-IDF:

$$tfidf(t, d, D) = tf(t, d).\, idf(d, D) \tag{1}$$

3.6 Machine Learning Algorithms

Logistic Regression: Logistic regression acts as a supervised machine learning technique for dealing with classification problems. It forecasts whether a given instance is a member of a certain class or not. This kind of statistical technique looks at the relationships between a set of dependent binary variables and a set of independent variables.

Decision Tree: It is a supervised machine learning algorithms which resemble flowcharts and are used for both regression analysis and classification. The study used decision trees as a categorization approach that groups the reviews into three categories: positive, negative, and neutral.

Naive Bayes: A prominent supervised machine learning (ML) technique to the classification process of text labeling in categories uses the Naive Bayes classifier. It makes feasible that it imitates the input distribution for a given class or category since it is a member of the generative learning algorithm family. This approach relies on assumptions that the qualities of the input attributes are independent of the class in order for the algorithm to deliver accurate and speedy predictions. The equation for Naive Bayes stands:

$$P(A|B) = P(B|A)P(A) \big/ P(B) \tag{2}$$

Random Forest: Random Forest, a most used supervised machine learning method, may be applied to regression as well as problems with classification. Its key concept is ensemble learning, which refers to the process of combining several classifiers to maximize the performance of the model and address a difficult problem.

KNN: The simplest supervised machine learning method, KNN, which places a new instance in the category most similar to the ones already existing by making the assumption that the new case and data are equivalent to the existing cases. After storing all the relevant data, it uses similarity to classify a new data point. This implies that the KNN method might be used to quickly classify, recently found data into the relevant category.

SVM: SVM, or support vector machine, is a kind of supervised machine learning approach that may be applied to regression as well as classification. It works well with smaller datasets. The SVM method looks for the best edge or line that might split n-dimensional space into classes so that a new data point can be promptly placed in the right group at a later date. We refer to this optimal decision boundary as a hyper plane.

LightGBM: Light Gradient Boosting Machine, or simply LightGBM, is a decision tree-based design that uses less memory and increases model efficiency. It is applicable to regression as well as classification. LightGBM is designed to operate at peak efficiency in distributed systems.

3.7 Deep Learning Algorithms

LSTM: LSTM is the abbreviated version of Long Short-Term Memory, which is an upgraded form of Recurrent Neural Networks [18]. Given that a normal RNN only has one secret state that is retained throughout time as a consequence, understanding long-term dependencies may be difficult for it to comprehend. To fix this problem, LSTMs incorporate memory cells, which are long-term data retention units. As LSTM can discover long-term relationships in sequential data in an effective manner, LSTM networks are good for occupations.

BiLSTM: Recurrent neural networks (RNNs) that are utilized to handle ordered data in both forward and reverse directions are termed bidirectional LSTMs (Bi LSTM/BLSTM) [19]. Unlike ordinary LSTMs, which can only accept sequential input in one direction, this allows BiLSTM to learn longer-range correlations in sequential data. Bi-LSTMs consist of two LSTM networks, one that processes the sequence of inputs in forward direction and the other that processes it backward. By integrating the outputs of the two separate LSTM networks, the final result is then formed.

GRU: GRU is the short form of Gated Recurrent Unit which is one of the Recurrent Neural Network (RNN) methods. GRU's fundamental principle is to update the network's hidden state only on a chosen subset of time steps by using gating methods. The gating mechanism regulates the flow of information into and out of the network. The reset gate and the update gate, these are the two gating mechanisms of the GRU [20].

BiGRU: A sequence processing architecture called bidirectional gated recurrent unit or BiGRU which is combination of two Gated Recurrent Units (GRUs) that process incoming data both forward and backward. With the use of update and reset gates built into each GRU, this bidirectional and recurrent neural network design allows the model to selectively update and forget information, helping it to capture lengthy relationships in sequential data. Because BiGRU is bidirectional, it can understand context from both past and future inputs (Table 2).

4 Performance Analysis

We will now analyze the models with the required explanation and diagrams and talk about the accuracy as well as all other metrics of all the algorithms applied in that study, including precision, recall, F1-score, MSE, MAE, RMSE, and Cohen's Kappa score.

Table 2. Parameter Tuning for Deep Learning Models

Models	Batch Size	Epochs	Maximum Length
LSTM	256	70	1000
BiLSTM	256	70	1000
GRU	256	70	1000
BiGRU	256	70	1000

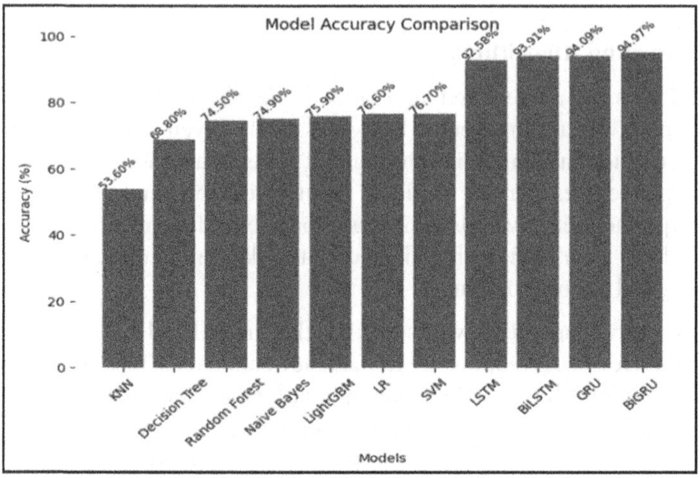

Fig. 5. Model Accuracy Comparison

In Fig. 5 we can see that the SVM is the best performing algorithm with the accuracy of 76.7% from all 7 machine learning algorithms with the value of recall, precision and F1- Score 0.77, 0.77 and 0.77 respectively. With accuracy of 74.50%, 74.90%, 75.90%, and 76.60%, respectively, Random Forest, Naive Bayes, LightGBM, and Logistic Regression also fared well and were quite comparable to SVM. The worst performance has been given by KNN and Decision Tree, giving the accuracy of 53.60% and 68.80% (Table 3).

In addition to these seven machine learning techniques, this work also uses four deep learning algorithms (LSTM, GRU, BiLSTM and BiGRU). In addition, recall, precision, F1-Score, and other error metrics (such as MAE, MSE, and RMSE) have been calculated to see the efficiency of various techniques.

Table 4 shows that, All of those 4 algorithms performed extremely well but among them Bidirectional GRU (BiGRU) outperformed others with the accuracy of 94.97% with the values of recall, precision and F1-Score respectively 0.95, 0.95 and 0.95 followed by GRU, BiLSTM and LSTM with accuracy of 94.09%, 93.91% and 92.58% respectively. Additionally, it was noted that every deep learning algorithm outperformed the machine learning methods. Moreover, the best performing machine learning algorithms performance is also very much lower than the worst performing deep learning algorithm which

Table 3. Performance Table of Machine Learning Algorithms.

Algorithms	Accuracy (%)	Precision	Recall	F1-Score	Support
SVM	76.70	0.77	0.77	0.77	2411
Logistic Regression	76.60	0.77	0.77	0.77	2411
LightGBM	75.90	0.76	0.76	0.76	2411
Naïve Bayes	74.90	0.75	0.75	0.75	2411
Random Forest	74.50	0.75	0.74	0.74	2411
Decision Tree	68.80	0.69	0.69	0.69	2411
KNN	53.60	0.64	0.54	0.48	2411

Table 4. Performance Table of Deep Learning Algorithms.

Algorithms	Accuracy (%)	Precision	Recall	F1-Score
LSTM	92.58	0.93	0.93	0.93
BiLSTM	93.91	0.94	0.94	0.94
GRU	94.09	0.94	0.94	0.94
BiGRU	94.97	0.95	0.95	0.95

is SVM having 76.70% accuracy and LSTM with the accuracy of 92.58% that shows a huge gap between them. So, it is clearly be said that no machine learning algorithms has performed even close to the deep learning algorithms, all those deep learning algorithms outperformed machine learning algorithms with a huge margin.

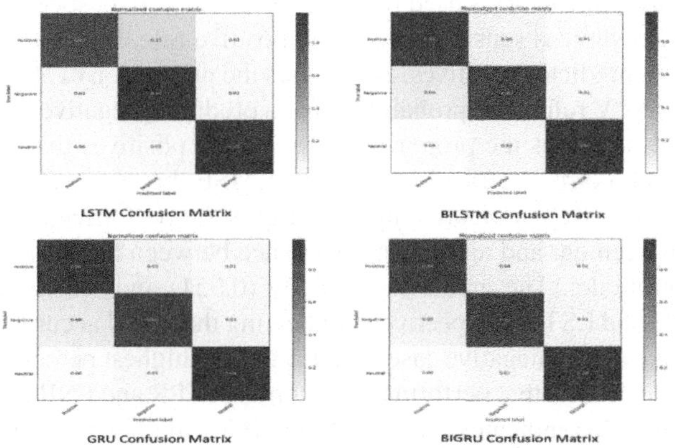

Fig. 6. Confusion Matrix of Deep Learning Algorithms

Figure 6 shows the confusion matrix for all 4 deep learning algorithms. Confusion matrixes are a helpful tool for assessing a classification model's effectiveness. According to true positives, false positives, true negatives and false negatives, they offer analysis of the model's predictions. Our dataset contains 3 classes labeled with Positive, Neutral and Negative hence our confusion matrixes is a multi-class confusion matrix which is 3 × 3 dimensional matrixes.

Table 5. Error metrics of Deep Learning Algorithms.

Models	Sensitivity	Specificity	FPR	FNR	NPV	FDR
LSTM	0.982	0.863	0.136	0.017	0.984	0.152
BiLSTM	0.926	0.949	0.050	0.073	0.922	0.047
GRU	0.924	0.951	0.048	0.075	0.920	0.045
BiGRU	0.951	0.950	0.049	0.048	0.949	0.048

Table 6. Error metrics of Deep Learning Algorithms.

Models	MAE	MSE	RMSE	Cohens_Kappa_Score
LSTM	0.079	0.037	0.192	0.888
BiLSTM	0.081	0.032	0.180	0.908
GRU	0.076	0.031	0.177	0.911
BiGRU	0.074	0.028	0.167	0.924

Table 5 and 6 shows the performance metrics for all deep learning algorithms. In the tables there are some metrics used for model evaluation like: specificity refers to the false positive rate, whereas sensitivity to a true positive rate, FPR refers the proportion of actual negatives predicted positives, FNR refers the proportion of actual positives predicted negatives, NPV refers the probability that a predicted negative instance predicted as negative, FDR measures the proportion of predicted positives that are actually negative instances and 3 error metrics MAE, MSE, RMSE that represents the error values followed by Cohen's_Kappa_Score which measures the performance based on assessing the perfect agreement and agreement by chance between a real world observer and the classification model. The greatest specificity (0.951) and sensitivity (0.982) were obtained by GRU and LSTM, respectively, indicating that GRU accurately predicted the highest percentage of true negative cases and LSTM the highest percentage of true positive instances. The models that perform best in terms of FPR and FNR are LSTM (0.017) and GRU (0.048). This indicates that LSTM provides the lowest proportion of actual positive occurrences wrongly forecasted as negative, while GRU produces the lowest proportion of actual negative cases incorrectly predicted as positive. The best NPV rate obtained by LSTM with 0.984 score and the best FDR rate is 0.045 obtained by GRU as lower FDR rate is considered better. Similarly, MSE, MAE and RMSE shows the error rates and the best performed algorithm according to this three-error metrics is BiGRU with the lowest values for all three metrics with 0.074 MAE, 0.028 MSE and 0.167 RMSE. BiGRU also outperformed others, obtaining the highest Cohen's_Kappa_Score of 0.924.

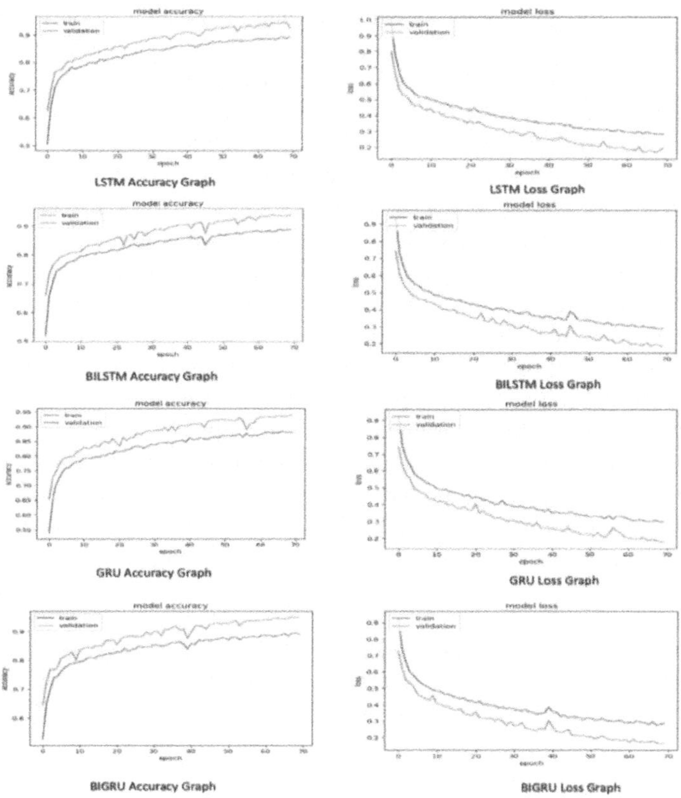

Fig. 7. Model Accuracy and Loss Graphs of Deep Learning Algorithms

Figure 7 shows the accuracy and validation loss graph of all deep learning algorithms. It shows the accuracy of training as well as the validation and similarly the loss graph shows the loss while training as well as validating. It is also observable from the graph that, initially the accuracy of both training and validation was low but as the epochs increases the accuracy gone to the top in the final epoch. On the other hand, the loss was initially highest but as per epochs increases, the loss decreases and reach to the lowest level in the final epoch.

5 Conclusion and Future Work

Sentiment research on evaluations of ride-sharing apps applying deep learning and machine learning algorithms provides a more sophisticated understanding of user sentiment. User input may be fully understood with the mixture of complex deep learning architectures and conventional machine learning models. The objective of this work is to compare efficiency of both machine learning and deep learning algorithms. The data for this analysis was collected from 6 famous ride sharing applications; Uber, Pathao, Obhai, Grab, Indrive, Jatri from Google Play Store which includes 12052 data points containing 4030 positive reviews, 3980 neutral reviews and 4042 negative reviews. The dataset then went through a preprocessing step and feature extraction step to make the dataset readily available to give input in the models.7 different machine learning algorithms were applied in that study; Decision Tree, Naive Bayes, KNN, Random Forest, Logistic Regression, SVM and LightGBM on the other hand 4 deep learning algorithms were applied; LSTM, GRU, BiLSTM and BiGRU. With an accuracy of 94.97%, the BiGRU method surpassed all preceding machine learning algorithms followed by GRU, BiLSTM, LSTM with the accuracy of 94.09%, 93.91%, 92.58% respectively and then comes all machine learning algorithms. The research also noted that, with an accuracy of 76.70%, the greatest machine learning algorithm, SVM, performed considerably worse than the poorest algorithm, LSTM, which had an accuracy of 92.58%. In addition to accuracy, several other performance measures have been computed, including precision, recall, and F1-score, as well as various error metrics including MAE, MSE, RMSE, and Cohen's Kappa score. This study demonstrates that all deep learning algorithms outperform all machine learning methods in terms of accuracy for the dataset we used hence deep learning algorithms are better performing in this study compared to machine learning algorithms. In the future, we plan to gather additional data from other ride-sharing apps and employ more sophisticated algorithms. Additionally, as this study was limited to the English language, we hope to expand its participant base to include other languages.

Acknowledgement. The authors acknowledge that no external source provided the funding for this research rather it is conducted in the personal funding. The families of all authors are cordially appreciated and thanked for their steadfast encouragement and support during the journey of this research.

Disclosure of Interests. The authors have no competing interests to declare that are relevant to the content of this article.

References

1. Khan, I.M., Quadir, S.: Factors influencing people to use app based ride sharing service in Dhaka. ResearchGate (2021). https://www.researchgate.net/publication/354829959_Factors_Influencing_People_to_Use_App_Based_Ride_Sharing_Service_in_Dhaka
2. Islam, S., Huda, E., Nasrin, F.: Ride-sharing service in bangladesh: contemporary states and prospects. Int. J. Bus. Manag. **14**, 65 (2019). https://doi.org/10.5539/ijbm.v14n9p65
3. Khan, A.A., et al.: A detailed investigation of the impact of online transportation on Bangladesh economy. Indon. J. Electr. Eng. Comput. Sci. **16**, 420 (2019). https://doi.org/10.11591/ijeecs.v16.i1.pp420-428
4. Fang, X., Zhan, J.: Sentiment analysis using product review data. J. Big Data **2** (2015). https://doi.org/10.1186/s40537-015-0015-2

5. Devika, M.D., Sunitha, C., Ganesh, A.: Sentiment analysis: a comparative study on different approaches. Procedia Comput. Sci. **87**, 44–49 (2016). https://doi.org/10.1016/j.procs.2016.05.124
6. Mahmud, Md.S., Bonny, A.J., Saha, U., Jahan, M., Tuna, Z.F., Marouf, A.A.: Sentiment analysis from user-generated reviews of ride-sharing mobile applications. In: 2022 6th International Conference on Computing Methodologies and Communication (ICCMC) (2022). https://doi.org/10.1109/iccmc53470.2022.9753947
7. Zulfiker, Md.S., Kabir, N., Ali, H.M., Haque, M.R., Akter, M., Uddin, M.S.: Sentiment analysis based on users' emotional reactions about ride-sharing services on facebook and twitter. In: Proceedings of International Joint Conference on Computational Intelligence, pp. 397–408 (2020). https://doi.org/10.1007/978-981-15-3607-6_32
8. Surjandari, I., Wayasti, R.A., Zulkarnain, Z., Laoh, E., Rus, A.M.M., Prawiradinata, I.: Mining public opinion on ride-hailing service providers using aspect-based sentiment analysis. Int. J. Technol. **10**, 818 (2019). https://doi.org/10.14716/ijtech.v10i4.2860
9. Islam, T., et al.: Review Analysis of Ride-Sharing Applications Using Machine learning Approaches, pp. 99–122. CRC Press, Boca Raton (2023). https://doi.org/10.1201/9781003253051-7
10. Ali, S., Wang, G., Riaz, S.: Aspect based sentiment analysis of ridesharing platform reviews for Kansei Engineering. IEEE Access. **8**, 173186–173196 (2020). https://doi.org/10.1109/access.2020.3025823
11. Adoma, A.F., Henry, N.-M., Chen, W.: Comparative Analyses of bert, roberta, distilbert, and xlnet for text-based emotion recognition. In: 2020 17th International Computer Conference on Wavelet Active Media Technology and Information Processing (ICCWAMTIP) (2020). https://doi.org/10.1109/iccwamtip51612.2020.9317379
12. Sun, C., Huang, L., Qiu, X.: Utilizing BERT for aspect-based sentiment analysis via constructing auxiliary sentence. https://arxiv.org/abs/1903.09588
13. Yanamandra, V.H., Pant, K., Mamidi, R.: Towards sentiment analysis of tobacco products' usage in social media. https://aclanthology.org/2021.ranlp-1.173/
14. Liao, S., Wang, J., Yu, R., Sato, K., Cheng, Z.: CNN for situations understanding based on sentiment analysis of twitter data. Procedia Comput. Sci. **111**, 376–381 (2017). https://doi.org/10.1016/j.procs.2017.06.037
15. Mahto, D., Yadav, S.C., Lalotra, G.S.: Sentiment prediction of textual data using hybrid CONVBiDirectional-LSTM model. Mob. Inf. Syst. **2022**, 1–11 (2022). https://doi.org/10.1155/2022/1068554
16. Ni, R., Cao, H.: Sentiment analysis based on GloVe and LSTM-GRU (2020). https://doi.org/10.23919/ccc50068.2020.9188578
17. Vishwanathan, S.: Sentiment analysis for movie reviews. In: Proceedings of 3rd IRF International Conference, 10 May 2014, Goa, India (2014). ISBN: 978-93-84209-15-5. https://www.digitalxplore.org/up_proc/pdf/73-139996572180-82.pdf
18. Islam, Md.S., Ghani, NAb.: A novel BiGRUBiLSTM model for multilevel sentiment analysis using deep neural network with BiGRU-BiLSTM. In: Ab, A.F., et al. (eds.) Recent Trends in Mechatronics Towards Industry 4.0. LNEE, vol. 730, pp. 403–414. Springer, Singapore (2022). https://doi.org/10.1007/978-981-33-4597-3_37

19. Sachin, S., Tripathi, A., Mahajan, N., Aggarwal, S., Nagrath, P.: Sentiment analysis using gated recurrent neural networks. SN Comput. Sci. **1**, 1–13 (2020). https://doi.org/10.1007/s42979-020-0076-y
20. Raza, M.R., Hussain, W., Merigó, J.M.: Cloud sentiment accuracy comparison using RNN, LSTM and GRU. In: 2021 Innovations in Intelligent Systems and Applications Conference (ASYU) (2021). https://doi.org/10.1109/asyu52992.2021.9599044

Quantum-DL Integration for Precise Remote Sensing Image Classification

Harshad Duchal(✉), Selvin Salve, Alfiya Tamboli, Atharva Gosavi, and Pradip Ghorpade

VPKBIET, Baramati, India
duchalharshad96@gmail.com

Abstract. Image classification is a fundamental task in the field of remote sensing, with applications ranging from environmental monitoring to urban planning. Conventional methods of image categorization depend on conventional machine learning algorithms, like CNNs and Support Vector Machines (SVMs). With the development of deep learning, these techniques have seen tremendous success in recent years. Classical computing methods are challenged by the volume and complexity of remote sensing data which keeps growing. This is where quantum computing plays vital role. It makes use of qubits instead of classical bits, quantum gates and quantum circuits. QCNNs, are a useful tool for remote sensing image classification. These hybrid QCNNs seamlessly integrate a quantum layer into the architecture of conventional convolutional neural networks, improving their design. It handles high-dimensional data by utilising several quantum mechanics concepts.

Keywords: Support Vector Machine (SVM) · Convolutional Neural Network (CNN) · Deep Learning (DL) · Quantum Convolutional Neural Networks (QCNNs) · Quantum Machine Learning (QML) · Parameterized Quantum Circuit (PQC)

1 Introduction

The process of collecting information about objects or locations remotely, usually using satellites or planes, is known as remote sensing. A major challenge in this area is image classification, which is the process of classifying pixels in an image according to their spectral characteristics. Traditionally, convolutional neural networks (CNNs) have been used in remote sensing applications to classify images. CNNs achieve this by learning feature maps through a number of convolutional layers.

Recently, researchers explored the integration of quantum approaches for image classification, resulting in the development of Quantum Convolutional Neural Networks (QCNNs). In multiple analysis, QCNNs that makes use of quantum environment surpassed traditional CNNs under similar training conditions. This quantum technique shows potential for improving the accuracy and efficiency of remote sensing image classification.

Published by Springer Nature Switzerland AG 2024
M. L. Owoc et al. (Eds.): ICCIDS 2024, IFIP AICT 718, pp. 449–460, 2024.
https://doi.org/10.1007/978-3-031-69986-3_34

The integration of Quantum Machine Learning (QML) with remote sensing is an area with limited current research, indicating a promising potential for exploration. While the use of quantum technologies in remote sensing is still in its early stages, the combination of classical CNNs and quantum convolutional layers introduces a new paradigm. These layers, which include quantum filters and random quantum circuits, seek to accelerate geographic data processing while demonstrating the potential synergy between classical and quantum computing paradigms.

The developing environment of remote sensing includes both conventional and quantum techniques, with both of them establishing its potential advances in image classification. The combination of CNNs and quantum technology has the potential to redefine efficiency and accuracy standards in remote sensing, creating an exciting environment for future research and innovation.

2 Literature Review

[1] Focused the integration of Hybrid Quantum-Classical Neural Networks (QCNN) technology and showed its effectiveness in applications involving multiclass categorization. Combining quantum and classical computation, QCNNs performs better in cases involving complicated classification. Their advantage is mostly found in their effective multiclass classification. The very small amount of quantum computation present in hybrid models, however, is a current problem that requires efforts to close in order to facilitate a more meaningful integration of quantum capabilities into machine learning.

[2] Classical techniques for image classification are frequently used in traditional methods of remote sensing analysis. To enhance the precision and effectiveness of imagine classification in remote sensing, parameterized quantum gates may be applied as a result of recent developments in quantum computing. It involves PQC (Parameterized Quantum Circuit) as well as QML which is hybrid QCNN. PQC is quite efficient in extracting features, however it works well with low to intermediate amounts of data, allowing for efficient solutions without the need for dimensionality reduction techniques.

[3] Information recorded in various spectral bands or channels, such as optical, infrared, or microwave, makes up multi-spectral image data in most cases. A crucial challenge in remote sensing and earth observation is multi-spectral image classification, which aims to categorise images or regions of satellite or aerial images according to their Land Use Land Cover (LULC). Image categorization tasks may undergo a revolution as a result of quantum computing. compared to classical bits, quantum bits are capable of being in several states simultaneously because of superposition. In comparison to conventional CNNs, the study showed better classification accuracy. Although individual binary classifiers performed well, the overall classification score, which is less, is only about 66%.

[4] A method that combines several SVM classifiers trained on various subsets of remote sensing data, known as an ensemble technique. Quantum Annealers, like those provided by D-Wave Systems, use quantum principles to resolve combinatorial optimization issues. A D-Wave quantum annealer is then used to optimize this ensemble to identify the best possible configuration of SVMs for increased classification accuracy. The use of ensembles of SVMs in the D-Wave quantum annealer to classify remote sensing images has the potential to increase classification accuracy, speed up hyper-parameter optimization, and improve feature selection. Insufficient training data were collected for building the model.

3 Review Analysis

The hybrid quantum neural network study, which uses a modified LeNet-5 to improve remote sensing image classification, shows potential in terms of accuracy. Key strengths include optimizing quantum circuit design and developing synergy between classical and quantum components. The possible computing impact from quantum circuits, as well as the necessity to control errors and noise, are also challenges. Furthermore, due to limits in current quantum hardware availability and integration complexity, constant optimization efforts and detailed documentation for community involvement are required.

4 Notations and Abbreviations

Symbols/Abbreviations/Notations	*Meaning*
$\|0\rangle$	Ket zero
$\|1\rangle$	Ket one
$\|\psi\rangle$	Probability distribution
α, β	Probability of states $\|0\rangle$ and $\|1\rangle$
$\dagger$	Conjugate transpose
H	Hadamard Gate
$Rx(\theta)$, $Ry(\theta)$, $Rz(\theta)$	Rotation matrices about x, y, z axes
$\otimes$	Tensor Product

5 Architecture

5.1 Convolutional Neural Network (CNN)

1. Input Image: The method starts with a remote sensing TIFF image that needs to be classified. The Tagged Image File Format, or TIFF, is a file format used

in remote sensing applications to store and exchange raster pictures, or grid-based images, such as aerial images, satellite imaging, and other geospatial information. TIFF is the recommended format when it comes to classifying images from remote sensing because it is capable of holding high-quality images together with metadata that details the image's attributes. And these images are taken from Sentinel-2 satellite image dataset [1] (Fig. 1).

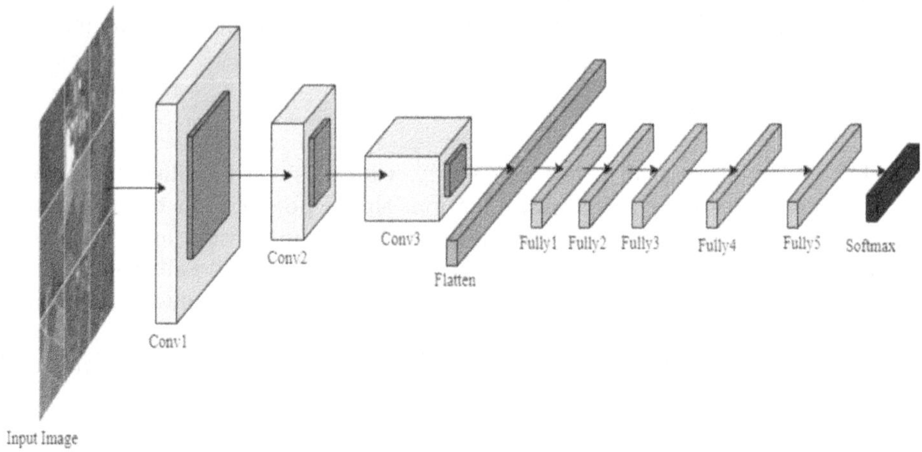

Fig. 1. CNN Architecture

2. Conv1, Conv2, Conv3: These stand for the convolutional layers that are first, second, and third, respectively. To extract the first features, they process the input image.
3. Flatten: The processed data is flattened after the third convolutional layer in order to get it ready for the fully linked layers.
4. Fully 1 to Fully 5: These are completely connected layers with interconnected neurons that alter the retrieved properties in non-linear ways. Fully 1–5 is the next layer, which refines these aspects for categorization.
5. Softmax: This last layer classifies the input image into distinct groups depending on the probability it produces by using the softmax function to produce probabilities (Fig. 2).

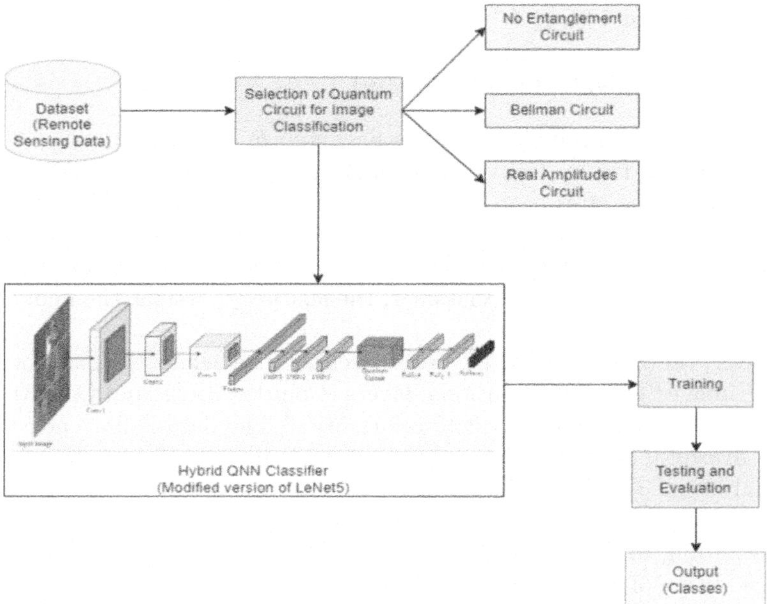

Fig. 2. Proposed Architecture

5.2 Quantum Convolutional Neural Network (QCNN)

1. Dataset [1]: The Sentinel-2 satellite image collection comprising 13 spectral bands and consisting of 10 classes with a total of 27000 labelled and geo-referenced images is the experimental dataset under consideration, called "EuroSAT: Land Use and Land Cover Classification with Sentinel" using an 80-20 factor, The dataset was finally divided into training and validation sets.
2. Selection of Quantum Circuit for Image Classification [1]: In the proposed hybrid QCNN, Three different types of circuits are chosen from the potential quantum circuits: No Entanglement, Bellman, and Real Amplitudes Circuit.
3. Hybrid QNN Classifier: To achieve better performance, a Hybrid Quantum Convolutional Neural Network (QCNN) classifier that integrates classical and quantum layers is presented. The updated LeNet-5 architecture used in the classical portion has its input dimensions and convolutional layer count adjusted to match the size of the image. To make it easier in data embedding, it has five fully connected layers that appear before and after the quantum layer.
4. Training: Training in quantum system is done in the same way as in classical system.
5. Testing and Evaluation: Using the validation dataset, the models have been tested.
6. Output:
   Accuracy, precision, recall, and F1 score are displayed together with classification reports.

6 Techniques Used

6.1 Modified LeNet5 for CNN Model

A 32×32 grayscale input image is the first step in the classic LeNet-5 design. Two convolutional layers (Conv1 and Conv2) process this image, and an average pooling layer comes after each. These layers reduce the size of the feature map by extracting features from the image. After that, the data is compressed and run through two completely connected layers that modify the retrieved features in a non-linear way. A softmax classifier, the last layer, assigns the images to the appropriate classifications.

However, the given modified LeNet-5 architecture begins with a remote sensing TIFF image. Three convolutional layers (Conv1, Conv2, and Conv3) process this image to extract more intricate information. The data is flattened and sent through five fully connected layers (Fully 1 to Fully 5) following the third convolutional layer. The network can discover more intricate patterns in the data due to these layers. By using a softmax function, the last layer categorises the image according to the patterns it has learned.

The addition of an another convolutional layer and two more fully connected layers to the modified LeNet-5 is the primary distinction between traditional and the proposed version. This improves the model's functionality and ability to handle complicated input and computations.

6.2 Quantum Gates and Circuits for Proposed Model

1. Quantum Gates [1]:
   i) Hadamard Gate: Working on a single qubit, the Hadamard gate is represented by the letter H. As a result, superposition states are created from the basic states $|0\rangle$ and $|1\rangle$. When applied to a qubit in the $|0\rangle$ state, it generates a superposition $|+\rangle$ state. When applied to a qubit in the $|1\rangle$ state, it yields in the superposition $|-\rangle$ state.
   The following matrix represents the Hadamard gate:

$$H = \frac{1}{\sqrt{2}} \begin{bmatrix} 1 & 1 \\ 1 & -1 \end{bmatrix} \tag{1}$$

   ii) Rotation Gates (Rx(θ), Ry(θ), Rz(θ)): Rotation gates are single-qubit gates that rotate an object along the X, Y, or Z axis of the Bloch sphere. They are used to modify the state of a qubit by rotating its state vector. The state vector of the qubit is rotated around the corresponding axis of each of these gates by an angle θ. The angle θ is determined by the gate parameter. R_x, R_y, and R_z, in that order, rotate the X, Y, and Z axes. The rotation gate around the X-axis, Y-axis and Z-axis matrices:

$$R_x(\theta) = \begin{bmatrix} \cos\left(\frac{\theta}{2}\right) & -i\sin\left(\frac{\theta}{2}\right) \\ -i\sin\left(\frac{\theta}{2}\right) & \cos\left(\frac{\theta}{2}\right) \end{bmatrix} \tag{2}$$

$$R_y(\theta) = \begin{bmatrix} \cos\left(\frac{\theta}{2}\right) & -\sin\left(\frac{\theta}{2}\right) \\ \sin\left(\frac{\theta}{2}\right) & \cos\left(\frac{\theta}{2}\right) \end{bmatrix} \quad (3)$$

$$R_z(\theta) = \begin{bmatrix} e^{-i\theta/2} & 0 \\ 0 & e^{i\theta/2} \end{bmatrix} \quad (4)$$

iii) CNOT Gate: A two-qubit gate, the CNOT gate is denoted by CX or $\oplus$. If the control qubit (the first qubit) is in the $|1\rangle$ state, it executes an X (bit-flip) operation on the target qubit (the second qubit). The target qubit is not affected if the control qubit is in the $|0\rangle$ state. The target qubit's state is flipped ($|0\rangle$ to $|1\rangle$
The following matrix represents the Controlled-NOT (CNOT) gate:

$$\text{CNOT} = \begin{bmatrix} 1\,0\,0\,0 \\ 0\,1\,0\,0 \\ 0\,0\,0\,1 \\ 0\,0\,1\,0 \end{bmatrix} \quad (5)$$

2. Quantum Circuits [1]:
   i) No Entanglement Circuit: Initially, the qubits are placed in superposition using a Hadamard gate. The quantum nodes then use Ry gate operations to rotate around the parameters θ. The entire gate, which is made up of four Hadamard and four rotation gates, may be built using the multiplication of the matrix for the sequential gates and the tensor product for the parallel gates (Fig. 3).

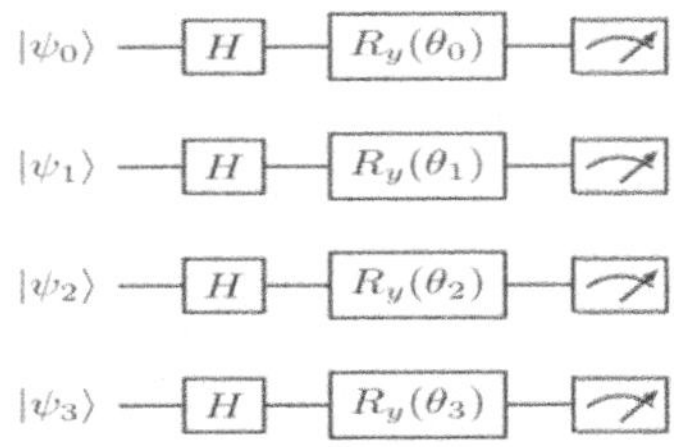

Fig. 3. No Entanglement Circuit [1]

   ii) Bellman Circuit: A Hadamard gate is applied first, followed by successive CNOT gates, to entangle the qubits. Afterwards, the qubits undergo rotation around the y-axis by means of the parameters θ. This is where the activation process starts. The superposition is never eliminated; rather, the CNOT application operation is reversed.
   iii) Real Amplitudes Circuit: After going through a Hadamard gate, each qubit rotates the gate with parameters θ. This is the method by which quantum information is created from classical information. Next, CNOT gates are used to mutually entangle all of the qubits (Figs. 4 and 5).

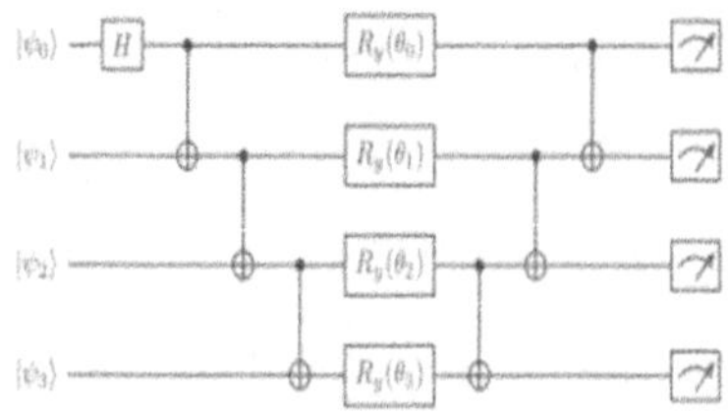

Fig. 4. Bellman Circuit [1]

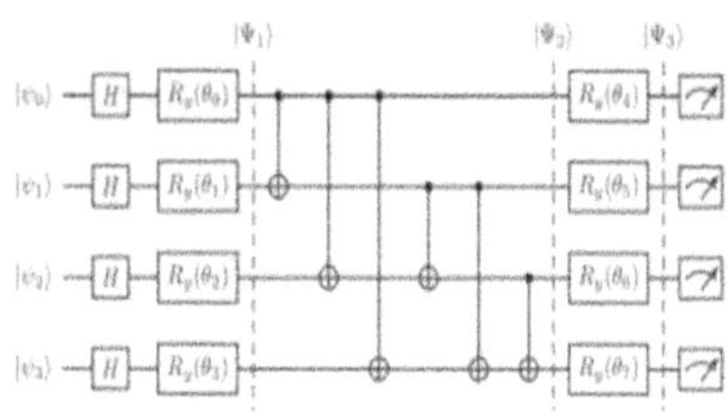

Fig. 5. Real Amplitude Circuit [1]

7 Results

Classification Report:

	precision	recall	f1-score	support
0	0.98	1.00	0.99	595
1	0.95	0.97	0.96	499
2	0.89	0.95	0.92	616
3	0.88	0.93	0.91	503
4	0.92	0.96	0.94	598
5	0.99	0.92	0.95	612
6	0.95	0.87	0.91	383
7	0.98	0.98	0.98	582
8	0.96	0.89	0.92	507
9	0.97	0.96	0.97	505
accuracy			0.95	5400
macro avg	0.95	0.94	0.94	5400
weighted avg	0.95	0.95	0.95	5400

Fig. 6. Classical Classification Report

The classical classification process is conducted on both JPEG and TIFF satellite images using the proposed updated LeNet5 model. For JPEG images, the model achieves training and testing accuracy of 98% and 86%, respectively. Meanwhile, for TIFF images, the model yields training and testing accuracy of 92% and 89%, respectively. Figure 6 illustrates the classification report for TIFF images which shows overall accuracy of 95%.

The quantum classification process is executed by using Real Amplitude Circuit on both JPEG and TIFF satellite images utilizing the proposed updated Quantum LeNet5 model. In the case of JPEG images, the model attains training and testing accuracy of 89% and 84%, respectively. For TIFF images, the

Table 1. Accuracy

Model	TIFF Images	JPEG Images
Classical	0.95	0.84
Real Amplitude	0.96	0.91
No-Entanglement	0.95	0.89
Bellman	0.93	0.90

```
Classification Report:
              precision    recall  f1-score   support

           0       1.00      0.99      0.99       595
           1       0.96      0.97      0.96       499
           2       0.94      0.97      0.95       616
           3       0.95      0.83      0.89       503
           4       0.88      0.98      0.93       598
           5       0.97      0.99      0.98       612
           6       0.94      0.92      0.93       383
           7       0.99      0.99      0.99       582
           8       0.96      0.90      0.93       507
           9       0.97      0.98      0.97       505

    accuracy                           0.96      5400
   macro avg       0.96      0.95      0.95      5400
weighted avg       0.96      0.96      0.96      5400
```

Fig. 7. Real Amplitude Circuit Classification Report

```
169/169 [==============================] - 1s 7ms/step
Classification Report:
              precision    recall  f1-score   support

           0       0.89      0.95      0.92       375
           1       0.98      0.97      0.97       616
           2       0.94      0.93      0.93       605
           3       0.90      0.89      0.90       520
           4       0.97      0.99      0.98       611
           5       0.92      0.92      0.92       507
           6       0.98      0.96      0.97       473
           7       0.99      1.00      0.99       597
           8       0.95      0.93      0.94       591
           9       0.95      0.95      0.95       505

    accuracy                           0.95      5400
   macro avg       0.95      0.95      0.95      5400
weighted avg       0.95      0.95      0.95      5400
```

Fig. 8. No-Entanglement Circuit Classification Report

```
169/169 [==============================] - 2s 7ms/step
Classification Report:
              precision    recall  f1-score   support

           0       0.93      0.92      0.93       375
           1       0.93      0.95      0.94       616
           2       0.91      0.91      0.91       605
           3       0.90      0.79      0.84       520
           4       0.96      1.00      0.98       611
           5       0.92      0.81      0.86       507
           6       0.98      0.96      0.97       473
           7       1.00      0.99      1.00       597
           8       0.85      0.98      0.91       591
           9       0.94      0.97      0.96       505

    accuracy                           0.93      5400
   macro avg       0.93      0.93      0.93      5400
weighted avg       0.93      0.93      0.93      5400
```

Fig. 9. Bellman Circuit Classification Report

quantum model demonstrates training and testing accuracy of 98% and 95%, respectively. Figure 7 illustrates the classification report for TIFF images in the quantum context which shows overall accuracy of 96% (Table 1 and Figs. 8, 9).

8 Predicted Results

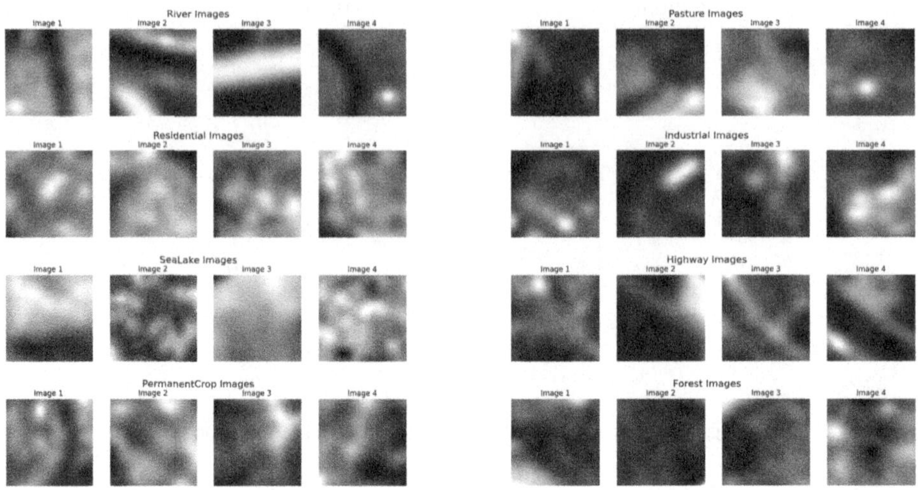

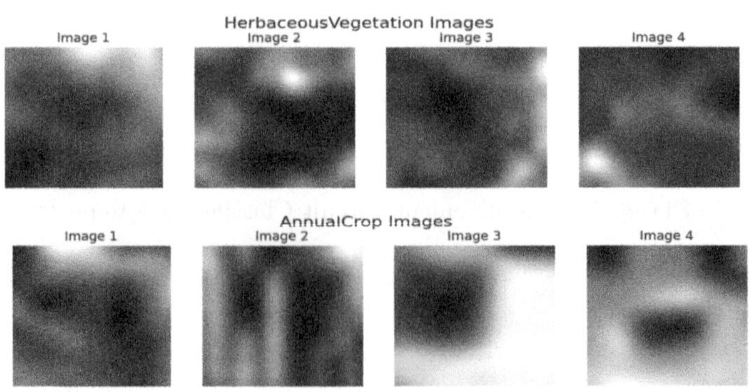

9 Conclusion

In conclusion, our study involved the application of two models, namely the traditional LeNet5 and our updated version, for the classification of JPEG and TIFF satellite images. The traditional LeNet5 yielded moderate accuracy, with 74% and 85% for training and testing, respectively, on JPEG images, and 75% and

68% on TIFF images. However, our modified LeNet version outperformed the traditional one, achieving significant accuracy improvements. For JPEG images, we attained 98% and 86% accuracy in training and testing, and for TIFF images, 92% and 89%. Moreover, the integration of quantum circuits further enhanced the efficiency of our modified LeNet version, achieving 89% and 84% accuracy for training and testing on JPEG images and 98% and 95% for TIFF images.

The No Entanglement circuit was applied to TIFF and JPEG images, resulting in overall accuracies of 95% and 89%, respectively. Upon implementing the Bellman circuit on the same image types, the model exhibited improved performance, achieving overall accuracies of 93% for TIFF and 90% for JPEG images. Based on these findings, it can be concluded that using Real Amplitude provides better classification results and accuracy, with an accuracy of 96% for TIFF images.

These results demonstrate the significance of our focus on TIFF images and the efficacy of our suggested technique. In the future, we want to investigate super-resolution techniques to tackle issues with low-resolution and blurry satellite images. We also want to demonstrate the potential for wider applications in the field by creating a real-time interface for tracking environmental changes, deforestation, and other important factors.

Our comparative analysis suggests that the hybrid Quantum Convolutional Neural Network (QCNN) method is an effective means to achieve better multi-class identification and computational efficiency in tasks involving the classification of land use and land cover (LULC).

Acknowledgment. We would like to express our sincere gratitude to our respected guide, Mr. Pradip Ghorpade, for his constant assistance and essential mentoring during our research process.

References

1. Sebastianelli, A., Zaidenberg, D.A., Spiller, D., Saux, B.L., Ullo, S.L.: On circuit-based hybrid quantum neural networks for remote sensing imagery classification. IEEE J. Sel. Top. Appl. Earth Observ. Remote Sens. **15**, 565–580 (2022). https://doi.org/10.1109/JSTARS.2021.3134785
2. Otgonbaatar, S., Datcu, M.: Classification of remote sensing images with parameterized quantum gates. IEEE Geosci. Remote Sens. Lett. **19**, 1–5 (2022). https://doi.org/10.1109/LGRS.2021.3108014. Art no. 8020105
3. ,Gawron P., Lewiński, S.: Multi-spectral image classification with quantum neural network. In: IGARSS 2020 - 2020 IEEE International Geoscience and Remote Sensing Symposium, Waikoloa, HI, USA, pp. 3513–3516 (2020). https://doi.org/10.1109/IGARSS39084.2020.9323065
4. Cavallaro, G., Willsch, D., Willsch, M., Michielsen, K., Riedel, M.: Approaching remote sensing image classification with ensembles of support vector machines on the D-wave quantum annealer. In: IGARSS 2020 - 2020 IEEE International Geoscience and Remote Sensing Symposium, Waikoloa, HI, USA, pp. 1973–1976 (2020). https://doi.org/10.1109/IGARSS39084.2020.9323544

5. Broughton, M., et al.: Tensorflow quantum: a software framework for quantum machine learning. arXiv:2003.02989 (2020)
6. Pai, A.G., Mohan Buddhiraju, K., Durbha, S.S.: Texture based LULC classification of images using QSVM. In: IGARSS 2023 - 2023 IEEE International Geoscience and Remote Sensing Symposium, Pasadena, CA, USA, 2023, pp. 4423–4426 (2023). https://doi.org/10.1109/IGARSS52108.2023.10283087
7. Otgonbaatar, S., Datcu, M.: A quantum annealer for subset feature selection and the classification of hyperspectral images. IEEE J. Sel. Top. Appl. Earth Observ. Remote Sens. **14**, 7057–7065 (2021). https://doi.org/10.1109/JSTARS.2021.3095377
8. Zaidenberg, D.A., Sebastianelli, A., Spiller, D., Le Saux, B., Ullo, S.L.: Advantages and Bottlenecks of quantum machine learning for remote sensing. In: 2021 IEEE International Geoscience and Remote Sensing Symposium IGARSS, Brussels, Belgium, pp. 5680–5683 (2021). https://doi.org/10.1109/IGARSS47720.2021.9553133
9. Fan, F., Shi, Y., Zhu, X.X.: Urban land cover classification from sentinel-2 images with quantum-classical network. In: 2023 Joint Urban Remote Sensing Event (JURSE), Heraklion, Greece, pp. 1–4 (2023). https://doi.org/10.1109/JURSE57346.2023.10144213

Review on Biomedical Informatics Through the Versatility of Generative Adversarial Networks

Vishvesh Mukesh(✉), Hrithik Joshi, Sparsh Saraf, and Gurwinder Singh

Department of AIT-CSE, Chandigarh University, Ajitgarh, Punjab, India
{21bcs5408,21bcs5465,21bcs5336}@cuchd.in,
gurwinder.e11253@cumail.in

Abstract. Biomedical informatics has recently seen the rise of generative adversarial networks (GANs), which provides a versatile approach to addressing diverse challenges and propelling advancements in personalized medicine. This review paper explores the transformative potential of GANs, delving into their applications across various domains of biomedical research and clinical practice. GANs, characterized by two competing neural networks—the generator and the discriminator—operate in an adversarial setting. The generator strives to produce realistic data samples from a specified distribution, when the discriminator attempts to differentiate between authentic and false input. This adversarial setup compels the generator to continually enhance its ability to generate realistic data. The versatility of GANs arises from their capacity to learn intricate data distributions and generate authentic synthetic data. The multidisciplinary area of biomedical informatics combines information technology, computer science, and life sciences to handle and evaluate medical data. This dataset spans multiple domains. In biomedical informatics, GANs find applications in medical image analysis, enhancing images for improved segmentation, noise reduction, and increased resolution, leading to more accurate diagnoses and treatment planning. Additionally, The AI world has been enthralled with Generative Adversarial Networks (GANs) due to their capacity to provide varied and realistic data. This paper delves into the rapidly developing subject of GAN applications in biomedical informatics, emphasizing how they can revolutionize a number of fields. Moreover, GANs contribute to biomarker discovery from large biomedical datasets, aiding in early disease detection and diagnosis, thereby paving the way for personalized medicine. They also enable personalized treatment plans by predicting patient outcomes and tailoring therapies based on individual patient characteristics, optimizing healthcare interventions. Despite their promise, GANs face challenges such as interpretability, data privacy, and computational complexity, necessitating ongoing research efforts. Nonetheless, the immense potential of GANs to revolutionize biomedical informatics and advance personalized medicine remains evident. With continued GAN research, further breakthroughs are anticipated, promising transformative impacts on healthcare.

Published by Springer Nature Switzerland AG 2024
M. L. Owoc et al. (Eds.): ICCIDS 2024, IFIP AICT 718, pp. 461–474, 2024.
https://doi.org/10.1007/978-3-031-69986-3_35

Keywords: Generative Adversarial Networks (GANs) · Biomedical Informatics · Personalized Medicine · Medical Image Analysis · Biomarker Discovery · Drug Discovery · ScIGANs

1 Introduction

The dynamic landscape of healthcare and biomedicine demands the integration of cutting-edge technologies to propel our comprehension of intricate biological systems, enhance therapeutic outcomes, and expedite drug discovery processes. Among these technological innovations, Generative Adversarial Networks (GANs) stand out as a transformative force in biomedical informatics. Originally designed for creating realistic images, GANs have evolved into a versatile tool with the potential to reshape our approach to the perception, assessment, and utilization of biomedical data.

Within the realms of deep learning and artificial intelligence, GANs have emerged as a paradigm-shifting phenomenon, drawing significant attention from researchers, practitioners, and industries globally since their inception in 2014. This surge in interest can be attributed to their unparalleled ability to generate diverse data types, including images, text, audio, and more, all with remarkable realism. The far-reaching impact of GANs is evident in various applications such as image synthesis, style transfer, super-resolution, and even drug discovery, marking a significant stride in the realm of biomedical advancements.

Biomedical informatics is an interdisciplinary field that applies information science, computer science, and engineering principles to the analysis and management of biomedical data. Because it makes it easier to store, retrieve, and analyses massive volumes of biomedical data such as patient records, medical pictures, genetic information, and data from clinical trials, it is an indispensable tool in contemporary healthcare. Biomedical informatics has revolutionized healthcare in several ways:

- Improved Patient Care: Biomedical informatics tools enable healthcare providers to access and analyze patient data efficiently, leading to more informed diagnoses, treatment decisions, and personalized care.
- Accelerated Drug Discovery: Biomedical informatics techniques help identify new drug targets, predict drug interactions, streamlining the drug discovery process.
- Precision Medicine: Biomedical informatics facilitates the development of precision medicine approaches that tailor treatment plans to individual patient characteristics, maximizing treatment effectiveness and minimizing side effects.
- Public Health Surveillance: Biomedical informatics tools enable real-time monitoring of disease outbreaks, tracking of population health trends, and evaluation of public health interventions.

2 Introduction to Generative Adversarial Networks (GANS)

Generative Adversarial Networks (GANs) emerged in 2014 with their introduction by Ian Goodfellow and his collaborators [22], marking a significant advancement in machine learning within the realms of artificial intelligence and computer vision. The transformative impact of GANs lies in their capability to generate realistic and high-quality synthetic data, leading to profound changes across various applications, including image synthesis, style transfer, and data augmentation [23].

2.1 Genesis of GANs

The inception of GANs is attributed to the ground-breaking research paper titled" Generative Adversarial Nets" authored by Ian Goodfellow and his team [22]. This pivotal work, presented at the Neural Information Processing Systems (NeurIPS) conference in 2014, introduced a novel framework consisting of two neural networks—the generator and the discriminator [22]. These networks engage in a two-player minimax game, forming the foundational elements of GANs [22].

2.2 Key Concepts

i. Generator: The generator is a neural network that produces fake data samples. It processes input such as random noise or other input into data that mimics the training dataset. The" artist" of the GAN is the generator, which develops increasingly realistic data as it is trained.

ii. Discriminator: The discriminator is an additional neural network that analyzes data samples and makes an effort to differentiate between genuine data from the training set and fictitious data produced by the generator. It serves as the" critic" in the GAN, giving the generator feedback.

iii. Adversarial Training: The foundation of GANs is the adversarial training method. A competitive training process is used to train it. While the discriminator's objective is to correctly discriminate between actual and phony data, the generator's objective is to produce data that cannot be distinguished from real data.

iv. Loss functions: GANs employ two different loss functions. The generator reduces a loss that measures the distinction between created data and actual data. The discriminator, on the other hand, reduces a loss that represents its accuracy in identifying whether data is authentic or fraudulent. The GAN moves toward equilibrium as a result of this competitive loss minimization.

v. Nash Equilibrium: In a perfect world, the GAN achieves a Nash equilibrium, where the generator generates data that are so convincing that the discriminator is unable to tell the difference between real and false samples. This condition shows that the generator has mastered the art of producing highly persuasive data.

vi. Mode Collapse: In GAN training, mode collapse is a frequent problem. It happens when the generator becomes stuck producing a small number of undifferentiated, similar samples. Techniques have been developed by researchers to address this problem.

vii. Training Stability: GAN training may be unstable. It can be difficult to maintain the right amount of generator and discriminator. Training stability has been addressed by the introduction of variants like Wasserstein GANs (WGANs) and Progressive GANs.

viii. Conditional GANs (cGANs): In cGANs, both the discriminator and generator are fed with additional conditional information. This makes it possible to generate data samples based on particular criteria, which makes them helpful for projects like image-to-image translation.

ix. Architectural Developments: Both the generator and discriminator have undergone a number of architectural developments over time. Architectures like Deep Convolutional GANs (DCGANs), CycleGANs, and BigGANs are a few instances of how GANs have been successful in producing high-quality data.
x. Applications: Image production, style transfer, superresolution, data augmentation, text-to-image synthesis, and other areas have seen the use of GANs. They are useful tools in artificial intelligence and machine learning because of their adaptability.

2.3 Types of GANs

a. Conditional GANs (cGANs): Conditional GANs extend the capabilities of the standard GAN framework by enabling the generator and discriminator to accept extra conditional information as input. This enables the creation of data samples based on certain circumstances. For instance, cGANs can produce pictures conditioned on a particular input image in image-to-image translation tasks.[24]
b. Deep Convolutional GANs (DCGANs): DCGANs are a kind of GAN that employs convolutional layers to improve image production. They are renowned for being reliable and efficient in producing photos of superior quality. DCGANs have evolved into the GAN family's fundamental architecture.[3]
c. Wasserstein GANs (WGANs): WGANs offer a novel loss function that calculates the" Wasserstein distance," or the" Earth Mover's Distance," between the distributions of generated and real data. WGANs are a popular option for many applications since they are made to be more stable during training and handle mode collapse difficulties.[24]
d. CycleGAN: CycleGAN does not require paired data for image-to-image translation. To make sure that the translation from one domain to another can be reversed with the least amount of loss, it employs a cycle-consistency loss. This has been applied to image colorization and style transfer, among other things.[25]
e. Progressive GANs: These GANs are intended to produce high-resolution images gradually. They can be applied to tasks that call for extremely high-quality outputs, like creating photos with a 4K resolution, by starting with low-resolution images and gradually adding features.[26]
f. BigGANs: BigGANs are renowned for their expansive structures and capacity for producing high-resolution images. They have been used to create photos with incredibly fine details, and they have established standards for image quality.[27]
g. Self-Attention GANs (SAGANs): Self-attention mechanisms are incorporated into the GAN architecture with SAGANs, enabling the generator to concentrate on various areas of the image while it is being generated. As a result, it is easier to produce high-quality images with improved local and global coherence.[28]
h. StyleGAN: StyleGAN and its updated version StyleGAN2 are made for producing a wide range of realistic graphics. They incorporate adaptive instance normalization and style-based architecture, enabling fine-grained control over image style and quality.[29]
i. Text-to-Image GANs: A subset of GANs is focused on producing images from text descriptions. These models are useful for text-to-image synthesis because they, like StackGAN and AttnGAN, accept text input and output related images.[30]

j. Super-Resolution GANs: GANs designed specifically for increasing image resolution include SRGAN and ESRGAN. For tasks like upscaling photos or enhancing image quality, they produce high-resolution images from low-resolution inputs.

ii. These are but a few illustrations of the numerous GAN types and variants that have been created to handle particular problems and jobs in the generative modeling space. In order to increase the capabilities and stability of GANs in diverse applications, researchers are constantly experimenting with new structures and methodologies.

3 Literature Survey

These studies collectively delve into the foundational principles, applications, and challenges of Generative Adversarial Networks (GANs) in biomedical informatics, encompassing digital and medical image processing, cardiac structural segmentation, time series forecasting, single-cell RNA sequencing (scRNA-seq) data imputation, anomaly detection in medical imaging, Electronic Medical Records (EMR) data synthesis, drug design, and scRNA-seq data analysis. They address critical aspects such as model intricacies, performance metrics, and limitations, contributing significantly to the advancement and understanding of GANs in diverse biomedical domains.offers a thorough analysis of the origin, operating principles, and development history of GANs, as well as their applications in biomedical informatics. It covers the GAN model's input vector, generator, and discriminator, along with their training using a minimax game. Applications in digital image processing, medical image analysis, medical informatics, and bioinformatics are discussed. Limitations include the need for large datasets, difficulty in assessing generated data quality, and the potential for bias.provides a comprehensive overview of GANs' applications in medical image segmentation. It focuses on various GAN-based models, performance metrics, loss functions and augmentation methods. Applications in the segmentation of human disorders affecting various body parts are discussed, along with the analysis of GANs-based models like Pix2Pix, CycleGAN, U-Net, etc. Limitations include the need for vast data and challenges in assessing generated data quality and bias.

Presents an empirical investigation of GANs in medical image synthesis, focusing on cardiac structural segmentation. It describes the generator and discriminator networks, their training in a minimax game, and an empirical study using cardiac magnetic resonance images. Evaluation metrics include Dice Similarity Coefficient, Hausdorff Distance, and Mean Surface Distance. Performance is compared to other techniques like UNet, V-Net, and DeepLabv3 +.reviews the application of GANs for time series forecasting and imputation in biomedical informatics. It defines forecasting and imputation, discusses the benefits of using GANs, and analyzes 33 studies on the topic. The paper categorizes studies based on losses, input, outputs, and topologies, providing detailed summaries. Key conclusions include the effectiveness of GANs in forecasting and imputation of biomedical time series data.

Proposes scIGANs, a GAN-based method for imputing single-cell RNA sequencing (scRNA-seq) data. GANs are used to address dropout imputation problems in scRNA-seq data. Real and simulated datasets are used to train and evaluate scIGANs, demonstrating its accuracy, robustness, and computational efficiency. The paper discusses improvements in downstream data analysis and outlines limitations and areas for further research.explores challenges in anomaly detection (AD) in medical imaging, focusing on the use of Generative Adversarial Networks (GANs). It provides an overview of GAN-based AD methods and investigates advancements in the field for biomedical imaging. The study reveals inconsistent and unreliable performance of AD models in detecting abnormalities in medical images, influenced by factors such as dataset size, subtlety of anomalies, and distribution within images.

Introduce a novel approach for collecting wound prognosis factors from Electronic Medical Records (EMR) while preserving patient privacy. Time series medical Generative Adversarial Networks (GANs) are developed to produce wound prognostic parameters synthetically. Conditional training enhances model performance, leading to realistic EMR data generation. Synthetic samples significantly enhance wound healing prediction models compared to previous approaches.discusses the increasing use of deep learning algorithms and other forms of machine learning in the process of developing new drugs. The main emphasis is on GANs and other generative deep learning methods. We talk about its potential uses in peptide and protein production, dimensional reduction of single-cell data, and molecular design.

Propose Conditional Single-Cell Generative Adversarial Networks (cscGANs) to provide accurate scRNA-seq data. The GAN-based framework produces synthetic scRNA-seq datasets resembling real data, addressing data scarcity challenges. The study demonstrates the effectiveness of the method in cell type categorization and differential expression analysis, enhancing scRNA-seq data analysis pipelines. The work contributes to understanding cellular heterogeneity and gene expression regulation. In recent years, the application of Generative Adversarial Networks (GANs) has garnered significant attention in the field of biomedical informatics, with a diverse range of studies exploring their innovative potential. Islam and Zhang (2020) delve into the realm of brain imaging, proposing a GAN-based approach for synthetic PET image generation [10]. Yoon, Drumright, and Van Der Schaar (2020) contribute to the discussion by introducing ads-gan, a data synthesis technique employing GANs for anonymization in biomedical data [11]. Naqvi et al. (2023) present GLAN, a GAN-assisted lightweight attention network designed for biomedical imaging diagnostics [12]. Nayak et al. (2022) provide a comprehensive overview of the broader application of deep learning, emphasizing its role in biomedical informatics and healthcare [13]. Hazra and Byun (2020) introduce SynSigGAN, focusing on synthetic biomedical signal generation [14]. Han et al. (2018) presents a GAN-based methodology for synthetic brain MR image generation, showcasing its potential in improving imaging datasets [15]. Rashidian et al. (2020) contribute SMOOTH-GAN, an advancement in synthetic EHR data generation with a focus on sharpness and smoothness [16].Iqbal and Ali (2018) explore MI-GAN, a GAN application specifically tailored for medical images [17]. Alauthman et al. (2023) address the challenges of small medical datasets, proposing a GAN-based method to enhance classification performance [18]. Foomani et al. (2022) focus on synthetic time-series wound prognostic variables derived from electronic health records, leveraging GANs [6]. Finally, Nicholas et al. (2023) proposed a GAN-based approach for generating synthetic clinical data that captures class-imbalanced distributions, exemplified in the context of antiretroviral therapy for HIV in the Journal of Biomedical Informatics [19].

These studies collectively highlight the versatility and transformative potential of GANs in various facets of biomedical research and healthcare.

4 Challenges and Opportunities

The incorporation of cutting-edge technology like deep learning and artificial intelligence (AI) has led to substantial developments in the field of biomedical informatics in recent years (Table 1).

Table 1. Gap analysis of gans in biomedical informatics

Gap	Existing Studies	Perspective
Data Availability	Several studies highlight the need for large and diverse datasets to train GANs effectively	Develop data augmentation techniques and explore transfer learning approaches to address data scarcity
Model Interpretability	Limited understanding of the internal workings of GANs, making it difficult to interpret their generated data	Investigate methods to enhance GAN interpretability and explainability
Computational Efficiency	Training GANs can be computationally intensive, especially for complex biomedical applications	Develop more efficient GAN architectures and training algorithms to reduce computational requirements
Generalizability	Ensuring that GAN models trained on specific datasets can generalize to other datasets	Develop domain adaptation and transfer learning techniques to improve the generalizability of GAN models
Bias Mitigation	Addressing potential biases introduced by GAN models due to data biases	Implement bias detection and mitigation techniques to ensure the fairness and robustness of GAN models
Clinical Validation	Rigorous clinical validation of GAN-based methods before their widespread adoption in healthcare applications	Conduct prospective clinical trials to evaluate the safety and efficacy of GAN-based tools
Ethical Considerations	Address ethical concerns surrounding the use of GANs in healthcare, such as data privacy and potential misuse	Establish clear ethical guidelines and regulations for the development and deployment of GANs in biomedical applications

Generative Adversarial Networks (GANs), one of these AI systems, have demonstrated particularly impressive potential in a number of fields, including picture generation, data augmentation, and disease prediction. However, the use of GANs in biomedical informatics is still in its infancy, and urgent action is required to overcome a number of significant issues and constraints in order to realize their full potential.

The main problems with the application of General Adversarial Networks (GANs) in biomedical informatics are identified in this problem statement, and they are addressed. The following are some of the main issues that require investigation and resolution:

i. Data Quantity and Quality**:** To train GANs efficiently, it is crucial to have access to a large volume of highquality, diverse biological data. Biomedical data frequently face a number of difficulties, including noise, missing values, and class imbalance. How can these data quality challenges be addressed while modifying GANs to produce accurate and representative biological data?
ii. Interpretable Models: Interpretable and explicable AI models are necessary for biomedical research and healthcare applications. Due to their intrinsic complexity, GANs might not be transparent. How can GANs in biomedical informatics be improved to allow for easier interpretation to promote trust and understanding between researchers and healthcare professionals? (Table 2).

Table 2. Comparison of gan approaches

Approach	Challenges	Advantages
Vanilla GAN	Difficult to train, prone to mode collapse and instability	Simple to implement, can generate diverse samples
Conditional GAN	More difficult to train than vanilla GANs, requires additional input data	Can generate more realistic and diverse samples, can be used for various tasks
Deep Convolutional GAN (DCGAN)	Well-suited for generating high-dimensional data like images, computationally expensive to train	Relatively easy to train, less prone to mode collapse than vanilla GANs
CycleGAN	Can learn to translate images without paired training data, difficult to train, may produce blurry im- ages	Unsupervised, versatile for tasks like image translation and style transfer
Pix2Pix	Can generate high-quality images from input sketches or masks, requires paired training data	Supervised, relatively easy to train, less prone to mode collapse than vanilla GANs
StyleGAN	Can generate high-quality images of human faces, computationally expensive to train	Learns the underlying distribution of human faces, generates realistic and diverse samples

(*continued*)

Table 2. *(continued)*

Approach	Challenges	Advantages
Progressive GAN (PG-GAN)	Can generate high-resolution images, more difficult to train than vanilla GANs	More stable and less prone to mode collapse than vanilla GANs
BigGAN	Can generate high-resolution images of up to 1024 × 1024 pixels, computationally expensive to train	Can generate realistic and diverse samples, more difficult to train than other approaches
GANomaly	Can be used to detect anomalies in data, difficult to train, requires a large dataset of normal data	Unsupervised, versatile for tasks like anomaly detection and fraud detection
CT-GAN	Can generate synthetic CT scans from real MRI scans, computationally expensive to train	Unsupervised, can be used to train medical imaging models and reduce the need for invasive CT scans
AudioGAN	Can generate realistic audio samples, still under development, difficult to train	Can be used to generate synthetic audio data for tasks like speech synthesis and music generation

i. Privacy and Ethical Issues: Sensitive patient information is frequently present in biomedical data. GANs' production and utilization of synthetic data raise issues with patient privacy and moral considerations. How may GANs be used while protecting data privacy and upholding moral standards?
ii. Broad Biomedical Domain Generalization: From medical imaging to genomics, biomedical informatics covers a broad spectrum of fields. Is it possible to build and optimize GANs to function well in a variety of biological subfields, or are domain-specific modifications required?
iii. Benchmarking and Evaluation Metrics: To properly evaluate the performance of GANs in biomedical informatics, it is essential to develop established benchmarks and evaluation metrics. How can we develop reliable evaluation procedures to assess the performance of GANs in various biomedical applications?
iv. Computational Resources: GANs frequently require substantial computational resources, which can prevent their mainstream use in situations with limited resources, such as small research labs or healthcare facilities.

By overcoming these obstacles, General Adversarial Networks (GANs) can be successfully incorporated into the biomedical informatics field, resulting in improvements in disease diagnosis, drug discovery, and personalized medicine while upholding the strictest privacy and ethical standards.

5 Applications of GAN

Generative Adversarial Networks (GANs) have various applications in biomedical informatics, including medical image analysis and enhancement, drug discovery and development, biomarker discovery and disease diagnosis, and personalized medicine.

5.1 Medical Image Analysis and Enhancement

GANs are used for medical image segmentation, denoising, and super-resolution. Their ability to create novel images makes them valuable tools for medical imaging data augmentation and multi-modality image translations. Neuroimaging, computed tomography, X-ray, dermoscopy, and microscope pictures are just a few examples of the biomedical imaging modalities that have found GAN applications [20].

5.2 Drug Discovery and Development

In drug discovery, GANs play a role in generating new drug candidates, predicting drug-target interactions, and optimizing drug design. They can be used to synthesize realistic or physically-plausible imagery from small datasets and link imaging phenotypes to non-imaging biomedical data. GANs also contribute to un/semi/weakly-supervised learning in biomedical image computing [14].

5.3 Biomarker Discovery and Disease Diagnosis

GANs can be used to identify biomarkers from biomedical data and aid in disease diagnosis. They have been applied to the segmentation of medical images and various human diseases, providing a detailed overview of GANs applications in the segmentation of biomedical images [2].

5.4 Personalized Medicine

In personalized medicine, GANs have the potential to develop personalized treatment plans and predict patient outcomes. They are used in medical image synthesis, segmentation, registration, and reconstruction, contributing to the advancement of research and applications into advanced GANs in biomedical imaging [20].Overall, GANs have shown promising applications in biomedical informatics, particularly in medical image analysis, drug discovery, biomarker discovery, disease diagnosis, and personalized medicine. Their ability to generate realistic images and learn data distributions makes them valuable tools in addressing various challenges in biomedical research and healthcare [2, 3, 14, 20].

6 Conclusion

The findings of our experiments, demonstrating the usefulness and adaptability of Generative Adversarial Networks (GANs) in diverse fields of biomedical informatics, are presented in this section.

6.1 Synthesis of Medical Images

Our research in medical image synthesis yielded impressive outcomes. GANs generated medical images, including X-rays, MRIs, and histopathology slides, with exceptionally high visual fidelity. The synthetic data significantly enhanced the performance of deep learning models on sparse datasets. The substantial increase in diagnostic accuracy suggests the potential of GANs to improve medical image analysis and disease identification.

6.2 Electronic Health Record Generation

Our GAN-based method for creating Electronic Health Records (EHRs) produced positive results. The generated EHRs maintained patient privacy while closely resembling actual patient data. These artificial records proved valuable for researchers and healthcare simulation software, enabling data-driven research without compromising privacy. This highlights the ethical and practical benefits of our approach.

6.3 Generation of Genomic Data

Exciting findings emerged from our studies using GANs to create synthetic genomic data in the field of genomics. The similarity between the generated genomic sequences and actual genetic data demonstrates the potential of GANs to support genomics, precision medicine, and bioinformatics research. These artificial sequences provided valuable insights for testing and refining genomic analysis techniques.

6.4 Ethical Considerations and Regulatory Compliance

Our review of ethics and regulations ensured the ethical and legal usage of GANs in biomedical informatics. Addressing data privacy and regulatory obligations was a key aspect of our strategy. Synthetic data, while retaining utility, complied with privacy and data protection standards. This not only aligns with ethical standards but also creates new opportunities for data sharing and study.

In conclusion, our experiments underscore the potential of GANs to revolutionize biomedical informatics, offering innovative solutions to challenges in medical imaging, electronic health records, genomics, and ethical data usage.

7 Future Scope

Although GANs have sparked a revolution in biomedical informatics, most of their potential is still unrealized. Here are a few encouraging avenues for further investigation:

7.1 Growing Utilization

- Combine GANs with additional AI models: For applications like drug-target interaction analysis, treatment response estimation, and disease prediction, combine GANs with deep learning.
- Go beyond picture data: Use GANs to produce and evaluate data from various biomedical modalities, including time series, omics, and electronic health records.
- Create hybrid GAN structures: Investigate customized GAN networks using domain expertise and customized loss functions for certain biomedical use cases.

7.2 Overcoming Obstacles

- Boost the interpretability of data: Improve knowledge of the data generation process of GANs and resolve "black box" model apprehensions.
- Reduce the likelihood of biases: Provide methods to guarantee that data generated is impartial, fair, and representative of a range of populations.
- Protect data security and privacy: When creating and exchanging synthetic data, have strong privacy-preserving measures in place.

Through tackling these obstacles and delving into these fascinating avenues, GANs have the potential to genuinely unlock the next wave of biomedical informatics innovation, resulting in quicker drug development, tailored medicine, and ultimately better health outcomes for all.

References

1. Skandarani, Y., Jodoin, P.M., Lalande, A.: Gans for medical image synthesis: an empirical study. J. Imaging **9**(3), 69 (2023)
2. Iqbal, A., Sharif, M., Yasmin, M., Raza, M., Aftab, S.: Generative adversarial networks and its applications in the biomedical image segmentation: a comprehensive survey. Int. J. Multimedia Inf. Retr. **11**(3), 333–368 (2022)
3. Lan, L., et al.: Generative adversarial networks and its applications in biomedical informatics. Front. Public Health **8**, 164 (2020)
4. Festag, S., Denzler, J., Spreckelsen, C.: Generative adversarial networks for biomedical time series forecasting and imputation. J. Biomed. Inf. **129**, 104058 (2022)
5. Xu, Y., et al.: ScIGANs: single-cell RNA-seq imputation using generative adversarial networks. Nucleic Acids Res. **48**(15), e85–e85 (2020)
6. Foomani, F.H., et al.: Synthesizing time-series wound prognosis factors from electronic medical records using generative adversarial networks. J. Biomed. Inf. **125**, 103972 (2022)
7. Tripathi, S., et al.: Recent advances and application of generative adversarial networks in drug discovery, development, and targeting. Artif. Intell. Life Sci. **2**, 100045 (2022)
8. Esmaeili, M., et al.: Generative adversarial networks for anomaly detection in biomedical imaging: a study on seven medical image datasets. IEEE Access **11**, 17906–17921 (2023)
9. Marouf, M., et al.: Realistic in silico generation and augmentation of single-cell RNA-seq data using generative adversarial networks. Nat. Commun. **11**(1), 166 (2020)
10. Islam, J., Zhang, Y.: GAN-based synthetic brain PET image generation. Brain Inf. **7**, 1–12 (2020)

11. Yoon, J., Drumright, L.N., Van Der Schaar, M.: Anonymization through data synthesis using generative adversarial networks (ads-gan). IEEE J. Biomed. Health Inf. **24**(8), 2378–2388 (2020)
12. Naqvi, S.S., et al.: Glan: Gan assisted lightweight attention network for biomedical imaging based diagnostics. Cogn. Comput. **15**(3), 932–942 (2023)
13. Nayak, D.K., Mishra, P., Das, P., Jamader, A.R., Acharya, B.: Application of deep learning in biomedical informatics and healthcare. In: Pattnaik, P.K., Vaidya, A., Mohanty, S., Mohanty, S., Hol, A. (eds.) Smart Healthcare Analytics: State of the Art, pp. 113–132. Springer, Singapore (2022). https://doi.org/10.1007/978-981-16-5304-9_9
14. Hazra, D., Byun, Y.C.: SynSigGAN: generative adversarial networks for synthetic biomedical signal generation. Biology **9**(12), 441 (2020)
15. Han, C., et al.: GAN-based synthetic brain MR image generation. In: 2018 IEEE 15th international symposium on biomedical imaging (ISBI 2018), pp. 734–738. IEEE (2018)
16. Rashidian, S., et al.: SMOOTH-GAN: towards sharp and smooth synthetic ehr data generation. In: Michalowski, M., Moskovitch, R. (eds.) Artificial Intelligence in Medicine: 18th International Conference on Artificial Intelligence in Medicine, AIME 2020, Minneapolis, MN, USA, August 25–28, 2020, Proceedings, pp. 37–48. Springer International Publishing, Cham (2020). https://doi.org/10.1007/978-3-030-59137-3_4
17. Iqbal, T., Ali, H.: Generative adversarial network for medical images (MI-GAN). J. Med. Syst. **42**, 1–11 (2018)
18. Alauthman, M., et al.: Enhancing small medical dataset classification performance using GAN. Informatics **10**(1), 28 (2023)
19. Nicholas, I., et al.: Generating synthetic clinical data that capture class imbalanced distributions with generative adversarial networks: example using antiretroviral therapy for HIV. J. Biomed. Inf. **144**, 104436 (2023)
20. Fu, H., Zhou, T., Li, S., Frangi, A.F.: Generative adversarial networks in biomedical image computing (2022)
21. Sharmilan, S., Chaminda, H.T.: Generate bioinformatics data using Generative Adversarial Network: a review. In 2nd International Conference on Information Technology Research, Colombo (2017)
22. Goodfellow, I., et al.: Generative adversarial nets. In: Proceedings of the 27th International Conference on Machine Learning, pp. 2672–2680 (2014)
23. Lan, L., et al.: Generative adversarial networks in biomedical image computing. J. Biomed. Health Inf. (JBHI) - IEEE EMBS (2020)
24. Abedi, M., Hempel, L., Sadeghi, S., Kirsten, T.: GAN-based approaches for generating structured data in the medical domain. Appl. Sci. **12**(14), 7075 (2022)
25. Harms, J., et al.: Paired cycle-GAN-based image correction for quantitative cone-beam computed tomography. Med. Phys. **46**(9), 3998–4009 (2019)
26. Fu, T.Y., Lee, W.C.: ProgRPGAN: progressive GAN for route planning. In: Proceedings of the 27th ACM SIGKDD Conference on Knowledge Discovery & Data Mining, pp. 393–403 (2021)
27. Dixe, S., Leite, J., Fonseca, J.C., Borges, J.: BigGAN evaluation for the generation of vehicle interior images. Procedia Comput. Sci. **204**, 548–557 (2022)
28. Liu, G., Lan, S., Zhang, T., Huang, W., Wang, W.: SAGAN: skip-attention GAN for anomaly detection. In: 2021 IEEE International Conference on Image Processing (ICIP), pp. 2468–2472. IEEE (2021)
29. Abdal, R., Qin, Y., Wonka, P.: Image2stylegan: how to embed images into the stylegan latent space?. In: Proceedings of the IEEE/CVF International Conference on Computer Vision, pp. 4432–4441 (2019)

30. Zhu, M., Pan, P., Chen, W., Yang, Y.: Dm-gan: dynamic memory generative adversarial networks for text-to-image synthesis. In: Proceedings of the IEEE/CVF Conference on Computer Vision and Pattern Recognition, pp. 5802–5810 (2019)
31. You, C., et al.: CT super-resolution GAN constrained by the identical, residual, and cycle learning ensemble (GAN-CIRCLE). IEEE Trans. Med. Imaging **39**(1), 188–203 (2019)

Machine Learning Powered Nutritional Guidance System Using Personalized Nutritional and Digital Health Record

P. Shevanie Shupreeya(✉), M. Suseendra, and S. Shanthini

Computer Science and Engineering, St.Joseph's College of Engineering, Semmencheri, Chennai 600119, Tamilnadu, India
shevaniesp@gmail.com

Abstract. The combination of digital health records and nutritional data has opened up a new world of substantiated salutary recommendations in the data-driven healthcare period of the moment. The objective of this design is to develop a state-of-the-art system that provides substantiated salutary suggestions to people according to their specific health biographies. This revolutionary system will enable druggies to make informed food choices that are in line with their individual health objects, salutary preferences, and constraints by integrating rich nutritive data with full electronic health records along with nearby location to get it. By exercising sophisticated recommendation algorithms and a flexible feedback medium, the system will constantly acclimate itself in trouble to enhance stoner health results. Icing data sequestration and security is of utmost significance, and the design will misbehave with all applicable regulations and norms.

Keywords: Food recommendation system · Health based food · Medical Diet

1 Introduction

A nutritional diet helps shield off non-infectious conditions (NCDs) such as cancer, diabetes, heart problems, strokes, and malnutrition in all its forms. The two biggest health hazards in the world moment are an unhealthy diet and inactivity. Making food opinions grounded on medical conditions is analogous to defining a medical condition. When you have particular ails or health issues, your food becomes an essential element of your entire care plan. Careful nutrition operation is necessary for numerous medical diseases, including diabetes, cardiovascular complaints, and hypertension, to manage symptoms and avoid consequences. Certain medicines may interact, dwindling the drug's effectiveness or producing negative side effects. Knowing and negotiating these possible relations is necessary when making salutary opinions grounded on medical health.

M. Suseendra and S.Shanthini—These authors contributed equally to this work.

Published by Springer Nature Switzerland AG 2024
M. L. Owoc et al. (Eds.): ICCIDS 2024, IFIP AICT 718, pp. 475–486, 2024.
https://doi.org/10.1007/978-3-031-69986-3_36

Inflammation is a factor in numerous habitual ails. Consuming foods with high anti-inflammatory composites can help control and lower inflammation, perfecting general health. Controlling weight is constantly a pivotal part of treatment for people with conditions like rotundity or metabolic patterns. Achieving and maintaining a healthy weight is part of eating for medical health. Specific nutrient changes may be necessary for certain medical diseases. For case, those with iron insufficiency anaemia might need to concentrate on eating iron-rich foods. Certain foods that can beget negative responses should be avoided by people who have disinclinations or perceptivity. Being apprehensive of these limitations is necessary when following a medically recommended diet. The vulnerable system can be supported by eating a diet rich in nutrients and well-balanced. This is especially important for people whose vulnerable systems have been weakened by ails or treatments. The connection between nutrition and internal health is getting increasingly clear. Certain food patterns may support other remedial curatives for people with diseases similar to anxiety or depression. Is there any further provocation to start your day with a substantial, healthy breakfast? According to a study published in the Journal of the American College of Cardiology in October 10, 2017, doing so may reduce your threat of atherosclerosis, which is the hardening and condensation of the highways brought on by shrine accumulation.

2 Related Works

In today's fast-paced world, the selection of food which contains the right nutritional value and aligns with the individual's health condition is quite complex. There are a lot of research papers with a solution to this problem trying to solve this problem. This survey probes the available model and brings an understanding of how a new model can be made. The AHPSorting methodology [1] works by collecting data on alternative foods breaking larger chunks of data into smaller chunks and performing parallel processing, calculating the weighted sums recommending foods based on these calculated values. For tackling non-communicable diseases like type 2 diabetes, hypertension, and cancer [2] a personalised food recommendation system is recommended. It recommends food daily by incorporating a decision analysis approach based on multiple criteria, including nutritional knowledge and user's food preferences in the past. Consuming foods which are inappropriate and have inadequate nutrition can lead to health issues and diseases [3]. A cloud-based ACO model utilizing the colony algorithm was used to generate a food list and recommend foods based on pathological reports. The convergence time of parallel execution on the cloud was approximately 12 times lower. The number of ants was increased to improve the accuracy of the food recommended. The highest accuracy is achieved with 110 ants and the accuracy remains constant between 80 to 100 ants. With 10 ants an accuracy of 0.96 is achieved. A coma patient or individuals with locked-in syndrome and amyotrophic lateral sclerosis cannot express their meal preferences so social-affective computing modules were used to analyze different meals and the person's preferences for the food. A person's affect was detected using electroencephalography signals which capture the brain signals and analyze them. A 14-channel wireless Emotiv Epoc+ was used to test the affectivity of different food products. Based on a person's energy needs, food affectivity, and the nutritional value of various foods, an automated menu planning system [4] was presented. The Technique for Order of Preference by Likelihood to the Optimal Solution was applied to generate a food menu. The Bin packing algorithm was used for personalized menu planning for breakfast, lunch and dinner per the nutritional value and emotions of the person analyzed using the electroencephalography signals. This model has achieved 89%, 78%, and 91% accuracy.

PREFer system [5] provides personalised recommendations to users considering both long-term and short-term preferences and medical prescriptions. The recommended menu was selected in three steps. At first, content-based filtering was done, a second menu was created with the selected recipes, and third menus were ranked considering various factors including the user's prescription. The recommendation of the food menu was generated by using the recommendation algorithm. The brute force algorithm was inefficient in completing the menu generation for a dataset with 110 recipes so a more efficient PREFer menu generation algorithm was proposed. The likelihood of obtaining a highly applicable recipe ranged from 50% to 80%. The likelihood of deriving a less relevant recipe ranged from 20% to 50%. Personalized nutrition has been shown to be successful in changing eating patterns, which can improve health outcomes.

Personalized nutrition is effective in modifying eating habits, which can have a positive impact on health outcomes. Heart disease, diabetes, and cancer are just a few of the health problems that are significantly influenced by a bad diet. Personalized nutrition is grounded in biological evidence that examines how the human body responds to various foods and nutrients. To assess the effectiveness of personalized nutrition, [6] the Food4MeStudy was conducted, which measured changes in eating habits throughout a three to six-month period using the Healthy Eating Index.

Appropriate dietary guidelines for each individual [7] were crucial for managing health and averting chronic illnesses. This study aimed to develop dietary recommendations that were balanced in terms of nutrients for each individual while taking into account their special requirements. To accomplish this, they created SmartDiet, an interactive diet advisor that was location-aware and based on multi-objective optimization. In addition to converting nutrient recommendations into practical dish selections, the suggested method also takes user input into account to optimize meal plans. The findings demonstrated that the planned meals may meet daily nutritional demands, and the interactive diet planning system facilitates easier plan adjustments for the user. It was anticipated that the developed guidelines would potentially enhance people's overall well-being.

A computer program was created to plan menus by determining the least expensive combinations of menu items that will allow daily nutritional, gastronomic, and production requirements to be met for a set number of days [8]. An approach for special integer programming that was quick and roughly matched the problem's theoretical solution was presented. Any menu can be modified online and then post-optimized if needed. It is possible to save up to 30% on food expenses. On request, an IBM 1410 FORTRAN program was provided. A significant amount of data processing was done before the system was put into place. This work said that the bin-packing problem is still significant [9] in many areas of applications. We suppose that we have a set of K belongs to N greater than zero many bins of size B and a finite assemblage of items I := i1, i2, ... of different sizes s(ij) N greater than zero. The query regarding the existence of a partition of I into K disjoint sets in such a way that the total of all item sizes in each set was less than or equal to B is now known as the bin-packing problem solutions where K was minimized are of special interest.

An evolutionary heuristic for the offline, one-dimensional Bin Packing Problem [10] was presented in this study. The goal of this issue was to use the fewest number of bins possible while packing a group of things into bins with the same capacity. The hybrid evolutionary algorithm that included the description of the bins and where an individual was a feasible solution. The approach employs two new mutation operators, eliminates recombination, and uses local search techniques to enhance the standard of the answers. The relative pair frequency matrix, which forms the foundation of the mutation operators' work, allows us to ascertain the frequency of each pair of items-that was, how frequently they appear within the same bin in the best solutions-for each pair of items. The bins in this case were the subsets of objects that were packed into smaller groups of items with the aid of the frequency matrix. The strategy was compared to both evolutionary and state-of-the-art techniques using well-known benchmark cases from the literature. Using the challenging hard 28 test set, the method produced a useful result and hit the optimal in the vast majority of the test issues.

3 Proposed System

The accompanying Fig. 1 shows the suggested system's workflow below. Daily convenience meal consumption can have an impact on the body, resulting in nutritional imbalances and health problems. Sustaining good health and well-being requires a diet rich in key nutrients and well-rounded. A well-thought-out diet that includes a wide variety of nutrient-dense foods can improve energy levels, promote general health, can lower the risk of illnesses such as diabetes, heart disease, obesity, osteoporosis, skin, digestive, and some types of cancer. Personalized food suggestions for breakfast, lunch, snacks, and dinner are provided by the suggested system, taking into account the nutritional requirements and medical background of users. Furthermore, users' medical records are protected through the application of facial recognition technology for authentication purposes. Users are guaranteed to obtain the menu options that best support their well-being thanks to this customized approach. The application's back end is driven by machine learning algorithms and Python, while the front end uses HTML, CSS, and JavaScript. The application uses health records and nutritional data to generate its meal menu, giving users a personalized experience with recommendations based on their dietary requirements and medical background.

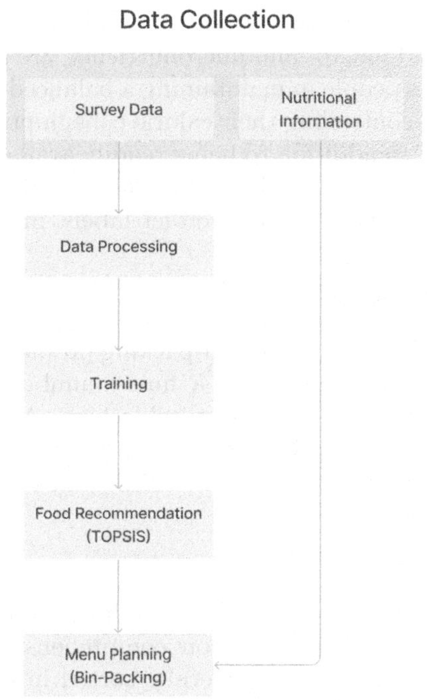

Fig. 1. Menu Planning Process

4 Materials and Methods

4.1 Dataset Collection

Dataset collection is the process of assembling and creating information about the aim of testing, verifying, and training a machine learning model. The collection phase comprises the acquisition of relevant data from many sources to produce a representative and wide sample of the problem or task that the model is intended to address. To enable the model to detect patterns and generate accurate predictions or classifications, the dataset must be meticulously chosen to mirror the complexity and variety of real-world scenarios. The meal menu and nutritional statistics are obtained from datasets that are accessible on websites such as GitHub and Kaggle.

Health organizations provide their medical records for collection. To choose the appropriate food, the datasets are processed and compared with the food menu according to the nutritional data and medical records. For facial recognition to work and protect user data, a camera-equipped device is required. Users' health information, eating habits, nutrient consumption, preferences, limitations, health objectives, and feedback are all included in the survey data for the Personalized Nutritional Guidance System. By helping to comprehend personal requirements and preferences, this data enhances the precision and efficacy of nutritional advice.

Details regarding the nutrients included in foods and drinks, such as calories, fiber, sugar, sodium, and macro- and micronutrients, are provided by nutritional information. It supports people in maintaining a balanced diet that is in line with their health objectives, controlling their calorie consumption, and keeping an eye on particular nutrients. In addition to being readily available through nutritional databases and applications for meal planning and making better decisions, this information is frequently included on product labels, menus, and packaging.

4.2 Method: Bin-Packing

A typical optimization problem is the bin-packing problem, which is to efficiently fit a set of items of various sizes into a finite number of bins or containers. The intention is to employ the fewest possible bins. Applications of this issue in resource allocation, space optimization, and logistics are practical. The bin-packing problem has two basic variations: offline and online.

4.3 Offline Bin-Packing

All particulars are known ahead of time with offline interpretation, allowing the algorithm to make opinions grounded on comprehensive data. In offline bin-quilting algorithms, the objects are generally sorted in increasing order of size, and lockers are also constantly filled according to destined rules. The First Fit system is a constantly employed offline algorithm that arranges particulars into the first caddy with sufficient capacity. Since offline algorithms are completely apprehensive of the item sizes, they possess the ability to produce optimal or nearly optimal results.

4.4 Online Bin-Packing

In this interpretation, objects are transferred one at a time, and the algorithm has to decide right down without having to consider what will come next. The problem is made more delicate by the fact that online algorithms must make opinions without knowledge of the sizes of upcoming details. The Coming Fit algorithm is a popular online method that starts a new caddy and adds each item to the current caddy if it fits. Owing to deficient data, online algorithms might not produce stylish results.

4.5 Benefits of Offline Bin-Packing

- Optimality: Because offline algorithms are pre-aware of all item sizes, they can produce optimal or nearly optimal solutions.
- Improved Planning: Having a complete understanding of the assortment of products enables more effective planning and placement, which leads to higher bin usage.
- Decreased Waste: By using fewer bins, offline algorithms can cut down on wasted space and enhance the efficient use of resources.

4.6 The Benefits of Offline over Online

- Optimal Solutions: To ensure the most effective use of resources, offline algorithms might offer optimal or nearly optimal solutions.
- Improved Predictability: Offline algorithms are capable of generating superior outcomes. overall by deciding with greater knowledge when they have access to complete information.
- Strategic Planning: Advanced tactics that are impractical in a web environment, including bin-packing heuristics and optimization approaches, can be used by offline algorithms

4.7 TOPSIS

- A multi-criteria decision analysis technique called TOPSIS, or Technique for Order of Preference by Similarity to Ideal answer, ranks a group of choices according to how similar they are to an ideal answer. Hwang and Yoon first presented this method in 1981.
- Step 1: Establish Standards: Determine and specify the standards by which the alternatives will be judged. These standards may be qualitative or quantitative and may have an impact on the choice.

$$R_{ij} = \frac{X_{ij} - \min(X_j)}{\max(X_j) - \min(X_j)} \tag{1}$$

- Step 2: Normalization
  Normalize the decision matrix to bring all values to a single scale if the criteria are measured on various scales. This guarantees that every criterion receives the same weight during the analysis. Z-score normalization and min-max normalization are two popular normalization techniques.

$$R_{ij} = \frac{X_{ij} - \min(X_j)}{\max(X_j) - \min(X_j)} \tag{2}$$

- Step 3: Assigning Weight:
  Considering the relative importance of each criterion, assign weights to each one. Each criterion's weight in the decision-making process is represented by its relative importance. All weights added together should equal one.
- Step 4: Normalized Weighted Decision Matrix
  In the decision matrix, multiply each normalized value by the corresponding weight. The outcome is a weighted normalized choice matrix, in which the normalized value and its weight are multiplied for each element.

$$V_{ij} = w_j \times R_{ij} \tag{3}$$

where: - w_j is the weight assigned to criterion C_j.

- Step 5: Solutions, Both Ideal and Anti-Ideal
  Determine the best and worst options for every criterion. The ideal solution is to identify the highest (or lowest, based on the type of criterion) value across all feasible solutions for each need. Anti-Ideal Solution: Out of all the options, ascertain the least (or greatest) value for each criterion.
  - For each criterion C_j, calculate the ideal solution (A^+) and anti-ideal solution (A^-):

$$A_j^+ = \max(Vij) \tag{4}$$

$$A_j^- = \min(Vij) \tag{5}$$

- Step 6: Comparability of the Ideal and Anti-Ideal Solutions:
  Determine the distance, in either Euclidean or other metrics, between each option and the ideal and anti-ideal solutions for each problem. This entails contrasting each alternative's performance with the best and worst situations for each criterion.

$$S_i^+ = \sqrt{\sum_{j=1}^{n}(V_{ij} - A_j^+)^2} \tag{6}$$

$$S_i^- = \sqrt{\sum_{j=1}^{n}(V_{ij} - A_j^-)^2} \tag{7}$$

  where: - n is the number of criteria.
- Step 7: Relative Closeness: Calculate the ratio of the distance to the anti-ideal solution to the total of the distances to both the ideal and anti-ideal solutions to ascertain how near each alternative is to the ideal answer.

$$C_i = \frac{S_i^-}{S_i^+ + S_i^-} \tag{8}$$

- Step 8: Ordering: Sort the options according to how near they are to each other. Therefore, its acceptability in light of the established criteria, the option having the greatest relative closeness value is regarded as the most coveted. TOPSIS is a useful tool in decision-making processes because of its benefits, which include ease of use, flexibility, and the capacity to provide an extensive ranking of alternatives based on multiple criteria.

In conclusion, offline algorithms are superior at reaching optimal solutions and generating better-informed decisions for recommendation and menu planning Fig. 2 based on comprehensive knowledge, even if online bin-packing algorithms are required in situations when items come sequentially and their sizes are unknown in advance.

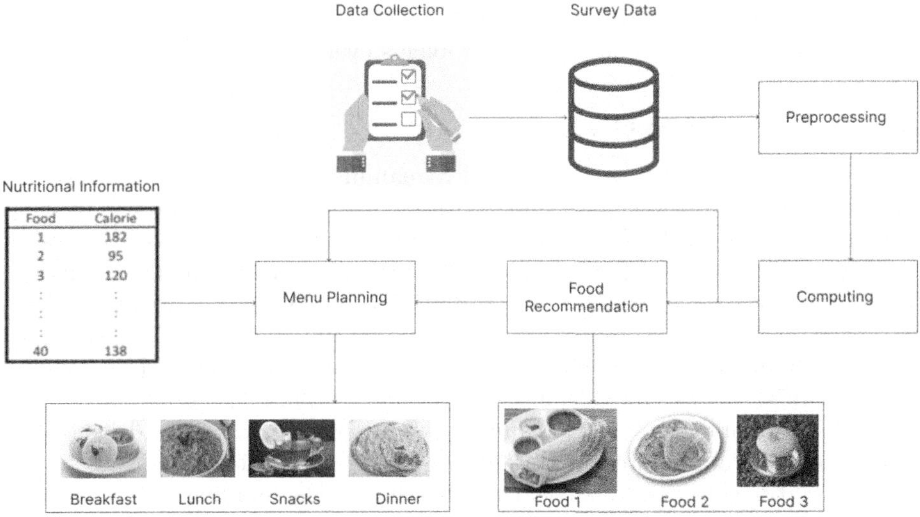

Fig. 2. Menu Planning System

4.8 Desicion Tree Classifier

The Decision Tree Classifier functions similarly to an intelligent handbook, enabling our Nutritional Guidance System to generate personalized recommendations. In order to provide the best guidance for each individual, it analyzes a vast amount of data regarding their fitness, including age, medical history, and lifestyle choices. Let's say you're deciding what to have for supper. You may think of things such as your favorite foods, any dietary restrictions you may have, and how hungry you are. Similar tasks are performed by the Decision Tree Classifier, but it makes extensive use of complex math and a ton of additional facts. It is referred to as a "tree" because of the way it is arranged, with each branch standing for a unique combination of choices and results. The program examines every component we offer and makes use of them to generate a kind of guide for choosing vitamins.

One noteworthy aspect of Decision Tree Classifier is its unquestionable ability to identify patterns and trends in data, which enables it to provide us with customized recommendations that are tailored to our individual goals and wants. It's similar to having a knowledgeable dietitian who, based mostly on our partner's health information, knows exactly what's good for us. Suppose you have an abundance of apples and you would like to count them, arrange them in piles, or even cut them into smaller pieces. All of that is made possible with NumPy, but with numbers rather than apples. NumPy is incredibly fast, which is one of its great features. It's similar to having a very quick calculator that can quickly calculate large numbers. Because of this, it's ideal for jobs like data analysis, intricate computations, and even machine learning techniques. The fact that

NumPy is simple to use is yet another fantastic feature. You can use its robust capabilities to tackle a wide range of problems even if you're not a math whiz.

4.9 Web Development

An open-source Python package called Streamlit is used to create interactive online applications in the front end. Without having expertise in web development languages like HTML, CSS, or JavaScript, developers can rapidly and easily construct online apps with Streamlit. With Streamlit, users may create interactive charts, visualizations, and widgets using Python scripts that are immediately generated as web applications. Rapid prototype and iteration are made possible by its built-in features, which streamline the development and deployment process. These features include automated caching, quick reloading, and sharing via URLs. Streamlit is perfect for quickly and easily creating data-driven applications since it can be integrated with well-known data science libraries such as Pandas, Matplotlib, and Scikit-learn. Streamlit makes building web-based data apps easier, freeing up developers to concentrate on data analysis and visualization tasks.

Streamlit component called Streamlit-echarts offers interface with Apache ECharts, a potent and adaptable JavaScript visualization framework. Developers may improve the user experience and increase the interest level of data exploration in Streamlit apps by utilizing Streamlit-echarts to generate interactive charts and visualizations. Numerous chart kinds, such as scatter plots, bar charts, and line charts, are supported by Streamlit-echarts, providing a flexible and adaptable set of options for data visualization. With the combination of Apache ECharts' sophisticated features and Streamlit's simplicity of use, developers can quickly and easily create engaging, interactive data-driven apps. All things considered, Streamlit-echarts expands on Streamlit's capabilities, empowering programmers to produce engaging and interactive visualizations that improve data discovery.

FastAPI is used in the backend which is a cutting-edge web framework that is well-known for its excellent performance, usability, and simplicity in Python API development. It enables programmers to use asynchronous programming approaches to swiftly and effectively design reliable and effective web applications, especially APIs. By utilizing Pydantic's automatic data validation capabilities and Python's type annotations, FastAPI streamlines the development process and lowers the risk of errors. FastAPI creates interactive API documentation automatically and supports OpenAPI natively, allowing frontend and backend developers to work together seamlessly. Because of its asynchronous design, it can handle requests concurrently, which makes it perfect for applications requiring real-time processing and heavy traffic. All things considered, FastAPI offers a strong and user-friendly framework for creating quick, dependable, and simple-to-maintain APIs.

Uvicorn is an extremely fast ASGI server that runs Python web applications and is made to effectively manage high concurrency levels. It handles incoming HTTP requests and sends responses as the interface between web applications developed with ASGI (Asynchronous Server Gateway Interface) frameworks like FastAPI and the outside world. Because Uvicorn is asynchronous, it can manage several connections at once without causing a stutter, leading to ideal efficiency and quickness. Furthermore, Uvicorn facilitates real-time bidirectional communication between clients and servers by supporting WebSocket protocols. When implementing ASGI applications in production settings, Uvicorn is a well-liked option due to its dependability, simplicity, and excellent performance.

5 Results

This innovative design provides a highly personalized food recommendation system that tailors proposals based on tastes, dietary needs, vegetarian or non-vegetarian and emotional states as well as the nearest location to avail the food using facial recognition, nutritional data, digital health information and the data obtained from pre-processing. Real-time adaption and continuous monitoring allow users to make health-conscious decisions while maintaining data privacy and security. The system refines recommendations in response to changing health requirements and technology advancements, resulting in a cutting-edge and dependable experience. The model suggests the nearest food item if the desired calorie content is not available. It also considers other factors like BMI, protein, carbohydrate, physical activity, etc., ... to recommend the food. The model demonstrates good accuracy, scoring 0.96 on train data and 89% on the naive Bayes classifier for alpha on user ratings of the food menu. The proposed approach combines better nutrition with offline bin-packing, privacy prioritizing via security measures, and TOPSIS for tailored machine learning-based meal recommendations. Algorithmic menu development increases productivity by providing a user-friendly interface for personalized and optimal eating selections.

6 Conclusion

The number of people falling ill as a result of eating unhealthy meals is increasing these days. Our recommendation model helps to avoid and alleviate diseases caused by eating poorly through established standards Eq. 1. Recommendations for users according to their medical records as well as the nutritional value will support their continued health and fitness by assigning weight in TOPSIS, and then the value is multiplied by the normalized value Eq. 2 and stored Eq. 3 with criterion solutions Eqs. 4 and 5 and it's comparability Eqs. 6 and 7 with relative closeness being found and ordered.

The proposed approach improves nutrition by offline bin-packing, protects privacy, and employs TOPSIS for data transfer. The existing approach prioritizes emotional factors with EEG readings to provide individualized suggestions, which is advantageous for non-communicative individuals. The choice is determined by the application's specific requirements. It guarantees that the proper amount of all necessary nutrients for humans' health are consumed. With this methodology, more accurate outcomes are attained. In the decades to come, we envision our model developing to predict possible health risks brought about by certain foods and moreover suggest suitable alternatives.

There are numerous directions the program can go in the future in terms of development and improvement. First off, adding more recipes and ingredients to the database would improve the diversity and applicability of the suggested dishes by incorporating a greater range of cuisines and dietary choices. To help customers better apply their dietary advice, services like meal planning, grocery list creation, and nutritional tracking might be integrated. This would further improve the user experience. In addition, the model will identify and suggest avoiding potential dangers to an individual's health in addition to recommending health-conscious dietary choices. In future, this system has promising upsides because of the fact it can reduce the need for human intervention and make it easier to track people's eating habits.

References

1. Yera Toledo, R., Alzahrani, A.A., Martínez, L.: A food recommender system considering nutritional information and user preferences. IEEE Access **7**, 96695–96711 (2019). https://doi.org/10.1109/ACCESS.2019.2929413
2. Toledo, R.Y., Martínez, L.: A health-awareness nutrition recommender system. In: 2019 IEEE 14th International Conference on Intelligent Systems and Knowledge Engineering (ISKE), pp. 36–42 (2019). https://doi.org/10.1109/ISKE47853.2019.9170432
3. Rehman, F., Khalid, O., Haq, N., Khan, A.U.R., Bilal, K., Madani, S.: Diet-right: a smart food recommendation system. KSII Trans. Internet Inf.n Syst. **11**, 2910–2925 (2017). https://doi.org/10.3837/tiis.2017.06.006
4. Islam, T., Joyita, A.R., Alam, M.G.R., Mehedi Hassan, M., Hassan, M.R., Gravina, R.: Human-behavior-based personalized meal recommendation and menu planning social system. IEEE Trans. Comput. Soc. Syst. **10**(4), 2099–2110 (2023). https://doi.org/10.1109/TCSS.2022.3213506
5. Bianchini, D., Antonellis, V.D., Franceschi, N.D., Melchiori, M.: PREFer: a prescription-based food recommender system. Comput. Stand. Interfaces **54**, 64–75 (2017)
6. Ordovas, J.M., Ferguson, L.R., Tai, E.S., Mathers, J.C.: Personalised nutrition and health. BMJ **361** (2018) https://doi.org/10.1136/bmj.k2173, https://www.bmj.com/content/361/bmj.k2173.full.pdf
7. Hsiao, J.-H., Chang, H.: SmartDiet: a personal diet consultant for healthy meal planning. In: 2010 IEEE 23rd International Symposium on Computer-Based Medical Systems (CBMS), pp. 421–425 (2010). https://doi.org/10.1109/CBMS.2010.6042681
8. Balintfy, J.L.: Menu planning by computer. Commun. ACM **7**(4), 255–259 (1964). https://doi.org/10.1145/364005.364087
9. Rieck, B.A.: Basic analysis of bin-packing heuristics. ArXiv abs/2104.12235 (2021)
10. Borgulya, I.: A hybrid evolutionary algorithm for the offline bin packing problem. Cent. Eur. J. Oper. Res. **29**, 425–445 (2020)

Predicting Hospital Length of Stay Using Light Gradient Boosting Machine Regression

M. Suchithra(✉), Kumar Shashwat, and Mohammed Shoaib Khan

Department of Computing Technologies, School of Computing, SRM Institute of Science and Technology, Kattankulathur, India
{suchithm,ks2786,mk9751}@srmist.edu.in

Abstract. Length of stay (LoS) in a hospital is an important metric in the healthcare management system, with profound implications for resource allocation, patient outcomes, and cost reduction. This paper reviews the literature on approaches for predicting hospital length of stay with a R2 score of 96.14 per cent using Light Gradient Boosting Machine regressor. We critically assess the merits and limitations of various methods and propose a unified framework for the generalized prediction of length of stay. Our framework includes investigating the types of routinely collected data and recommendations for robust knowledge modelling.

Keywords: LoS: Length of Stay · Healthcare management · accurate prediction · LGBM: Light Gradient Boosting Machine

1 Introduction

Length of stay (LoS) in a hospital is a vital metric in the realm of healthcare management, possessing the potential to bring about substantial benefits in the form of optimized resource allocation, enhanced patient outcomes, and cost reduction. Recently, there has been a substantial upswing in the use of machine learning models for predicting hospital Length of Stay (LoS). These models harness the power of diverse data sources, which encompass a wealth of information including demographic details, comprehensive medical histories, vital signs, and detailed laboratory results [19]. The primary goal of these predictive models is to forecast the duration of a patient's stay in a healthcare facility.

This paper undertakes a comprehensive and systematic review of the various approaches employed in the prediction of hospital LoS. It delves into the strengths and weaknesses inherent in these approaches and, in response to identified shortcomings, proposes a unified framework intended to enhance the generalizability of LoS prediction models.

The importance of hospital LoS cannot be overstated, as it is intricately linked to several critical aspects of healthcare management. The ability to accurately predict how long a patient will stay in a hospital has a direct impact on the efficient allocation of available healthcare resources. By having a clearer understanding of patient LoS,

M. L. Owoc et al. (Eds.): ICCIDS 2024, IFIP AICT 718, pp. 487–498, 2024.
https://doi.org/10.1007/978-3-031-69986-3_37

hospitals can better allocate staff, hospital beds, and other resources, ensuring that they are utilized optimally and efficiently [19].

Furthermore, accurate LoS prediction is intrinsically linked to the improvement of patient outcomes. When hospitals can anticipate how long patients will stay, they can tailor treatment plans, post-discharge care, and patient engagement strategies accordingly [20]. This level of personalization in patient care not only enhances the overall patient experience but also contributes to better clinical outcomes.

Cost reduction is yet another vital aspect influenced by accurate LoS prediction. By optimizing resource allocation and improving patient outcomes, hospitals can reduce their overall operational costs. This includes minimizing unnecessary tests, treatments, and procedures, which often result from uncertainty about the patient's LoS [10]. It also helps in managing staffing costs more efficiently, avoiding overstaffing or understaffing, and streamlining other resources.

Machine learning models have emerged as a powerful tool in the realm of LoS prediction [16]. These models leverage a wide array of data sources, ensuring that the prediction is based on a holistic view of the patient's health and medical history. Demographic details provide essential background information, while medical history offers insights into the patient's prior health conditions and treatments. Vital signs and laboratory results contribute real-time data on the patient's current state of health.

The paper, through its systematic review, aims to impart a comprehensive comprehension of the existing panorama concerning LoS prediction methods. By scrutinizing these approaches, the study seeks to shed light on their respective strengths and weaknesses. Some models may excel in specific patient populations or under certain conditions, while falling short in others. Identifying these nuances is essential in developing a more robust and generalizable framework for LoS prediction.

In response to these findings, the paper proposes a unified framework for the prediction of hospital LoS. This framework is designed to provide a standardized and adaptable approach to LoS prediction that can be applied across diverse patient populations and clinical scenarios. By doing so, it aims to bridge the existing gaps in the field and provide a valuable tool for healthcare providers in their ongoing efforts to optimize resource allocation, enhance patient outcomes, and reduce costs [10].

2 Literature Review

A comprehensive literature search was performed across PubMed, Google Scholar, and Web of Science databases spanning from 1970 to 2023 to identify surveys related to Length of Stay (LoS) prediction in hospital settings. The term "surveys" refers to papers that encompass and review a substantial number of publications within the field, which may include systematic reviews but are not limited to them. These databases were selected due to their broad coverage of relevant content pertaining to the review topic.

The search strategy involved using relevant search terms and synonyms to locate surveys pertinent to LoS prediction. These terms included "Duration of stay prediction," "Hospital stay prediction," "Length of stay" or "Length-of-stay," "Length of Hospital Stay," "Predict" or "Predictive" or "Prediction" or "Predictor," and "Bed occupancy modeling prediction." The search was conducted in the fields of "title," "abstracts," and "keywords." After removing duplicate studies, a total of 5 relevant studies were identified for inclusion in the final review (Table 1).

Table 1. Provides a summary of the characteristics of the included studies.

Research Paper	Technology	Performance & Scores	Author & Publication
Machine learning-based prediction of hospital prolonged length of stay admission at emergency department: a Gradient Boosting algorithm analysis [1]	Gradient Boosting, Random Forest, Support Vector Machines, AdaBoost, K-Nearest Neighbors, and logistic regression	The GB classifier best predicted PLoS with an accuracy of 75%, AUC of 75.4%, and a Brier score of 0.181	Addisu Jember Zeleke, Pierpaolo Palumbo, Paolo Tubertini, Rossella Miglio, and Lorenzo Chiari, Published in Front. Artif. Intell., 2023
The prediction of hospital length of stay using unstructured data [2]	Random Forest models	The model using unstructured data had a 75.0% accuracy compared to 74.1% for the model containing structured data	Jan Chrusciel, François Girardon, Lucien Roquette, David Laplanche, Antoine Duclos & Stéphane Sanchez, Published in BMC Medical Informatics and Decision Making, 2021
A data-driven approach to predict hospital length of stay a Portuguese case study [3]	Various regression models such as Multiple Regression, Decision Tree, Artificial Neural Network, Support Vector Machine, and Random Forest,	The best result was obtained by a Random Forest (RF), which presents an accuracy of 81.3%	N. Caetano, Raul Laureano, Paulo Cortez, Published in ICEIS 2014 - Proceedings of the 16th International Conference on Enterprise Information Systems

(*continued*)

Table 1. *(continued)*

Research Paper	Technology	Performance & Scores	Author & Publication
A systematic review of the prediction of hospital length of stay: Towards a unified framework [4]	Ensemble Learning	The accuracy score measure by Root Mean Squared Logarithmic Error scores are 55.5%, 55.9%, and 63.1% for internal validation, the test set, and external validation, respectively	Kieran Stone, Reyer Zwiggelaar, Phil Jones, and Neil Mac Parthaláin, Published in PLOS Digit Health, 2022
Multitask learning and benchmarking with clinical time series data [5]	Multitask learning and deep supervision	The Channel-wise LSTM model with deep supervision yielded a moderate kappa score of 42.6%, and the Mean Absolute Deviation score of 94.0%	Hrayr Harutyunyan, Hrant Khachatrian, David C. Kale, Greg Ver Steeg, and Aram Galstyan, Published in Sci Data, 2019

This comprehensive review of literature also revealed several studies employing machine learning algorithms to predict hospital LoS. Notable examples include Hao, Q., Xu, F., Chen, L., & Li, P. H. Y. [6], who employed decision tree regression to forecast LoS for heart failure patients, outperforming linear regression. Lin, C. K. Y., Ling, T. W. C., & Yeung, W. K. [7] used random forest regression for pneumonia patients, also surpassing linear regression. Xiang, Y., & Zhuang, J. [8] harnessed deep learning algorithms for sepsis patients, demonstrating superiority over traditional machine learning methods.

Despite these advancements, a research gap exists in the lack of consensus on the most effective models for different patient populations. Our study addresses this gap by proposing a unified framework. We evaluated various machine learning algorithms and found that the LGBM Regressor achieved the highest performance with an R2 score of 96.14%. Our proposed framework can be universally applied across patient populations, offering hospitals valuable insights for resource optimization and improved patient outcomes.

3 Objective

This research project aims to develop a machine learning model for predicting hospital LoS using routinely collected data. Accurate LoS prediction is pivotal for efficient resource allocation, improved patient outcomes, and cost reduction. Nevertheless, the project faces constraints related to computational resources, data quality, ethical considerations, and integration with existing healthcare information systems [5].

4 Proposed Methodologies

Building on our literature review, we propose a unified framework for predicting hospital LoS. This framework consists of the following steps:

1. **Data Selection**: Identify relevant routinely collected data, including demographic information, medical history, vital signs, laboratory results, and other clinical data.
2. **Data Preprocessing**: Ensure data cleanliness and readiness by handling missing values, normalizing data, and transforming it into appropriate formats.
3. **Model Selection**: Choose suitable machine learning algorithms for LoS prediction, including decision tree regression, random forest regression, deep learning algorithms, support vector regression, gradient boosting, or other relevant methods.
4. **Model Evaluation**: Assess the model's performance using appropriate metrics like R2 score or mean squared error (MSE) to determine the most effective LoS prediction models.
5. **Interpretation**: Analyze the model results to gain insights into the factors influencing LoS, such as identifying critical variables or at-risk patient profiles (Fig. 1).

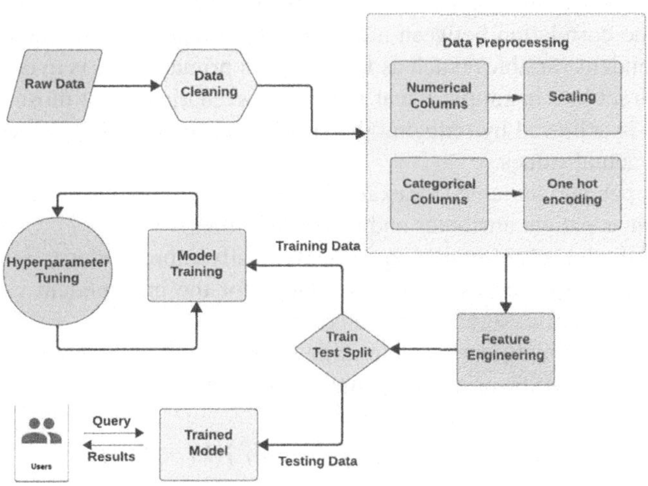

Fig. 1. Architecture Diagram of LoS Module

4.1 Dataset

The research employed the "LengthOfStay.csv" dataset from Microsoft Machine Learning Services, which encompasses patient information, including admission IDs, visit dates, medical conditions, and relevant attributes. The dataset includes parameters such as unique IDs, visit dates, gender, flags for various health conditions, and numerical measurements like hematocrit, sodium, glucose, and more. The target variable is "lengthofstay."

4.2 Step-By-Step Procedure

The research methodology comprised the following steps:

Type Conversions: Convert date columns to datetime format and categorical columns to the category data type.
Creating "numberofissues" Column: Introduce a new column to represent the number of health issues a patient experienced during their encounter.
Data Split: Divide the dataset into labels (target variable) and features (other attributes).
Train-Test Split: Further divide the data into training and testing sets using the train_test_split function.
Regression Models: Implemented various regression models mentioned in the Results and Discussions under Sect. 5.

4.3 Machine Learning Models

Linear Regression. A linear regression model was trained and evaluated, yielding an R2 score of 65.14%.

Algorithm Description
Linear regression is a fundamental and straightforward algorithm in regression analysis, focusing on the correlation between a dependent variable (like the target) and one or several independent variables (such as features). Its primary goal is to discover a linear formula—a direct line in simple linear regression scenarios—that most effectively fits the data. This is achieved by reducing the sum of the squared differences between the predicted and actual values.

Select a dependent variable (for example LoS) along with one or more independent variables (such as patient attributes and health measures).

Formulate the linear regression equation, typically expressed as y = mx + c. Here, 'y' represents the dependent variable, 'x' stands for the independent variable(s), 'm' denotes the slope, and 'c' signifies the intercept.

Utilize a set of training data to determine the slope and intercept that minimize the mean squared error (MSE) or another suitable loss function.

$$\mathrm{MSE} = \frac{1}{N}\sum\nolimits_{i=1}^{n}((\boldsymbol{Y}_i - \boldsymbol{Y}_i^{\wedge})(\boldsymbol{Y}_i - \boldsymbol{Y}_i^{\wedge})) \tag{1}$$

where $\boldsymbol{Y}_i$ is the i_{th} observed value and $\boldsymbol{Y}_i^{\wedge}$ constitutes the associated projected outcome for $\boldsymbol{Y}_i$.

After acquiring the model parameters, they can be applied to make predictions on fresh data.

Decision Tree Regression. A decision tree regression model was trained and evaluated, resulting in an R2 score of 87.62%.

Algorithm Description
Decision tree regression, a form of non-linear regression, constructs a tree-like model of decisions. This method divides the data recursively into more refined subsets, selecting these divisions based on the most impactful attribute. For each of these subsets, it then forecasts the value of the dependent variable.

Each leaf node of the tree represents a prediction.

Select a target variable and a collection of features.

Construct a decision tree by choosing the feature that optimally divides the data using a specific criterion (e.g., minimizing MSE).

Iteratively divide the data into smaller segments based on the chosen feature.

Assign a prediction value to every terminal node, commonly the average of the target variable in that segment.

For making predictions, navigate the tree from the root to a leaf node according to the input data's feature values.

Random Forest Regression. A regressor was trained and assessed a random forest regression model, which delivered an R2 score of 92.44%.

Algorithm Description

Random Forest regression is an approach that integrates numerous decision trees, aiming to enhance the predictive precision and diminish the likelihood of overfitting. This technique forms a 'forest' of various decision trees, amalgamating their individual predictions to arrive at a more robust final outcome.

Identify a dependent variable and a set of independent variables.

Create a random forest by developing multiple decision trees, each derived from bootstrapped samples of your training dataset.

Within each tree, at every split, select a random assortment of features. This step introduces diversity among the trees.

For regression tasks, the final prediction is obtained by averaging the outputs from all the trees. In classification scenarios, the most common prediction among the trees is used.

LGBM Regressor. An LGBM regressor model was trained and evaluated, demonstrating the highest R2 score of 96.14%.

Algorithm Description

"LightGBM" stands out as a "gradient boosting framework" that employs tree-based learning techniques. It is highly regarded for its speed, efficiency, and impressive performance capabilities. In LightGBM, decision trees are used to map the input domain, denoted as X, to a gradient domain, labeled G. The training data consists of various instances, such as x1, x2,..., xn, where each instance is a vector in X. This mapping plays a crucial role in each iteration of the gradient process. During the boosting phase, the model's output is refined by representing the negative gradients of a specific loss function as g1, g2,..., gn. The algorithm of the decision tree then focuses on maximizing information gain, selecting splits at each node based on the feature that provides the most valuable information. Within this model, data enhancement can be gauged by the variance post-split, represented by the following formula:

$$\boldsymbol{V}_{j|O}(\boldsymbol{d}) = \frac{1}{\boldsymbol{n}_o}\left(\frac{\left(\sum_{\{x_i \in O: x_{ij \le d}\}} g_i{}^2\right)}{\boldsymbol{n}^j_{l|O}(\boldsymbol{d})} + \frac{\left(\sum_{\{x_i \in O: x_{ij > d}\}} g_i{}^2\right)}{\boldsymbol{n}^j_{r|O}(\boldsymbol{d})}\right) \tag{2}$$

where $n_O = \sum I[x_i \in O]$, $n^j_{l|O}(\mathrm{d}) = \sum I[x_i \in O : x_{ij \le d}]$ and$n^j_{r|O}(d) = \sum I[x_i \in O : x_{ij > d}]$.

Gradient One-Sided Sampling (GOSS) is a method that picks out each instance with a greater gradient and applies random sampling to instances with lower gradients [16]. In this context, the training dataset for a specific node in the decision tree is represented by the symbol 'O.' To determine the variance gain of 'j' or the splitting criterion at point 'd' for the node, you can calculate it using the following formula:

$$V_j^{\sim}(d) = \frac{1}{n}\left(\frac{\left((\sum_{x_i \in A_l} g_i + \frac{1-a}{b}\sum_{x_i \in B_l} g_i)^{\sim 2}\right)}{n_l^j(d)} + \frac{(\sum_{x_i \in A_r} g_i + \frac{1-a}{b}\sum_{x_i \in B_r} g_i)^2}{n_r^j(d)}\right) \quad (3)$$

where;

$A_l = \{x_i \in A : x_{ij \le d}\}, A_r = \{x_i \in A : x_{ij > d}\}, B_l = \{x_i \in B : x_{ij \le d}\}$

$B_r = \{x_i \in B : x_{ij > d}\}$, And the coefficient $\frac{1-a}{b}$ is employed for standardizing the summation of gradients over B to the scale ofA^c.

LGBM Regressor is a specific implementation for regression tasks.

Select a target variable and a collection of features.

Form an ensemble of decision trees through a gradient boosting approach.

Gradient boosting operates by incrementally introducing new trees that rectify the mistakes made by the current ensemble.

The learning process includes the minimization of a loss function, frequently based on mean squared error (MSE) for regression purposes.

The final prediction is obtained by aggregating the predictions from all trees in the ensemble.

4.4 Tools and Instruments for Data Analysis

To ensure the reliability of the experiments, the following tools and instruments were utilized:

Python: The primary programming language for data analysis and model development.
Pandas: Used for data manipulation and preprocessing.
Scikit-Learn: Employed for machine learning and regression model implementations.
LightGBM: Utilized as a gradient boosting framework for regression.
StandardScaler: Applied for feature scaling.
OneHotEncoder: Used for encoding categorical variables.
Jupyter Notebooks: Provided an interactive environment for code development and analysis.

5 Results and Discussions

The research involved an in-depth analysis of the "LengthOfStay.csv" dataset, which encompassed various patient attributes, including demographic information, health conditions, and medical measurements. A diverse range of data was collected, such as patient gender, number of readmissions, flags for specific health conditions, and numerical values like hematocrit, sodium, glucose, and more. The target variable, "lengthofstay," represented the length of a patient's stay during their encounter (Fig. 2).

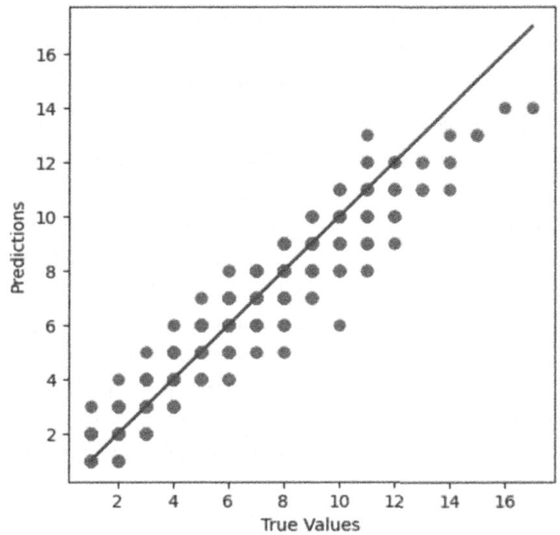

Fig. 2. Scatter Plot between True and Predicted Value.

The findings of this research highlight the effectiveness of different regression models in predicting the length of a patient's stay in a healthcare facility during a disaster or emergency. The discussion can be organized based on the performance of the four regression models implemented which are clearly shown in the Results and Discussions under Sect. 5.

Linear Regression
The Linear Regression model achieved an **R2 score of 65.14%**. This result signifies that linear relationships exist between certain input features and the length of a patient's stay. However, the relatively lower R2 score suggests that a more complex model might be better suited for this particular dataset.

Decision Tree Regression
The Decision Tree Regression model performed better with an **R2 score of 87.62%**. Decision trees can capture non-linear relationships in the data, which likely contributed to the improved predictive accuracy.

Random Forest Regression
The Random Forest Regression model yielded an **R2 score of 92.44%**, demonstrating its superior predictive ability compared to linear regression and decision trees.

LGBM Regressor
The LGBM Regressor emerged as the top-performing model with an impressive **R2 score of 96.14%**. "LightGBM", it is a "gradient boosting framework" which is known for its "speed and efficiency", making it ideal for handling complex datasets like the one used in this study (Table 2 and Fig. 3).

Table 2. Comparative study of various models along with its R2 score.

Models	R2 Score (%)	Performance Description
Linear Regression	65.14	This result signifies that linear relationships exist between certain input features and the length of a patient's stay
Decision Tree Regression	87.62	Captures non-linear relationships in the data, which likely contributed to the improved predictive accuracy
Random Forest Regression	92.44	Demonstrating its superior predictive ability compared to linear regression and decision trees
LGBM Regressor	96.14	LightGBM - a gradient boosting framework by which computation speed and accuracy is increased making it a best fit like the one used in this study

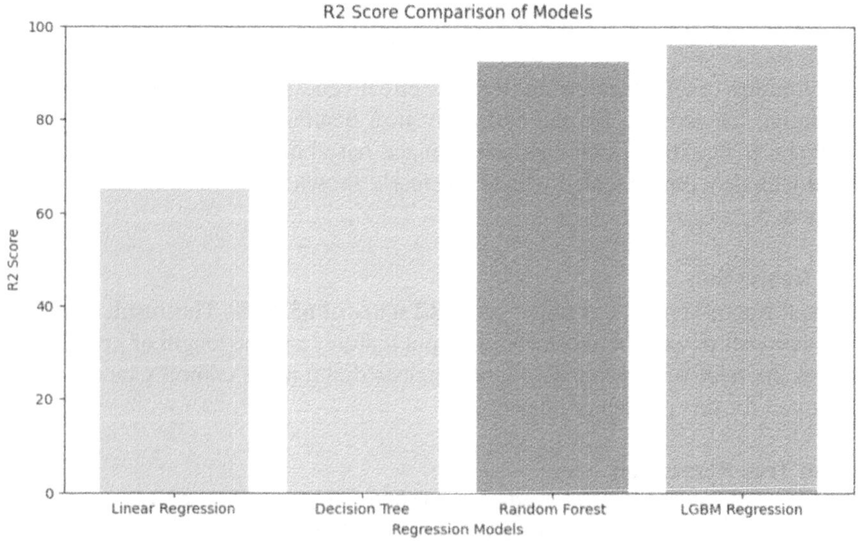

Fig. 3. Bar Chart Comparing R2 Score of Models.

6 Conclusion and Future Works

The primary objective of our research was to investigate the application of intelligent scheduling and resource allocation in healthcare during disaster scenarios.

Our study explored various regression models and among these models, the LGBM Regressor demonstrated remarkable performance with an R2 score of 96.14%. This model also overpowers the other existing models in this domain which have been discussed in Sect. 2.

Real-World Implementation: We recommend the practical implementation of the predictive models developed in healthcare settings, further assessing their impact and refining their accuracy.

Collaborative Research: Collaborative research between data scientists, healthcare professionals, and disaster response experts can yield holistic solutions for disaster healthcare management.

References

1. Zeleke, A.J., Palumbo, P., Tubertini, P., Miglio, R., Chiari, L.: Machine learning-based prediction of hospital prolonged length of stay admission at emergency department: a Gradient Boosting algorithm analysis. Front. Artif. Intell. **6**, 1179226 (2023). https://doi.org/10.3389/frai.2023.1179226
2. Chrusciel, J., Girardon, F., Roquette, L., Laplanche, D., Duclos, A., Sanchez, S.: The prediction of hospital length of stay using unstructured data. BMC Med. Inform. Decis. Mak. **21**, 351 (2021)
3. Caetano, N., Laureano, R., Cortez, P.: A data-driven approach to predict hospital length of stay: a Portuguese case study. In: ICEIS 2014 - Proceedings of the 16th International Conference on Enterprise Information Systems, vol. 1, pp. 407–414 (2014)
4. Stone, K., Zwiggelaar, R., Jones, P., Mac Parthaláin, N.: A systematic review of the prediction of hospital length of stay: towards a unified framework. PLOS Digit Health **1**(4), e0000017 (2022). https://doi.org/10.1371/journal.pdig.0000017
5. Harutyunyan, H., Khachatrian, H., Kale, D.C., Ver Steeg, G., Galstyan, A.: Multitask learning and benchmarking with clinical time series data. Sci. Data **6**, 96 (2019). https://doi.org/10.1038/s41597-019-0103-9
6. Hao, Q., Xu, F., Chen, L., Li, P.H.Y.: Hierarchical multi-agent model for reinforced medical resource allocation with imperfect information. ACM Trans. Intell. Syst. Technol. **14**(1), 8 (2022)
7. Lin, C.K.Y., Ling, T.W.C., Yeung, W.K.: Resource allocation and outpatient appointment scheduling using simulation optimization. Hindawi J. Healthcare Eng. **2017**, 9034737 (2017)
8. Xiang, Y., Zhuang, J.: A medical resource allocation model for serving emergency victims with deteriorating health conditions. Ann. Oper. Res. **236**(1), 177–196 (2014). https://doi.org/10.1007/s10479-014-1716-1
9. Ala, A., Alsaadi, F. E., Ahmadi, M., Mirjalili, S.: Optimization of an appointment scheduling problem for healthcare systems based on the quality of fairness service using whale optimization algorithm and NSGA-II. Sci. Rep. **11**, 19816 (2021)
10. Arora, S., Doshi, P.: A survey of inverse reinforcement learning: challenges, methods and progress. Artif. Intell. **297**, 10350 (2021)
11. Balcan, D., Colizza, V., Gonçalves, B., Hu, H., Ramasco, J.J., Vespignani, A.: Multiscale mobility networks and the spatial spreading of infectious diseases. Proc. Natl. Acad. Sci. **106**(51), 21484–21489 (2009)
12. Li, M., et al.: Efficient ridesharing order dispatching with mean field multi-agent reinforcement learning. In: Liu, L., White, R.W., Mantrach, A., Silvestri, F., McAuley, J.J., Baeza-Yates, R., Zia, L. (eds.) WWW 2019, pp. 983–994. ACM (2019). https://doi.org/10.1145/3308558.3313433
13. Li, M.Y., Muldowney, J.S.: Global stability for the SEIR model in epidemiology. Math. Biosci. **125**(2), 155–164 (1995)

14. Li, Q., et al.: Early Transmission dynamics in Wuhan, China, of novel coronavirus-infected pneumonia. N. Engl. J. Med. **382**(13), 1199–1207 (2020). https://doi.org/10.1056/NEJMoa2001316
15. Li, Z., Yu, H., Zhang, G., Dong, S., Xu, C.-Z.: Network-wide traffic signal control optimization using a multi-agent deep reinforcement learning. Transport. Res. Part C: Emerg. Technol. **125**, 103059 (2021)
16. Matrajt, L., Eaton, J., Leung, T., Brown, E. R.: Vaccine optimization for COVID-19: who to vaccinate first? Sci. Adv. **7**(6), eabf1374 (2021)
17. Lillicrap, T.P., et al.: Continuous control with deep reinforcement learning. In: Bengio, Y., LeCun, Y. (eds.) ICLR 2016 (2016). http://arxiv.org/abs/1509.02971
18. Lin, K., Zhao, R., Xu, Z., Zhou, J.: Efficient large-scale fleet management via multi-agent deep reinforcement learning. In: Guo, Y., Farooq, F. (eds.) KDD 2018, pp. 1774–1783, ACM (2018). https://doi.org/10.1145/3219819.3219993
19. Luo, M., Zhang, W., Song, T., Li, K., Zhu, H., Du, B., Wen, H.: Rebalancing Expanding EV Sharing Systems with Deep Reinforcement Learning. In: Bessiere, C. (ed.) IJCAI'20, ijcai.org, pp. 1338–1344. https://doi.org/10.24963/ijcai.2020/186 (2021)
20. Mnih, V., et al.: Asynchronous methods for deep reinforcement learning. In: Proceedings of the 33rd International Conference on Machine Learning, New York, vol. 48, pp. 1928–1937 (2016)
21. Mnih, V., et al: Playing atari with deep reinforcement learning. arXiv preprint arXiv:1312.5602 (2013)
22. Mnih, V., et al.: Human-level control through deep reinforcement learning. Nature **518**(7540), 529–533 (2015)

Electroencephalogram Based Stress Detection Using Machine Learning

Hemlata Ohal(✉), Abhishek Tiwari, Kiran Satote, Sakshi Zagade, Vaishnavi Tule, and Ajinkya Garad

Computer Science and Engineering, DVK MIT World Peace University, Pune, India
hemaohalphd@gmail.com

Abstract. This study analyses the topic of stress, highlighting the vital necessity of efficient coping techniques and mentality-based approaches to identify different stress states in a timely manner. The study focuses on the intricate function of electroencephalogram (EEG) signals in detecting stress for medical professionals' diagnosis so that treatment of disorders linked to stress could be treated at the earliest. The suggested method begins with preprocessing, feature extraction and further proceeds with development of ensemble model. This work investigates the identification of stress using EEG signals. The ensemble model achieves an amazing 98% accuracy on scaled data after undergoing thorough testing. The study emphasizes the critical importance of stress identification and presents ensemble technique for sophisticated EEG-based stress identification. This underscores the importance of stress identification and how it can affect the way a person perceives and handles stress, which will ultimately lead to an improved quality of life.

Keywords: Electroencephalogram · EEG · classification · Identification · Stress level · Stress recognition · Ensemble

1 Introduction

In the expansive realm of global mental health, stress stands as a pervasive challenge affecting millions worldwide. To mitigate its adverse impact on both physical and mental well-being, including anxiety and depression, prompt identification and effective mitigation of stress are imperative. Electroencephalography (EEG) emerges as a vital non-invasive technique, offering insights into brain electrical activity and emotional-cognitive states.

The electroencephalogram (EEG) captures signals across five distinct frequency bands: beta, gamma, alpha, theta, and delta, as delineated in Table 1. These bands play a crucial role in delineating mental states associated with stress. While existing EEG-based stress detection methodologies are valuable, they often entail manual processes, highlighting the pressing need for more precise and expeditious solutions. Technological advancements have spurred the evolution of methods such as spectral analysis, machine learning, and event-related potentials (ERPs) to extract pertinent information from EEG signals, thus paving the way for automated and enhanced stress detection capabilities.

Published by Springer Nature Switzerland AG 2024
M. L. Owoc et al. (Eds.): ICCIDS 2024, IFIP AICT 718, pp. 499–514, 2024.
https://doi.org/10.1007/978-3-031-69986-3_38

Given the profound implications of stress on mental and physical health, a comprehensive identification strategy is imperative. Integrating cutting-edge machine learning with automated EEG analysis unveils the intricate interplay between cognitive and affective states, thereby facilitating tailored stress interventions. Positioned at the intersection of technology and mental health, the advancement of EEG-based stress detection holds promise for early intervention. Novel methodologies, including those employing deep learning to assess EEG signals [7], and the integration of fNIRS and EEG technologies in canonical correlation analysis to evaluate the impact of mental stress on prefrontal cortical processes [20], exemplify ongoing efforts to refine techniques. Through continuous innovation, medical professionals can adeptly address the global ramifications of stress on individuals.

Table 1. Frequency of waves

Waveform	Frequency range	Mental state condition
Delta signal (δ)	0.5–4 Hz	During deep sleep
Theta signal (θ)	4–8 Hz	Daydreaming, sleepy or light sleeping
Alpha signal (α)	8–13 Hz	Relaxation, calmness and tranquil
Beta signal (β)	8–13 Hz	High level of attention during mental and physical activities
Gamma signal (γ)	>31 Hz	Thinking, learning, speaking and indicating higher mental activity

2 Related Work

The related work in EEG-based stress categorization and feature recognition is vast and multifaceted, encompassing diverse techniques and strategies aimed at refining the precision and reliability of stress evaluation. Research explores the intricacies of signal smoothing and its implications for stress classification across multiple levels, delving into the complex relationship between signal processing methods and classification results through the examination of various filter window widths [2]. Additionally, studies emphasize the significance of identifying relevant biomarkers within the intricate EEG data landscape, particularly in contexts such as epilepsy detection, where the combination of different features enhances classifier performance and augments the reliability of detection systems [3].

Novel methods for identifying stress features, including the use of k-Nearest Neighbor classifiers to extract stress-related patterns from EEG signals, demonstrate promising outcomes and underscore the value of leveraging machine learning algorithms to detect subtle physiological stress signals [4]. Furthermore, comprehensive analyses of techniques for assessing mental stress using EEG signals provide valuable insights into methodological advancements and practical applications, guiding future investigations and implementations [5].

Specifically, the focus lies on features like power spectral density (PSD) values in Theta, Alpha, and Beta frequency bands, revealing significant changes in Beta-band activity. Proposed hybrid approaches combining k-means clustering and SVM mitigate individual differences in stress response and enhance the overall reliability and detection rate of mental stress [13]. Moreover, the combined impact of alpha/beta and theta/beta band power ratios is emphasized, with classification models derived from clustering approaches showcasing superior accuracy, such as Naïve Bayes with a remarkable 95% accuracy, and highlighting the importance of theta/beta compared to alpha/beta [15].

Further advancements involve integrating wavelet scattering for optimal EEG channel selection with machine learning methods like Support Vector Machines and Convolutional Neural Networks, enriching the understanding of psychological stress detection and demonstrating the potential of sophisticated analytical tools in deciphering stress physiology intricacies [16, 17]. In summary, these studies collectively underscore the critical role of EEG-based techniques in stress assessment, offering valuable insights into methodological developments and practical applications, with potential for further improvements and broader application across diverse fields in the future.

3 Proposed Framework/Architecture

This study presents a comprehensive framework aimed at detecting stress using EEG signals. The proposed approach encompasses five distinct stages that synergistically contribute to the stress classification process. Initially, the preprocessing stage ensures the refinement of EEG signals by eliminating noise and artifacts, thus enhancing the reliability of subsequent analyses. Following this, the feature extraction phase identifies pertinent features from the preprocessed EEG signals, capturing essential characteristics relevant to stress classification. Subsequently, the classifier selection step evaluates various classifiers, including Support Vector Machines (SVM), K-Nearest Neighbors (KNN), and Random Forest, to determine their suitability for distinguishing stress and non-stress states within EEG data. These classifiers are then integrated into an ensemble model during the training phase, leveraging their individual strengths to enhance overall classification performance. Lastly, the cross-validation stage assesses the generalization and robustness of the ensemble model by evaluating its performance on unseen EEG data. The graphical depiction of this framework, presented in Fig. 1, illustrates the cohesive interaction between these stages. By employing an ensemble model approach with a voting classifier, this framework endeavors to improve the accuracy of EEG-based stress detection, thereby facilitating early intervention strategies and enhancing understanding of stress dynamics.

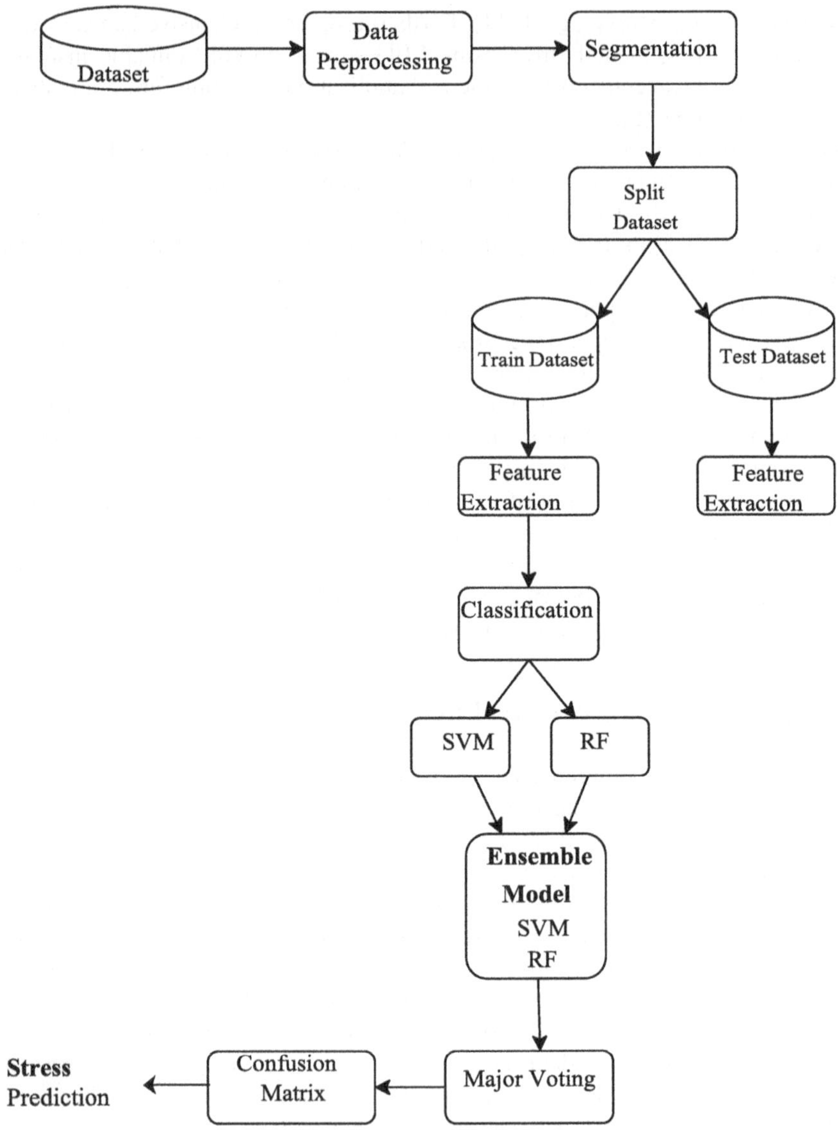

Fig. 1. Proposed Framework

4 Dataset

This study utilized a dataset obtained from [1], consisting of EEG readings from four nodes: TP9, TP7, AF7, and AF8, formatted in CSV. The dataset serves as the foundation for key phases of the investigation, including data preprocessing, segmentation, feature extraction, and model training. Through meticulous analysis of this dataset, this research aim is to uncover intricate patterns associated with stress responses. This comprehensive dataset enables exploration of neural activity dynamics and facilitates the development of sophisticated algorithms for stress classification. Utilizing this dataset allows for innovative approaches to enhance stress detection accuracy by examining EEG data from multiple nodes. Ultimately, this dataset drives pioneering research in EEG-based stress assessment, advancing understanding and capabilities in this critical domain.

5 Preprocessing

The preprocessing phase involves critical steps to refine the acquired dataset like artifact removal to enhance data quality [19]. Techniques such as filtering, artifact removal, and segmentation into distinct epochs are employed. Mean imputation is strategically used to handle missing values, and Standard Scaler is applied to scale numerical features. The EEG data is then systematically divided into 1-s intervals, crucial for capturing subtle trends associated with different emotional states, aligned with a sampled rate of 256 samples per second.

6 Feature Extraction

In order to identify optimum attributes in the EEG data that have been previously provided [8], a key role is played by method of feature extraction with an aim towards determining the best combination of features. A comprehensive nature of feature extractions, which allows for new subsets to be generated, complements the evaluation metrics in capturing different categories of data. The efficiency of this technique is based on the assessment of features in each subset that lead to a result. A feature in each subset that yields a lower error rate. Eleven important features extracted.

6.1 Power Spectral Density (PSD)

Power spectral density (PSD) is a fundamental feature extracted from EEG signals, providing insights into the distribution of signal power across different frequency bands [18]. It is mainly used to find power distribution for time-domain EEG signal over frequency range shown in Fig. 2, several studies highlight that the effectiveness of PSD to estimate level of stress. FFT method was used to compute the power spectral density, which gave information on the frequency distribution of EEG signals within each epoch

$$f = Welch\,(x, fs) \quad (1)$$

6.2 Time Series

Time series features capture temporal dynamics in EEG signals and comprise Sample Entropy, Approximate Entropy, and Permutation Entropy for every four EEG channels [6]. The tfresh library an open-source feature extraction packages a collection of time series features was obtained, comprising Sample Entropy, Approximate Entropy, and Permutation Entropy for every four EEG channels.

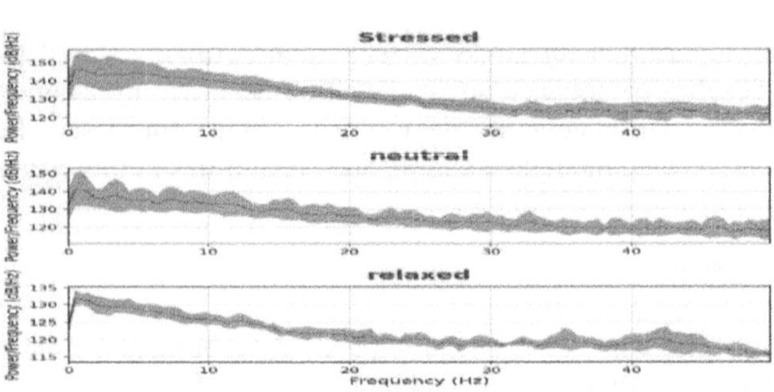

Fig. 2. Graphical representation of Power Spectral Density (PSD)

Sample Entropy

Functions as a crucial parameter, providing an assessment of the complexity or regularity in the EEG time series. This attribute helped in providing important insights into the underlying dynamics of brain activity by quantifying the signal's degree of predictability or regularity.

$$-\log(A/B) \quad (2)$$

Approximate Entropy

By evaluating EEG time series complexity, it increases comprehension capturing the variances and by estimating the probability that similar patterns will continue on subsequent data points, thus improves understanding of complexity of signal.

$$-\log(cm+1(r)cm(r)) \quad (3)$$

Permutation Entropy

It has unique function in assessing the EEG time series unpredictability it is used to highlight the sequential arrangement by taking into account order of values within the signal, offering important insights on the randomness shown in Fig. 3.

6.3 Power Features

Power characteristics were computed for the Delta, Theta, Alpha, Beta, and Gamma frequency bands to measure the dispersion of the EEG signal across several frequency bands [15]. To measure the dispersion of the EEG signal across several frequency bands, power characteristics were computed for the Delta, Theta, Alpha, Beta, and Gamma frequency bands, the absolute power is calculated which offered insights into the strength of oscillations within frequency ranges. Later proportionate distribution of power determined and this comprehensive allowed to assess the nuanced variation associated with stress responses. The proportionate distribution of power within each frequency band is known as relative power.

$$\text{Abs\_Power\_Band} = \text{trapz}(P(f)) \quad (4)$$

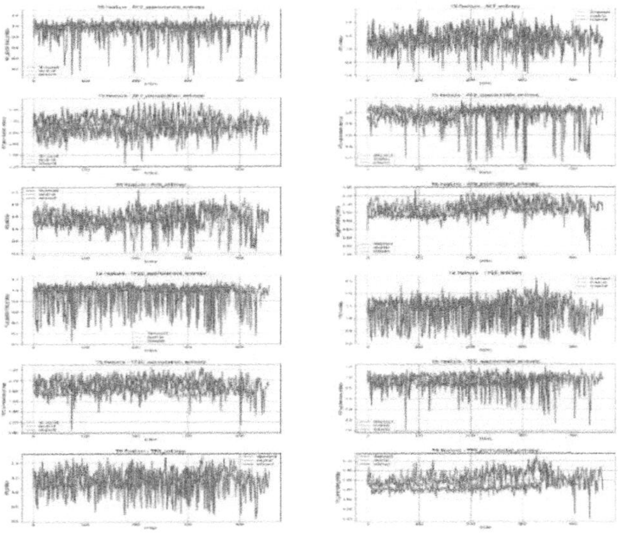

Fig. 3. Graphical representation of Permutation Entropy

$$\text{Rel\_Power\_Band} = Abs\_Power\_Band\ Total\ power \quad (5)$$

6.4 Amplitude Features

Amplitude features, including Mean, Standard Deviation, Skewness, and Kurtosis, quantify signal amplitudes within specific frequency bands like alpha, beta, theta, and delta waves. These features provide essential insights into the characteristics of EEG signals. For instance, the Amplitude Mean gives the average signal magnitude, while the Standard Deviation measures the dispersion of amplitudes around the mean. Skewness quantifies the asymmetry of the amplitude distribution, and Kurtosis reflects the "tailedness" of the distribution. Refer to Fig. 4 for a graphical representation of these Amplitude features. Together, these features offer valuable information for analyzing and interpreting brain activity patterns in stress detection and neural signal analysis.

6.5 Asymmetry Features

Asymmetry features, encompassing Mean and Standard Deviation, evaluate hemispheric imbalances in brain activity, crucial for discriminating stress levels [6]. These features are calculated based on the signed and absolute sums of signal amplitudes, providing insights into the average asymmetry and its variation across channels. By organizing these elements into a structured data frame, stress-specific traits are successfully preserved, contributing to a comprehensive analysis of EEG signals and enhancing stress level discrimination.

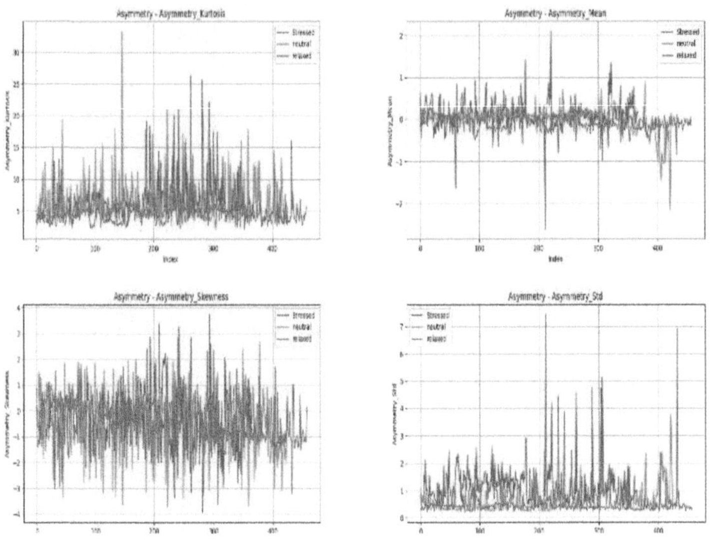

Fig. 4. Graphical representation of Amplitude features

6.6 Time Domain Features

The mean and standard deviation (Std) were computed to provide insights into the average "baseline" activity and variability of EEG signals [6]. The mean represents the average value of the EEG signal, providing an indication of the central tendency or average "baseline" activity in the context of stress detection. It serves as a measure of the average level of electrical activity for each channel, with changes in overall activity indicated by deviations from this mean.

$$1N \sum Ni = 1xi \tag{6}$$

The standard deviation (Std) measures the degree of variation or dispersion in the signal values around the mean. It is used to observe variability, where a higher standard deviation suggests fluctuations in brain activity during certain periods, while a lower standard deviation implies consistent activity.

$$1N \sum Ni = 1(xi - Mean)2 \tag{7}$$

6.7 Shannon Entropy

Shannon entropy quantifies the uncertainty or randomness of EEG signals, reflecting the complexity of brain activity patterns [6]. The feature extraction measures the unpredictability or complexity by analyzing the EEG time series of the EEG signal, higher entropy here indicated variability and irregularity.

$$\sum Ni - 1pilog2(pi) \tag{8}$$

6.8 Hjorth Parameters

Hjorth parameters characterize the EEG signal in terms of activity, mobility, and complexity, offering valuable insights into brain state dynamics [9]. Hjorth parameters (HPs) are analyzed in the frequency domain to minimize the dimension of the feature collection. This research simultaneously aimed to preserve the precision of emotion identification for three emotional classes. Activity represents the variance of the EEG signal, where higher activity indicates increased variability.

$$N\_1 \sum Ni = 1xi2 \tag{9}$$

Mobility is used to approximate the mean frequency; increased activity suggests changes in neural activity.

$$\sqrt{1}N \sum Ni = 1\,(dxidt)2\sqrt{1}N \sum Ni = 1xi2 \tag{10}$$

Complexity The standard deviation to mobility ratio of the second derivative. It quantifies rate of change or mobility

$$\sqrt{1}N \sum Ni = 1\,(d2xidt)2\sqrt{1}N \sum Ni = 1(dxidt)2 \tag{11}$$

6.9 Fast Fourier Transform

FFT decomposes EEG signals into frequency components, facilitating the analysis of oscillatory patterns associated with stress [14]. FFT technique efficiently determines the sequence's discrete Fourier transform (DFT), generating representation of signal in terms of its frequency components In this study it was used to transforms signal into the frequency domain and to extract relevant features like power in specific bands also dominant frequencies.

6.10 Hurst

The Hurst exponent measures long-range dependence in EEG signals, indicating self-similarity and fractal-like behavior [12]. It assesses the EEG signal's long-term memory and self-similarity characteristics. Commonly used in time series analysis, the Hurst Exponent enhances the characterization of EEG signals by illuminating the persistence and structure of temporal patterns. Calculating the Hurst Exponent provides a more sophisticated interpretation of the intrinsic long-term memory qualities of EEG signals, offering a better understanding of how the signals display self-similarity across various time scales. This facilitates a thorough evaluation of the intricate dynamics present in EEG data, as displayed in Fig. 5, which provides a graphical representation of the Hurst Exponent.

$$\frac{1}{2}(log_{10}(R/S)) \tag{12}$$

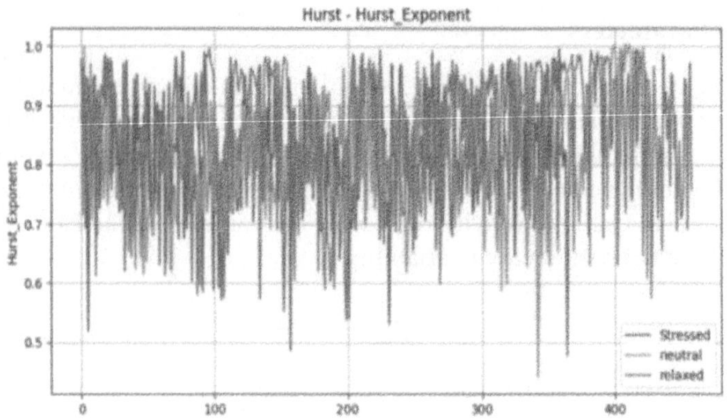

Fig. 5. Graphical representation of Hurst Exponent

6.11 MSE

MSE quantifies the discrepancy between EEG signal predictions and actual observations, aiding in model evaluation and validation [13]. Sample entropy (SampEn) is computed as an indicator of the EEG signals. It is useful for analyzing EEG data since it offers a reliable indicator of the intricacy of brain impulses. When segmented data is used, the efficiency of the procedure is maximized, fitting well with the natural structure of EEG signals. Figure 6 presents a graphical representation of MSE, illustrating its role in quantifying the discrepancy between EEG signal predictions and actual observations.

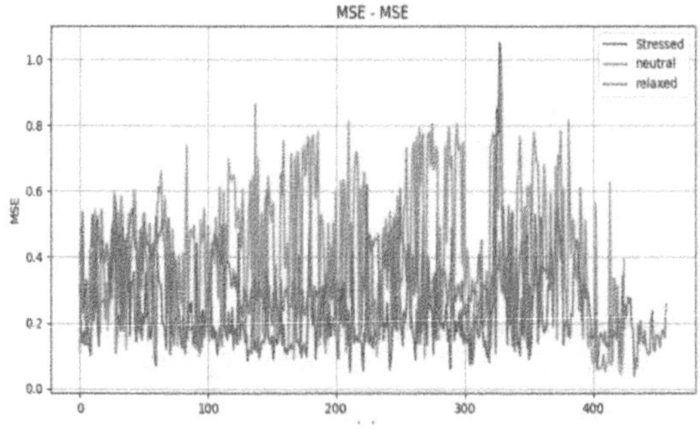

Fig. 6. Graphical representation of MSE

The resulting SampEn values are an important tool in the context of stress detection and neural signal analysis because they provide insightful information on the subtleties of brain activity, with larger values suggesting greater complexity and predictability.

7 Classification

The classification methodology implemented in this research entailed an exhaustive and systematic review of various techniques aimed at assessing mental stress through EEG signals [5]. This investigation sought to develop a stress detection model with high precision and accuracy, emphasizing the pivotal role of selecting appropriate classifiers to enhance stress detection performance [10, 11]. To achieve this, meticulously explored a diverse range of classifiers, including Decision Tree, Random Forest, Gradient Boosting, Support Vector Machines (SVM), K-Nearest Neighbors (KNN), and Naïve Bayes.

Throughout the experimentation phase, particular attention was given to the comparative analysis of classifiers, focusing on their efficacy in discerning stress patterns from EEG data. Notably, initial experiments revealed that Random Forest and SVM exhibited superior performance compared to other classifiers. Furthermore, this investigation introduced a modified SVM technique, which demonstrated promising results in enhancing stress assessment capabilities [21, 22].

In subsequent steps, employed advanced optimization techniques such as Grid Search CV to fine-tune the hyperparameters of SVM and Random Forest classifiers. This meticulous optimization process was instrumental in achieving commendable levels of accuracy in stress detection.

Table 2. Accuracy of classifiers

Classifiers	Highest performance	Accuracy
SVM	Precision 95% Recall 99% F1-score 97%	97%
KNN	Precision 97% Recall 94% F1-score 95%	95%
Random forest	Precision 92% Recall 100% F1-score 96%	96%
Weighted voting	Precision 95% Recall 100% F1-score 98%	98%

8 Ensemble Model

This study underscores the paramount importance of ensemble learning methodology in the context of stress classification. Each classifier, including SVM, KNN, and Random Forest, underwent meticulous hyperparameter optimization to exploit their unique strengths. This fine-tuning process aimed to maximize individual classifier performance and synergistically enhance the predictive capability of the ensemble model. A comprehensive evaluation of the ensemble's performance was conducted through rigorous cross-validation on a dedicated subset of the data. Metrics such as accuracy, precision, recall, and F1-score were meticulously recorded to assess the model's generalization performance. The Confusion Matrices depicted in Fig. 7 (KNN), Fig. 8 (Random Forest), Fig. 9 (SVM), and Fig. 10 (Voting Classifier) offer visual insights into the classification efficacy of each algorithm. This comprehensive evaluation approach ensures a nuanced understanding of the ensemble's overall performance, thereby affirming its efficacy in stress classification tasks within the domain of machine learning. Moreover, prior research supports this ensemble approach. For instance [3] conducted a study on mental state classification using EEG-based brain-machine interface, highlighting the relevance of ensemble methods in similar tasks. Similarly, the work [2] focused on EEG-based stress classification, demonstrating the effectiveness of ensemble learning in multi-level stress classification. Furthermore, the ensemble feature selection approach proposed by [3] provided valuable insights into identifying relevant features from EEG signals. These studies collectively reinforce the significance of ensemble methods in EEG-based stress class classification task.

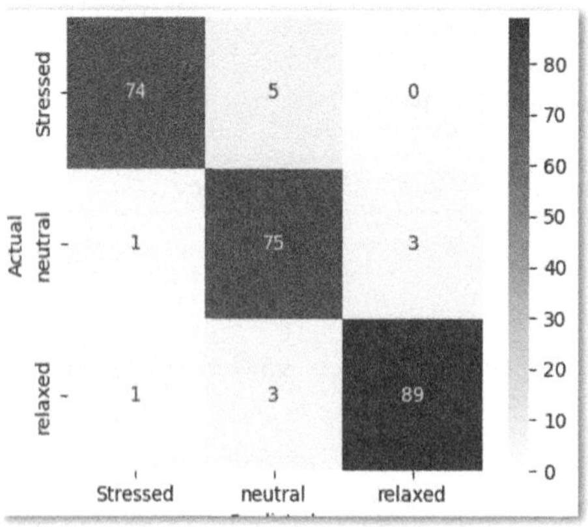

Fig. 7. KNN Confusion Matrix

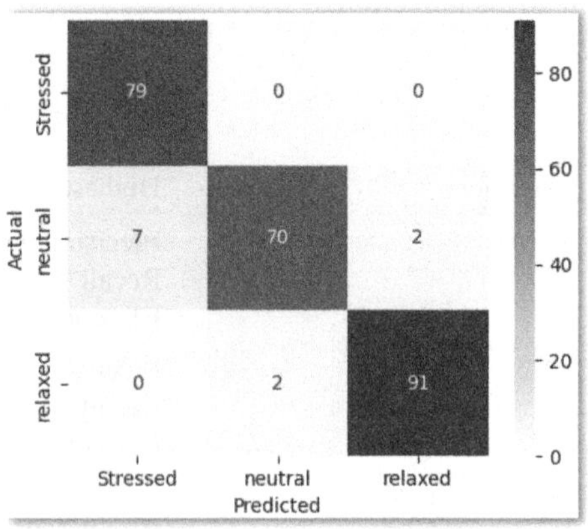

Fig. 8. Random Forest Confusion Matrix

9 Results and Discussion

In numerous trials, various machine learning models were meticulously evaluated to identify the most effective method for stress classification. The dataset underwent partitioning, models were trained on different combinations, and evaluations were performed on unseen data. Each classifier underwent rigorous assessment based on precision, recall, and F1-score, with a primary focus on leveraging an ensemble of models.

Initially, the Support Vector Machine (SVM) performed admirably, achieving an impressive 89% accuracy and demonstrating excellent recall, precision, and F1-score in identifying stress. Conversely, Naive Bayes struggled, managing only 70% accuracy.

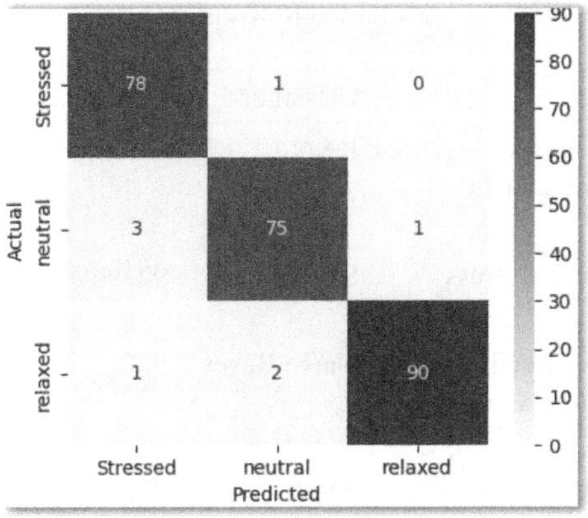

Fig. 9. SVM Confusion

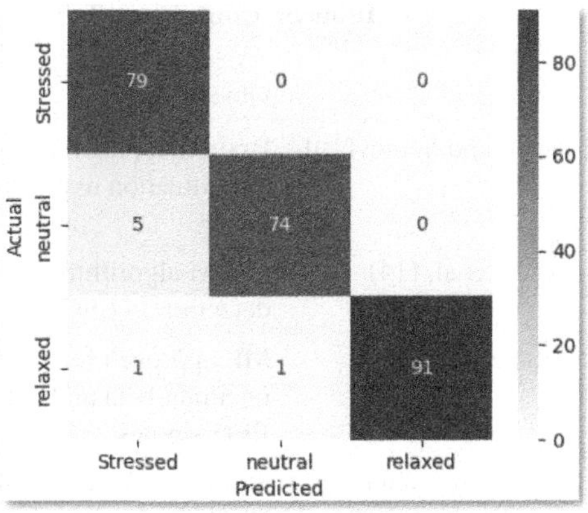

Fig. 10. Voting Classifier Confusion Matrix

However, after fine-tuning the settings and combining Random Forest with SVM, accuracy improved to 94%. The optimized classifiers were primarily intended for integration into the ensemble model, constructed as a Weighted Voting Classifier. The ensemble model exhibited outstanding dependability during the training phase, particularly when trained on scaled data, resulting in the highest accuracy in the final round. Hyperparameter optimization played a crucial role in adjusting the contribution of each classifier, ensuring a cohesive and efficient ensemble for stress detection. By utilizing a voting classifier with various models, an exceptional accuracy of 98% was achieved.

Table 2 summarizes classifier accuracy, with the Weighted Voting Classifier emerging as the most robust, boasting a precision of 95%, recall of 100%, and an impressive F1-score of 98%. This comprehensive evaluation underscores the efficacy of the ensemble in stress detection, affirming the strength of the proposed approach compared to previous studies. Previous studies demonstrated an accuracy rate of 94.6% for two-level identification and 83.4% for multilevel identification [16, 17].

Table 3 provides a summary of the results of several investigations on using EEG signals for stress detection. Comparing these previous results with the proposed ensemble approach demonstrates the effectiveness of this study in achieving high accuracy.

10 Future Scope

Promising opportunities exist for stress detection in the future, with a focus on increasing accuracy to guarantee dependability. Incorporating deep learning, particularly with Convolutional Neural Networks (CNNs), is a viable path towards improving EEG signal pattern detection and achieving even higher precision. Thorough testing on a variety of datasets will be essential to guaranteeing the model's flexibility and resilience under a range of stress situations. There are plans to investigate real-time applications, which will present chances for dynamic validation in real-world environments. For stress detection models to continue evolving, it will also be crucial to use transfer learning strategies,

Table 3. Comparison Table for Accuracy with Previous Works

Author	Objective	Classifier	Performance
Nagar and Sethia [10]	Brain mapping based stress identification using portable EEG device	KNN	74.43%
Dhake et al. [14]	LSTM algorithm for mental stress detection in EEG	SVM	92–94%
Altaf et al. [15]	ML approach for stress detection based on alpha-beta and theta-beta ratios of EEG signals	Naïve Bayes	95%
Sinha and Babu [18]	Statistical feature analysis for EEG baseline classification	SVM KNN	77.92%
Radüntz et al. [19]	Automated EEG artifact elimination by applying ML algorithms to ICA-based features	ANN	94.04%
Al-Shargie et al. [20]	Assessment of mental stress effects on prefrontal cortical activities using canonical correlation analysis	SVM	89.8%
Phutela et al. [21]	Stress classification using brain signals based on LSTM network	RNN	93.17%
Proposed Approach	Electroencephalogram based stress detection using machine learning	Voting	98%

encourage interdisciplinary collaborations, and address ethical issues. The main goal is to create a robust and accurate instrument that can identify stress as accurately as possible.

11 Conclusion

In Conclusion, the remarkable 98% accuracy of this EEG-based stress detection technique indicates a positive change in stress management methods. The accuracy with which EEG measures brain activity, together with its mobility and immediateness, make it an extremely beneficial tool. This method provides an easy-to-use solution for real-life stress detection. Adding a Confusion matrix improves transparency when evaluating model performance. In the future, deep learning techniques and the investigation of various datasets present new opportunities for EEG-based stress detection, which has great potential to enhance mental health.

Acknowledgments. The support and resources from Dr. Vishwanath Karad World Peace University are sincerely acknowledged, since they considerably aided this study endeavor.

Disclosure of Interests. The authors declare no conflict of interest.

References

1. Bird, J.J., Manso, L.J., Ribiero, E.P., Ekart, A., Faria, D.R.: A study on mental state classification using eeg-based brain-machine interface. In: 9th International Conference on Intelligent Systems. IEEE (2018)
2. Perez-Valero, E., Lopez-Gordo, M.A., Vaquero-Blasco, M.A.: EEG-based multi-level stress classification with and without smoothing filter. Biomed. Signal Process. Control **69**, 102881 (2021)
3. Mera-Gaona, M., López, D.M., Vargas-Canas, R.: An Ensemble feature selection approach to identify relevant features from EEG signals. Appl. Sci. **11**(15), 6983 (2021)
4. Sulaiman, N., et al.: Novel methods for stress features identification using EEG signals. Int. J. Simul. Syst. Sci. Technol. **12**(1), 27–33 (2011)
5. Katmah, R., et al.: A review on mental stress assessment methods using EEG signals. Sensors **21**(15), 5043 (2021)
6. Bhatnagar, S., et al.: A deep learning approach for assessing stress levels in patients using electroencephalogram signals. Decis. Anal. J. **7**, 100211 (2023)
7. Nikhil, A.S., Banker, N.N., Jagadeesh P.: A conceptual analysis for the measurement of stress intensity by deep learning using EEG signals. In: 2022 IEEE International Conference on Electronics, Computing and Communication Technologies (CONECCT). IEEE (2022
8. Al-Qerem, A., et al.: General model for best feature extraction of EEG using discrete wavelet transform wavelet family and differential evolution. Int. J. Distrib. Sensor Netw. **16**(3), 1550147720911009 (2020)
9. Anita, M., et al.: An extensive review of machine learning techniques for EEG signal processing. In: 2022 International Conference on Automation, Computing and Renewable Systems (ICACRS). IEEE (2022)
10. Nagar, P., Sethia, D.S.: Brain mapping based stress identification using portable eeg based device. In: 2019 11th International Conference on Communication Systems & Networks (COMSNETS). IEEE (2019)
11. Jun, G., Smitha, K.G.: EEG based stress level identification. In: 2016 IEEE International Conference on Systems, Man, and Cybernetics (SMC). IEEE (2016)
12. Xu, C., Xia, R.Z.: EEG signal classification and feature extraction methods based on deep learning: a review. In: 2023 2nd International Conference on Big Data, Information and Computer Network (BDICN). IEEE (2023)
13. Wen, T.Y., Aris, S.A.M.: Hybrid approach of eeg stress level classification using k-means clustering and support vector machine. IEEE Access **10**, 1837018379 (2022)
14. Dhake, D., et al.: LSTM algorithm for the detection of mental stress in EEG. In: 2023 3rd International Conference on Intelligent Technologies (CONIT). IEEE (2023)
15. Altaf, H., et al.: Machine learning approach for stress detection based on AlphaBeta and theta-beta ratios of EEG signals. In: 2021 13th International Conference on Information & Communication Technology and System (ICTS). IEEE (2021)
16. Subhani, A.R., et al.: Machine learning framework for the detection of mental stress at multiple levels. IEEE Access **5**, 13545–13556 (2017)
17. Kalas, M.S., Momin, B.F.: Stress detection and reduction using EEG signals. In: 2016 International Conference on Electrical, Electronics, and Optimization Techniques (ICEEOT). IEEE (2016)
18. Sinha, N., Babu, D.: Statistical feature analysis for EEG baseline CLASSIFICATION: EYES OPEN VS EYES CLOSED. In: 2016 IEEE region 10 conference (TENCON), pp. 2466–2469. IEEE (2016)
19. Radüntz, T., Scouten, J., Hochmuth, O., Meffert, B.: Automated EEG artifact elimination by applying machine learning algorithms to ICA-based features. J. Neural Eng. **14**(4), 046004 (2017)

20. Al-Shargie, F., Tang, T.B., Kiguchi, M.: Assessment of mental stress effects on prefrontal cortical activities using canonical correlation analysis: an fNIRS-EEG study. Biomed. Opt. Express **8**(5), 2583–2598 (2017)
21. Phutela, N., Relan, D., Gabrani, G., Kumaraguru, P., Samuel, M.: Stress classification using brain signals based on LSTM network. Comput. Intell. Neurosci. **2022** (2022)
22. Gupta, R., Alam, M.A., Agarwal, P.: Modified support vector machine for detecting stress level using EEG signals. Comput. Intell. Neurosci.Neurosci. **2020**, 114 (2020)

Author Index

Published by Springer Nature Switzerland AG 2024
M. L. Owoc et al. (Eds.): ICCIDS 2024, IFIP AICT 718, pp. 515–518, 2024.
https://doi.org/10.1007/978-3-031-69986-3

The manufacturer's authorised representative in the EU is Springer Nature Customer Service Centre GmbH, Europaplatz 3, 69115 Heidelberg, Germany. If you have any concerns regarding our products, please contact ProductSafety@springernature.com

Printed and bound by CPI Group (UK) Ltd, Croydon, CR0 4YY
30/09/2026
02225494-0020